METACOMMUNITIES

METACOMMUNITIES

Spatial Dynamics and Ecological Communities

EDITED BY

Marcel Holyoak, Mathew A. Leibold,
and Robert D. Holt

The University of Chicago Press

Chicago and London

MARCEL HOLYOAK is associate professor of environmental science and policy at the University of California, Davis.

MATHEW A. LEIBOLD is associate professor of integrative biology at the University of Texas at Austin.

ROBERT D. HOLT is professor of ecology at the University of Florida.

The University of Chicago Press, Chicago 60637
The University of Chicago Press, Ltd., London
© 2005 by The University of Chicago
All rights reserved. Published 2005
Printed and bound by CPI Group (UK) Ltd, Croydon, CR0 4YY

14 13 12 11 10 09   2 3 4 5

ISBN 978-0-226-35063-9 (cloth)
ISBN 978-0-226-35064-6 (paper)

Library of Congress Cataloging-in-Publication Data

Metacommunities : spatial dynamics and ecological comunities / edited by Marcel Holyoak,
Mathew A. Leibold, and Robert D. Holt
p. cm.
Includes bibliographical references and index.
ISBN 0-226-35063-0 (cloth : alk. paper) — ISBN 0-226-35064-9 (pbk. : alk. paper)
1. Spatial ecology. 2. Biotic communities. 3. Biogeography. 4. Population biology.
I. Holyoak, Marcel. II. Leibold, Mathew A. III. Holt, Robert D.
QH541.15.S62M48 2005
577—dc22
2005002196

Contents

Preface

Community ecology as a field is concerned with explaining the patterns of distribution, abundance and interaction of multiple species. Since at least the 1960s when MacArthur and Wilson published *The Theory of Island Biogeography,* ecologists have questioned the extent to which spatial processes are responsible for producing community patterns. Interest in spatial community ecology has grown extensively during the last decade, fueled by a realization that there were gaps in the theory that logically would be required to explain the functioning of communities connected by dispersal. For example, in their current forms, island biogeography, traditional community ecology, and metapopulation ecology cannot fully explain how both local and regional species diversity are maintained, such that local diversity can contribute to the maintenance of regional diversity and vice-versa. Similarly, and of relevance to conservation, we do not have a theory to predict how food web structure and dynamics will change during habitat fragmentation; these gaps in theory might hinder prediction of what will happen to communities when spatially continuous habitat is divided into well connected fragments, and eventually becomes small, isolated fragments. Ecologists have used the term "metacommunity" to describe groups of communities that are connected by dispersal, and the term is being used with increasing frequency (e.g., over twenty times in published papers during 2003). This book came about through a symposium held at the Ecological Society of America annual meeting in Madison, Wisconsin (August 2001) and through a working group on metacommunities at the National Center for Ecological Analysis and Synthesis (NCEAS). Other authors who were working on key pieces of the story to date were also invited to contribute chapters.

This book aims to bring the metacommunity concept to a broad audience. It attempts to do so by synthesizing current empirical and theoretical knowledge, and through novel contributions in a broad range of areas. The greatest challenge in putting together this book was to integrate theory and empirical knowledge about spatial community ecology. We have therefore chosen a variety of theoretical approaches that we believe are amenable to empirical investigation, and we selected empirical contributors who were both doing appropriate work and were capable of making strong theoretical connections. (Naturally, we were not able to include all of the people who are doing relevant and excellent work.) Contributors were encouraged to make theoretical-empirical connections through the initial invitation, two rounds of revisions with reviews coming from other chapter authors, and editorial comments from members of an NCEAS working group on metacommunities, including ourselves.

Our target audience is graduate students and researchers in ecology, evolution, and related applied areas of environmental management and conservation. Most of the chapters are accessible to a broad audience, but a few chapters discuss more complex modeling techniques and multivariate statistical methods. For these more complex parts we have also tried to ensure that the purpose and outcome of the methods is clear without having to understand all of the details.

Chapter 1 serves as an introduction to the metacommunity concept and describes the layout of this volume. The book is organized into an overview (chapter 1), Core Concepts (chapters 2–3), Empirical Perspectives (chapters 4–9), Theoretical Perspectives (chapters 10–13), Emerging Areas and Perspectives (chapters 14–20) and a final coda section. Each of the major sections has its own introduction that gives overviews of the content, in each case placing chapters into a larger ecological framework. Our greatest hope for this book is that it will fuel further investigation and increased rigor in untangling the role of spatial dynamics and patterns in community ecology.

Acknowledgments

This book would not have been possible without the hard work of its many contributors. The authors of the chapters in this volume did an amazing job in getting thorough reviews and revisions to us in a timely fashion. We also thank our colleagues for their encouragement and support, for helpful discussions of the science, and for editorial feedback on manuscripts. This book greatly benefited from the editorial assistance and expertise of Christie Henry and Jennifer Howard at the University of Chicago Press. A graduate discussion group at the University of California at Davis (ECL290-053) read earlier drafts of most of the book manuscript and gave comments that greatly improved its logic, organization, and clarity: we thank Liz Chamberlain, Sarah Elmendorf, Melanie Gogol-Prokurat, Brett Harvey, Michelle Hornberger, Ivan Kauter, Richard Lankau, Kim Preston, Matthew Schlesinger, Aaron Setran, Jenna Shinen and Sam Veloz. Several chapter authors contributed more in ideas and / or reviews than is apparent from their authorship in chapters of the book, and in this respect we wish to thank Priyanga Amarasekare, Karl Cottenie, Martha Hoopes, Jamie Kneitel, and David Tilman. The book plan and book each benefited from two anonymous reviews. We thank the reviewers for their helpful suggestions.

This book was facilitated by financial and logistic support for the Metacommunity Working Group at the National Center for Ecological Analysis and Synthesis, a center funded by the National Science Foundation (DEB-9421535), University of California at Santa Barbara, and the state of California. The Ecological

Society of America facilitated the holding of a symposium on the topic of meta-communities (during August 2001) that helped to get this project started.

Marcel Holyoak would like to also thank Sharon P. Lawler for her patient support and understanding throughout this project. Helpful feedback and criticism were provided by Jun Bando, Amber Mace, Denise Piechnik, Drew Talley, Theresa S. Talley, and those acknowledged above in the ECL290 class. I thank Matthew Schlesinger for providing additional helpful comments on, and improvements to, several chapters. M. H. was supported by NSF DEB-0213026. Mathew Leibold and Marcel Holyoak thank Joe Travis for not interrupting Bob Holt when he needed to finish his chapters.

Mathew Leibold would like to thank all of those who provided comments, NCEAS and acknowledge support from NSF, the University of Chicago, and University of Texas at Austin.

Robert D. Holt would like to thank Lynne Holt and all of those who provided feedback on this book. He thanks NSF, NIH, NCEAS, and the University of Florida for support contributing to his involvement in this project.

CHAPTER 1

Metacommunities

A Framework for Large-Scale Community Ecology

Marcel Holyoak, Mathew A. Leibold, Nicolas Mouquet,
Robert D. Holt, and Martha F. Hoopes

A primary goal of ecology is to measure, understand, and predict patterns of biodiversity, including the numbers of kinds of organisms and their genetic and phenotypic or functional diversity. Patterns in the distribution and abundance of species are often striking, inspiring awe of nature and fostering a desire to conserve biodiversity. Understanding such patterns has crucial practical utility, for instance as part of the ongoing quest to understand the role of biodiversity in ecosystem functioning (e.g., maintaining water quality, atmospheric CO_2 levels, or primary production; Naeem et al. 1999; Loreau 2000). Dealing with anthropogenic global change provides a strong motivation to articulate the mechanisms creating and maintaining biodiversity.

Biodiversity is structured by processes operating at several hierarchical scales, including populations of individual species, interacting populations of different species (predators and prey, competitors, etc.), and whole communities and ecosystems (e.g., indirect interactions, levels of ecosystem functioning). The patterns of biodiversity that we seek to understand are also innately spatial, scaling from local ecosystems to landscapes and entire biogeographic regions (e.g., Wiens 1989; Levin 1992; Holt 1993; Rosenzweig 1995; Maurer 1999; Hubbell 2001; Chase and Leibold 2003). Surprisingly, there are many gaps in the empirical and theoretical knowledge that could logically explain the dynamics of entire communities in spatially structured habitats (e.g., collections of fragments). This book aims at filling some of these gaps by highlighting the emergence of a new focus for ecologists working at this level of organization—what is known as "metacommunity ecology."

The kinds of patterns we seek to explain are established by existing empirical studies, for instance, by studies that measure species diversity locally (α-diversity), among-localities (β-diversity) and regionally (γ-diversity; Whittaker 1960; Magurran 1988). Especially interesting are cases where such patterns of diversity are linked to changes in composition and related to environmental factors such as environmental gradients. We argue that a complete theory for species diversity would have the potential, as is appropriate for the study system, to explain the following:

• How diversity varies at different scales ranging from that of single point samples, through elements of spatial and temporal turnover over different scales, to the regional scale. Most community theory is focused on the "local" scale and much less thought has gone into explaining diversity at other scales (e.g., Pimm 1982; Polis and Winemiller 1996; Morin 1999).

• How diversity is related to other major features of communities and ecosystems such as trophic structure (allocation of biomass into different functional groups), and rates of flow of materials through food webs and ecosystems. Again, much of "diversity theory" (e.g., Whittaker 1960; MacArthur and Wilson 1967; Magurran 1988) is focused on explaining patterns in the number of species without relating this to the ways these species participate in other important ecological processes.

• How patterns of diversity at different scales are related to processes involving dispersal as it affects either colonization rates (rates of introduction of novel species into communities where they were previously absent) or population dynamics per se (frequently involving *mass effects, rescue effects* or *source-sink* relations among different local communities—see table 1.1 for definitions of italicized words). While much recent work has explored the effects of dispersal in a piecemeal fashion there is still much to do to understand the full spectrum of dispersal-mediated dynamics that might occur among interacting species at different spatial and temporal scales. This question builds on the foundations of island biogeography and metapopulation theory (e.g., MacArthur and Wilson 1967; Hanski and Gilpin 1997; Hanski and Gaggiotti 2004).

   This book aims to begin to provide a body of work that can address all of these questions in a unified way. We encourage readers to think broadly about the kinds of empirical, theoretical and synthetic work that can contribute to understanding species diversity and the spatial structure of communities, coalescing around the theme of metacommunities.

   During the last few years ecologists have increasingly questioned whether the existing conceptual framework of community ecology is adequate for describing the dynamics of communities that are connected across space. The metacommunity concept has emerged as a new and exciting way to think about spatially-extended communities. It leads us to ask novel questions about the mechanisms that structure ecological communities and that create emergent patterns, such as patterns of species diversity and distribution. A *metacommunity* can be defined as a set of local communities that are linked by dispersal (Hanski and Gilpin 1991; Wilson 1992; table 1.1). In turn, a *community* may be defined as a collection of species occupying a particular *locality* or habitat. These definitions describe a hierarchy of scales and emphasize the ways in which processes occurring at smaller scales interact with those at larger scales (e.g., Levins and Culver 1971; Vandermeer 1973; Crowley 1981; Law et al. 2000; Mouquet and Loreau 2002). It is these interactions among processes at different spatial scales that are central to meta-community thinking and that form the core of this book.

This introductory chapter has five purposes. First, we elaborate on the motivation for studying metacommunities. Second, we flesh out the metacommunity concept by building on the definitions above. Third, we provide a set of definitions in table 1.1 that facilitate discussion. Fourth, we describe four conceptual models that help to simplify thinking about metacommunities (following Leibold et al. 2004). Fifth, we highlight the variety of ways in which metacommunities are being studied by introducing the rest of this book.

The Need for the Metacommunity Concept

This book arose because of empirical and theoretical gaps in the ecological literature that could limit the success of both pure and applied ecology. This section describes some problems that indicate the need to consider the spatial dynamics of communities.

A good example of a classical community ecology concept that has been misleading because of our failure to explicitly consider space is the intermediate disturbance hypothesis (IDH) (Connell 1978). The IDH is the most frequently cited nonequilibrium mechanism of species coexistence (Wilson 1990), and predicts that species diversity will be greatest at intermediate levels of disturbance. In thirty-six empirical studies (Shea et al. 2004), what counted as "intermediacy" of disturbance was defined in terms of intensity (seventeen cases), frequency (thirteen cases), time since disturbance (three cases), extent (two cases), and duration (one case). However, of twenty-seven published empirical tests of the IDH, only ten (37%) showed the predicted relationship of maximum species diversity at intermediate disturbance (Holyoak, unpublished data). Furthermore, Roxburgh et al. (2004) pointed out that disturbance per se is not the coexistence mechanism involved in the IDH. Instead the *storage effect* and *relative nonlinearity* (see table 1.1 and Hoopes et al., chapter 2 for further explanation) are the mechanisms of coexistence; these may be independent of disturbance in many systems. The lack of congruency with the IDH in many empirical tests is therefore not surprising because disturbance is not necessarily the mechanism of coexistence even when disturbance influences communities! Roxburgh et al. (2004) made possible the identification of coexistence mechanisms by searching for indicators of relative nonlinearities and the storage effect within spatially explicit models. In a lucid review of empirical studies, Shea et al. (2004) began the search for such mechanisms and clarified the role of disturbance in nature. Our initial ideas about the IDH (e.g., Connell 1978), and their recent reinterpretation (Roxburgh et al. 2004; Shea et al. 2004), are a good example of where taking a closer look at spatial dynamics has led to important new insights.

A second motivation for studying metacommunities comes from our desire to

conserve biodiversity in landscapes experiencing fragmentation. Habitat fragmentation creates patchy landscapes in which dispersal may be required for persistence, and is acknowledged to be an important factor driving the loss of biodiversity (e.g., Wilcove et al. 2000). However, fragmentation studies typically use
empirical trends to predict how communities will change during fragmentation
because we lack a general metacommunity theory to guide us in how to measure
and analyze natural fragmented communities. In a recent book on forest fragmentation and management, Lindenmayer and Franklin (2002) recount a large
number of examples where fragmentation produced largely unexpected effects
either on individual species or biodiversity. Experimental studies of fragmentation also frequently produce "surprising" effects (Debinski and Holt 2000). Unexpected effects took a variety of forms, but commonly observed phenomena
were that fragmentation responses were influenced by the nature of the habitat
"matrix" between patches (see also Davies et al., chapter 7), and by changes in
habitat within patches (e.g., edge effects). Empirical work on fragmentation often
investigates the ability of species' traits to predict responses to fragmentation, but
rarely attempts to explicitly deal with community structure (metacommunity
studies, such as those in this book, are exceptions to this generalization). Metapopulation models provide a motivation for studying species interactions within
communities. Single species are equivalent to noninteracting species and specialist predators and prey or competitors exemplify interacting species. In single species metapopulation models, the subdivision of habitat that results from fragmentation can only be detrimental—as fragmentation proceeds, previously
stable populations in large undivided habitats become increasingly small and isolated, making them vulnerable to local extinction through demographic stochasticity, but with a reduced capacity for patches to be to be recolonized (Harrison
and Taylor 1997). For interacting pairs of species, where a species can drive another locally extinct, persistence and diversity can actually be enhanced by fragmentation (subdivision). This may occur as formerly extinction-prone interacting populations in large areas of habitat become fragmented and various *spatial
dynamics* (e.g., colonization-competition trade-offs, see table 1.1) become possible that can enhance persistence and diversity (Harrison and Taylor 1997;
Hoopes et al., chapter 2). The degree to which species negatively interact could
therefore be critical to the way in which species respond to fragmentation. It is an
open question whether the responses of biodiversity to fragmentation are best
predicted using community-level theory (metacommunities) or species-level
theory (metapopulations), and the answer is likely to depend on the degree to
which species interact, on how such interactions are modified by spatial dynamics, and by how such pair-wise interactions are embedded in more complex multispecies communities.

The absence of a theory that provides mechanisms for responses to fragmentation potentially limits both our ability to predict how communities will change

under altered circumstances and our ability to effectively manage communities and metacommunities by manipulating habitat factors at landscape scales. Since the most general goal of conservation efforts is to maintain biodiversity, it is worrying that we at present attempt this without a complete theory that can explain the maintenance of biodiversity over ecologically relevant periods of time. These deficiencies in knowledge also carry over to managing fisheries through protecting areas in marine reserves, to restoring habitats where placement of restoration sites is an issue, to managing invasive (and spreading) species, to predicting the impacts of climate change, and to managing ecosystem properties that are linked to biodiversity.

A specific example helps provide motivation for studying the role of community-level mechanisms and especially species interactions in understanding responses to habitat fragmentation (Allan et al. 2003; LoGiudice et al. 2003). Forest fragmentation and habitat destruction in Dutchess County (NY, USA) have been shown to reduce mammalian species diversity and to elevate population densities of white-footed mice (*Peromyscus leucopus*). Fragmentation is also expected to cause an increase in the human exposure to Lyme disease because the disease's vector, black-legged ticks (*Ixodes scapularis*), are more likely to be infected with the Lyme bacterium (*Borrelia burgdorferi*) after feeding on mice compared to other vertebrate hosts. The frequency of tick infection declined linearly as fragment area increased, while mammalian species diversity increased, and mice density declined (Allan et al. 2003). Different vertebrate species have been shown to harbor different numbers of ticks, leading to different survival rates of ticks and different infection rates of ticks with the Lyme bacterium. White-footed mice are overwhelmingly the greatest producers of infected ticks, and the ability of mice to produce ticks is different for the various vertebrate hosts (figure 1.1; LoGiudice et al. 2003). Squirrels (*Sciurus carolinensis* and *Tamiasciurus hudsonicus*) are estimated to have the greatest combined effects in reducing the potential for Lyme disease (figure 1.1) (because of mechanisms like competition between vertebrate host species and tick preference for different hosts). Several questions follow from these observations, and all of them are likely to require spatial answers: (1) What are the implications of the differences among vertebrate hosts and differences in the sequence of community assembly for the occurrence of Lyme disease (LoGiudice et al. 2003)? Community assembly in fragments results from the movement of species between fragments (a spatial dynamic). (2) Are vertebrate hosts responding directly to habitat change or are they undergoing indirect changes caused by interactions with other species? This question is also central to testing the "species sorting perspective" of metacommunities sketched later in this chapter. (3) How do species interactions between the Lyme bacterium, the tick and vertebrate hosts operate? Predator-prey and host-disease metapopulation models show the potential for these interactions to be strongly influenced by spatial dynamics (see Hoopes et al. chapter 2, and Holt and Hoopes, chapter 3).

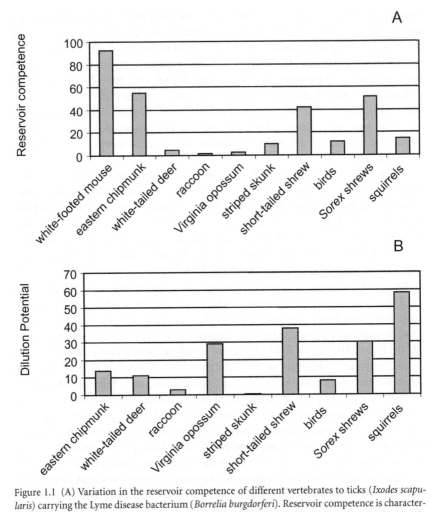

Figure 1.1 (A) Variation in the reservoir competence of different vertebrates to ticks (*Ixodes scapularis*) carrying the Lyme disease bacterium (*Borrelia burgdorferi*). Reservoir competence is characterized by three components: susceptibility of the host to the infection when bitten by an infected tick, the ability of the pathogen to magnify and persist in the host, and the efficiency of transmitting the bacterium to ticks. Reservoir competence was measured in the field. (B) The ability each species to reduce the effect of white-footed mice (the most competent reservoir) on the prevalence of infected ticks (measured as a percentage dilution) in a two-host community consisting of mice plus the focal species and a focal community in which mice are the only potential host. Data from LoGiudice et al. (2003).

These examples illustrate substantial gaps in our knowledge that require examination of the role of spatial structure and dynamics in ecological communities. This book provides many further examples of problems that motivate the study of metacommunities. The caveats introduced into various pieces of work show that we are just beginning on a journey of discovery. This volume is intended to provide for a broad community of basic and applied ecologists the essential conceptual building blocks for further exploration of metacommunities.

Defining Metacommunities

Earlier we defined a *metacommunity* as a set of local communities that are linked by dispersal (Hanski and Gilpin 1991; Wilson 1992), and a *community* as a collection of species occupying a particular *locality* or habitat (table 1.1). This set of definitions works well for conceptualizing metacommunities, but is often complicated by the complex nature of real metacommunities. This section discusses some of these complexities, first for communities then for metacommunities.

Defining Local Communities

Local communities can be defined in various ways. A theoretical approach is to define a community as encompassing an area within which all individuals are equally likely to interact, precluding any spatial heterogeneity in distribution or abundance. This highly simplified view in effect assumes that mass action and mean field conditions are adequate descriptors of dynamics, as seen in population dynamic models such as the classic Lotka-Volterra equations and their extensions (e.g., May 1973, Pimm and Lawton 1978, McCann et al. 1998). There are at least three more practical approaches, which are not mutually exclusive. The first is to select systems with relatively discrete boundaries, such as lakes and ponds. Similarly, in metapopulation studies discrete habitat *patches* are often delineated (Hanski and Simberloff 1997). Extending this idea to communities is difficult because species making up these communities often vary in their scales of movement (Schoener 1983; reviewed by Clobert et al. 2001). A second approach is to define communities based on structurally dominant organisms (usually plants); hence, we might discuss oak woodlands or tallgrass prairies (Clements 1936). The utility of this approach depends on the relative influence of physical conditions, dispersal, stochastic events, species interactions, and succession (e.g., Gleason 1926, Clements 1936). This approach is useful for habitat types and sets of species that are known to be repeatable in occurrence and sufficiently permanent relative to the duration of typical ecological studies to permit field studies. A third approach is to define a community as including all of the species present at a selected locality. This common working definition of a community has a suite of problems that are also relevant to the habitat-based approach, including identifying the relevant spatial scale at which to describe a community (is there a nonarbitrary

Table 1.1 Terms used in defining metacommunities

Term	Definition
Ecological scales of organization	
Population	All individuals of a single species within a habitat patch.
Metapopulation	A set of local populations of a single species that are linked by dispersal (after Hanski and Gilpin 1997).
Community	The individuals of all species that potentially interact within a single patch or local area of habitat.
Metacommunity	A set of local communities that are linked by dispersal of multiple potentially interacting species (Wilson 1992).
Descriptions of space	
Patch	A discrete area of habitat. Patches have variously been defined as microsites or localities (Levins 1969; Tilman 1994; Hanski and Simberloff 1997; Amarasekare and Nisbet 2001; Mouquet and Loreau 2002).
Microsite	A site that is capable of holding a single individual. Microsites are nested within localities.
Locality	An area of habitat encompassing multiple microsites and capable of holding a local community.
Region	A large area of habitat containing multiple localities and capable of supporting a metacommunity. This corresponds to the mesoscale (Holt 1993).
Types of dynamics	
Spatial dynamics	Spatial changes in the distributions or abundances of individuals or species. Different types of mechanisms are discussed by Holyoak and Ray (1999) and Hoopes et al. (chapter 2).
Mass effect	A mechanism for spatial dynamics in which there is net flow of individuals created by differences in population size (or density) in different patches (Shmida and Wilson 1985).
Rescue effect	A mechanism for spatial dynamics in which there is the prevention of local extinction of species by immigration (Brown and Kodric-Brown 1977).
Source-sink effect	A mechanism for spatial dynamics in which there is the enhancement of local populations by immigration in sink localities due to migration of individuals from other localities where emigration results in lowered populations.
Storage effect	Subadditivity in a species (usually the poorer competitor's) response to competition in good and poor environments. This response is reflected at the population level through the presence of buffering mechanisms (seed banks, diapause), which allow the species to store resources during times of relative harshness and yet reemerge in the population at other times (after Shea et al. 2004; see Hoopes et al., chapter 2 for further explanation).
Relative nonlinearity	A persistence mechanism where the population growth rates of competing species respond differently and nonlinearly to competition (or resource availability). For example, if one responds linearly to increasing competition while the other responds in a nonlinear fashion and is negatively affected by high competition (after Shea et al. 2004).
Colonization	A mechanism for spatial dynamics in which populations become established at sites from which they were previously absent.
Dispersal	Movement of individuals from a site (emigration) to another (immigration).
Stochastic extinction	A mechanism whereby established local populations become extinct for reasons that are independent of other species present or of any deterministic change in patch quality. Possible mechanisms include stochastic components associated with small populations and extinctions due to stochastic environmental changes (i.e., disturbances) that can affect large populations.

Term	Definition
Deterministic extinction	A mechanism whereby established local populations of component species become extinct due to deterministic species interactions or aspects of patch quality.
Metacommunity dynamics	The dynamics that arise within metacommunities. Logically, these consist of spatial dynamics, community dynamics (multispecies interactions or the emergent properties arising from them within communities), and the interaction of spatial and community dynamics. Care needs to be taken to use the term only when another, existing, term would not suffice.

Types of model population or community structure

Classic (Levins) metapopulation	A group of identical local populations with finite and equal probabilities of extinction and recolonization—no rescue effects occur.
Source-sink system	A system with habitat specific demography such that some patches (source habitats) have a finite growth rate of greater than unity and produce a net excess of individuals that migrate to sink patches. Populations in sink habitats have finite growth rates of less than one and would decline to extinction without immigration from sources (Holt 1985; Pulliam 1988).
Mainland-island system	A system with variation in local population size that influences the extinction probability of populations. Systems are usually described as consisting of extinction-resistant mainland populations and extinction-prone island populations (Boorman and Levitt 1973).
Open community	A community which experiences immigration and / or emigration.
Closed community	A community that is isolated, receiving no immigrants and giving out no emigrants.
Patch occupancy model	A model in which patches contain either individuals or populations of one or more species and where local population sizes are not modeled.
Spatially explicit model	A model in which the arrangement of patches or distance between patches can influence patterns of movement and interaction.
Spatially implicit model	A model in which the arrangement of patches and / or individuals does not influence the dynamics of the system. Movement is assumed equally likely between all patches.
Spatial dependence	Implies that the response variable is spatially structured because it depends on explanatory (e.g., physical) variables that are themselves spatially structured by their own generating processes (Legendre et al. 2002). Spatial dependence between abundance and spatial habitat factors is implied most strongly by the species sorting perspective, and is assumed absent in the neutral perspective.
Spatial autocorrelation	Implies that the value of response variable y at site j is assumed to result from some dynamic process within variable y itself. Spatial autocorrelation actually refers to the lack of independence among the error components of field data, as a function of geographic distance among the sites (Legendre et al. 2002).

Metacommunity perspectives

Patch dynamic perspective	A perspective that assumes that patches are identical and that each patch is capable of containing populations. Patches may be occupied or unoccupied. Local species diversity is limited by dispersal or by species interactions. Spatial dynamics are dominated by local extinction and colonization.
Species sorting perspective	A perspective emphasizing that resource gradients or patch types cause sufficiently strong differences in the local demography of species and the outcomes of local species' inter-actions that patch quality and dispersal jointly affect local community composition. This perspective emphasizes spatial niche separation above and beyond spatial dynamics. Dispersal is important because it allows compositional changes to track changes in local environmental conditions.

Table 1.1 continued

Term	Definition
Mass effects perspective	A perspective that focuses on the effect of immigration and emigration on local population dynamics. In such a system species can be rescued from local competitive exclusion in communities where they are bad competitors by immigrating from communities where they are good competitors. This perspective emphasizes that spatial dynamics affect local population densities.
Neutral perspective	A perspective in which all species are similar in their competitive ability, movement, and fitness (Hubbell 2001; Chave 2004). The dynamics of species diversity are then derived both from probabilities of species loss (extinction, emigration) and gain (immigration, speciation). Community composition drifts through time, termed "ecological drift."

way of defining the spatial boundaries of the community? Wiens 1989; Levin 1992) and the appropriate criteria for including tourist species that are present for only part of the time (e.g., Abramsky and Safriel 1980; Cousins 1990).

The problem of defining communities is simpler in experimental and some natural model systems, such as artificial laboratory systems (Lawler 1998; Jessup et al. 2004), natural phytotelmata (Kitching 2000), and other systems where patches can be sampled in their entirety (e.g., Worthen 1989). The practicality of this approach depends on whether the metacommunity can be adequately described by focusing on particular patch types and ignoring the habitat matrix and other resources in the environment. It is also often not clear whether the results of such investigations are applicable in all generality to more complex systems at larger spatial scales (e.g., Naeem 2001). An alternative approach is to develop statistical and analytical techniques for defining community boundaries in a wider range of natural systems. Although invaluable and indeed often inescapable, this tactic runs the risk that the original objective—understanding diversity—may be subsumed in the need to grapple with a welter of multivariate complexities.

Defining Metacommunities and Metacommunity Dynamics

A metacommunity is easiest to conceptualize when all of the interacting species utilize the same set of discrete habitat patches and have local populations that use resources at the same within-patch scale. However, many communities lack discrete boundaries, and indeed many populations are regulated over multiple spatiotemporal scales. In addition, the exact spatial placement of habitats can influence metacommunity dynamics, and species can differ in the extent to which they disperse. Interpatch dispersal at sufficiently low rates can lead to variation in species composition from patch to patch. We believe that although existing models of metacommunities (e.g., Hubbell 2001; Mouquet and Loreau 2002, 2003) give a highly simplified representation of the spatial complexities of natural assemblages of organisms, the insights that have come from them are useful in developing theories for more complex metacommunity scenarios.

In this book, we will frequently refer to *metacommunity dynamics*. Metacommunity dynamics logically consist of either the spatial dynamics (table 1.1) or regional properties (resulting from dynamics) of communities occupying two or more interconnected patches. To be useful, the term *metacommunity dynamics* should be distinct from existing definitions. In contrast to metapopulation dynamics, metacommunity dynamics should involve more than two interacting species. The term should also be distinguished from the term *community dynamics;* the metacommunity concept would be most useful if it pertained to a system in which the dynamics of individual species were altered both by species interactions (of more than two species) and dispersal. This book describes a wide range of dynamics that arise because of metacommunity structure. Care should be taken to use preexisting terms such as *population dynamics, community dynamics, spatial and metapopulation dynamics* as complementary concepts to metacommunity dynamics.

In the remainder of this section we discuss three problems that must be addressed when considering metacommunity dynamics. First is the extent to which species can interact and the system still can be usefully considered a metacommunity. Second, we address the influence of dispersal on metacommunity structure. Third, we discuss the representation of spatial and temporal dynamics in metacommunities. We discuss the choice of how many spatial scales to recognize, whether spatial arrangement of habitat is explicitly considered or not, and whether habitat is treated as static or instead has its own dynamics.

SPECIES INTERACTIONS

A central community and metacommunity question is the extent to which species interact (Laska and Wootton 1998; Berlow et al. 1999). We can distinguish three limiting cases that encompass most possible metacommunities. The first case provides a null model for predicting expected patterns of assemblages of independent species. It consists of a multipatch system containing multiple independent metapopulations of different species, such that there are no interspecific species interactions with population dynamical consequences. Each metapopulation is a "set of local populations which interact via individuals moving among populations" (Hanski and Gilpin 1991; table 1.1). In such a case, no particular added understanding is gained from either the community or metacommunity perspective, and an equal understanding could be obtained by separately considering the metapopulation dynamics of each species. The second case consists of communities where all species interact in a factorial manner. If interactions are strong enough to cause extinctions of species from local communities (e.g., McCann et al. 1998) and there is a high between-species variance in competitive or predatory ability, then a single dominant (keystone) species can have a disproportionate impact on the rest of the community. In this scenario, local and regional dynamics of all of the species can be predicted by using metapopulation models for either a single (dominant) species or two interacting species (the dominant and each prey or inferior competitor in turn). In the third case, species interactions are weaker but still influence population dynamics and between-species variance in competitive and predatory abilities is smaller. This third case is novel in its simultaneous consideration of the spatial dynamics and interactions of more than two species, which may be more complex than those considered by current metapopulation models. The metacommunity examples in this book, together with metapopulation models (Hoopes et al., chapter 2) and work on food web modules (McCann et al. 1998; Holt and Hoopes, chapter 3), leads us to predict that variation in the extent to which species interact has profound consequences for metacommunity dynamics.

HOW MUCH DISPERSAL?

Another question that arises when defining metacommunities is how much dispersal (as measured, for example, by numbers of individuals and species, and distances moved) is required for a system to be deemed a metacommunity? This question has broad ecological and evolutionary consequences (Ims and Yoccoz 1997; Clobert et al. 2001). Dispersal may influence both local and regional dynamics. In the context of metacommunities, *spatial dynamics* can be defined as the regional dynamics that arise in multiple patches that are linked by movement (table 1.1). The mechanisms of spatial dynamics broadly include colonization-extinction dynamics, rescue effects, habitat-specific demography, and the dynamics of habitat patches themselves (table 1.1; reviewed by Holyoak and Ray 1999; Hoopes et al., chapter 2). Movement may also alter local demography, species interactions, and the aggregate properties of communities. In many patchy systems, it is likely that multiple species are mobile and that dispersal varies among species. Different species are likely to have their own rates of movement that represent a combination of evolved abilities and responses to their environment (e.g., Holling 1986, Clobert et al. 2001, Rodríguez 2002).

The question of how much movement is best answered using mathematical models. The influence of dispersal in this volume is reviewed extensively by Hoopes et al. in chapter 2, Holt and Hoopes in chapter 3, Mouquet et al. in chapter 10, and Loreau et al. in chapter 18. Techniques where the influence of dispersal can be analyzed are presented by Mouquet et al. in chapter 10, Law and Leibold in chapter 11 and Chesson et al. in chapter 12. Both theory and observation paint a rich tapestry of movement-related life-history patterns (e.g., Roff 2001, Kneitel and Chase 2004). A well-known example of such a life-history pattern is the competition-colonization trade-off, where a negative correlation between colonization (dispersal) and competitive ability allows many species to coexist (Kneitel and Chase 2004; Hoopes et al., chapter 2; Mouquet et al. chapter 10). To date, other life-history patterns involving movement have generally not been incorporated into metacommunity models.

REPRESENTATION OF SPATIAL AND TEMPORAL DYNAMICS IN METACOMMUNITIES

Up to this point, this chapter has considered only local and regional scales. It is also possible to imagine species interactions occurring over more than two scales (e.g., Kolasa and Romanuk, chapter 9). The number of spatial scales that are required to represent the dynamics of real metacommunities is not yet clear; indeed, a continuum of scales may in the end prove to be involved in determining the spatial dynamics of communities. Many current models of metacommunity dynamics, especially those inspired by work on sessile taxa, are based on a three-level hierarchy of scales (table 1.1). At the smallest scale, *microsites* can hold a single individual. Microsites are nested within *localities* that hold local communities similar to those in conventional species interaction models. In turn, local communities are connected to other such communities as part of a metacommunity occupying a *region*. In much of the literature, localities are equivalent to habitat patches. This distinction becomes blurred in patch occupancy models (Hoopes et al., chapter 2; Mouquet et al., chapter 10), where patches can be viewed either as microsites or localities holding individuals, populations, or communities. We use the term *patch* as equivalent to a locality capable of holding a local population or community (table 1.1).

In considering how to measure and represent metacommunities it is also necessary to decide whether the explicit spatial location of individual communities or patches should be recorded and tracked, or not (spatially implicit analyses; table 1.1). There is a cost to explicitly representing space, since a great deal of extra information needs to be recorded and analyzed, and mathematical models frequently become analytically intractable (e.g., Durrett and Levin 1994a, 1994b, Pacala et al. 1996). There are a wide variety of kinds of dynamics that can be represented in spatially explicit models, but not in spatially implicit models (see Hoopes et al., chapter 2, and Holt et al., chapter 20, for examples). Ignoring explicit space may also lead us to view metacommunities (and metapopulations) as being simpler than they actually are (Hanski and Gaggiotti 2004). For example, mass effects (see table 1.1 and the next section) are more likely with localized dispersal and when populations of particular species in adjacent areas differ greatly in size (Hoopes et al., chapter 2; Mouquet et al. chapter 10). Mass effects could easily be overlooked in spatially implicit analyses using either field or model data. Consequently, we ignore spatially explicit patch arrangement at our peril. Nonetheless, spatially implicit models are valuable because of their analytical tractability. They also allow spatially implicit field community data to be analyzed even if not all of the communities within a landscape were sampled. There are a growing number of statistical methods for dealing with spatially explicit data and these could be valuable for metacommunity analyses (e.g., Borcard and Legendre 2002; Legendre et al. 2002).

In most of this book we treat habitat patches as permanent, whereas they may have their own dynamics (transient patches are treated by Miller and Kneitel in chapter 5, and Chesson et al. in chapter 12). Creation of habitat patches (e.g., by disturbance) may produce new opportunities for colonization, and patch destruction (e.g., by succession) can eliminate communities within patches. Changes within patches, such as through succession, can also alter patch quality, thereby altering local demography (e.g., through source-sink dynamics; Pulliam 1988; table 1.1). For ecosystems, Holling and colleagues have developed links between processes at multiple scales, including processes that create and destroy habitat patches (Holling 1994; Peterson et al. 1998; Allen and Holling 2002). Resilience is generated by diverse, but overlapping, functions within a scale and by apparently redundant species that operate at different scales, thereby reinforcing ecosystem functioning across scales (see also Loreau et al., chapter 18). The distribution of functional diversity within and across scales enables regeneration and renewal to occur following ecological disruption over a wide range of scales (Peterson et al. 1998). A key insight is that the relationship of organisms and processes at different scales interacts with renewal (patch creation) processes. Community ecology has tended to take a more simple view of the dynamics of habitat, such as simply creating empty patches by disturbance (e.g., Connell 1978). There is also a growing literature about how the destruction and creation of habitat patches influences metapopulation dynamics (e.g., Ellner and Fussman 2003, Hastings 2003). While few contributions in this book directly address intrinsic habitat dynamics, the studies mentioned above demonstrate that such dynamics can have strong effects on populations, communities, and ecosystems. Such studies provide interesting areas for future research, especially when considered together with species' traits, life histories, and evolution (see also Leibold et al., chapter 19).

Four Perspectives on Metacommunities

We present four conceptual models to describe metacommunities, and each model illuminates different aspects of spatial community dynamics (table 1.2). Because several factors differ between the models (see also Chase et al., chapter 14), deciding which is more appropriate for a particular study system should not be the main aim. Rather, studies should investigate the mechanisms driving dynamics (e.g., the factors in tables 1.2 and 14.1). The integration of the different metacommunity models with one another is ongoing (e.g., Mouquet et al., chapter 10), as are more detailed investigations of the population dynamic mechanisms (Law and Leibold, chapter 11; Chesson et al., chapter 12).

To date, theoretical and empirical work on metacommunities largely falls along four broad perspectives that we refer to as the "patch dynamic," "species sorting," "mass effects" and "neutral" perspectives (table 1.2). Below we present the theory, but reserve discussion of empirical examples until later in this chapter.

The Patch Dynamic Perspective

The first perspective extends metapopulation models for patch dynamics to more than two species. Because it considers multiple species it also can be considered to build on the equilibrium theory of island biogeography (MacArthur and Wilson 1967). This approach assumes the existence of multiple identical patches (islands) that undergo both stochastic extinctions (as in standard single species metapopulation dynamics) and deterministic extinctions (like metapopulation models for interacting species; Harrison and Taylor 1997; Hoopes et al., chapter 2). Dispersal counteracts these extinctions by providing a source of colonization into empty patches. For coexistence to occur, dispersal rates must be limited so that dominant species cannot drive their competitors or prey to regional extinction. Because all patches are identical and there are no permanent refuges for species, it is likely that local within-patch species composition and diversity will change through time.

The equilibrium theory of island biogeography also assumes a prominent role for extinction and colonization in setting levels of biodiversity on islands. However, in the equilibrium theory species from a fixed pool of mainland species randomly colonize islands (patches), so that mainland species diversity determines the regional (all-island) species diversity. Empirical evidence, including that discussed by MacArthur and Wilson (1967), clearly indicates that this is not always realistic because many systems do not have a large mainland with a fixed species composition. The equilibrium theory of island biogeography considers only the number of species in a community and does not include community (trophic) structure, species identities, or niche differentiation (Chase and Leibold 2003; Holt et al., chapter 3). Metapopulation models and the patch dynamics perspective differ from the equilibrium theory in that they recognize that spatial dynamics (reviews: Holyoak and Ray 1999; Hoopes et al., chapter 2) can enhance persistence and that the number of species in a region might emerge from an agglomeration of dynamics within many interlinked patches.

Metapopulation models typically contain only simplified food web structure, which often arises from considering only one or two species, or species within a single trophic level or pairs of levels (reviews: Hanski and Gilpin 1997; Hoopes et al., chapter 2; Mouquet et al., chapter 10; examples of multispecies models include: Hastings 1980; Hassell et al. 1994; Tilman 1994; Amarasekare and Nisbet 2001; Mouquet and Loreau 2002, 2003). Three approaches have been used to model these kinds of dynamics, which we describe for competing species and then extend to consumer-resource systems.

Models based on patch dynamics often utilize occupancy formalisms in which patches are either vacant or are occupied by populations at equilibrium (usually point equilibrium density, but possibly under other nonpoint equilibria). This formalism is consistent with an assumption that local dynamics occur on a faster timescale than does dispersal or the regional dynamics from colonization and extinction. The simplest version of this model considers only regional coexistence of competing species (Levins and Culver 1971) and does not explicitly consider local dynamics. For competitive metacommunities in a homogeneous environment, regional coexistence is possible given an appropriate trade-off between competitive ability and dispersal. Yu and colleagues (Yu and Wilson 2001; Yu et al. 2001) have considered a trade-off between fecundity and dispersal, and Adler and Mosquera (2000) examine a trade-off between mortality and competitive ability in interference competition, all with similar conclusions.

This classic two-level (local versus regional) approach has been rescaled by Tilman (1994), and briefly by Hastings (1980), who considered a single community divided into single-resource patches that contain at most an individual (microsites rather than localities in our terminology; table 1.1). Because microsites hold single individuals, extinction rates are reinterpreted as mortality rates, and colonization as birth and movement. The results of this approach are essentially the same as the above approach: coexistence is possible given an appropriate trade-off between competitive and colonization abilities (or fecundity).

A third type of formalism simulates localities containing populations with local dynamics, also represented by Lotka-Volterra equations and, patches linked by diffusive dispersal (e.g., Case 1991). This model can produce rather more complex results than the previous two formalisms (e.g., Hoopes et al., chapter 2; Mouquet et al., chapter 10), but many similar results emerge.

The effect of predator-prey interactions on regional persistence has been considered in patch occupancy models with patches containing individuals or populations (e.g., Caswell 1978, McCauley et al. 1993) and also in models with explicit local dynamics (e.g., Crowley 1981). Adding predators or competitors that are capable of causing local extinctions of other species to these models leads to constraints on the dispersal rates at which regional persistence is possible. For instance, prey species must colonize patches faster than they are driven extinct, and more rapidly than predators, and persistence may only be possible at intermediate dispersal rates (reviews: Kareiva 1990; Taylor 1990; for other examples see Hassell et al. 1994, Hess 1996).

The Species Sorting Perspective

The second approach builds on theories of community change over environmental gradients (Whittaker 1972) and considers the effects of local abiotic features on population vital rates and species interactions (Leibold 1998). In this perspective, the ensemble of local patches is heterogeneous in some local factors, and the outcome of local population dynamics (species interactions and individual species' responses) depends on these spatially varying aspects of the abiotic environment. Like many patch dynamics models, this approach assumes a separation of time scales between local population dynamics and colonization-extinction dynamics. Populations are assumed to be able to reach their equilibrium behavior (stable points, oscillations, or complex attractors) between the time of colonization and when environmental perturbations might cause local extinction. Colonization is assumed to occur frequently enough that local assembly trajectories reach their endpoint states (Law and Morton 1993), but not so often that mass effects occur (see the next section). Such endpoint states are often communities that are uninvasible by any of the other species in the metacommunity. However, endpoints can sometimes consist of cyclical patterns of compositional change among a given set of compositional states (Law and Morton 1993, Steiner and Leibold 2004). The net effect is that the species present within a community are determined by the abiotic conditions in local patches. Consequently, local species diversity and composition are expected to be relatively constant or bounded through time. Dispersal is restricted to colonization and does not extend to allowing species to persist in sink habitats.

This species sorting perspective has much in common with traditional theory on niche separation and coexistence (Dobzhansky 1951; MacArthur 1958; Pianka 1966). Indeed, in this traditional view, local abiotic conditions determine community composition. The main differences are that a metacommunity perspective forces us to think about the links between local and regional diversity, and about the role of regional diversity in making local communities appear saturated (Shurin and Srivastava, chapter 17). The result is that species distributions are closely linked to local conditions and (unlike the patch dynamic perspective) are largely independent of unrelated purely spatial effects (Cottenie and de Meester, chapter 8; Leibold and Norberg 2004). Different model formalisms become more appropriate as dispersal increases and mass effects exert a dominant influence on metacommunity dynamics.

The Mass Effects Perspective

While the patch dynamic and species sorting perspectives assume a separation of time scales between local dynamics and colonization-extinction dynamics, important regional dynamics may also emerge when local population dynamics are quantitatively affected by dispersal. The mass effects perspective (Shmida and Wilson 1985) represents a multispecies version of source-sink dynamics (Holt 1985, 1993; Pulliam 1988) and rescue effects (Brown and Kodric-Brown 1977; definitions are given in table 1.1). Differences in population density (or mass) at different locations, or asymmetric dispersal, can drive both immigration and emigration between local communities. Immigration can supplement birth rates and enhance densities of local populations beyond what might be expected in closed communities, and emigration can similarly enhance the loss rates of local populations. Such mass effects due to dispersal can have potentially strong influences on the relationships between local conditions and community structure (Holt 1993; Mouquet et al., chapter 10).

It should be noted that mass effects can occur in the absence of habitat heterogeneity from patch to patch, but are more predictable when habitats are heterogeneous and thus also are expected to fit the species sorting perspective. As noted by Holt and Hoopes (chapter 3), one basic difference between the two perspectives has to do with the emphasis placed on the strength of local exclusion due to interspecific interactions and abiotic conditions, relative to the magnitude of flux rates due to dispersal. In the absence of explicit patch differences, there are two versions of the model for competing species: a pure competitive, weighted lottery model (Chesson 1985; Chesson and Huntly 1989; Iwasa and Roughgarden 1986; Mouquet and Loreau 2002) and one based on the classical MacArthur model of species competition (Levin 1974; Amarasekare 2000; Amarasekare and Nisbet 2001). The two approaches introduce a constraint of regional similarity between coexisting species that adds some complexity to the predictions, but provides an important finding from these two approaches (Amarasekare and Nisbet 2001; Mouquet and Loreau 2002). Coexistence in such a metacommunity is obtained through a regional balance of local competitive abilities. As a consequence, species are locally different but regionally similar in their competitive abilities (Mouquet and Loreau 2002; Mouquet et al., chapter 10). Mass effects allowing local coexistence are constrained in complex ways (Amarasekare and Nisbet 2001) because coexistence requires spatial variance in fitness, which cannot be maintained at high levels of dispersal among patch types. This basic idea could be extended to consider other kinds of species interactions besides competition. Marked spatial heterogeneity in patch types is likely to reduce species turnover within local communities and to fix local communities in space, if enough time has elapsed for locally superior species to arrive and exert local dominance (Hoopes et al., chapter 2; Mouquet et al., chapter 10).

The Neutral Perspective

All of the above approaches assume that species differ significantly from each other either in their niche relations with local factors, and / or in their abilities to disperse or avoid local extinctions. The resulting dynamics depend on differences among species or the trade-offs that emerge from these assemblages, with multiple consequences at local and regional scales. In the absence of any such differences among species, the behavior of metacommunities can be dramatically different from models with trade-offs or species-specific differences (Caswell 1978; Hubbell 2001; Chave 2004). Neutral models predict a gradual loss of all competing species via a potentially slow process of random walks. The resultant temporal change in species composition was termed ecological drift by Hubbell (2001). Thus, in contrast to the other views described above, neutral models alone cannot explain how differences in local and regional diversity are maintained. Informative reviews of neutral models are provided by Bell (2001) and Chave (2004). Hubbell (2001) has explored a neutral model in situations assuming a time scale over which speciation counteracts the extinction process due to drift, and he points out that even slow speciation rates can sustain high levels of diversity in such metacommunities. Under slow speciation rates, the neutral model has its own metacommunity dynamics predominantly influenced by slow random patterns of compositional change in space and time. Even if Hubbell's model is described as being closer to a continent-island system than to a metacommunity, in which sets of communities are linked through immigration and emigration, it can be interpreted as the endpoint of a continuum of coexistence mechanisms within the metacommunity framework (just as it is in metapopulation theory, Hanski and Gyllenberg 1993). Neutral models also currently lack trophic or other community structure. Despite this limitation, the "neutral" view can be regarded as a null hypothesis for the other three views described above (cf. Bell 2000). However, it may also literally describe dynamics of some communities where species are close to being equivalent, or where transient dynamics are very long. Hubbell and Bell's neutral community models merit close attention because they predict a surprising number of community and metacommunity patterns (Bell 2001; Hubbell 2001). Neutral models and ways to test them are discussed further in Chase et al. (chapter 14).

Applying These Conceptual Models to Real Metacommunities

Real ecological communities are probably subject to both habitat variability and to local stochastic or nonequilibrium dynamics, limiting the explanatory power of each of these paradigmatic metacommunity models. A synthetic perspective on metacommunities would be a great improvement in understanding how communities are structured by the joint action of processes operating at both local and regional scales. Clearly all four of the perspectives outlined above capture interesting aspects of metacommunity dynamics. Further, it is unlikely that all of the species interacting in a given metacommunity will uniformly conform to any one of these perspectives. Instead, it is likely that each of these sets of processes will play interactive roles in structuring real metacommunities. The extent to which real metacommunities will conform to the predictions sketched above will depend on how well each system conforms to the assumptions of the models (table 1.2).

These models make at least two types of direct assumptions. First, the models differ in their assumptions about the nature of differences among localities. In the case of the patch dynamic and neutral models, the assumption is that local sites do not differ systematically in any respect except for the species composition that exists at any given moment in time (table 1.2). By contrast, the mass effects and species sorting perspectives assume that there are intrinsic, persistent differences among local sites in their attributes, so that different species might be favored at different sites. Second, these models differ in the amount of interpatch movement, which is assumed to be limited in neutral and patch dynamics models, but could be greater in the species sorting and mass effects perspectives (table 1.2). A

Table 1.2 A comparison of four conceptual models of metacommunities

Characteristic	Patch dynamics	Species sorting	Mass effects	Neutral models
(1) Patch similarity	Similar	Dissimilar	Dissimilar	Similar
(2) Interpatch movement	Low rate	Not specified; needs to be sufficiently high for species to be present in suitable patches but too low for mass effects	Higher and may be regional	Localized (i.e., not global)
(3) Species similarity	Similar or dissimilar. Competitive models require trade-offs for regional coexistence	Species must differ in their ability to perform under different conditions	Species must differ in their ability to perform under different conditions	All individuals have identical fitness
(4) Local and regional species composition	Local varies through time, regional is more constant	Local and regional are more constant	Local and regional are more constant through time, assuming that (2) is constant	Local and regional vary through time
(5) Spatial synchrony	At least some asynchrony	Not specified	Synchronous because of (2)	At least some asynchrony (but not specified)
(6) Equilibrium of local community dynamics	Not reached because of (2)	Assumed to be at an equilibrium condition	Not at local dispersal-free equilibrium because of high movement (2), but could reach a new equilibrium with dispersal	Absent because of drift

further assumption implied by spatially implicit models discussed in the previous sections is that patches are uniformly distributed over space (isotropy) and equally linked by dispersal. However, such a restriction would be unnecessary if a spatially explicit approach were taken.

There are several implicit assumptions in metacommunity models. The four models differ in the assumptions they make about the ecological traits of species involved in the metacommunity. The neutral model assumes that there is no variation (and hence no covariation) in ecological traits that influence net fitness. In the patch dynamics models for competitive metacommunities, the assumption is that competitive ability varies and that covariance with dispersal is sufficiently negative to permit regional coexistence (see Hoopes et al., chapter 2 and Mouquet et al., chapter 10). In the mass effects and species sorting models, the assumptions are that there are trade-offs in the abilities of species to perform well under different habitat conditions. Another way to think about this is as a form of *spatial dependence* (as opposed to *spatial autocorrelation;* see table 1.1) between species performance and the spatial environment (Legendre et al. 2002). Spatial dependence between abundance and spatial habitat factors is implied by the species sorting perspective, but mass effects would shift patterns in abundance away from strict spatial dependence on local conditions. Dispersal limitation (as in the patch dynamics perspective) may also prevent strong spatial dependence arising between habitats and populations. The neutral perspective by contrast assumes no spatial dependence between populations and habitat factors. Unlike spatial dependence, spatial autocorrelation could arise in any of the perspectives from spatially correlated variation arising due to similarity of environmental factors or populations that are closer together in space, and spatially localized dispersal.

A second implicit assumption is the emergent effects on species composition. Differences in patch conditions and corresponding species responses are likely to lead to more fixed species compositions in local communities under the species sorting perspective than with the mass effects perspective: the neutral and patch dynamics perspectives should lead to the most variability in composition (table 1.2). It is likely that these differences in local composition will also carry across to controls on regional composition.

Third, variation among these perspectives in assumptions about movement rates is likely to lead to differences in the synchrony of population fluctuations in different patches, as described in table 1.2. (This also assumes that there are some differences in local conditions that can alter demography.)

A final difference between the perspectives is whether local communities are at their theoretical equilibria, which would result from all species having arrived and interactions having played out through time in all patches (table 1.2). With limited dispersal, local dynamics are likely to have caused extinction on a rapid timescale compared to the time between colonization events. This makes it more likely that local communities do not contain the full complement of species that is theoretically possible in both the patch dynamics and neutral perspectives. In the idealized species sorting view communities are expected to be at their theoretical within-locality equilibrium because dispersal occurs at a rate sufficient to "seed" all communities with all potential occupants, but insufficient to otherwise perturb local dynamics. However, dispersal rates are sufficiently large that species arrive in localities more frequently than they go locally extinct. In the mass effects perspective, both source-sink dynamics for individual species and mass effects across species will perturb local communities from the theoretical equilibrium expected from closed communities, and a new regional equilibrium may or may not result (e.g., because sedentary competitive dominants in a local community are excluded by a high rate of "spillover" of less effective competitors from other communities; Holt et al. 2003).

Undoubtedly there are other logical differences between the different perspectives that could be drawn out (e.g., those in table 14.1). Ultimately, a theoretical synthesis of the different perspectives is required (Mouquet et al., chapter 10; Chase et al., chapter 14). There are also many factors not included in these four perspectives that are likely to influence metacommunity dynamics, such as local dynamics (Hoopes et al., chapter 2); synthesizing insights from ongoing and future empirical studies of metacommunities will doubtless reveal unexpected effects. These existing models are a starting point, rather than a complete framework, for metacommunity ecology. One major class of factors that is only briefly touched on in this book is evolutionary processes (but see Leibold et al., chapter 19, McPeek and Gomulkiewicz, chapter 15, and Holt et al., chapter 20). The dynamics of actual metacommunities may depend strongly on how the species pool has evolved (Shurin et al. 2000; Shurin and Srivastava, chapter 17). The integration of species' traits and life history relationships with metacommunity ideas, taking into account both microevolutionary dynamics and macroevolutionary processes, remains a major challenge.

A Roadmap for This Book:
The Variety of Ways to Think about Metacommunities

Making progress in understanding metacommunities is amenable to many approaches. The contents of this book are framed around four of these approaches: empirical perspectives consisting of both observational and manipulative studies, conceptual syntheses, theoretical approaches, and emerging areas and perspectives. Perhaps the greatest challenge in studying metacommunities is to integrate these approaches in productive ways. Below we discuss how these approaches map onto our proposed framework for metacommunity ecology.

Core Concepts

The next two chapters draw on traditional community ecology, metapopulation studies, and island biogeography to make the case that the metacommunity concept addresses key gaps in ecological understanding. Hoopes et al. (chapter 2) compare and contrast the predictions from spatial models for single noninteracting species versus pairs of interacting species (competitors, predators and prey, mutualists). Hoopes et al. (chapter 2), like Holt and Hoopes (chapter 3), demonstrate the central role that interactions between species and between species and the environment plays in metacommunities. They also emphasize that the interaction between dispersal and spatial structure is critical to metacommunity structure and composition. Holt and Hoopes (chapter 3) discuss the ways that predictions for food web modules (of three to four interacting species) are relevant to metacommunities. The approach is valuable because it allows ecologists to draw on a very large literature on species interactions within local communities, including evaluations of stability and trophic control (Murdoch and Oaten 1975;

Holt 1977; Kuno 1987; Abrams and Walters 1996; Holt and Polis 1997). Holt and Hoopes also discuss how island biogeographic thinking can be expanded to include simple trophic structure and discuss the consequences of this for community assembly.

Empirical Studies

This section illustrates a variety of empirical studies that have addressed the relevance of metacommunity concepts to particular systems.

Chapters 4 and 6 involve examples where, following on from the ideas in chapter 3, food web modules or community structure have been studied. Van Nouhuys and Hanski (chapter 4), describe patch dynamics in a Finnish system consisting of hundreds of patches containing a food web consisting of up to three plants, two butterfly species, five primary parasitoids, and two hyperparasitoids. Unlike in the next example, chapter 4 considers relatively uniform patches. Pitcher plants form temporary patches of aquatic habitat, requiring dispersal of at least some inhabitants, which range from bacteria to insects (Miller and Kneitel, chapter 5). Miller and Kneitel study a variety of community properties, such as community assembly and the response to nutrient enrichment. The pitcher plant system represents an example (like Resetarits et al., chapter 16) where the landscapes consists of patches that vary in position both spatially and temporally, and in local site quality. The dynamics of such systems may be strongly dependent on traits related to spatial dynamics such as dispersal and dormancy (Harrison and Taylor 1997; McPeek and Kalisz 1998).

Chapter 6 by Gonzalez considers a model empirical system that is useful for examining the influence of movement on patterns of species diversity. Carpets of epilithic moss containing a species-rich assemblage of microarthropods represent readily manipulable microlandscapes. The system illustrates the consequences of dispersal for various aspects of local and regional species diversity.

In chapter 7, Davies et al. study the most elusive form of metacommunity—systems in which habitats are permanent but patch boundaries are less distinct. In their study of a landscape containing *Eucalyptus* forest fragmented by the planting of nonnative pine woodland, they show that assemblages of ground dwelling beetles are characterized by lower temporal population variability in fragments than in more spatially continuous habitat. They suggest that this is due to the influence of species that use the habitat matrix. In such systems, the degree to which spatial dynamics are relevant is likely to vary with the degree of habitat specialization, which influences the organisms' perception of habitat size and isolation (Harrison 1997).

This book includes two chapters addressing systems with patches that appear heterogeneous but permanent and where, like the previous chapter, movement appears to play a strong role in controlling local species diversity. Cottenie and De Meester (chapter 8) describe zooplankton in an interconnected system of ponds, some with fish predators, and some without. They use a variety of multivariate statistics to analyze variation in density and composition that is related to the habitat (indicating species sorting) and spatial position (indicating mass effects).

The complexity of empirical perspectives on metacommunities reaches its height in Kolasa and Romanuk's chapter 9, which describes a rock pool system with a wide range of invertebrate taxa; these authors suggest that physical conditions are critical to organizing the communities and creating a hierarchy of scales.

Theoretical Approaches

By contrast to the cases in chapters 2 and 3 with low dispersal, Mouquet et al. (chapter 10), consider a wider range of dispersal rates, including those that are high enough to cause mass and rescue effects. They do so in the context of metacommunities of competitors and integrate patch dynamic and mass effects models by showing the effects of different levels of dispersal in the presence and absence of patch heterogeneity. These mass and rescue effects modify both species abundance (e.g., source-sink dynamics; Pulliam 1988) and species interactions (Holt 1985; Holt et al. 2003; Danielson 1991).

The next two chapters consider techniques for studying metacommunity dynamics. Law and Leibold (chapter 11) demonstrate how patch occupancy models can be used to create a link to permanence as a measure of persistence. The chapter illustrates this technique by using an example of intransitive competition. The technique is applicable to any case where an assembly map can be drawn (Warren et al. 2003) and there is a separation of local and regional timescales (a theme that also arises in chapter 10).

In chapter 12, Chesson et al. discuss a powerful modeling technique, scale transition theory, which uses models fitted to empirical data to partition out spatial, temporal, and spatiotemporal elements of community structure. Scale transition theory could be used to model a very large range of population and community problems. The completeness of the framework presented makes it an attractive technique for considering all aspects of spatiotemporal dynamics. The technique is complex, and to aid in making it more accessible and highlight its utility, a companion empirical chapter by Melbourne et al. (chapter 13) describes its application to a variety of empirical problems.

Emerging Areas and Perspectives

All four of these perspectives on metacommunities that we have described are admittedly incomplete and present challenges: to empiricists to evaluate their relevance to real systems, and to theoreticians to synthesize their viewpoints and to elucidate mechanisms. In this regard, chapter 14 by Chase et al. is interesting because it describes a variety of empirical patterns and the (incomplete) explanations for them that are based on the four metacommunity perspectives and niche theory. The authors concentrate on competitive metacommunities and describe a full range of testable hypotheses (e.g., table 14.1) that come from contrasting the four perspectives presented above, and especially from thinking about testing ideas from neutral models.

Neutral models are fascinating in part because of the simplicity of the assumptions that they make. In particular, the complete absence of competitive differences and niche differences between species is something that makes many ecologists scratch their heads in puzzlement. In a provocative chapter 15, McPeek and Gomulkiewicz describe the relationships between population genetics and Hubbell's (2001) neutral theory, and the apparent relationship between these theories and McPeek's own studies of damselflies. Chapter 15 provides a potential example of neutral dynamics and a nice example of a case where neutral theory has made us question our perception of an empirical system.

Resetarits et al. (chapter 16) and colleagues use a system of temporary and permanent ponds to explore the potential importance of habitat selection behavior for local and regional community structure. This is a new and exciting area that could integrate behavior into metacommunity dynamics. The ideas have broad interaction with the evolutionary ideas in chapters 15 and 19.

Shurin and Srivastava's chapter 17 returns to considering patterns of species diversity. A pattern that has long intrigued both community ecologists and island biogeographers is the relationship between local and regional species diversity. Ecologists have inferred a regional influence on local communities based on nonasymptotic relationships between local and regional diversity. Classically, the shape of this relationship has been used to infer whether or not local communities are saturated (i.e., susceptible to invasions). The authors discuss possible interpretations of this relationship and highlight the importance of the area from which a species pool is drawn (experimentally or by dispersal).

Just as ecologists have linked species diversity to ecosystem functioning (Kinzig et al. 2002) it is useful to consider how metacommunity diversity is linked to the regional functioning of ecosystems. In chapter 18, Loreau et al. describe a metaecosystem as "a set of ecosystems connected by spatial flows of energy, materials, and organisms across ecosystem boundaries." This idea builds naturally on the metacommunity concept. It is complementary to ideas about ecosystem subsidy (Polis, Anderson, et al. 1997; Polis, Power, et al. 2004), which simplifies thinking about flows across ecosystem boundaries by treating fluxes as donor controlled rather than dynamic. The metaecosystem concept includes both physical and biotic drivers of ecosystem functions.

The book ends with two prospectus chapters that discuss some necessary future work and a short summary of important findings. Leibold et al. (chapter 19) consider the relevance of evolution to the four metacommunity perspectives presented in this chapter and describe the potential for metacommunities to be complex adaptive systems (see also Leibold and Norberg 2004). Holt et al. (chapter 20) summarize some emerging directions from this book and discuss a broad range of topics that are inadequately covered by this book and that would benefit from future attention. The most significant insights from this book are summarized in a short, final coda.

Conclusions

In this introductory overview, we have argued that metacommunity approaches can substantially change the ways in which we interpret ecological phenomena, both at local and metacommunity scales. We have proposed a definition for metacommunities and have reviewed four simplistic approaches to modeling them. It is clear that any synthesis linking these four approaches to each other would greatly facilitate empirical work and provide a much more realistic framework for understanding large-scale ecological processes. The four perspectives considered show how the metacommunity concept leads us to identify the important roles of habitat and movement in modifying community and metacommunity diversity and abundance. While this chapter draws on much classic work such as island biogeography and the study of vegetation patterns along environmental gradients, novel insights are coming from two forms of integration. First, this approach provides a testable framework for what we believe are the main factors influencing local and regional community structure and dynamics. Second, the novel integration of spatial dynamics with community ecology approaches that have conventionally been limited to the local scale is providing exciting new insights into large-scale community processes.

Acknowledgments

We thank Jun Bando, Kim Preston, Drew Talley, Theresa S. Talley, Melanie Gogol-Prokurat, and two anonymous referees for helpful comments.

Literature Cited

Abrams, P. A., and C. J. Walters. 1996. Invulnerable prey and the paradox of enrichment. Ecology 77:1125–1133.

Abramsky, Z., and U. Safriel. 1980. Seasonal patterns in a Mediterranean bird community composed of transient wintering and resident passerines. Ornis Scandinavica 11:201–216.

Adler, F. R., and J. Mosquera. 2000. Is space necessary? Interference competition and limits to biodiversity. Ecology 81:3226–3232.

Allan, B. F., F. Keesing, and R. S. Ostfeld. 2003. Effect of forest fragmentation on Lyme disease risk. Conservation Biology 17:267–272.

Allen, C. R., and C. S. Holling. 2002. Cross-scale structure and scale breaks in ecosystems and other complex systems. Ecosystems 5:315–318.

Amarasekare, P. 2000. The geometry of coexistence. Biological Journal of the Linnean Society 71:1–31.

Amarasekare, P., and R. M. Nisbet. 2001. Spatial heterogeneity, source-sink dynamics, and the local coexistence of competing species. American Naturalist 158:572–584.

Bell, G. 2000. The distribution of abundance in neutral communities. American Naturalist 155: 606–617.

———. 2001. Neutral macroecology. Science 293:2413–2418.

Berlow, E. L., S. A. Navarrete, C. J. Briggs, M. E. Power, and B. A. Menge. 1999. Quantifying variation in the strengths of species interactions. Ecology 80:2206–2224.

Boorman, S. A., and P. R. Levitt. 1973. Group selection on the boundary of a stable population. Theoretical Population Biology 4:85–128.

Borcard, D., and P. Legendre. 2002. All-scale spatial analysis of ecological data by means of principal coordinates of neighbour matrices. Ecological Modelling 153:51–68.

Brown, J. H., and A. Kodric-Brown. 1977. Turnover rates in insular biogeography: Effect of immigration on extinction. Ecology 58:445–449.

Case, T. J. 1991. Invasion resistance, species build-up and community collapse in metapopulation models with interspecies competition. Biological Journal of the Linnean Society 42:239–266.

Caswell, H. 1978. Predator mediated co-existence: A non-equilibrium model. American Naturalist 112:127–154.

Chave, J. 2004. Neutral theory and community ecology. Ecology Letters 7:241–253.

Chase, J. M., and M. A. Leibold. 2003. Ecological niches: Linking classical and contemporary approaches. University of Chicago Press, Chicago, IL.

Chesson, P. L. 1985. Coexistence of competitors in spatially and temporally varying environments: A look at the combined effects of different sorts of variability. Theoretical Population Biology 28:263–287.

Chesson, P., and N. Huntly. 1989. Short-term instabilities and long-term community dynamics. Trends in Ecology and Evolution 4:293–298.

Clements, F. E. 1936. Nature and structure of climax. The Journal of Ecology 24:252–284.

Clobert, J., E. Danchin, A. A. Dhondt, and J. D. Nichols. 2001. Dispersal. Oxford University Press, Oxford, UK.

Connell, J. H. 1978. Diversity in tropical rainforests and coral reefs. Science 199:1302–1310.

Cousins, S. H. 1990. Countable ecosystems deriving from a new food web entity. Oikos 57:270–275.

Crowley, P. H. 1981. Dispersal and the stability of predator-prey interactions. American Naturalist 118:673–701.

Danielson, B. J. 1991. Communities in a landscape: The influence of habitat heterogeneity on the interactions between species. American Naturalist 138:1105–1120.

Debinski, D. M., and R. D. Holt. 2000. A survey and overview of habitat fragmentation experiments. Conservation Biology 14:342–355.

Dobzhansky, T. G. 1951. Genetics and the origin of species. Columbia University Press, New York, N.Y.

Durrett, R., and S. A. Levin. 1994a. The importance of being discrete (and spatial). Theoretical Population Biology 46:363–394.

Durrett, R., and S. A. Levin. 1994b. Stochastic spatial models: A user's guide to ecological applications. Philosophical Transactions of the Royal Society of London, Series B 343:329–350.

Ellner, S. P., and G. Fussmann. 2003. Effects of successional dynamics on metapopulation persistence. Ecology 84:882–889.

Gleason, H. A. 1926. The individualistic concept of the plant association. Bulletin of the Torrey Botanical Club 53:7–26.

Hanski, I. and O. Gaggiotti. 2004. Ecology, genetics, and evolution of metapopulations. Academic Press, New York, NY.

Hanski, I., and M. Gilpin. 1991. Metapopulation dynamics: Brief history and conceptual domain. Biological Journal of the Linnean Society 42:3–16.

———. 1997. Metapopulation biology: Ecology, genetics and evolution. Academic Press, San Diego, CA.

Hanski, I., and M. Gyllenberg. 1993. Two general types of metapopulation models and the core-satellite species hypothesis. American Naturalist 142:17–41.

Hanski, I., and D. Simberloff. 1997. The metapopulation approach, its history, conceptual domain, and application to conservation. Pages 5–26 *in* I. P. Hanski, and M. E. Gilpin, eds. *Metapopulation dynamics: Ecology, genetics, and evolution.* Academic Press, San Diego.

Harrison, S. 1997. How natural habitat patchiness affects the distribution of diversity in Californian serpentine chaparral. Ecology 78:1898.

Harrison, S., and A. D. Taylor. 1997. Empirical evidence for metapopulation dynamics: A critical review. Pages 27–42 *in* I. Hanski, and M. E. Gilpin, eds. *Metapopulation dynamics: Ecology, genetics and evolution.* Academic Press, San Diego.

Hassell, M. P., H. N. Comins, and R. M. May. 1994. Species coexistence and self-organizing spatial dynamics. Nature 370:290–292.

Hastings, A. 1980. Disturbance, coexistence, history, and competition for space. Theoretical Population Biology 18:363–373.

———. 2003. Metapopulation persistence with age-dependent disturbance or succession. Science 301:1525–1526.

Hess, G. 1996. Disease in metapopulation models: Implications for conservation. Ecology 77:1617–1632.

Holling, C. S. 1986. The resilience of ecosystems: Local surprise and global change, Pages 292–317 *in* W. C. Clark, and M. R. E., eds. *Sustainable development of the biosphere.* Cambridge University Press, Cambridge, U.K..

———. 1994. Simplifying the complex—the paradigms of ecological function and structure. Futures 26:598–609.

Holt, R. D. 1977. Predation, apparent competition, and the structure of prey communities. Theoretical Population Biology 12:237–266.

———. 1985. Population dynamics in two-patch environments: Some anomalous consequences of an optimal habitat distribution. Theoretical Population Biology 28:181–208.

———. 1993. Ecology at the mesoscale: The influence of regional processes on local communities. Pages 77–88 *in* R. E. Ricklefs, and D. Schluter, eds. *Species diversity in ecological communities: Historical and geographical perspectives.* University of Chicago Press, Chicago, IL.

Holt, R. D., M. Barfield, and A. Gonzalez. 2003. Impacts of environmental variability in open populations and communities: "Inflation" in sink environments. Theoretical Population Biology 64:315–330.

Holt, R. D., and G. A. Polis. 1997. A theoretical framework for intraguild predation. American Naturalist 149:745–764.

Holyoak, M., and C. Ray. 1999. A roadmap for metapopulation research. Ecology Letters 2:273–275.

Hubbell, S. P. 2001. *The unified neutral theory of biodiversity and biogeography.* Princeton University Press, Princeton, N.J.

Ims, R. A., and N. G. Yoccoz. 1997. Studying transfer processes in metapopulations: Emigration, migration and colonization. Pages 247–265 *in* I. Hanski, and M. E. Gilpin, eds. *Metapopulation dynamics: Ecology, genetics and evolution.* Academic Press, New York, N.Y.

Iwasa, Y., and J. Roughgarden. 1986. Interspecific Competition among Metapopulations with Space-Limited Subpopulations. Theoretical Population Biology 30:194–214.

Jessup, C. M., R. Kassen, S. E. Forde, B. Kerr, A. Buckling, P. B. Rainey, and B. J. M. Bohannan. 2004. Big questions, small worlds: Microbial model systems in ecology. Trends in Ecology and Evolution 19:189–197.

Kareiva, P. 1990. Population dynamics in spatially complex environments: Theory and data. Philosophical Transactions of the Royal Society of London, Series B 330:175–190.

Kinzig, A. P., S. W. Pacala, and D. Tilman. 2002. *The functional consequences of biodiversity: Empirical progress and theoretical extensions.* Princeton University Press, Princeton, N.J.

Kitching, R. L. 2000. *Food webs and container habitats: The natural history and ecology of phytotelmata.* Cambridge University Press, New York, N.Y.

Kneitel, J. M., and J. M. Chase. 2004. Trade-offs in community ecology: Linking spatial scales and species coexistence. Ecology Letters 7:69–80.

Kuno, E. 1987. Principles of predator-prey interaction in theoretical, experimental, and natural population systems. Advances in Ecological Research 16:250–331.

Laska, M. S., and J. T. Wootton. 1998. Theoretical concepts and empirical approaches to measuring interaction strength. Ecology 79:461–476.

Law, R., U. Dieckmann, and J. A. J. Metz. 2000. Introduction. Pages 1–6 *in* U. Dieckmann, R. Law, and J. A. J. Metz, eds. *The geometry of ecological interactions: Simplifying spatial complexity.* University of Chicago Press, Chicago, IL.

Law, R., and R. D. Morton. 1993. Alternative permanent states of ecological communities. Ecology 74:1347–1361.

Lawler, S. P. 1998. Ecology in a bottle: Using microcosms to test theory. Pages 236–253 *in* W. J. Resetarits, Jr., and J. Bernardo, eds. *Experimental ecology: Issues and perspectives.* Oxford University Press, New York, N.Y.

Legendre, P., M. R. T. Dale, M.-J. E. Fortin, J. Gurevitch, M. Hohn, and D. Myers. 2002. The consequences of spatial structure for the design and analysis of ecological field surveys. Ecography 25:601–615.

Leibold, M. A. 1998. Similarity and local co-existence of species in regional biotas. Evolutionary Ecology 12:95–110.

Leibold, M. A., M. Holyoak, N. Mouquet, P. Amarasekare, J. M. Chase, M. F. Hoopes, R. D. Holt, J. B. Shurin, R. Law, D. Tilman, M. Loreau, and A. Gonzalez, A. 2004. The metacommunity concept: A framework for multi-scale community ecology. Ecology Letters 7:601–613.

Leibold, M. A., and J. Norberg. 2004. Biodiversity in metacommunities: Plankton as complex adaptive systems? Limnology and Oceanography 49:1278–1289.

Levin, S. A. 1974. Dispersion and population interactions. American Naturalist 108:207–228.

———. 1992. The problem of pattern and scale in ecology. The MacArthur award lecture. Ecology 73:1943–1967.

Levins, R. 1969. Some demographic and genetic consequences of environmental heterogeneity for biological control. Bulletin of the Entomological Society of America 15:237–240.

Levins, R., and D. Culver. 1971. Regional coexistence of species and competition between rare species. Proceedings of the National Academy of Science, USA 68:1246–1248.

Lindenmayer, D. B., and J. F. Franklin. 2002. *Conserving forest biodiversity: A comprehensive multi-scaled approach.* Island Press, Washington, D.C.

LoGiudice, K., R. S. Ostfeld, K. A. Schmidt, and F. Keesing. 2003. The ecology of infectious disease: Effects of host diversity and community composition on Lyme disease risk. Proceedings of the National Academy of Sciences of the United States of America 100:567–571.

Loreau, M. 2000. Biodiversity and ecosystem functioning: Recent theoretical advances. Oikos 91:3–17.

MacArthur, R. H. 1958. Population ecology of some warblers of northeastern coniferous forests. Ecology 39:599–619.

MacArthur, R. H., and E. O. Wilson. 1967. *The theory of island biogeography.* Princeton University Press, Princeton, N.J.

Magurran, A. E. 1988. *Ecological diversity and its measurement.* Princeton University Press, Princeton, N.J.

Maurer, B. A. 1999. *Untangling ecological complexity: The macroscopic perspective.* University of Chicago Press, Chicago, IL.

May, R. M. 1973. *Stability and complexity in model ecosystems.* Princeton Monographs in Biology. Princeton University Press, Princeton, NJ.

McCann, K., A. Hastings, and G. R. Huxel. 1998. Weak trophic interactions and the balance of nature. Nature 395:794–798.

McCauley, E., W. G. Wilson, and A. M. de Roos. 1993. Dynamics of age- and spatially-structured predator-prey interactions: Individual-based models and population level formulations. American Naturalist 142:412–442.

McPeek, M. A., and S. Kalisz. 1998. On the joint evolution of dispersal and dormancy in metapopulations. Advances in Limnology 10:33–51.

Morin, P. J. 1999. *Community ecology.* Blackwell Science Inc., Oxford, U.K.

Mouquet, N., and M. Loreau. 2002. Coexistence in metacommunities: The regional similarity hypothesis. American Naturalist 159:420–426.

Mouquet, N., and M. Loreau. 2003. Community patterns in source-sink metacommunities. American Naturalist 162:544–557.

Murdoch, W. W., and A. Oaten. 1975. Predation and population stability. Advances in Ecological Research 9:1–31.

Naeem, S. 2001. Experimental validity and ecological scale as criteria for evaluating research programs. Pages 223–250 *in* R. H. Gardner, W. M. Kemp, V. S. Kennedy, and J. E. Petersen, eds. *Scaling relations in experimental ecology.* Columbia, New York, N.Y.

Naeem, S., F. S. Chapin, III, R. Costanza, P. R. Ehrlich, F. B. Golley, D. U. Hooper, J. H. Lawton, R. V. O'Neill, H. A. Mooney, O. E. Sala, A. J. Symstad, and D. Tilman. 1999. Biodiversity and ecosystem functioning: Maintaining natural life support processes. Issues in Ecology. Washington, Ecological Society of America.

Pacala, S. W., C. Canham, J. Saponara, J. Silander, R. Kobe, and E. Ribbens. 1996. Forest models defined by field measurements. II. Estimation, error analysis, and dynamics. Ecological Monographs 66:1–44.

Peterson, G., C. R. Allen, and C. S. Holling. 1998. Ecological resilience, biodiversity, and scale. Ecosystems 1:6–18.

Pianka, E. R. 1966. Latitudinal gradients in species diversity: A review of concepts. American Naturalist 100:33–46.

Pimm, S. L. 1982. *Food webs.* Chapman and Hall, London, UK.

Pimm, S. L., and J. H. Lawton. 1978. On feeding on more than one trophic level. Nature 275:542–544.

Polis, G. A., W. B. Anderson, and R. D. Holt. 1997. Toward an integration of landscape and food web ecology: The dynamics of spatially subsidized food webs. Annual Review of Ecology and Systematics 28:289–316.

Polis, G. A., M. Power and G. Huxel. 2004. *Food webs at the landscape level.* University of Chicago Press, Chicago, IL.

Polis, G. A., and K. O. Winemiller. 1996. *Food webs: Integration of patterns and dynamics.* Chapman and Hall, New York, N.Y.

Pulliam, H. R. 1988. Sources, sinks, and population regulation. American Naturalist 132:652–661.

Rodríguez, M. A. 2002. Restricted movement in stream fish: The paradigm is incomplete, not lost. Ecology 83:1–13.

Roff, D. A. 2001. *Life history evolution.* Sinauer Associates, Inc., New York, N.Y.

Rosenzweig, M. L. 1995. *Species diversity in space and time.* Cambridge University Press, Cambridge, U.K.

Roxburgh, S. H., K. Shea, and J. B. Wilson. 2004. The intermediate disturbance hypothesis: Patch dynamics and mechanisms of coexistence. Ecology 85:359–371.

Schoener, T. W. 1983. Rates of species turnover decreases from lower to higher organisms: A review of the data. Oikos 41:372–377.

Shea, K., S. H. Roxburgh, and E. S. J. Rauschert. 2004. Moving from pattern to process: Coexistence mechanisms under intermediate disturbance regimes. Ecology Letters 7:491–508.

Shmida, A., and M. V. Wilson. 1985. Biological determinants of species diversity. Journal of Biogeography 12:1–20.

Shurin, J. B., J. E. Havel, M. A. Leibold, and B. Pinel-Alloul. 2000. Local and regional zooplankton species richness: A scale-independent test for saturation. Ecology 81:3062–3073.

Steiner, C. F., and M. A. Leibold. 2004. Cyclic assembly trajectories and scale-dependent productivity-diversity relationships. Ecology 85:107–113.

Taylor, A. D. 1990. Metapopulations, dispersal, and predator-prey dynamics: An overview. Ecology 71:429–433.

Tilman, D. 1994. Competition and biodiversity in spatially structured habitats. Ecology 75:2–16.

Vandermeer, J. H. 1973. On the regional stabilization of locally unstable predator-prey relationships. Journal of Theoretical Biology 41:161–170.

Warren, P. H., R. Law, and A. J. Weatherby. 2003. Mapping the assembly of protist communities in microcosms. Ecology 84:1001–1011.

Whittaker, R. H. 1960. Vegetation of the Siskiyou Mountains, Oregon and California. Ecological Monographs 30:279–338.

Whittaker, R. H. 1972. Evolution and measurement of species diversity. Taxon 21:213–251.

Wiens, J. A. 1989. Spatial scaling in ecology. Functional Ecology 3:385–397.

Wilcove, D. S., D. Rothstein, J. Dubow, A. Phillips, and E. Losos. 2000. Leading threats to biodiversity: What's imperiling U.S. species? Pages 239–254 in B. A. Stein, L. S. Kutner, and J. S. Adams, eds. Precious heritage: The status of biodiversity in the United States. Oxford University Press, Oxford, U.K.

Wilson, J. B. 1990. Mechanisms of species coexistence: Twelve explanations for Hutchinson's "paradox of the plankton": Evidence from New Zealand plant communities. New Zealand Journal of Ecology 13:17–42.

Wilson, D. S. 1992. Complex interactions in metacommunities, with implications for biodiversity and higher levels of selection. Ecology 73:1984–2000.

Worthen, W. B. 1989. Effects of resource density on mycophagous fly dispersal and community structure. Oikos 54:145–153.

Yu, D. W., and H. B. Wilson. 2001. The competition-colonization trade-off is dead: Long live the competition-colonization trade-off. American Naturalist 158:49–63.

Yu, D. W., H. B. Wilson, and N. E. Pierce. 2001. An empirical model of species coexistence in a spatially structured environment. Ecology 82:1761–1771.

CORE CONCEPTS

Marcel Holyoak

It is remarkable how rich and intricate the effects of introducing space can be on community dynamics. The chapters in this section build on two major areas that have contributed to our interest in metacommunities, and which provide building blocks for fuller consideration of the role of space in communities.

The first introductory chapter of part 1 by Hoopes et al. (chapter 2) reviews the effects of space on coexistence of pairs of species interacting through competition, predation and mutualism. The metapopulation and related literature covered in this chapter provide a logical starting point from which to consider metacommunities. Another difficult issue is how to scale up from pairwise interactions to whole food webs in a way that includes spatial structure. Holt and Hoopes (chapter 3) investigate the role of spatial processes in food web dynamics, a poorly studied problem in community ecology. First, they explore how spatial flows and dispersal influence community modules, recurrent structures consisting of a few species engaged in a defined set of interactions. Second, by extending island biogeographic theory to incorporate multitrophic interactions, Holt and Hoopes illustrate how one may move beyond community modules in integrating food web ecology into the framework of metacommunity dynamics. These chapters together with chapter 1 pave the way for consideration of empirical evidence for metacommunities in the next section of the book.

Part I

CORE CONCEPTS

CHAPTER 2

The Effects of Spatial Processes on Two Species Interactions

Martha F. Hoopes, Robert D. Holt, and Marcel Holyoak

Introduction

Species interactions take place in a spatial context. That context affects organism encounters, yet the most common and familiar models in ecology ignore spatial processes and assume that interactions between groups of individuals, populations, and species are well represented by the interactions of the average individual. Here we will discuss the role of spatial processes in population and community dynamics with a focus on models that address the role of spatial processes in species coexistence and the maintenance of diversity. Recent interest in the role of spatial ecology has led to diverse literature on this topic (e.g., Caswell and Cohen 1991; Hanski 1997; Tilman and Kareiva 1997; Chesson 2000a; Holt 2002) that makes it impossible to summarize the full scope of this work in a single chapter. Instead, we will use illustrative examples that reveal the influence of space on coexistence in competition, mutualism, and consumer-resource interactions.

This chapter examines how spatial processes affect species diversity and the persistence of species interactions in general, not just at or near equilibrium. Often ecologists are interested in examining the factors affecting the coexistence or distribution of organisms in relation to particular other species or environmental factors. Equilibrium outcomes of spatial models may not provide the best answers for these questions. Many models display long periods of transient dynamics, in which populations and occupancies are not only not at equilibrium levels but are even on qualitatively different trajectories (Hastings and Higgins 1994; Ovaskainen and Hanski 2002; but see Labra et al. 2003). Over a given time period, these transient dynamics may dictate metapopulation occupancy and dynamics more accurately than will the equilibrium outcomes of spatial models (Holt 1992; Ruxton and Doebeli 1996). This is particularly true with metapopulations at high occupancy, which slows dynamics (Kinzig et al. 1999), and when spatial structure is combined with temporal heterogeneity (Gonzalez and Holt 2002). Temporal fluctuations may be on a time scale faster than the period required to reach equilibrium outcomes, so that spatial populations are always experiencing disturbances or always reacting to changing quality. Transient dynamics are ubiquitous in natural systems, and must be considered in understanding multispecies persistence and coexistence.

We begin with single species spatial models, which form the basis for multi-species models, then proceed through competitive, consumer-resource, and mutualism models. Because the literature we review is extensive and complex, we provide table 2.1 as a synoptic overview. Other perspectives on some of the topics we cover are provided in this volume by Holt and Hoopes in chapter 3, Mouquet et al. in chapter 10, and Chesson et al. in chapter 12.

Some pairwise species interactions do not fall neatly into any of the categories that we cover. The most notable of these is disease dynamics. Although consumer-resource models encompass many elements of spatial disease models, disease organisms experience hosts as separate (and heterogeneous) patches, thereby experiencing additional levels of spatial hierarchy (within individual hosts, among hosts, and among host populations) beyond the other systems that we discuss. Excellent reviews of spatial disease models include works by Sattenspiel and Simon (1988), Andreasen and Christiansen (1989), Bolker and Grenfell (1995), Hess (1996), Lloyd and May (1996), Grenfell and Harwood (1997), Rhodes and Anderson (1997), Grenfell et al. (2001), Rodriguez and Torres-Sorando (2001), McCallum and Dobson (2002), and Park et al. (2002).

Spatial Models for Single Species
Models without Local Dynamics

One of the simplest spatial models for a single species is the patch-occupancy metapopulation formulation (Levins 1969, 1970), in which organisms exist in an infinite array of discrete populations or patches. The model assumes global dispersal, so that all patches are equally accessible from all other patches. It also assumes that patches have only two states, empty or occupied, and so ignores local population dynamics. The proportion of patches occupied at any time changes according to

$$\frac{dp}{dt} = cp(1 - p) - mp, \qquad\qquad 2.1$$

where p is the proportion of currently occupied patches, c is the rate at which empty patches are colonized, and m is the rate at which occupied patches go extinct. The nontrivial equilibrium occupancy, $p^* = 1 - m/c$, is globally stable when it is feasible, implying persistence if the colonization rate exceeds the extinction rate. Immediately after colonization, local population sizes are assumed to reach a carrying capacity so that continued dispersal has no effect on local dynamics. Dynamics within patches are, thus, assumed to happen on a faster time scale than dynamics among patches.

COMPARISON WITH NONSPATIAL MODELS

Because global dispersal creates a well-mixed system in which all patches are equally accessible, this metapopulation model is only implicitly spatial, meaning that patches are separate but there is no defined spatial distance among them. This model is analogous to a logistic model in which each patch represents a site occupied by a single individual (Gilpin and Hanski 1991; Caughley 1994; Amarasekare 1998a). Hence, factors that influence the persistence of a single population can have corresponding impacts on the persistence of a metapopulation. For example, metapopulations can experience demographic stochasticity and reduced probability of persistence at low occupancy (Stacey et al. 1997). Similarly, metapopulations can experience Allee effects in which a critical occupancy is required for colonization or persistence (Amarasekare 1998a). Harding and McNamara (2002) discuss a generalized Levins model with nonlinear extinction and colonization terms, comparable to incorporating nonlinear density dependence in local population models.

FACTORS AFFECTING PERSISTENCE

For a metapopulation to persist longer than a single local population, patches must act independently to a degree so that the entire metapopulation has a different extinction probability than a single population (Quinn and Hastings 1987), meaning that extinctions are not completely synchronized (Harrison and Quinn 1990). Equation 2.1 assumes that extinctions are independent among patches. In real-world systems or more complex models, synchrony is affected by both the degree of spatial correlation in the environment as well as dispersal distances and rates (figure 2.1; Hill et al. 2002; Ovaskainen et al. 2002). We will find that the synchrony of dynamics across space is an important subtheme as we add species interactions to our spatial models.

One of the simplest ways to include the factors that affect synchrony (isolation and correlation) is to use an incidence function model, which links patch-specific size (or quality) and isolation to extinction and colonization rates (Hanski 1994). Incidence function models are generally still patch occupancy models that do not follow local dynamics, and moreover they require an equilibrium between colonization and extinction. The assumption that patch size or quality matters in an incidence function model, however, is an implicit acknowledgment that emigration and immigration may be functions of population size and that population size may be affected by habitat quality.

Models with Local Dynamics
IMPLICIT SPACE

The simplest single species models that follow local dynamics in more than one local population do not include explicit spatial distances or heterogeneous

Table 2.1 An overview of the effects of adding spatial dynamics to different types of species interactions

Features of spatial dynamics	Simplest type of model	Examples	Comments
Single species or noninteracting species			
Asynchronous dynamics lead to a regional average occupancy despite local fluctuations, most basic statistical averaging	Patch occupancy (no local dynamics), spatially implicit	Levins 1969	• Formally equivalent to the logistic equation at a regional level. • Patches are independent and identical so that patch colonization and extinction probabilities are fixed. • Population dynamic features, e.g., Allee effects and demographic stochasticity can be represented at a metapopulation scale.
Factors leading to asynchrony are explicit; statistical averaging	Patch occupancy, spatially implicit incidence function models	Hanski 1994	• Recognize the importance of independence (asynchrony), or differences among patches in prolonging regional persistence time relative to population models; local dynamics are *statistically averaged* to give persistent regional dynamics. • Between-patch differences act through influencing colonization and extinction probabilities, e.g., through population/ patch size or quality.
Rescue effects, source-sink dynamics, pseudosink effects	Models with local population dynamics with implicit space	Hastings 1983; Comins and Noble 1985; Holt 1985; Watkinson and Sutherland 1995; Amarasekare 1998b; Amarasekare and Nisbet 2001	• Local population sizes are explicitly represented. • Variation in patch quality can promote source and sink dynamics with strong rescue effects, where sink populations are maintained by immigration. • Immigration can create density-dependent reductions in local population growth (termed pseudosink effects by Watkinson and Sutherland 1985). • Density independent emigration may cause declines in local population size.

Pattern formation / aggregation	Models with local population dynamics and explicit space.	Durrett and Levin 1994a, 1994b; Bolker and Pacala 1997	• Local population dynamics coupled with larger scale dispersal can lead to pattern formation, and patterns can influence both local and regional dynamics.
Competitive interactions Competitive dominance, refuge effect or rescue effect	Spatially implicit models with habitat heterogeneity: (1) permanent habitat heterogeneity	Horn and MacArthur 1972; Comins and Noble 1985	• Habitat heterogeneity can alter within-patch competitive ability so that different species are competitively dominant in different localities. • Mechanism is equivalent to a refuge, which is more typically considered in predator-prey models.
Nonlinear, nonadditive responses to temporal heterogeneity	(2) Temporal habitat heterogeneity	Chesson and Warner 1981; Warner and Chesson 1985; Chesson 1994, 2000a	• If habitat quality varies through time, coexistence is possible via a temporal *storage effect* if: i. long-lived life stages are present to permit survival through bad periods; and ii. species have different responses to environmental conditions so that maximum growth occurs at different times.
Nonlinear, nonadditive responses to spatial heterogeneity	(3) Spatial habitat heterogeneity	Chesson 1985, 2000a	• Variation in density across space can allow species to persist in habitats where they are poor competitors through source and sink dynamics. This is termed the *spatial storage effect*. • Source and sink dynamics, and the rescue / mass effects they entail are required and dispersal (across space) has the same function as long-lived life stages in the temporal storage effect (involving dispersal across time).
Competition-dispersal trade-off leading to regional equilibrium (statistical averaging)	Spatially implicit models for homogeneous habitats, with species' trade-offs	Levins and Culver 1971; Hastings 1978; Tilman 1994	• A trade-off between competitive ability and dispersal ability can promote regional coexistence.

Table 2.1 continued

Features of spatial dynamics	Simplest type of model	Examples	Comments
Pattern formation / aggregation	Spatially explicit models for homogeneous habitats	Neuhauser 1998; Bolker and Pacala 1999; Law and Diekmann 2000	• Inferior competitors can repeatedly escape from competition by finding empty habitat not yet colonized by the superior competitor. • As habitat is destroyed superior competitors are lost first because of their poor dispersal abilities. • Clustering, associated with pattern formation, coupled with local interactions altered competitive effects and could change both local and regional competitive dominance. • At least three competitive strategies of space-use exist: tolerance (ability to resist invasion and take over space), dispersal, and exploitation (speed of utilizing space).
Competition-dispersal trade-off and pattern formation / aggregation	Models that combine competitive shifts in dominance (heterogeneous habitats) with species' trade-offs	Amarasekare and Nisbet 2001; Mouquet and Loreau 2002	• There are threshold levels of dispersal that allow species to coexist: If species dispersal is too high, then the advantage of niche partitioning is lost, whereas with too little dispersal, inferior competitors cannot survive.
Consumer-resource interactions Statistical averaging	Models with many independent interacting entities, e.g. multiple patches or explicit space with locally interacting individuals	Allen 1975; Crowley 1981; Reeve 1988; de Roos et al. 1991, 1998; McCauley et al. 1993, 1996; Wilson 1998	• Although local populations may fluctuate in density and are extinction-prone, regional population densities are relatively constant because of averaging population densities over fluctuating subpopulations. • Asynchrony is critical to the independence

Feature	Model	References	Spatially dynamic features
Density dependence induced by dispersal: refuge effects, rescue effects, and source-sink dynamics	Models with patch heterogeneity	Crowley 1981; Sih 1987; Reeve 1988; Murdoch et al. 1992; Holt and Hassell 1993; Hochberg et al. 1996	of the local populations across which the regional density is averaged. • Predator and prey dispersal can have different effects on synchrony. • Any supply of individuals from high- to low-density areas may promote density dependence (the refuge effect), which extends persistence through a rescue effect. • This process is similar to source-sink maintenance of inferior competitors in the competition models. • Local heterogeneity can be fixed or from spatiotemporal variation, as we found in spatial models of competition.
Nonlinear, nonadditive responses to heterogeneity	Patch, lattice, and reaction-diffusion models with spatial heterogeneity in densities and nonlinearities in predator-prey responses.	McCauley et al. 1996; De Roos et al. 1998; Gurney and Veitch 2000	• This feature is similar to a spatial storage effect (Chesson 1985, 1990; Warner and Chesson 1985). • Prey show strong patterning / aggregation in all models.
Mutualistic interactions			
Nonlinear, nonadditive responses to heterogeneity	Obligate mutualisms with implicit space	Hutson et al. 1985; Hastings and Wolin 1989	• Mutualistic effects that vary nonlinearly spatially allow obligate mutualists to span a broader range of densities and still persist; this is similar to a spatial storage effect.
Spatial heterogeneities, refuge or rescue effects, density dependence induced by dispersal	Facultative mutualisms	Addicott 1981; Ringel et al. 1996	• Facultative mutualists improve the quality of habitat where they coexist, leading to spatial heterogeneities and mass effects that always increase the persistence of two species interactions. • A mutualist as a third species can alter the balance of competition or consumer-resource interactions and can create the spatial heterogeneities listed in previous sections of the table.

Note: The table lists the spatially dynamic features that emerge from models and includes only models that exemplify these features.

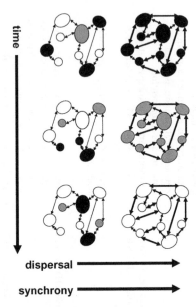

Figure 2.1 Relationship between dispersal and synchrony in metacommunities. Each row of figures represents a metacommunity through three successive temporal snapshots. Different colors represent different species combinations, resulting in different local communities. Arrow thickness represents dispersal rate. Note that the top row has more dispersal between patches as well as higher rates of dispersal. Synchrony of local communities is higher in the top metacommunity than in the bottom metacommunity.

habitats. Differences in local population sizes allow these models to reveal rescue effects and costs of dispersal (Brown and Kodric-Brown 1977; Amarasekare 1998b). A *rescue effect* occurs if emigrants from high population density patches prevent low population density patches from declining to extinction (table 1.1). Rescue effects in homogeneous habitats are only possible if dynamics are asynchronous (Doebeli 1995).

With differences in patch quality, however, a broader range of dynamics appears even with implicit space (Hastings 1983; Comins and Noble 1985; Amarasekare and Nisbet 2001). Coupling habitats with different qualities can lead to source-sink dynamics (Holt 1985; Pulliam 1988; Pulliam and Danielson 1991). Strictly speaking, *sources* are areas in which species experience positive population growth, and sinks are areas in which they experience negative population growth at all densities; populations in sinks are only maintained by immigration from source areas (Holt 1985; Pulliam 1988). With strong mass effects (Shmida and Ellner 1984; see table 1.1 for a definition), such that immigration exceeds emigration, populations can be held above carrying capacity in lower quality areas even if this area is strictly speaking a source (Holt 1983, 1985). Without immigration, populations decline in these elevated areas, but not to extinction. This makes it hard to tell sources from sinks in practice (Watkinson and Sutherland 1995). For the purposes of this chapter, the term *source-sink* denotes a class of models that examine movement of individuals from high quality to low quality areas with effects on local population dynamics.

If patches differ in quality, rescue effects are more likely; for example immigration from source patches may rescue sink patches. Such rescue effects explain some of the stabilizing influence of habitat heterogeneity in spatial predator-prey models (Crowley 1981; Holyoak and Lawler 1996). Movement from high quality to low quality areas also has associated costs (Amarasekare 1998b). Individuals may die while dispersing or end up in lower habitat quality. For organisms free to move and capture resources according to their availability (i.e., populations with ideal-free distributions; Fretwell and Lucas 1970), individuals are only found in areas of low habitat quality when moving suboptimally (Hastings 1983; Holt 1985), meaning that they do not select habitats to maximize their individual fitness (Delibes, Ferreras, et al. 2001; Delibes, Gaona, et al. 2001b). Species cannot necessarily optimize their fitness, however, because they cannot always cue in on habitat factors that differentiate between high and low habitat quality (Pulliam and Danielson 1991). Source-sink models have received wide attention in conservation circles because they demonstrate that the addition of low quality habitat can increase the regional abundance of a species by maintaining individuals not only in high quality areas but also in low quality areas where they would not otherwise survive (Hoopes and Harrison 1998). For a single species in a temporally invariant environment, however, sink habitat only increases population persistence and abundance if populations do not have an ideal free distribution; population increases only occur when individuals disperse passively or use low-quality habitat exclusively after high-quality habitat is depleted (Holt 1985; McPeek and Holt 1992; Donahue et al. 2003). In the latter case sink habitats do not drain reproductive individuals away from areas where they can contribute to population growth. When organisms cannot distinguish between habitats, dispersal into low-quality or sink habitat can decrease population abundance and persistence (Holt 1985; Donovan and Thompson 2001; Gundersen et al. 2001). We will see later that this trend extends to multispecies models where dispersal does not always increase community stability or diversity if it overexposes rare populations to increased predation or competition or decreased facilitation.

EXPLICIT SPACE

Single-species models that include explicit space range from extremely detailed spatially explicit population models (SEPMs) based on empirical spatial and movement data to much more general reaction-diffusion models that combine basic population growth equations (reaction) with diffusive movement (diffusion). SEPMs offer excellent predictive abilities about specific habitats but are computationally intensive and difficult to analyze (Pacala et al. 1996). Their use has increased with recent computational advances. Reaction-diffusion models are well developed and tractable but include nonzero rates of infinite dispersal and may not reflect realistic movement patterns (Okubo 1980). Despite these simplifications, reaction-diffusion models are the basis of many advances in spatial ecology and are still essential in the field.

Models with explicit space include the range of behavior found in spatially implicit models but with a greater range of impediments or aids to movement. Rescue effects are less frequent because immigrants cannot come from infinite distances. In this way, spatially explicit models with local dynamics are somewhat parallel to incidence functions in patch occupancy models; where incidence functions make some patches more or less likely to be occupied, explicit space in models with local dynamics increases the differences between population sizes and makes some populations more likely to be rescued than others.

These differences can induce heterogeneities in homogeneous habitats. Models with explicit local interactions—meaning models in which population dynamics are followed explicitly at scales smaller than the dispersal distance—frequently lead to pattern formation (Durrett and Levin 1994b). For example, lattice formulations, which follow populations or individuals distributed across a grid of spatial habitats are followed in discrete time, lead to clustering of individuals (Durrett and Levin 1994a). Similarly, point-process models, which examine populations in continuous space (point) and follow the dynamics (process) of population density in continuous time, also can lead to clustering pattern formation (Bolker and Pacala 1997). Although the habitat is homogeneous, these emergent patterns are themselves spatial heterogeneities. As with implicit space models, heterogeneities, including spatial patterns, can alter the persistence of interacting species in models with explicit space (De Roos et al. 1998; Bolker and Pacala 1999; Gurney and Veitch 2000).

Spatial Competition Models

Spatial models of competition generally address the maintenance of diversity by examining equilibrium conditions for species coexistence, or nonequilibrium conditions for the persistence of two or more competitors (Hastings 1980; Tilman 1980, 1994; Chesson 1985, 2000a; Pacala 1986). The use of the term *coexistence* is somewhat confusing in the literature because it can refer to coexistence within a locality or an entire region. Here we use it to mean coexistence within a region regardless of whether coexistence is possible in individual localities (Amarasekare and Nisbet 2001; Mouquet et al. 2002).

Spatial competition models often use as a reference point the expected outcomes of Lotka-Volterra models of direct competition. For example, adding space to models of competition can allow inferior competitors to persist (Hastings

1980; Tilman et al. 1994; Amarasekare and Nisbet 2001; Mouquet and Loreau 2002), can shift relative dominance as local abundances change (Bolker and Pacala 1999; Loreau and Mouquet 1999), and can lead to the exclusion of competitive dominants (Law and Diekmann 2000). Here we will only summarize some of the major spatial processes responsible for these shifts. A more thorough review of spatial competition models is provided by Mouquet et al. in chapter 10.

Although there are several mechanisms that can maintain species diversity (see Chesson [2000b] for an excellent review), the two most common mechanisms in spatial models are (1) habitat heterogeneity that shifts the local balance of competition so that different species are competitively dominant in different localities, and (2) a trade-off between competitive ability and dispersal ability (figure 2.2). The first mechanism basically offers a refuge (Horn and MacArthur 1972; Comins and Noble 1985); this refuge need not be constant in space but can emerge from environmental fluctuations in either space or time (Chesson 1985, 2000a; Warner and Chesson 1985). In the second mechanism, inferior competitors repeatedly escape from competition by finding empty habitat not yet colonized by the superior competitor (Levins and Culver 1971; Hastings 1978; Tilman 1994). These two mechanisms sound remarkably similar when stated this way. The biological concepts underpinning the two ideas approach each other, but there is a fundamental difference in the model formulation for a shift in competitive superiority (first mechanism) and a constant chase (second mechanism). We will first detail models that rely on habitat heterogeneity to maintain diversity, then examine models with a trade-off between competitive ability and dispersal ability, and finally discuss ways in which these two mechanisms can interact.

Heterogeneous Habitats and Switches in Competitive Dominance

A simple way to allow two species to coexist in a region is to have the region include qualitatively different habitats in which each species specializes so that there is niche differentiation. Intuitively, if species switch competitive superiority with habitat type, then a region can support as many species as there are habitat types as long as each species can maintain a balance of births and deaths (or extinctions and recolonizations) within its preferred habitat (Horn and MacArthur 1972). If the quality of a spatially uniform environment changes through time, then species may partition dominance through time in a similar way, allowing multiple species to coexist (Grubb 1977; Sale 1977; Chesson and Warner 1981; Comins and Noble 1985). This has been called the "storage effect" because species with long-lived life stages and overlapping generations can store up the benefit from good years by reproducing in positive conditions and surviving through negative conditions (Chesson and Warner 1981; Warner and Chesson 1985; Chesson 1994). Species need to have different responses to variation so that there are species-specific responses to the environment that allow niche partitioning (Chesson 2000a). The storage effect requires subadditivity, in which the covariance between competitive processes and the environment decreases with density (Chesson and Huntly 1989; Chesson 1990, 1994). This means that, with a range of densities and environments, a species experiences lower competitive effects on average than if all the individuals were at the mean density and experiencing the mean environment and level of competition. This effect arises from Jensen's inequality, which states that the mean of a decelerating function (concave down, 2nd derivative negative) is less than the function at the mean (Ruel and Ayres 1999; see also figure 12.1 and box 12.2). For populations spread across space

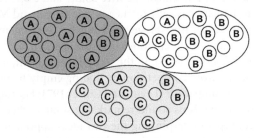

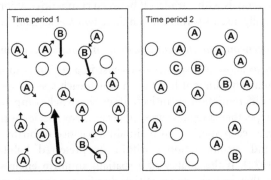

Figure 2.2 Regional competitive coexistence through (a) heterogeneity in the competitive hierarchy or (b) a competition-dispersal trade-off. Species can coexist regionally through either competitive spatial mechanism. Circles represent local communities with A, B, and C representing different species or community types. In the top panel (a) different colors represent different habitat types. Each species competes best in a particular habitat type, and all three species persist in the system. Notice that each species is also found, through source-sink dynamics, in habitats in which it is an inferior competitor. In the lower panel (b) arrows represent dispersal, with arrow size depicting dispersal distance and rate. Locally, species A outcompetes species B, which outcompetes species C; however inferior competitors disperse better and persist in the system.

this systematic difference from the behavior expected from mean conditions can be extremely important for persistence. Jensen's inequality and the nonadditive responses to heterogeneity that result from it not only play a role in competitive coexistence in the storage effect but also play a role in spatial consumer resource and mutualism models.

Chesson (1985, 2000a) has pointed out that storage effects can also be spatial, meaning that species survive in habitats where they are extremely poor competitors because the environment is favorable for them in other patches. Spatial storage effects are, in fact, more likely than temporal storage effects, as they can come from pure spatial variation as well as from spatiotemporal variation, which affects different parts of the environment at different times (Chesson 1985). As with the source-sink effect described in the single species section above, maintenance in low quality areas results from mass effects, or movement of individuals from high quality areas (see also Mouquet et al., chapter 10; Chesson et al., chapter 12; Amarasekare and Nisbet 2001; Mouquet and Loreau 2002).

Homogeneous Habitats and a Competition Colonization Trade-off

When there is no location in which the competitive hierarchy switches, poor competitors can nonetheless survive by dispersing to take advantage of empty habitat. The competition-dispersal trade-off appears most frequently in patch occupancy models similar to the metapopulation approach in equation 2.1 (e.g., Levins and Culver 1971; Hastings 1980; Nee and May 1992; Tilman 1994; Mouquet and Loreau 2002). These models generally assume hierarchical competition, meaning that species have fixed competitive rankings. When a superior species colonizes a site occupied by an inferior competitor, the latter is supplanted. Sites available for colonization, therefore, consist of all sites not occupied by the species itself or a superior competitor. The dynamics of the ith species can be represented as

$$\frac{dp_i}{dt} = c_i p_i \left(1 - \sum_{j=1}^{i} p_j \right) - m_i p_i - \left(\sum_{j=1}^{i-1} c_j p_j p_i \right). \qquad 2.2$$

Here p_i is the fraction of the habitat occupied by species i, c_i and m_i are the colonization rate and extinction rate, respectively, of species i, and the competitive superiority of species decreases with increasing i. Patches can be occupied by single individuals or populations. This model permits an infinite number of species to potentially coexist, provided inferior local competitors are superior at reaching empty patches, and may explain how multiple species can coexist on a single limiting resource (Hastings 1980; Tilman 1994). The "R^*-rule" in resource competition suggests that—on a single limiting resource—only the species that can reduce the resource to the lowest level while maintaining a positive growth rate should survive (Tilman 1980, 1982). More generally, nonspatial models suggest that the equilibrium number of species in closed habitats should match the number of limiting resources. Patch occupancy models help to bridge the discrepancy between R^* predictions and the diversity found in real communities by considering the effect of space on species interactions.

Note that space is still implicit in models like equation 2.2, meaning that there are no explicit distances between patches or sizes of patches. Adler and Mosquera (2000) have pointed out that models without explicit space but with interference competition and frequency dependence also predict a potentially infinite number of coexisting competitors. Note that a spatially implicit model is similar to Adler and Mosquera's nonspatial model with frequency dependence because both models look at the proportion of a system in particular states (2000). In patch occupancy models the states are the occupying species, and in frequency dependent models the states are the competing species.

Models based on equation 2.2 offer qualitative insight into the identity of coexisting species. The competitive hierarchy dictates dispersal requirements for coexisting inferior competitors. Competitive dominants always leave a portion of the habitat unoccupied, leaving room for an inferior competitor to invade, but only an inferior competitor with high colonization rates is capable of finding and using the vacant habitat (Horn and MacArthur 1972; Hastings 1980; Tilman 1994). Taking these insights to more quantitative levels can be misleading, though, because superior competitors can always invade, regardless of their dispersal rates (Kinzig et al. 1999). Some researchers have derived quantitative explanations for limits to similarity of competing species by looking at expressions for successive invaders based on the invasion criteria from equation 2.2 (Pacala and Tilman 1994; Tilman 1994). These quantitative predictions are not reliable because, as superior competitors invade, they change the amount of unoccupied habitat and the subsequent quantitative dispersal requirements for coexistence of inferior competitors (Horn and MacArthur 1972; Kinzig et al. 1999).

These patch occupancy models for competition shed light on relative species abundance and the effects of habitat alteration, fragmentation, or destruction (Hastings 1980; Nee and May 1992; Tilman et al. 1994). As habitat quality declines or habitats are destroyed, superior competitors with lower dispersal rates should suffer more than inferior competitors that are better at dispersal (Tilman et al. 1994). The models suggest that, because of competitive release, weedy species may increase in abundance in disturbed habitats (Nee and May 1992); diversity (Hastings 1980) and extinction probabilities (Tilman et al. 1994; Huxel and Hastings 1998, 1999) may be hard to predict with initial loss of superior competitors (Tilman et al. 1994). These qualitative insights are instructive, but models based on equation 2.2 may not provide appropriate tools for examining specific outcomes at high diversity levels because in that region the dynamics are so slow as to be pathological (meaning they lead to internal inconsistencies or contradictions of model assumptions; Kinzig et al. 1999).

Models with Explicit Space

Models with explicit space allow researchers to examine the effects of extremely local interactions and pattern formation, as discussed in the section on single species models, and may be particularly useful for examining questions with an applied twist (e.g., effects of habitat destruction on coexistence). In a cellular automata model, Neuhauser (1998) found that local interactions altered competitive effects and could change both local and regional competitive dominance. The proximity of neighbors affected the strength of interactions and led to superior competitors excluding inferior competitors from their immediate vicinity. Habitat destruction did not necessarily affect superior competitors more, and actual outcomes depended on the spatial structure imposed by species interactions.

Other competition models with explicit spatial structure have also found that local interactions and pattern formation can change regional dominance. In a spatially continuous model with a homogeneous environment, Law and Diekmann (2000) used moment closure to examine the effects of local dispersal and competition. Unlike neighborhood competition models, which assume a uniform distribution of competitive effects over some local distance (Pacala 1986), the moment-closure approach allows a probability distribution or kernel of dispersal and competitive effects (Bolker and Pacala 1997, 1999; Law and Diekmann 2000). Law and Diekmann found that an inferior competitor with superior dispersal could actually regionally exclude a superior competitor. As the superior competitor eliminated the inferior competitor from its local vicinity, the two species formed clusters; the superior competitor suffered from increased intraspecific competition while the inferior competitor escaped from interspecific effects (Law and Diekmann 2000).

Bolker and Pacala (1999) used a similar moment closure approach to isolate the spatial and competitive trade-offs that contribute to coexistence. They identified three mechanisms involved in competitive use of space: (1) tolerance, which in their definition ranged from the ability to take over occupied space to the ability to resist encroachment, (2) dispersal, which in this definition described the distance moved, and (3) exploitation, which involved the speed with which an organism could actually make use of space to reproduce. Although most models separate tolerance from the other two mechanisms, most spatially implicit models combine the second and third mechanisms into a single process. The explicit spatial structure of the model allowed the authors to separate the ability to reach habitat (dispersal) from the ability to reproduce (exploitation). In patch dynamic models, colonization combines both finding empty localities and instantly reaching carrying capacity in the colonized area (Hanski 1983). Bolker and Pacala found that coupling long-range dispersal with exploitation was unlikely to maintain inferior competitors in the system unless superior competitors were far more clumped than empirically demonstrated for plant species (1999). Instead, they found that strategies combining short dispersal with either high exploitation or tolerance allowed coexistence. Short dispersal and resistance to competitors created what they called a "phalanx effect" in which clusters of conspecifics experienced lower competition. One might expect this dynamic from a species able to change its competitive regime through allelopathy or other processes that affect conspecifics less than heterospecifics (Bolker and Pacala 1997). Although individuals in the phalanx might individually be inferior competitors, when aggregated in clusters they could limit the effects of the superior competitor and change the balance of interspecific and intraspecific competition (Bolker and Pacala 1997). Note that combining both exploitation and the phalanx effect would create a superior competitor.

Similar clustering effects appear in cellular automata models, which follow local interactions in discrete space (Molofsky 1994; Molofsky et al. 2001). Despite a homogeneous environment, organisms in cellular automata models create spatial patterns that alter the quality of local habitats or competitive neighborhoods, just as we saw in spatially explicit models for single species. Consideration of pattern formation in models for competing species shows that, as spatial structure becomes more explicit, spatial processes increasingly affect and in turn are affected by species interactions.

Interaction between Competition-Dispersal Trade-offs and Heterogeneity Induced Shifts in Dominance

As spatial structure in model formulations increases, the two basic mechanisms for competitive coexistence in effect interact. Our initial discussion of heterogeneity in the spatial or temporal environment pointed out that average competitive values may change with spatial or temporal fluctuations, allowing switches in local or short-term competitive dominance and the maintenance of multiple species in a region (Warner and Chesson 1985). With a competition-dispersal trade-off, the competitive hierarchy is fixed so such switches are not possible; yet models that look at this trade-off in explicit space with local interactions find that the average competitive effect actually experienced by individuals across the landscape can be altered by spatial patterns that form in response to this competition-dispersal trade-off. Models that attempt to examine the effects of both habitat heterogeneity and trade-offs have found that there are threshold levels of dispersal that allow species to coexist (Amarasekare and Nisbet 2001; Mouquet and Loreau 2002). If species dispersal is too high, then the advantage of niche partitioning is lost. With too little dispersal, inferior competitors cannot survive. A similar threshold is apparent in Bolker and Pacala's explicit space model, where high levels of dispersal of the inferior competitor did not generally promote coexistence (1999). In chapter 10, Mouquet et al. more thoroughly review models combining source-sink dynamics and life history trade-offs.

Spatial Consumer-Resource Models

Nonspatial consumer-resource models with living resources tend to be unstable, meaning that one or both species goes extinct or both species oscillate. This lack of stability affects persistence and the maintenance of diversity. Nonspatial predator-prey models have generally been based on either a Lotka-Volterra formulation in continuous time (Lotka 1925; Volterra 1926a, 1926b, 1931) or a Nicholson-Bailey formulation in discrete time (Nicholson 1933; Nicholson and Bailey 1935). The most basic nonspatial formulations of each model include only the two interacting species; neither model leads to local stability without the addition of direct density dependence in some fashion (May 1976). The Lotka-Volterra model leads to neutral stability in which populations do not return to equilibrium after a perturbation, but instead oscillate around the equilibrium with amplitude cycle set by the amplitude of the perturbation. The Nicholson-Bailey model produces diverging oscillations until one species goes extinct (May 1976). Stabilizing factors for both models include the addition of a carrying capacity or density dependence for either or both species and an immune stage or age class for the prey (which acts as a refuge; Murdoch and Oaten 1975; Kuno 1987). In the Lotka-Volterra formulation the addition of a Type III functional response in the predator also offers the prey a refuge at low density (Murdoch and Oaten 1975). In the Nicholson-Bailey formulation, adding a negative binomial attack rate, which simulates aggregation in the predator attacks, helps to stabilize the model by offering prey a partial refuge away from clumps of predators (May 1978). We will show that some of the spatial factors that stabilize spatial predator-prey models and assist persistence mimic the same features required for stability in these nonspatial models.

Because of their inherent instability, consumer-resource interactions have received considerable attention from theoreticians and have a broader response to spatial processes than other two-species models. To avoid unnecessary complexity, we focus on stabilizing features of spatial processes in two-species predator-prey models. Following Briggs and Hoopes (2004), we have classified spatial processes that stabilize dynamics into three categories: (1) spatial averaging of fluctuations, frequently referred to as "statistical stability" (De Roos et al. 1991; McCauley et al. 1993, 1996), (2) density dependence induced by dispersal (Murdoch et al. 2003), and (3) nonadditive, nonlinear responses to heterogeneity in densities (Briggs and Hoopes 2004). These three mechanisms are not completely independent, and many models combine aspects of more than one of them. Below we provide an explanation of each mechanism and some examples. We do not specifically discuss spatially explicit versus spatially implicit formulations as we did for single species and competitive models because such divisions do not as succinctly divide model outcomes as they do in the other models. The first and third stabilizing mechanisms below are found more frequently in spatially explicit formulations, but all three stabilizing mechanisms can be found in both implicit and explicit spatial models. There are a number of good reviews of spatial consumer-resource models available for the reader wanting more depth on this subject (Murdoch and Oaten 1975; Briggs and Hoopes 2004; Murdoch et al. 2003).

Statistical Stability, Homogeneous Space

The first of these mechanisms, spatial averaging or statistical stability (De Roos et al. 1991), is seen most frequently in models with many patches or a lattice with local interactions (Allen 1975; De Roos et al. 1991, 1998; McCauley et al. 1996; Wilson 1998). In these models, although local patches may continue to fluctuate, regional population densities are relatively constant. This constancy comes from averaging population densities over fluctuating subpopulations. This mechanism does not require homogeneous space, but it can be swamped by the other two mechanisms in models with heterogeneous space. In models with explicit space there is a characteristic spatial scale below which local dynamics continue to fluctuate as in isolated, nonspatial predator-prey models and there is often pattern formation that induces spatial heterogeneities (De Roos et al. 1991; McCauley et al. 1993, 1996; Wilson et al. 1993). In patch models with small numbers of identical patches, this sort of statistical stability generally requires patches in out-of-phase limit cycles (Adler 1993; Jansen 2001). With more patches, patch models can have more variability in patch densities because the law of large numbers still leads to relatively constant average regional densities (Allen 1975; Crowley 1981). In order for regional densities to remain fairly constant, local populations must fluctuate asynchronously, which generally requires low to intermediate levels of dispersal (Allen 1975; Crowley 1981; De Roos et al. 1991; Adler 1993; but see Jansen 2001). Predator and prey dispersal can have different effects (McCauley et al. 1993), but this result comes partially from heterogeneities in local interactions, which is part of the third mechanism discussed below.

Density Dependence Induced by Dispersal: Heterogeneities in Space or Initial Conditions

The second stabilizing mechanism can come from any of a number of different spatial processes that result in density dependence in local dynamics. The simplest of these processes is a refuge or a constant number of dispersers (Crowley 1981; Sih 1987; Reeve 1988). As local densities vary, this movement leads to smaller changes in large populations and larger changes in small populations, resulting in density dependence that usually stabilizes local dynamics (Holt 1993; Briggs and Hoopes 2004; Murdoch et al. 2003), though immigration can at times be destabilizing (Holt 2002). Heterogeneities in initial conditions can lead to density-dependent interactions in a similar fashion although very specific dispersal parameters are required to maintain such heterogeneities (Crowley 1981). A propagule pool can similarly cause local density dependence. If all dispersers move into a pool from which they are evenly distributed, then immigration is higher per capita in low density patches than in high density patches (Weisser et al. 1997). Similarly, a time lag in dispersal—which is somewhat like a temporal dispersal pool—can be stabilizing as it disconnects dispersal from local dynamics (Gourley and Britton 1996; Neubert et al. 2002). Density dependence can also come from constant per capita immigration or emigration rates among heterogeneous patches (Crowley 1981; Holt 1984; Murdoch et al. 1992; Holt and Hassell 1993; Hochberg et al. 1996). This process is similar to source-sink maintenance of inferior competitors in the competition models. Local heterogeneity can be fixed or from spatiotemporal variation, as we found in spatial models of competition (Crowley 1981). Predation that is focused on areas where prey have high growth rates can be particularly stabilizing (Holt 1993; Holt and Hassell 1993). Conversely, as in the competition and single-species models, any dispersal that overemphasizes areas where populations are doing poorly will destabilize the dynamics and reduce persistence (Murdoch et al. 1992; Holt and Hassell 1993). For example, if predators aggregate too well on prey, they can synchronize patches and destabilize dynamics (Murdoch et al. 1992).

Nonlinear, Nonadditive Responses to Heterogeneity

The third stabilizing mechanism comes only with spatial heterogeneity in densities and nonlinearities in predator-prey responses. This feature is similar to a spatial storage effect (Chesson 1985, 1990; Warner and Chesson 1985) in that prey benefit more in areas where they are at high density than they lose in areas where they are at low density; this nonlinearity leads to subadditivity of density and predation effects (figure 12.1; box 12.2; Chesson 1990). The predators, therefore, still control the prey but without the overcompensation effects that would cause cycles. This sort of effect has been demonstrated with patch models (De Roos et al. 1998), lattice models (McCauley et al. 1996), and reaction-diffusion models (Gurney and Veitch 2000). In all of these models prey show distinct spatial patterns. In the patch model example, predators with a Type II functional response were spread evenly across the patches; prey dispersal combined with the nonlinear predator functional response led to some patches devoid of prey and others at high density (De Roos et al. 1998). In the individual-based lattice model example, predator functional responses were nonlinear because of age-structure; adult and juvenile predators had similar attack rates, but juveniles did not reproduce (McCauley et al. 1996). Local dispersal of juveniles led to prey clustering, creating the necessary spatial heterogeneities; this clustering combined with the nonlinear predation stabilized the system and increased persistence (McCauley et al. 1996). In the reaction-diffusion model, Gurney and Veitch (2000) found that low predator dispersal and immobile prey could form spiral waves in a continuous model and clusters in a discrete analogue; with Rosenzweig-MacArthur predation

(Rosenzweig and MacArthur 1963), both spatial patterns could stabilize the system.

Combining the Three Stabilizing Mechanisms

These three stabilizing mechanisms in consumer-resource interactions are not independent and may not only act together but even be indistinguishable in both models and empirical systems. For example, the formation of stable spatial patterns promotes statistical stability, creates differences in local densities that can lead to density dependent dispersal, and creates heterogeneities that may interact with predator nonlinearities. In many models in which spatial patterns do not stabilize dynamics, they nonetheless lead to increased persistence (Hassell et al. 1991; Comins et al. 1992; Rohani et al. 1996; Gurney et al. 1998). Note that transient dynamics can contribute to persistence (Hastings 2001). This is particularly true for statistical stability, which is basically a series of very local systems with transient dynamics that are constantly buffeted by the transient dynamics of their neighbors. On the other hand, models with all of these mechanisms may lead to extinction, cycling, or complex dynamics. No mechanism guarantees stability. Instead each mechanism *can* promote stability and persistence, given appropriate model structure, rates and modes of dispersal, and system size. The functional form of predator responses and prey growth can strongly affect model outcomes (Rohani and Ruxton 1999) with differences between comparable discrete and continuous models (Wilson 1996, 1998; Gurney and Veitch 2000). Dispersal is equally crucial. Too much can synchronize systems so that dynamics are similar to those of nonspatial models (Crowley 1981; Adler 1993; McCauley et al. 1993; Doebeli 1995), and too little dispersal can fail to separate dynamics enough (McCauley et al. 1993; Jansen 2001). The size of the system and boundary conditions can affect spatial pattern formation and stability (Wilson 1996; Donalson and Nisbet 1999), and in small systems even or odd numbers of patches can affect the ability to attain out of phase limit cycles (Adler 1993). We have here only provided a brief outline or explanation and one or two examples. There are also examples of similar models with slightly different combinations of functional responses or dispersal distances that lead to very different outcomes. However, we suggest that any system that is stable or persists will include one of these three mechanisms.

Spatial Mutualism Models

Although mutualistic interactions appear to be common in nature (Crepet 1983; Boucher 1985; Schwartz and Hoeksema 1998), there are far fewer models that address the population dynamic consequences of mutualisms than there are for competition or predation (Wolin 1985; Bronstein 1994). Simple models suggest that mutualistic interactions are unstable with a tendency toward extinction and unbounded population growth; however obligate and facultative mutualisms display different dynamics (May 1974; Vandermeer and Boucher 1978). Obligate mutualisms are those in which each partner requires the other for survival. Examples include some plant-pollinator systems, such as figs with fig wasps and yuccas with yucca moths (Bronstein 1988; Pellmyr and Huth 1994), mycorrhizae and some plant species (Hoeksema and Bruna 2000); and coelenterates and zoo-xanthellae (Goreau et al. 1979). Facultative mutualisms are those in which each partner can survive without the other. Plants with multiple dispersal pathways and pollinating insects with a range of food resources are good examples of facultative mutualists. Because there are fewer mutualism models, let alone spatial mutualism models, we briefly summarize nonspatial models, examine the effects of spatially implicit processes on obligate mutualisms, and then extend these results qualitatively to facultative mutualisms.

Nonspatial Mutualism Models

The simplest models of obligate mutualisms generally consist of Lotka-Volterra competition equations with positive, rather than negative, constant interaction coefficients to indicate linear, positive effects from species interactions (Gause and Witt 1935; Vandermeer and Boucher 1978). These models often assume a negative or zero carrying capacity to meet the assumption that the two species do not survive in the absence of the other. As with competition models, these mutualism models produce intersecting linear isoclines. The intersection of the isoclines is an unstable equilibrium point with a tendency toward extinction below the equilibrium point and limitless growth above it, an unrealistic prediction that has hampered the study of mutualisms in communities. Nonlinear interaction coefficients between the two species (e.g., saturating benefits, DeAngelis et al. 1986) though, can give curved isoclines (figure 2.3); when these isoclines intersect so that there are two interior equilibria (i.e., the curves create an eye by crossing twice), the second intersection is a locally stable equilibrium, and the species persists as long as densities stay above the lower intersection or saddle equilibrium point (May 1976; Vandermeer and Boucher 1978; DeAngelis et al. 1986). Because these more realistic models predict that both mutualistic partners go extinct when densities drop below this threshold lower equilibrium point, they suggest that obligate mutualisms should only persist in relatively stable environments where population densities are unlikely to experience extreme reductions (May 1974; Travis and Post 1979). Other researchers have suggested that the diversity of obligate mutualists in nature indicates instead that these models fail to capture the full range of factors affecting the population dynamics of mutualists (Heithaus et al. 1980; Boucher et al. 1982; Wolin 1985; Ringel et al. 1996).

Models of facultative mutualisms are far easier to stabilize than obligate mutualisms (Wolin and Lawlor 1984; Wolin 1985). Because facultative partners can survive without the other species, their populations grow when small and do not tend toward extinction at low densities (Vandermeer and Boucher 1978; Travis

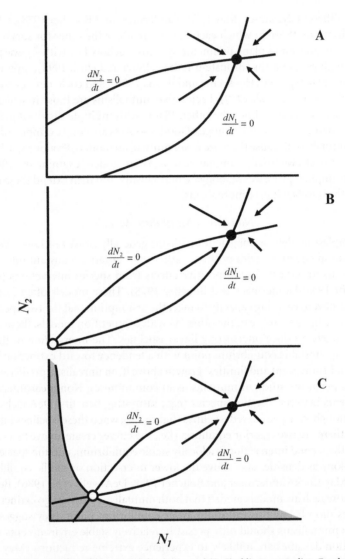

Figure 2.3 Isoclines and equilibria for (A) facultative and (B, C) obligate mutualists. Facultative mutualisms can be very stable although obligate mutualisms may not be. Solid circles represent stable equilibria and open circles represent unstable equilibria. In panel A each species can exist without the other, making the species facultative mutualists. Notice that each species exists at a higher density with its mutualistic partner than alone. In panels B and C, the two species can only exist in the presence of the other, making them obligate mutualists. In panel C the unstable equilibrium is a saddle point with a stable manifold that acts as a threshold. At densities below this manifold (in the shaded region), the species cannot persist (after Vandermeer and Boucher 1978; Dean 1983).

and Post 1979; Wolin and Lawlor 1984). Mutualistic effects may also be incorporated in multiple ways in models of facultative mutualisms. For example, the presence of a facultative mutualistic partner may increase carrying capacities or alter the speed at which populations approach carrying capacity (Dean 1983; Wolin 1985; Wright 1989). The effects of facultative mutualists can also vary with the density of either partner (Wolin and Lawlor 1984). In general mutualistic effects at low density or decreasing with density lead to stable equilibria (Dean 1983; Wolin and Lawlor 1984; Wright 1989); however all facultative mutualisms, even those that do not contain stable equilibria, increase the persistence of the mutualistic partners (Boucher et al. 1982; Wolin 1985).

Adding Space to Models of Obligate Mutualism

One of the missing factors of realism in obligate mutualism models is spatial structure. Hutson et al. (1985) point out that nonspatial approaches to modeling obligate mutualisms assume that populations are so evenly spread across an environment that modeling average dynamics is equivalent to modeling complete dynamics. If, instead, one assumes that populations are spread out across space with different densities in different locations, then drops in local population density will not necessarily take even an obligate mutualist below a threshold density (Hutson et al. 1985). In fact, if one can describe density across linear space with a function, then all points above that function should be dynamically equivalent to average population densities that uniformly approach a stable equilibrium point (figure 2.4; Hutson et al. 1985). There should be, therefore, a vast array of distributions for which the spatial mutualistic interaction is extremely resilient to perturbations (Hutson et al. 1985). Similarly, Hastings and Wolin (1989) examined a structured population model to look at obligate mutualists in spatial subpopulations. The model is an age-structured model in which local density is assumed to increase with patch age (time since colonization). If extinction rates decrease with density, then there is an equilibrium distribution of numbers within patches, and all equilibria are stable (Hastings and Wolin 1989). This outcome is similar to the dynamic equilibrium found in patch occupancy single species and competition models in which local occupancy may vary but the fraction of occupied patches is constant. This outcome is also similar to the statistical averaging found in consumer-resource models with increased persistence or regional stability arising from statistical stability. Space, therefore, can stabilize models of obligate mutualisms. These spatial models suggest not only that obligate mutualisms may persist in nature, but also that the consideration of spatial structure may be particularly important in community dynamics involving such mutualists.

Adding Space to Models of Facultative Mutualism

Because even nonspatial models suggest that facultative mutualisms not only persist in nature but can increase the persistence of each species beyond their indi-

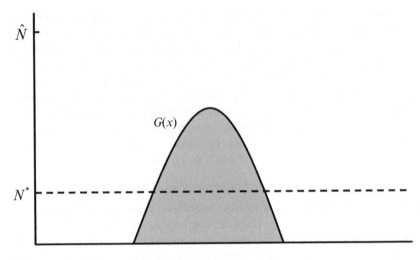

one dimensional space (x)

Figure 2.4 One dimensional spatial representation of an obligate mutualist. Obligate mutualisms may be far more stable in space than nonspatial models suggest. The density of each species in figure 2.3 could be represented in one dimensional space. If the species were uniformly distributed across space, then at the unstable equilibrium the density at each point x would be N^*, and any density below the dashed line would bring the interaction below the threshold stable manifold in figure 2.3C, leading to the extinction of the two obligate mutualists. As suggested by Hutson et al. (1985), though, a variety of potential functions, $G(x)$, could describe the distribution of the species in space. As long as $N(x)$ is greater than $G(x)$, meaning the distribution does not lie in the shaded region, the species should head toward the stable equilibrium, and the mutualists should persist. This result suggests that obligate mutualists can persist despite very low densities in some areas, and that these interactions might be quite stable.

vidual expected persistence time, spatial models of facultative mutualisms offer more insight into the mechanisms of single-species spatial persistence or spatial processes in multispecies models. The nonuniform distribution of a facultative mutualist can create higher quality habitats that lead to mass effects for focal single species. Similarly, mutualists can stabilize larger community dynamics and food web modules by creating habitat heterogeneity that shifts the balance of competition and allows inferior competitors to persist or that offers a refuge from predators (Heithaus et al. 1980; Addicott 1981; Boucher et al. 1982; Ringel et al. 1996). Mutualisms are not always direct interactions and are not always without costs; they represent one end of a continuum of interactions from predatory or parasitic to facilitative (Abrams and Matsuda 1996; Abrams et al. 1998). Many researchers have pointed out the likelihood of evolution of mutualistic interactions from parasitic ones (Axelrod and Hamilton 1981; Toft and Karter 1990; Nee 2000; Bronstein 2001; Westerbergh and Westerbergh 2001), and many empirical and modeling studies have examined interactions that vary spatially or temporally from positive to negative (Vandermeer et al. 1985; Dodds 1988; Abrams and Matsuda 1996; Ringel et al. 1996; Abrams et al. 1998). The stabilizing influence of mutualistic interactions, therefore, may be extremely common and hard to detect in communities, and mutualists may play an important and poorly understood role in metacommunity dynamics.

Conclusion

This chapter has outlined some basic spatial processes in models of two species interactions, with an emphasis on key features that help to increase stability or persistence in single and two-species models. The first feature is asynchrony in local dynamics, particularly in models with homogeneous space. This feature is crucial for single species models but is also important in statistical averaging found in competition-dispersal trade-off competitive models, statistical stability in consumer resource models, and the stable distribution of population abundances found in Hastings and Wolin's spatially structured mutualism model (1989).

With heterogeneity in habitat quality, the potential for persistent interactions or communities increases by creating refuges or disparities in habitat quality that increase differences in local population abundances. With dispersal these disparities lead to mass effects that allow rescue effects and source-sink dynamics in single species and competitive models, and create induced density-dependent stability in predator-prey interactions. Facultative mutualistic interactions can increase such spatial heterogeneities and contribute to the persistence not only of the mutualistic partners but also of their predators.

Spatial heterogeneity and nonlinearities in response to that heterogeneity can also increase persistence in all three types of two-species models. Nonlinear, nonadditive responses to other species can create thresholds as well as relative spatial refuges. They can interact with spatial or temporal heterogeneity to change average regional parameter values from those expected for a mean environment as in temporal and spatial storage effects in competitive interactions, nonlinear stabilization in predator-prey interactions, and the potential maintenance of obligate mutualists with nonuniform distributions.

As discussed in the consumer-resource section, these mechanisms can interact and are not always completely separable. Pattern formation in homogeneous landscapes creates heterogeneities that influence persistence, so that organisms in homogeneous landscapes may experience not only statistical averaging but also mass effects from local changes in habitat quality due to altered interaction strengths or barriers to dispersal. Also, if interacting species disperse at different scales or use different resources, they may experience very different environments, even so far as one species perceiving a homogeneous environment while the other perceives a heterogeneous one. This is frequently the case in predator-prey models.

We have emphasized modeling approaches that highlight the maintenance of diversity, yet community dynamics can just as frequently be affected by spatial responses to species interactions that limit or reduce diversity. As we have discussed in each section, a balance between too little and too much dispersal is important in order to maintain asynchrony between local communities. At higher levels of dispersal, spatial subdivision disappears and the stabilizing or persistence increasing effects of spatial processes may be diluted or may disappear altogether. Facultative mutualisms, for example, may decrease persistence of other species by increasing abundance and dispersal between local communities so that synchrony increases (Ringel et al. 1996). Dispersal in heterogeneous landscapes can permit exclusion that otherwise would not be observed, and at times even destabilize predator-prey interactions.

With all formulations the addition of space can stabilize dynamics, increase persistence of species, and offer some explanations for the maintenance of species diversity. These spatial dynamics are crucial to metacommunities. We will see in the next chapter that additional interacting species add heterogeneity and can distinctly alter dynamics as we discussed briefly in regard to mutualistic interactions. Additional species may alter local habitat quality leading to variation in spatial effects that may allow inferior competitors in one area to dominate in another, that may offer alternative prey resources or shelter from predation, or that may facilitate growth or reproduction in ways that alter the outcomes of other species interactions. As with space, species interactions can have both facilitative and negative effects. In many cases the addition of species or interactions to the models we have outlined may destabilize stable interactions or may change spatial or temporal heterogeneity just enough to remove a fugitive species from the system and lead to reductions in local and metacommunity diversity.

Acknowledgments

This manuscript benefited from comments by Brett Melbourne and Richard Law. This work was conducted as part of the Metacommunity Working Group at the National Center for Ecological Analysis and Synthesis (NCEAS), a center funded by the National Science Foundation (DEB-9421535), University of California Santa Barbara, and the state of California. MFH was supported by NSF grant DEB-9806635 and NIH grant R01 ES12067-01 to Cheryl J. Briggs. MH was supported by NSF DEB-0213026. RDH was supported by NSF, NIH, and the University of Florida Foundation.

Literature Cited

Abrams, P. A., R. D. Holt, and J. D. Roth. 1998. Apparent competition or apparent mutualism? Shared predation when populations cycle. Ecology 79:201–212.

Abrams, P. A., and H. Matsuda. 1996. Positive indirect effects between prey species that share predators. Ecology 77:610–616.

Addicott, J. F. 1981. Stability properties of two-species models of mutualism: Simulation studies. Oecologia 49:42–49.

Adler, F. R. 1993. Migration alone can produce persistence of host-parasitoid models. American Naturalist 141:642–650.

Adler, F. R., and J. Mosquera. 2000. Is space necessary? Interference competition and limits to biodiversity. Ecology 81:3226–3232.

Allen, J. C. 1975. Mathematical models of species interactions in space and time. American Naturalist 109:319–342.

Amarasekare, P. 1998a. Allee effects in metapopulation dynamics. American Naturalist. 152:298–302.

———. 1998b. Interactions between local dynamics and dispersal: insights from single species models. Theoretical Population Biology 53:44–59.

Amarasekare, P., and R. M. Nisbet. 2001. Spatial heterogeneity, source-sink dynamics, and the local coexistence of competing species. American Naturalist 158:572–584.

Andreasen, V., and F. B. Christiansen. 1989. Persistence of an infectious-disease in a subdivided population. Mathematical Biosciences 96:239–253.

Axelrod, R., and D. W. Hamilton. 1981. The evolution of cooperation. Science 211:1390–1396.

Bolker, B., and B. Grenfell. 1995. Space, persistence and dynamics of measles epidemics. Philosophical Transactions of the Royal Society of London, series B, 348:309–320.

Bolker, B. M., and S. W. Pacala. 1997. Using moment equations to understand stochastically driven spatial pattern formation in ecological systems. Theoretical Population Biology 52:179–197.

———. 1999. Spatial moment equations for plant competition: Understanding spatial strategies and the advantages of short dispersal. American Naturalist 153:575–602.

Boucher, D. H. 1985. *The biology of mutualism*. Oxford University Press, New York.

Boucher, D. H., S. James, and K. H. Keeler. 1982. The ecology of mutualism. Annual Review of Ecology and Systematics 13:315–347.

Briggs, C. J., and M. F. Hoopes. 2004. Stabilizing effects in spatial parasitoid-host and predator-prey models: A review. Theoretical Population Biology 65:299–315.

Bronstein, J. L. 1988. Limits to fruit production in a monoecious fig: Consequences of an obligate mutualism. Ecology 69:207–214.

———. 1994. Our current understanding of mutualisms. Quarterly Review of Biology 69:31–51.

———. 2001. The costs of mutualism. American Zoologist. 41:825–839.

Brown, J. H., and A. Kodric-Brown. 1977. Turnover rates in insular biogeography: Effect of immigration on extinction. Ecology 58:445–449.

Caswell, H. and J. E. Cohen. 1991. Disturbance, interspecific interaction, and diversity in metapopulations. Biological Journal of the Linnean Society 42:193–218.

Caughley, G. 1994. Directions in conservation biology. Journal of Animal Ecology 63:215–244.

Chesson, P. L. 1985. Coexistence of competitors in spatially and temporally varying environments: A look at the combined effects of different sorts of variability. Theoretical Population Biology 28:263–287.

———. 1990. Geometry, heterogeneity and competition in variable environments. Philosophical Transactions of the Royal Society of London, series B, 330:165–173.

———. 1994. Multispecies competition in variable environments. Theoretical Population Biology 45:227–276.

———. 2000a. General theory of competitive coexistence in spatially-varying environments. Theoretical Population Biology 58:211–237.

———. 2000b. Mechanisms of maintenance of species diversity. Annual Review of Ecology and Systematics 31:343–366.

Chesson, P., and N. Huntly. 1989. Short-term instabilities and long-term communities dynamics. Trends in Ecology and Evolution 4:293–298.

Chesson, P. L., and R. R. Warner. 1981. Environmental variability promotes coexistence in lottery competitive systems. American Naturalist 117:923–943.

Comins, H. N., M. P. Hassell, and R. M. May. 1992. The spatial dynamics of host-parasitoid systems. Journal of Animal Ecology 61:735–748.

Comins, H. N., and I. R. Noble. 1985. Dispersal, variability, and transient niches: Species coexistence in a uniformly variable environment. American Naturalist 126:706–723.

Crepet, W. L. 1983. The role of insect pollination in the evolution of angiosperms. Pages 29–50 *in* L. Real, ed. *Pollination biology.* Academic Press, Orlando, FL, USA.

Crowley, P. H. 1981. Dispersal and the stability of predator-prey interactions. American Naturalist 118:673–701.

Dean, A. M. 1983. A simple model of mutualism. American Naturalist 121:409–417.

DeAngelis, D. L., W. M. Post and C. C. Travis. 1986. *Positive feedback in natural systems.* Springer-Verlag, New York.

Delibes, M., P. Ferreras, and P. Gaona. 2001a. Attractive sinks, or how individual behavioural decisions determine source-sink dynamics. Ecology Letters 4:401–403.

Delibes, M., P. Gaona, and P. Ferreras. 2001b. Effects of an attractive sink leading into maladaptive habitat selection. American Naturalist 158:277–285.

De Roos, A. M., E. McCauley, and W. G. Wilson. 1991. Mobility versus density-limited predator-prey dynamics on different spatial scales. Proceedings of the Royal Society of London, series B. Biology 246:117–122.

———. 1998. Pattern formation and the spatial scale of interaction between predators and their prey. Theoretical Population Biology 53:108–130.

Dodds, W. K. 1988. Community structure and selection for positive or negative species interactions. Oikos 53:387–390.

Doebeli, M. 1995. Dispersal and dynamics. Theoretical Population Biology 47:82–106.

Donahue, M. J., M. Holyoak, and C. Feng. 2003. Patterns of dispersal and dynamics among habitat patches varying in quality. American Naturalist 162:302–317.

Donalson, D. D., and R. M. Nisbet. 1999. Population dynamics and spatial scale: Effects of system size on population persistence. Ecology 80:2492–2507.

Donovan, T. M., and F. R. Thompson. 2001. Modeling the ecological trap hypothesis: A habitat and demographic analysis for migrant songbirds. Ecological Applications 11:871–882.

Durrett, R., and S. A. Levin. 1994a. The importance of being discrete (and spatial). Theoretical Population Biology 46:363–394.

———. 1994b. Stochastic spatial models: A user's guide to ecological applications. Philosophical Transactions of the Royal Society of London, series B, 343:329–350.

Fretwell, S. D., and H. L. Lucas. 1970. On territorial behavior and other factors influencing habitat distribution in birds. I. Theoretical development. Acta Biotheoretica 19:16–36.

Gause, G. F., and A. A. Witt. 1935. Behavior of mixed populations and the problem of natural selection. American Naturalist 69:596–609.

Gilpin, M., and I. Hanski. 1991. *Metapopulation dynamics: Empirical and theoretical investigations.* Academic Press, London, UK.

Gonzalez, A., and R. D. Holt. 2002. The inflationary effects of environmental fluctuations in source-sink systems. Proceedings of the National Academy of Sciences 99:14872–14877.

Goreau, T. F., N. I. Goreau, and T. J. Goreau. 1979. Corals and coral reefs. Scientific American 241:124–136.

Gourley, S. A., and N. F. Britton. 1996. A predator-prey reaction-diffusion system with nonlocal effects. Journal of Mathematical Biology 34:297–333.

Grenfell, B., and J. Harwood. 1997. (Meta)population dynamics of infectious diseases. Trends in Ecology & Evolution 12:395–399.

Grenfell, B. T., O. N. Bjornstad, and J. Kappey. 2001. Travelling waves and spatial hierarchies in measles epidemics. Nature 414:716–723.

Grubb, P. J. 1977. The maintenance of species-richness in plant communities: The importance of the regeneration niche. Biological Reviews of the Cambridge Philosophical Society 52:107–145.

Gundersen, G., E. Johannesen, H. P. Andreassen, and R. A. Ims. 2001. Source-sink dynamics: How sinks affect demography of sources. Ecology Letters 4:14–21.

Gurney, W. S. C., and A. R. Veitch. 2000. Self-organization, scale and stability in a spatial predator-prey interaction. Bulletin of Mathematical Biology 62:61–86.

Gurney, W. C. S., A. R. Veitch, I. Cruickshank, and G. McGeachin. 1998. Circles and spirals: Population persistence in a spatially explicit predator-prey model. Ecology 79:2516–2530.

Hanski, I. 1983. Coexistence of competitors in patchy environment. Ecology 64:493–500.

———. 1994. A practical model of metapopulation dynamics. Journal of Animal Ecology 63:151–162.

———. 1997. Metapopulation dynamics: From concepts and observations to predictive models. Pages 69–72 in I. Hanski and M. E. Gilpin, eds. Metapopulation biology: Ecology, genetics, and evolution. Academic Press, San Diego, CA.

Harding, K. C. and J. M. McNamara. 2002. A unifying framework for metapopulation dynamics. American Naturalist 160:173–185.

Harrison, S., and J. F. Quinn. 1990. Correlated environments and the persistence of metapopulations. Oikos 56:293–298.

Hassell, M. P., H. N. Comins, and R. M. May. 1991. Spatial structure and chaos in insect population dynamics. Nature 353:255–258.

Hastings, A. 1978. Spatial heterogeneity and the stability of predator-prey systems: Predator-mediated coexistence. Theoretical Population Biology 14:380–395.

———. 1980. Disturbance, coexistence, history, and competition for space. Theoretical Population Biology 18:363–373.

———. 1983. Can spatial variation alone lead to selection for dispersal? Theoretical Population Biology 24:244–251.

———. 2001. Transient dynamics and persistence of ecological systems. Ecology Letters 4:215–220.

Hastings, A., and K. Higgins. 1994. Persistence of transients in spatially structured ecological models. Science 263:1133–1136.

Hastings, A., and C. L. Wolin. 1989. Within-patch dynamics in a metapopulation. Ecology 70:1261–1266.

Heithaus, E. R., D. C. Culver, and A. J. Beattie. 1980. Models of some ant-plant mutualisms. American Naturalist 116:347–361.

Hess, G. 1996. Disease in metapopulation models: Implications for conservation. Ecology 77:1617–1632.

Hill, F. M., A. Hastings, and L. W. Botsford. 2002. The effects of small dispersal rates on extinction times in structured metapopulation models. American Naturalist. 160:389–402.

Hochberg, M. E., G. W. Elmes, J. A. Thomas, and R. T. Clarke. 1996. Mechanisms of local persistence in coupled host-parasitoid associations: The case model of Maculinea rebeli and Ichneumon eumerus. Philosophical Transactions of the Royal Society of London, series B, 351:1713–1724.

Hoeksema, J. D., and E. M. Bruna. 2000. Pursuing the big questions about interspecific mutualism: A review of theoretical approaches. Oecologia 125:321–330.

Holt, R. D. 1984. Spatial heterogeneity, indirect interactions, and the coexistence of prey species. American Naturalist 124:377–406.

———. 1985. Population dynamics in two-patch environments: Some anomalous consequences of an optimal habitat distribution. Theoretical Population Biology 28:181–208.

———. 1992. A neglected facet of island biogeography: The role of internal spatial dynamics in area effects. Theoretical Population Biology 41:354–371.

———. 1993. Ecology at the mesoscale: The influence of regional processes on local communities. Pages 77–88 *in* R. Ricklefs and D. Schluter, eds., *Species Diversity in Ecological Communities*. University of Chicago Press: Chicago.

———. 2002. Food webs in space: On the interplay of dynamic instability and spatial processes. Ecological Research 17:261–273.

Holt, R. D., and M. P. Hassell. 1993. Environmental heterogeneity and the stability of host-parasitoid interactions. Journal of Animal Ecology 62:89–100.

Holyoak, M., and S. P. Lawler. 1996. The role of dispersal in predator—prey metapopulation dynamics. Journal of Animal Ecology 65:640–652.

Hoopes, M. F., and S. Harrison. 1998. Metapopulation, source-sink and disturbance dynamics. Pages 135–151 *in* W. J. Sutherland, ed. *Conservation Science and Action*. Blackwell Science, Oxford, UK.

Horn, H. S., and R. H. MacArthur. 1972. Competition among fugitive species in a harlequin environment. Ecology 53:749–752.

Hutson, V., R. Law, and D. Lewis. 1985. Dynamics of ecologically obligate mutualisms—effects of spatial diffusion on resilience of the interacting species. American Naturalist 126:445–449.

Huxel, G. R., and A. Hastings. 1998. Population size dependence, competitive coexistence and habitat destruction. Journal of Animal Ecology 67:446–453.

———. 1999. Habitat loss, fragmentation, and restoration. Restoration Ecology 7:309–315.

Jansen, V. A. A. 2001. The dynamics of two diffusively coupled predator-prey populations. Theoretical Population Biology 59:119–131.

Kinzig, A. P., S. A. Levin, J. Dushoff, and S. Pacala. 1999. Limiting similarity, species packing, and system stability for hierarchical competition-colonization models. American Naturalist 153:371–383.

Kuno, E. 1987. Principles of predator-prey interactions in theoretical, experimental, and natural populations. Advances in Ecological Research 16:250–331.

Labra, F. A., N. A. Lagos, and P. A. Marquet. 2003. Dispersal and transient dynamics in metapopulations. Ecology Letters 6:197–204.

Law, R., and U. Diekmann. 2000. A dynamical system for neighborhoods in plant communities. Ecology 81:2137–2148.

Levins, R. 1969. Some demographic and genetic consequences of environmental heterogeneity for biological control. Bulletin of the Entomological Society of America 15:237–240.

———. 1970. Extinction. Pages 75–107 *in* M. Gerstenhaber, ed. *Some mathematical problems in biology*. American Mathematical Society, Providence, RI, USA.

Levins, R., and D. Culver. 1971. Regional coexistence of species and competition between rare species. Proceedings of the National Academy of Sciences 68:1246–1248.

Lloyd, A. L., and R. M. May. 1996. Spatial heterogeneity in epidemic models. Journal of Theoretical Biology 179:1–11.

Loreau, M., and N. Mouquet. 1999. Immigration and the maintenance of local species diversity. American Naturalist 154:427–440.

Lotka, A. J. 1925. *Elements of physical biology*. Williams and Wilkins, Baltimore, MD, USA.

May, R. M. 1974. *Stability and complexity in model ecosystems*. Princeton University Press, Princeton, NJ. USA.

———. 1976. Models for two interacting populations. Pages 78–104 *in* R. M. May, ed. *Theoretical ecology: Principles and applications*. Blackwell Scientific, Oxford, UK.

———. 1978. Host-parasitoid systems in patchy environments: A phenomenological model. Journal of Animal Ecology 47:833–844.

McCallum, H., and A. Dobson. 2002. Disease, habitat fragmentation and conservation. Proceedings of the Royal Society of London, series B. Biological Sciences 269:2041–2049.

McCauley, E., W. G. Wilson, and A. M. de Roos. 1993. Dynamics of age-structured and spatially structured predator-prey interactions: Individual-based models and population-level formulations. American Naturalist 142:412–442.

———. 1996. Dynamics of age-structured predator-prey populations in space: Asymmetrical effects of mobility in juvenile and adult predators. Oikos 76:485–497.

McPeek, M. A., and R. D. Holt. 1992. The evolution of dispersal in spatially and temporally varying environments. American Naturalist 140:1010–1027.

Molofsky, J. 1994. Population dynamics and pattern formation in theoretical populations. Ecology 75:30–39.

Molofsky, J., J. D. Bever, and J. Antonovics. 2001. Coexistence under positive frequency dependence. Proceedings of the Royal Society of London, series B, 268:273–277.

Mouquet, N., and M. Loreau. 2002. Coexistence in metacommunities: The regional similarity hypothesis. American Naturalist 159:420–426.

Mouquet, N., J. L. Moore, and M. Loreau. 2002. Plant species richness and community productivity: Why the mechanism that promotes coexistence matters. Ecology Letters 5:56–65.

Murdoch, W. W., C. J. Briggs, and R. M. Nisbet. 2003. *Consumer-resource dynamics*. Princeton University Press, Princeton, NJ.

Murdoch, W. W., C. J. Briggs, R. M. Nisbet, W. S. C. Gurney, and A. Stewart-Oaten. 1992. Aggregation and stability in metapopulation models. American Naturalist 140:41–58.

Murdoch, W. W., and A. Oaten. 1975. Predation and population stability. Advances in Ecological Research 9:1–131.

Nee, S. 2000. Mutualism, parasitism and competition in the evolution of coviruses. Philosophical Transactions of the Royal Society of London, series B. Biological Sciences 355:1607–1613.

Nee, S., and R. M. May. 1992. Dynamics of metapopulations: Habitat destruction and competitive coexistence. Journal of Animal Ecology 61:37–40.

Neubert, M. G., P. Klepac, and P. v. d. Driessche. 2002. Stabilizing dispersal delays in predator-prey metapopulation models. Theoretical Population Biology 61:339–347.

Neuhauser, C. 1998. Habitat destruction and competitive coexistence in spatially explicit models with local interactions. Journal of Theoretical Biology 193:445–463.

Nicholson, A. J. 1933. The balance of animal populations. Journal of Animal Ecology 2:132–178.

Nicholson, A. J., and V. A. Bailey. 1935. The balance of animal populations. Part I. Proceedings of the Zoological Society of London 3:551–598.

Okubo, A. 1980. Diffusions and ecological problems: Mathematical models. Springer-Verlag, New York.

Ovaskainen, O., and I. Hanski. 2002. Transient dynamics in metapopulation response to perturbation. Theoretical Population Biology 61:285–295.

Ovaskainen, O., K. Sato, J. Bascompte, and I. Hanski. 2002. Metapopulation models for extinction threshold in spatially correlated landscapes. Journal of Theoretical Biology 215:95–108.

Pacala, S. W. 1986. Neighborhood models of plant population dynamics. Single-species and multi-species models of annuals with dormant seeds. American Naturalist 128:859–878.

Pacala, S. W., C. Canham, J. Saponara, J. Silander, R. Kobe, and E. Ribbens. 1996. Forest models defined by field measurements: Estimation, error analysis and dynamics. Ecological Monographs 66:1–44.

Pacala, S. W., and D. Tilman. 1994. Limiting similarity in mechanistic and spatial models of plant competition in heterogeneous environments. American Naturalist 143:222–257.

Park, A. W., S. Gubbins, and C. A. Gilligan. 2002. Extinction times for closed epidemics: The effects of host spatial structure. Ecology Letters 5:747–755.

Pellmyr, O., and C. J. Huth. 1994. Evolutionary stability of mutualism between yuccas and yucca moths. Nature 372:257–260.

Pulliam, H. R. 1988. Sources, sinks, and population regulation. American Naturalist 132:652–661.

Pulliam, H. R., and B. J. Danielson. 1991. Sources, sinks, and habitat selection: A landscape perspective on population dynamics. American Naturalist 137:S50–S66.

Quinn, J. F., and A. M. Hastings. 1987. Extinction in subdivided habitats. Conservation Biology 1:198–208.

Reeve, J. D. 1988. Environmental variability, migration, and persistence in host-parasitoid systems. American Naturalist 132:810–836.

Rhodes, C. J., and R. M. Anderson. 1997. Epidemic thresholds and vaccination in a lattice model of disease spread. Theoretical Population Biology 52:101–118.

Ringel, M. S., H. H. Hu, G. Anderson, and M. S. Ringel. 1996. The stability and persistence of mutualisms embedded in community interactions. Theoretical Population Biology 50:281–297.

Rodriguez, D. J., and L. Torres-Sorando. 2001. Models of infectious diseases in spatially heterogeneous environments. Bulletin of Mathematical Biology 63:547–571.

Rohani, P., R. M. May, and M. P. Hassell. 1996. Metapopulations and equilibrium stability: The effects of spatial structure. Journal of Theoretical Biology 181:97–109.

Rohani, P., and G. D. Ruxton. 1999. Dispersal-induced instabilities in host-parasitoid metapopulations. Theoretical Population Biology 55:23–36.

Rosenzweig, M. L., and R. H. MacArthur. 1963. Graphical representation and stability conditions of predator-prey interactions. American Naturalist 97:209–223.

Ruel, J. J., and M. P. Ayres. 1999. Jensen's inequality predicts effects of environmental variation. Trends in Ecology and Evolution 14:361–366.

Ruxton, G. D., and M. Doebeli. 1996. Spatial self-organization and persistence of transients in a metapopulation model. Proceedings of the Royal Society of London, series B. Biological Sciences 263:1153–1158.

Sale, P. F. 1977. Maintenance of high diversity in coral reef fish communities. American Naturalist 111:337–359.

Sattenspiel, L., and C. P. Simon. 1988. The spread and persistence of infectious-diseases in structured populations. Mathematical Biosciences 90:341–366.

Schwartz, M. W., and J. D. Hoeksema. 1998. Specialization and resource trade: Biological markets as a model of mutualisms. Ecology 79:1029–1038.

Shmida, A., and S. Ellner. 1984. Coexistence of plant species with similar niches. Vegetatio 58:29–55.

Sih, A. 1987. Prey refuges and predator-prey stability. Theoretical Population Biology 31:1–13.

Stacey, P. B., V. A. Johnson, and M. L. Taper. 1997. Migration within metapopulations: The impact upon local population dynamics. Pages 267–291 in I. A. Hanski and M. E. Gilpin, eds. Metapopulation biology: Ecology, genetics, and evolution. Academic Press, San Diego, CA.

Tilman, D. 1980. Resources: A graphical-mechanistic approach to competition and predation. American Naturalist 116:362–293.

———. 1982. Resource competition and community structure. Princeton University Press, Princeton, NJ, USA.

———. 1994. Competition and biodiversity in spatially structured habitats. Ecology 75:2–16.

Tilman, D., and P. M. Kareiva. 1997. Spatial ecology: The role of space in population dynamics and interspecific interactions. Princeton University Press, Princeton, NJ, USA.

Tilman, D., R. M. May, C. L. Lehman, and M. A. Nowak. 1994. Habitat destruction and the extinction debt. Nature 371:65–66.

Toft, C., and A. J. Karter. 1990. Parasite-host coevolution. Trends in Ecology and Evolution 5:326–329.

Travis, C. C., and W. M. Post. 1979. Dynamics and comparative statics of mutualistic communities. Journal of Theoretical Biology 78:553–571.

Vandermeer, J., B. Hazlett, and B. Rathcke. 1985. Indirect facilitation and mutualism. Pages 326–343 in D. H. Boucher, ed. The biology of mutualism. Oxford University Press, New York.

Vandermeer, J. H., and D. H. Boucher. 1978. Varieties of mutualistic interactions in population models. Journal of Theoretical Biology 74:549–558.

Volterra, V. 1926a. Fluctuations in the abundance of a species considered mathematically. Nature 188:558–560.

———. 1926b. Variazioni e fluttuazioni del numero d'individui in specie animali conviventi. Memoria Academia de Lincei 2:31–113.

———. 1931. Variation and fluctuations of the number of individuals in animal species living together. Pages 409–448 *in* R. N. Chapman, ed. *Animal ecology.* McGraw-Hill, New York.

Warner, R. R., and P. L. Chesson. 1985. Coexistence mediated by recruitment fluctuations: A field guide to the storage effect. American Naturalist 125:769–787.

Watkinson, A. R., and W. J. Sutherland. 1995. Sources, sinks and pseudo-sinks. Journal of Animal Ecology 64:126–130.

Weisser, W. W., V. A. A. Jansen, and M. P. Hassell. 1997. The effects of a pool of dispersers on host-parasitoid systems. Journal of Theoretical Biology 189:413–425.

Westerbergh, A., and J. Westerbergh. 2001. Interactions between seed predators / pollinators and their host plants: A first step towards mutualism? Oikos 95:324–334.

Wilson, W. G. 1996. Lotka's game in predator-prey theory: Linking populations to individuals. Theoretical Population Biology 50:368–393.

———. 1998. Resolving discrepancies between deterministic population models and individual-based simulations. American Naturalist 151:116–134.

Wilson, W. G., A. M. de Roos, and E. McCauley. 1993. Spatial instabilities within the diffusive Lotka-Volterra system: Individual-based simulation results. Theoretical Population Biology 43:91–127.

Wolin, C. L. 1985. The population dynamics of mutualistic systems. Pages 248–269 *in* D. H. Boucher, ed. *The biology of mutualism.* Oxford University Press, New York.

Wolin, C. L., and L. R. Lawlor. 1984. Models of facultative mutualism: Density effects. American Naturalist 124:843–862.

Wright, D. H. 1989. A simple, stable model of mutualism incorporating handling time. American Naturalist 134:664–667.

CHAPTER 3

Food Web Dynamics in a Metacommunity Context

Modules and Beyond

Robert D. Holt and Martha F. Hoopes

Introduction

Most natural metacommunities are also food webs, with producers and consumers at different trophic levels. What is the relationship between metacommunity processes and food webs?

The "ur-theory" of metacommunities is surely the theory of island biogeography developed by MacArthur and Wilson (1967), in which dispersal is asymmetrical between a continental source and island recipient communities. This famous monograph and the rich literature it spawned almost entirely focused on the "horizontal" structure of communities, such as the number of species within a taxon as a function of island size, and largely ignored food web interactions (Whittaker 1992; but see Lomolino 1984 and Spencer and Warren 1996). Likewise, a vast literature documented the importance of "vertical" forces in communities, ranging from reciprocal controls of predator diversity and prey diversity, to indirect impacts of predators on plant production (e.g., Holt and Lawton 1994; Pace et al. 1999; Estes et al. 2001; Persson et al. 2001; Chase et al. 2002). Food webs are a basic organizing theme in studies of many core ecological issues (Pimm 1982; Lawton 1989; Warren 1994; Cohen et al. 2003), yet until relatively recently ecologists paid scant attention to how spatial processes might influence food web structure and dynamics (Schoener 1989; Holt 1993; Polis, Holt, et al., 1996; Polis, Power, et al. 2004).

An important challenge in community ecology, and the theme of this chapter, is to weave together traditional food web ecology and metacommunity dynamics. A growing body of evidence points to the importance of space in food web ecology. Consider for instance food chain length (Post 2002; Holt and Post, MS). There are suggestive hints from surveys of connectance webs that food chains are longer in larger ecosystems (e.g., Rey and McCoy 1979; figure 5 in Schoener 1989; figure 7.2 in Holt 1993). Stable isotope analyses show that trophic rank of the top predator in lakes of the northeastern United States is strongly correlated with lake volume, with little residual effect of productivity (Post et al. 2000). A further line of evidence for metacommunity effects on food web structure comes from species-area relationships, which can reflect the impact of regional processes on species richness (Holt 1993; Rosenzweig 1995; Rosenzweig and Ziv 1999). For instance, on patches of clover and vetch in agricultural landscapes in central Europe, parasitoid species richness increases much more rapidly with area than does the richness of their hosts (Kruess and Tscharntke 2000), leading to lower rates of parasitism on smaller patches (Kruess and Tscharntke 1994). In European calcareous grasslands, habitat fragmentation affects species of higher trophic rank and trophic specialists particularly strongly (Steffan-Dewenter and Tscharntke 2002). Habitat fragmentation differentially impacts species at different trophic ranks (Didham et al. 1998; Holyoak 2000; Davies et al. 2001). Vertebrate carnivores seem particularly vulnerable to extinction in small fragments (e.g., Crooks and Soule 1999), which can have devastating consequences for the remainder of the community (Terborgh et al. 2001). Spencer et al. (1999) examined the trophic structure of arthropod assemblages in temporary ponds in Israel, and observed that the fraction of species that are predators increases strongly with pond area. Oceanic island biotas are often particularly poor in predator species (Rosenzweig 1995; Schoener et al. 1996) and so have short food chains (Schoener 1989).

These examples provide tantalizing evidence that ecosystem size and distance from source pools have profound effects on food web structure. For reasons discussed below, these effects could partly reflect metacommunity dynamics. (Further discussion of metacommunities and ecosystem properties can be found in Loreau et al., chapter 18.) Chapter 1 (see also Leibold et al. 2004) outlines four different perspectives on metacommunities: patch dynamics, species sorting, mass effects, and neutrality. All of these perspectives could pertain to food webs.

Strictly neutral models have not been developed for food webs, and in any case, in the development of neutral food web theory some constraints must be surely be placed on the system (e.g., one cannot have predators in a persistent food web without also including their prey). If species have roughly equivalent resource requirements, and are experiencing similar sets of predators, then sometimes they are lumped in food web analyses (e.g., functional groups). These may be candidate pieces of the full web where neutral models could apply.

Species sorting involves classic and familiar issues in community ecology (rules of dominance, invasibility, and exclusion due to the combined impact of abiotic factors and local interactions; e.g., Chase and Leibold 2003). Food web models often predict a plethora of alternative stable equilibria, with the one being realized depending on initial conditions. If there is dispersal, and occasional local disturbances that reinitiate local community assembly, then one of these states tends to dominate regionally (Shurin et al. 2004). Alternative food web states are most likely to be observed if alternative communities can sort out along environmental gradients (Shurin et al. 2004).

The mass effect has recently received a great deal of attention from food web ecologists under the rubric of "spatial subsidies" (Polis, Anderson, et al. 1997;

Polis, Power, et al. 2004). Mass effects have many potentially important impacts on local food webs, ranging from stabilization of otherwise unstable interactions (e.g., Huxel and McCann 1998), to generating reversals of local competitive dominance (Holt 2004).

Finally, patch dynamics occur if species in local communities often go extinct, but can persist overall because of colonization from a regional ensemble of local communities. If food webs experience frequent strong disturbances, or if local trophic interactions are quite unstable, such extinctions are likely. Below, we will consider in more details some models for simple metacommunities fitting the assumptions of patch dynamic theory.

Food web ecology is an enormous subdiscipline of ecology, and there are many legitimate approaches to studying food webs (Pimm 1982; Cohen, Beaver, et al. 1993; Cohen, Jonnson, et al. 2003; Polis and Winemiller 1996; Polis et al. 2004). One approach is to focus on entire, fully articulated webs, addressing issues such as connectance, patterns of interaction strength, the relationship between diversity and stability, and rigid circuit patterns. There are many challenges to developing adequate empirical characterizations of any but the simplest food webs (e.g., Cohen et al. 1993; Polis 1994). This is due in large measure to the large number of species in most webs, and the complex, reticulate, and variable network of interactions among these species. For these same reasons, theoretical models for the dynamics of entire webs are often built on highly simplified and unrealistic assumptions about interspecific relationships. A continuing challenge in both empirical and theoretical studies of food webs is to develop approaches to surmount this "curse of dimensionality" (Cohen et al. 1993).

In this chapter, we use a complementary approach to whole web analyses to address the interplay of food web and metacommunity dynamics. A conceptual way station between the relative simplicity of single-species population dynamics and the almost overwhelming richness of full food webs is the analysis of "community modules" (Holt 1997b; Persson 1999). The basic insight is that food webs contain recurrent structures that involve a small number of species (e.g., three to six) engaged in a defined pattern of interactions. At times, empirical systems may closely match the structure of a given module. Systems with strong interactors and well-defined functional groups often fit simple modules; this seems particularly true in the simplified communities of agroecosystems and other anthropogenic landscapes (e.g., Evans and England 1997; Muller and Brodeur 2002). Moreover, modules are basic building blocks of more complex communities. Analyses of modules can provide a handle for grappling with processes believed to be general drivers of community dynamics.

Van Nouhuys and Hanski (2002; chapter 4) provide a nice overview of real-world metacommunity dynamics for a number of modules, centered on the Glanville fritillary metapopulation in the Åland Islands off the south coast of Finland. After a brief discussion of some general issues, we consider several familiar community modules, embedded in a metacommunity context—pairwise trophic interactions, food chains, and shared predation. In the final section we sketch some thoughts on how to go beyond modules in relating food web ecology to metacommunity dynamics, and we present a novel, simple model extending island biogeography theory to multiple trophic levels. Further discussion of food web issues, particularly in the context of landscape ecology, can be found in chapter 20 by Holt et al.

Conceptual Overview

In an influential review, Kareiva highlighted how population dynamics may be fundamentally influenced by the fact that individuals disperse as well as interact (1990). The magnitude of the influence of dispersal on interactions depends on the spatial scale of environmental variation, relative to dispersal rates. If dispersal rates are very high, the metacommunity is just a well-mixed soup of interactions, creating one large spatially-distributed community. Conversely, if dispersal rates are vanishingly small, the only species expected to be present are those that can persist based on local environmental conditions and interspecific interactions; in this case, for all practical purposes communities are closed and could potentially be described by existing food web theory. Although Kareiva was concerned with single species and pairs of interacting species, his general point pertains to modules and indeed to entire food webs. The importance of metacommunity dynamics relative to local interactions in explaining food web structure should reflect the interplay of dispersal rates and the scale of patchiness and spatial heterogeneity (e.g., Hoopes et al., chapter 2). Moreover, the importance of dispersal must be gauged against the strength of local interactions (Holt 2004; see also below).

One of the key ways that dispersal and species interactions can come together is in the process of community assembly. Dispersal constraints define the species pool available for colonization into a local community (Belyea and Lancaster 1999), whereas local food web interactions can determine which colonists actually become established. For instance, Shurin (2001) experimentally demonstrated that predators attacking a zooplankton community facilitated invasion by competitor and prey species from a regional species pool, and that predator impact depended on community openness: predators reduced local species richness in closed communities, but enhanced richness in open communities. The likelihood of exclusion can itself have an implicit spatial dimension; for example, exclusion may be more likely in a small than in a large patch, because the latter may be more likely to have refuges from predation or competition.

Historical contingencies (e.g., priority effects) can arise because of the interplay of dispersal constraints and local web interactions. With strong negative interactions, low rates of extinction and low rates of dispersal, it is relatively easy to generate alternative community compositions in food web models (Luh and

Pimm 1993; Law and Morton 1996). Such alternative states get blurred at higher invasion rates (Lockwood et al. 1997), and are less likely to persist regionally if there are frequent local extinctions and global dispersal (Shurin et al. 2004). However, alternative states could be important contributors to metacommunity diversity at landscape or regional scales if disturbances are infrequent and dispersal is localized.

Community Modules and Metacommunities

Pairwise Trophic Interactions

Predator-prey interactions are the core building blocks out of which food webs are built, so their general features can influence the properties of the entire system. In chapter 2, Hoopes et al. consider in some detail the mechanisms by which spatial dynamics and spatial structure can lead to the regional persistence of predator-prey interactions, and so here we simply note key insights that pertain more broadly to multispecies food web interactions.

In pairwise predator-prey interactions, a necessary condition for a stable equilibrium is that at least one species experience direct density dependence. Similarly, in multispecies systems, direct, negative density dependence (measured by the trace—the sum of nonzero elements along the diagonal of the community matrix) is a necessary condition for a stable equilibrium (May 1973). Movement among habitats can create an "induced" form of local density dependence (Holt 1993); if a population of size N receives I immigrants into a population, the per capita effect on growth is I/N, a term which declines with increasing N. This negative density dependence can stabilize otherwise unstable local interactions. This effect helps explain the stabilizing impact of source-sink dynamics and spatial refuges in both pairwise and multispecies predator-prey interactions (Holt 1984, 1985, 1993; Nisbet et al. 1993; Huxel and McCann 1998; Briggs and Hoopes 2004).

Many food webs contain specialist predators and parasitoids. Specialist enemies impact their prey more when those prey are more common (an idea that stems back at least to Janzen 1970; for formal treatments see Armstrong 1989 and Grover 1997). This leads to density-dependent mortality, which frees space and resources for other species, thus promoting local diversity. If metapopulation dynamics promote persistence of a specialist predator-prey interaction (Hanski 1999; Bonsall et al. 2002), this indirectly facilitates the persistence of other species sharing that prey's resources. More broadly, if keystone species dominate local community structure, their dynamics also loom large in the metacommunity. For instance, a keystone predator may experience metapopulation dynamics because of recurrent extinctions unrelated to its impact on its food base (Britton et al. 2001; Shurin and Allen 2001). This sets up a parallel dynamic in the prey community, since local predator extinctions unleash competitive interactions among prey species and these lead to further local extinctions. Variation in the abundance and distribution of a keystone species due to dispersal should thus have reverberating effects on the rest of the community.

Competitive Modules

Other chapters in this book deal with competitive interactions (e.g., Mouquet et al., chapter 10), so here we only touch on this important topic. In any food web, if consumers overlap in their diet, exploitative competition may occur. Spatial dynamics may help explain the coexistence of consumers competing for shared resources. For instance, Ruxton and Rohani (1996), building on an earlier coupled lattice model of Hassell et al. (1994), showed that coexistence between two parasitoids competing for a single host species can occur in a metacommunity, given a trade-off between local attack rates and ability to move among local populations. Such coexistence was robust to varying assumptions about spatial and temporal heterogeneity. Shurin and Allen (2001) explore a metacommunity model in which a predator permits competing prey species to coexist locally, when the competitors cannot coexist in the absence of the predator (for related models see Caswell 1978 and Britton et al. 2001). The model splices together metapopulation models for competing prey (e.g., Levins and Culver 1971) and metapopulation models for predator-prey interactions (e.g., Holt 1997a; May 1994). Shurin and Allen (2001) found that predators generally promoted regional coexistence, but could have positive or negative effects on mean local diversity. They suggest that with multiple generalist predators, each with different impacts on their prey, one could observe a positive correlation between the local diversity of predators and prey. Below we will develop a quite different model that leads to a similar conclusion.

Spatial Determinants of Food Chain Length: Metacommunity Perspectives

The food chain module describes a tritrophic interaction of a basal resource (e.g., a plant), sustaining a consumer (e.g., a herbivore), which in turn supports another consumer (e.g., a predator). Interpreted literally, an unbranched food chain arises from interlocked trophic specializations, leading to stacked specialists (Holt 1993). Theoretical studies of food chains are central to the hypothesis of exploitation ecosystems (e.g., Oksanen, Fretwell, et al. 1981; Oksanen, Oksanen, et al. 1992; Oksanen, Schneider, et al. 1999), a hypothesis that emphasizes the interplay of top-down and bottom-up forces in community organization (Leibold 1996; Sinclair et al. 2000). Here we address several questions about this module. What factors determine food chain length, both in the absence and presence of dispersal? How do tritrophic interactions respond to spatial flows among different habitats?

Traditional explanations of the factors limiting food chain length emphasize ecological energetics and the stability of local interactions (Pimm 1982; Post

2002). Schoener (1989) extended the energetic hypothesis to a "productive space" hypothesis, which is that food chain length is governed by the total energy available to a given trophic level (productivity per unit area or volume, times area or volume). However, productivity alone does not at present seem to be a good predictor of food chain length, whereas habitat area or volume can influence chain length (Post 2002; though see Rosenzweig 1995 and Vander Zanden et al. 1999). This area effect could arise from metacommunity dynamics (Holt and Post, MS).

Given tight trophic specialization, spatial effects influencing the persistence of basal resource species are automatically transmitted to higher-ranked species (Holt 1993; Holt et al. 1999; Van Nouhuys and Hanski 2002). Colonization-extinction dynamics in a metacommunity can constrain food chain length. We illustrate this with a simple "donor-controlled" model. By donor control, we mean that a resource population has extinction and colonization dynamics that are independent of top-down effects of consumer populations. However, we assume that consumers can only colonize a patch if their required resource is already present, and if the resource goes extinct, so too does the consumer, so there are strong bottom-up effects.

For a species of trophic rank j in this donor-controlled food chain, a standard metapopulation model (Holt 1996, 1997a, 1997b) is

$$\frac{dp_j}{dt} = c_j p_j (h_j - p_j) - e_j p_j, \tag{3.1}$$

where p_j is the fraction of patches occupied by species j, h_j is the fraction of the landscape suitable for species i, c_j is the per patch colonization rate, and e_j is the rate of extinction. The basal species in the chain persists only if $h_1 > e_1/c_1$, and if it persists its equilibrial occupancy is $p_1^* = h_1 - e_1/c_1$.

What about the species of rank 2? Because it requires the prior presence of species 1, suitable habitat for species 2 is the current fraction of the landscape containing species 1. If species 1 is at equilibrium, we can set $h_2 = p_1^*$ in equation (3.1); the equilibrial occupancy of species 2 is $p_2^* = h_2 - e_2/c_2 = h_1 - e_1/c_1 - e_2/c_2$, hence species 2 persists only if $h_1 > e_1/c_1 + e_2/c_2$. (Note that p_i^* is also the equilibrial fraction of patches that have a food chain of length i.) Similarly, habitat patches suitable for the top predator contain both the intermediate consumer and the basal species, so the top predator persists only if $h_1 > e_1/c_1 + e_2/c_2 + e_3/c_3$. By induction, for a donor-controlled food chain of length n the criterion for persistence of the top-ranked species is

$$h_1 > \sum_{j=1}^{n} e_j/c_j. \tag{3.2}$$

As one ascends the food chain, by inspection of expression 3.2 it is clear that there are increasingly stringent criteria for persistence of the top-ranked species (whose presence determines chain length). Sparse habitats, which have small values for

h_1, are particularly unlikely to sustain food chains comprised of specialists (Holt 1997a, 1997b, 2002; Melian and Bascompte 2002). Moreover, the basal species is unlikely to sustain a long food chain if it has a low maximal occupancy (e.g., because of its own high extinction or low colonization rates).

If a species at any given trophic rank goes extinct on a patch, so do all higher-ranked species that depend on it, so extinction rates must stay the same or increase with trophic rank. A striking example comes from fragmented boreal forest, where specialist food chains of a bracket fungus, a tineid moth herbivore, and a specialist tachinid fly parasitoid become increasingly truncated with increasing time since fragmentation (Komonen et al. 2000). The presence and abundance of fungal fruiting bodies is highly variable through time (Hanski 1989), which makes it harder for this resource to sustain a chain of specialist consumers. Another example is provided in chapter 4 by Van Nouhuys and Hanski, who argue that metacommunity effects are the dominant factor explaining the restricted distribution of a specialist parasitoid with limited dispersal abilities in the Åland Islands; this system matches an assumption of the model, which is that the specialist parasitoid experiences donor control (Van Nouhuys and Tay 2001).

Alternative Stable States in Food Chain Length at the Landscape Scale

More generally, predators will influence prey colonization and/or extinction rates. For instance, if predation reduces average prey population size, or generates strongly unstable dynamics with fluctuations to low levels, predators can elevate prey extinction rates. Relaxing the assumption of donor control (allowing top-down effects to occur) leads to models with a more complex algebraic structure, but does not change the fundamental conclusion that food chain length can be constrained due to coupled metapopulation dynamics (Holt 1997a, 1997b). This is particularly true if extinction rates always increase with a lengthening of the food chain in a patch. However, interesting novel effects can arise in tritrophic metacommunities with top-down effects that *enhance* local stability, such as alternative, stable landscape states with different food chain lengths. Holt (1997a) generalizes the model (equation 3.1) to include such effects, and Holt (2002) presents examples of alternative states. Rather than describe this model in all its algebraically complex glory, we here attempt to give the reader a flavor for why alternative states can arise if top-down effects are sufficiently strong.

A tritrophic predator-prey model described by May (1973) reveals that local dynamics can be stabilized by a top predator, which can lead to alternative stable states for food chain length on a landscape. Unstable dynamics in a two species system can lead to low densities of the basal prey species, leading to its possible extinction, followed by extinction of the intermediate predator. Such extinctions in a metacommunity context can imply low occupancies for the intermediate predator—too low for the top predator to increase when rare. However, if the top predator is sufficiently common, it may reduce extinction rates in the patches it occupies, and colonization from these occupied patches can permit the stable persistence of the entire food chain in the metacommunity.

An alternative scenario for tritrophic interactions in a metacommunity was explored by Jansen (1995). In contrast to the above model, Jansen assumed that for consumers (either the herbivore or top predator), dispersal occurred solely due to local extinction of their required resource. Such dispersal can be strongly destabilizing. The reason is that dispersal permits a delay in the response by the consumer population to declining resource levels, allowing consumers to push resources even lower across the landscape and thus increasing the time required for resource recovery. The model also permits alternative states, with both stable three-species equilibria and limit cycles emerging in a given environment, but from different initial conditions.

Shared Predation and Apparent Competition

Top predators can stabilize the dynamics of other species, and so facilitate prey persistence. Conversely, top predators can attack prey at sufficient rates to exclude some prey species from communities. In particular, generalist predators can lead to the extinction of prey species due to the maintenance of the predator by alternative prey. This indirect interaction, called apparent competition, has been studied extensively theoretically (e.g., Holt 1977, 1984) and has received a considerable amount of empirical attention as well (Chaneton and Bonsall 2000). Indirect exclusion of prey due to shared predation can occur in a metacommunity, even among prey species that are never found together. An experimental demonstration of this effect is provided by Bonsall and Hassell (1997, 1998). In their system, each of two moth hosts (*Plodia interpunctella* and *Ephestia kuehniella*) for a parasitoid (*Venturia canescens*) was maintained in a separate laboratory arena, so in effect each species occupied a distinct habitat, with no interspecific competition. Each host species could persist for long time periods when coexisting with the parasitoid alone. However, when both host species were present, and the parasitoid (but not either host) was permitted to move freely between the habitats, one host species was rapidly excluded due to the spillover of parasitoids moving between habitats. This exclusion arose because parasitoids were produced in sufficient numbers by the host with higher intrinsic rate of increase (*P. interpunctella*) to drive the exclusion of the other host.

A simple two-patch metacommunity model (Holt 1997a; for similar models see Swihart et al. 2001; Melian and Bascompte 2002) illustrates that for a predator that feeds on two prey species, predator mobility is a critical determinant of prey coexistence. For simplicity, assume the two prey species use distinct resources in different patches, and so do not directly compete (as in the experiment just described). The potential for indirect competitive exclusion in metacommunities is illustrated by the following model. We show the equations just for prey species 1 and predators occupying patches with that prey (a similar pair of equations describes prey 2 with subscripts 1 and 2 reversed and the predator occupying patches with prey 2):

$$\frac{dp_1}{dt} = c_1 p_1 (h_1 - p_1 - q_1) - e_1 p_1 - p_1 (c_{11} q_1 + c_{12} q_2),$$ (3.3)

$$\frac{dq_1}{dt} = p_1 (c_{11} q_1 + c_{12} q_2) - e_{1q} q_1.$$ (3.4)

In equation 3.3, h_i is the fraction of the landscape with habitat suitable for prey species i. The fraction of the landscape occupied by prey species i alone is p_i. The fraction of the landscape occupied simultaneously by prey i and the predator is q_i. We assume that predators can only colonize patches in which one or the other prey species already resides. The parameter c_i scales colonization by prey i of patches of type i; e_i is the extinction rate of prey i, in the absence of the predator; c_{ij} is the rate of colonization by predators into patch type i, drawn from patch type j. Finally, e_{iq} is the rate at which predators drive prey (and thus themselves) extinct within patches. The model assumes that predators have a very strong effect on local prey abundance, making those prey in patches with predators essentially irrelevant to prey colonization into empty patches; successful prey colonization depends on dispersers emitted by predator-free patches.

A key feature of this model is that alternative prey species occupy mutually exclusive habitats, and so do not directly interact. The predator, however, can colonize across as well as within the two habitats, and so provides a conduit of indirect negative interaction between prey species. This can lead to apparent competitive exclusion in the metacommunity. If prey i is present alone, it persists if $h_i > e_i / c_i$. The predator can persist on prey i alone if $c_i (h_i - e_{iq}/c_{ii}) - e_i > 0$. We assume this is true. Coexistence requires that each prey species be able to increase when rare, given that the other prey species and predator are at equilibrium, implying the following joint condition for coexistence:

$$\frac{c_1 (h_1 - e_{1q}/c_{11}) - e_1}{c_1 + c_{11}} < \frac{c_2 h_2 - e_2}{c_{21}},$$ (3.5)

and

$$\frac{c_2 (h_2 - e_{2q}/c_{22}) - e_2}{c_2 + c_{22}} < \frac{c_1 h_1 - e_1}{c_{12}}.$$ (3.6)

Expressions 3.5 and 3.6 imply that if the predator has little cross-habitat colonization, prey coexistence is assured; if for each prey species, cross-habitat colonization by the predator is less than within-habitat colonization, there is a range of parameters permitting coexistence; and, there is a range of habitat availabilities that implies the indirect exclusion of the prey species requiring that habitat, which would suffice for that prey to persist together with the predator, were they alone. If the inequalities in equations 3.5 and 3.6 are reversed, one expects prey species exclusion. The model suggests that prey species may be vulnerable to exclusion from a metacommunity for many reasons: vulnerable species may be specialized to rare habitat types, have lower intrinsic rates of colonization, have higher intrinsic rates of extinction (independent of predation), or be more vulnerable to extinction when confronted by the predator.

This model shows how apparent competitive exclusion in a metacommunity can arise because of predator dispersal. Were such exclusion to occur, one is likely to miss the mechanism in observational field studies, since at equilibrium the predator will be absent from any patch without prey!

Community Modules in Spatially Explicit Landscapes

These metacommunity models for modules of interacting species assume global dispersal; patch arrangement is ignored. In spatially explicit metacommunity models with localized dispersal, spatial patterns may arise that are important in determining persistence (see also Hoopes et al., chapter 2). Spatiotemporal dynamics can produce dynamics that are consistently out of phase in different parts of the landscape; dispersal between populations at peaks and those at low abundances can help rescue local populations from extinction.

Consider the two modules we have discussed above: food chains and apparent competition due to shared predation. Wilson et al. (1998) examined a stochastic tritrophic model in a cellular lattice with nearest-neighbor dispersal and strongly unstable local interactions, and showed that lattice size (a measure of metacommunity "size") had a strong effect on the persistence of the food chain. Small lattices did not permit the simultaneous existence of local populations in sufficiently different phases to generate the stability rescue effect of dispersal, which could have prevented local (and regional) extinction. To persist, a three-species system required lattices an order of magnitude larger in area than did a two-species host-parasitoid system. These area effects on food chain length were particularly pronounced with large differences among species in dispersal rates. In effect, expected food chain length should increase with lattice size, because larger lattices permit regional mechanisms of persistence to operate more effectively. It is likely that this effect contributes to the observed influence of ecosystem size on food chain length in some natural systems (Post 2002; Holt and Post, MS).

The model for apparent competition explored above (equations 3.3 and 3.4) assumes global dispersal for all species. With spatially explicit interactions and local dispersal, in a metacommunity one can observe coexistence under shared predation that would otherwise not occur. This is illustrated by a model studied by Bonsall and Hassell (2000), who examined apparent competition between two hosts species sharing a parasitoid in a lattice. Within cells, parasitism is described by a Nicholson-Bailey model. Dispersal is among nearest-neighbor cells. Within a single closed patch the dynamics are unstable, and host coexistence does not occur. In a homogeneous, well-mixed system, the theoretical expectation is that the host species with the higher value of the intrinsic growth rate, scaled against the attack rate, should tend to displace the alternative host species (Holt and Lawton 1993).

In the model of Bonsall and Hassell (2000) there are no within-patch mechanisms permitting coexistence, but dispersal is limited. The parasitoid inflicts parasitism evenhandedly on the two hosts, and one host has a higher intrinsic growth rate. The model predicts apparent competitive exclusion in a wide range of circumstances (as expected from the results of Holt and Lawton 1993). But it also shows that coexistence can occur in a metacommunity, and for two distinct reasons. The inferior host species could persist if it is a fugitive species, with a higher dispersal rate than the superior species (an analogue for apparent competitive interactions of the familiar colonization-competition trade-off; see Hoopes et al., chapter 2). More surprisingly, the inferior species may also persist if it has a much *slower* rate of dispersal! The interesting finding that sluggish inferior prey can persist reflects phenomena that arise only in a spatially structured metacommunity with limited dispersal. If the superior host and parasitoid are both dispersing, but dispersal is localized, parasitoid numbers tend to be highest in patches temporarily containing the superior host. In effect, the sedentary behavior of the inferior host means it will be left behind by waves of parasitoids tracking the superior host over space, so the inferior host enjoys transient refuges (often found in the troughs of the spiral waves these models can generate on the lattice). In chapter 2, Hoopes et al. describe parallel spatial mechanisms of escape in systems of directly competing species.

Beyond Modules

The module approach, although useful (and indeed we would argue essential) as a tool for analyzing the structure and dynamics of complex communities, is not sufficient for understanding all aspects of food web structure. As the number of species being considered explicitly grows, the number of possible module configurations grows much faster. As an example, Sinclair et al. (2000) in reviewing tritrophic dynamics with just three components note that there are twenty-seven possible configurations of interactions (including direct density dependence). One way to circumvent the issue of dimensionality is to lump species into broad functional groups. However, ignoring heterogeneity within nodes of lumped food webs must be done cautiously. Seemingly slight differences in the web of interactions can at times profoundly influence dynamics. For instance, Persson et al. (2001) experimentally enriched aquatic food webs in tanks, and found that the detailed structure of the system (e.g., the presence of inedible as well as edible producers) was essential for interpreting impacts of enrichment. Similarly,

Abrams (1993) in studies of food web models observed that disparate responses of biomass to increased productivity arose between models with slight differences in the configuration of food web interactions (e.g., presence or absence of omnivory).

Despite these cautionary remarks, relatively simple effects may emerge when one considers shifts in diversity in food webs with well-defined trophic levels. Caswell and Cohen (1993) superimposed disturbance regimes on patch dynamic models of competing species, and found that species richness tended to be maximal at intermediate levels of disturbance. Wootton (1998) considered how disturbance influenced species diversity in a community with multiple species at several trophic levels. His model consisted of MacArthur's resource-consumer equations, with superimposed density-independent mortality and immigration from an external source. Depending on the details (e.g., which species immigrates) immigration could either enhance or eliminate the effect of disturbance on coexistence. The latter effect was particularly likely when top consumers were mobile. Wootton concludes that "the surprisingly different effects of immigration... suggest that its effects on more complex situations also merit further exploration" (1998). In the following section, we examine communities with well-defined trophic levels by using an approach that deliberately ignores the detailed pattern of trophic interactions among species to examine how ecosystem size influences species richness at different trophic levels.

Trophic Island Biogeography: A Step toward Generality

The stacked specialist models for food chains discussed above provide a first step toward a generalization of island biogeography and metapopulation theory to food webs. Yet these models are limited, because they assume tight trophic specialization, which is not necessarily the norm for predators. Developing comparable models for trophic generalists that keep track in detail of each possible community configuration and transitions amongst them leads to models of daunting complexity. An alternative approach we explore here is to radically simplify the problem by assuming a minimal set of assumptions about the likely relationships between trophic diversity on adjacent levels. Our aim is to develop a qualitative theory predicting how species richness at various trophic ranks scales with area (e.g., of islands, or habitat patches).

Assume that multiple species can co-occur at each trophic level (either regionally, locally, or both), but that broad, qualitative constraints define coexistence. General ecological theory (e.g., Whittaker 1975) predicts that a more diverse resource base should support a more diverse consumer base, given that many consumers are relatively specialized in their diets; there is suggestive support for this hypothesis from the plant and arthropod communities of Cedar Creek, Minnesota (Siemann 1998). We develop a "minimalist" island biogeographic model for two trophic levels, where we deliberately ignore many details of trophic interactions. Let P denote the number of predator species present on an island of size A, and S denote the number of prey species. We assume that the number of species at each trophic level is determined by colonization from a source pool, and extinctions. Moreover, we assume that trophic interactions are donor-controlled, so that colonization-extinction dynamics of the prey level are not driven by changes in the predator community. However, the converse will not be true; an increase in the number of prey species present should affect colonization and extinction rates in the predator trophic level.

Following MacArthur and Wilson (1967), prey species dynamics are described by

$$dS/dt = C - E = (c - sS) - eS. \qquad (3.7)$$

Here, C is the total rate of colonization of new prey species into the community (colonization entails establishment of viable populations), and E is the total rate of extinction of resident, established prey species. To make the model algebraically transparent (as did MacArthur and Wilson), we make these rates depend in a simple linear manner on species richness. The parameter c is the rate at which new prey species successfully colonize empty islands, s describes the reduction in rate of colonization with increasing island richness, and e is the rate of extinction, per resident species. At equilibrium, we have $S^* = c/(s + e)$.

Let a power law, $S = qA^z$, describe among-island variation in prey species richness, where A is island area, z describes the strength of the species-area relationship, and q is a taxon-specific parameter. After taking natural logs and differentiating S^* with respect to natural log of area we can form the identity

$$z = \partial \log S^* / \partial \log A$$
$$= (1/c)\partial c/\partial \log A - (1/(s + e))(\partial s/\partial \log A + \partial e/\partial \log A). \qquad (3.8)$$

In principle, any of the parameters c, s, and e describing community dynamics could vary with island area. For instance, a larger area provides a larger target (larger c), holds more species when saturated (smaller s), and has a lower extinction rate of resident species (lower e). The two terms in the right-hand parenthesis are thus negative, so $z > 0$.

In like manner, the dynamics of the predator community can be described by colonization and extinction:

$$dP/dt = C' - E' = (c' - s'P) - e'P. \qquad (3.9)$$

Here the symbols match those for the prey. Equilibrial richness of predators is $P^* = c'/(s' + e')$.

Again, we would like to know how predator species richness scales with island area. We assume that predator colonization and extinction rates are determined not directly by area, but rather by the number of prey species present. There may still be emergent area effects on predator richness arising indirectly via area effects on prey richness.

One expects predator colonization to increase with prey species richness S (i.e., $\partial c'/\partial \log S > 0$). If a predator is a specialist, to successfully colonize its required prey species must be present. It is reasonable to hypothesize that in general, a particular prey species is more likely to be present if the total number of prey species is larger. For generalist predators, colonization success may also increase with increasing prey species for several distinct reasons. First, if total food supply scales with prey species richness, colonization should be more likely if there are more prey species resident. Second, if different prey provide different limiting nutrients (the obligate-generalist case of Holt et al. 1999), it is more likely the predator can colonize into a richer prey community.

With more prey species, there is also a greater chance that predators can have sufficiently distinct diets that competition is moderated. Even if there is no competition among predators, a greater diversity of prey permits bet-hedging in the face of temporal variability. So, the number of predators that can be sustained in a saturated community should increase with prey richness ($\partial s'/\partial \log S < 0$), and the extinction rate of predators already present will be lower with more prey species present ($\partial e'/\partial \log S < 0$). Ritchie (1999) presents evidence for one system (prairie dog colonies sustained by herbaceous plant communities) where local extinction rates decline with increasing prey species richness.

Using these inequalities, and with an application of the chain rule to the expression for equilibrial predator richness, we have

$$z' = \partial \log P^*/\partial \log A = (\partial \log S^*/\partial \log A)$$
$$\times \; [(1/c')\,\partial c'/\partial \log S^* - (1/(s' + e'))(\partial s'/\partial \log S^* + \partial e'/\partial \log S^*)] \quad (3.10)$$

or compactly,

$$z' = zQ, \quad\quad\quad (3.11)$$

where Q is the right-bracketed expression in (3.10). The quantity Q describes the strength of the species-area relationship in the predator community, relative to that in the prey community on which they depend. With our assumptions, an increase in prey species richness should increase predator colonization rates (higher c' and/or lower s'), and reduce predator extinctions (lower e'). Hence, Q is positive, so predator richness should always scale positively with island area. However, for predator species richness to scale *more strongly* with area than does the prey (predators have a higher z-value), we must also have Q greater than 1.

It is likely that the magnitude of Q will depend on whether or not the predators in question are specialists, or generalists. Several distinct processes could make Q lower for generalist predators than for specialists, making it more likely z would not always increase with increasing trophic rank. Consider first colonization dynamics.

Generalist consumers may be able to readily colonize, given only a small subset of the resident prey community. Moreover, initial colonization should not strongly depend on the richness of the resident prey community (lowering $\partial c'/\partial \log S$). A specialist consumer by contrast requires that a particular prey species be present, before it can colonize. By chance, many species-poor communities (e.g., on small areas) will lack its required prey, whereas species-rich communities will harbor that prey. This automatically increases the dependency of c' on prey species richness for specialists, compared to generalists. So considering just the first term above suggests that it is reasonable that Q should be lower for generalists.

If all consumers are specialists, their extinction rates can be no lower than the extinction rates of their required prey (and may be higher). If a generalist can subsist on various subsets of the prey it can utilize, there should be a reduction in the dependency of extinction rates on prey species richness for generalist predators, compared to specialists (hence, a decrease in the magnitude of $\partial e'/\partial \log S$). This should also reduce Q for generalists.

Finally, if there were no local extinctions, the island predator community would equilibrate at $K(S) = c'/s'$, which we might consider to be the "saturation" richness of the community. An increase in the number of prey species may not greatly increase the number of generalist species, compared to specialists, because of the opportunity for overlap in diet, competitive interactions, and intraguild predation. In the above model, this could be described by decreasing the magnitude of $\partial s'/\partial \log S$, again lowering Q, and hence z'.

These observations suggest that generalist predators should have lower values of z than do specialists, and possibly even lower values than that of their prey (figure 3.1). Holt et al. (1999) reviewed empirical relationships between trophic rank and the species-area relationship, and observed instances of both $Q > 1$, and $Q < 1$. In systems dominated by trophic specialists (e.g., the parasitoids on habitat patches studied by Kruess and Tscharntke [2000]), one observes stronger species-area relationships at higher trophic ranks, so $z' > z$. However, in other systems (e.g., invertebrate consumers on islands in the Gulf of California; Holt et al. 1999; G. A. Polis, pers. comm.), predators (e.g., scorpions) are highly generalized and have lower z-values than do some lower-ranked trophic levels (e.g., plants). This suggests that in these systems $z' < z$; area has a stronger effect on species richness at low trophic levels. Steffen-Dewenter and Tscharntke (2000) showed that the predicted effect of trophic generalization on the magnitude of z is evident in butterflies differing in dietary breadth on habitat fragments; the regression coefficient of log(species) versus log(area) (the z-value) increases monotonically from butterflies, which are extreme generalists, to those which are oligophagous, to those which are tight specialists on a single host plant. Given that butterflies often show metapopulation dynamics (Hanski 1999), it would be interesting to know the relative contribution of colonization and extinction (as mediated through host species richness) to this pattern.

Several cautionary remarks are in order.

First, we assumed that predator dynamics depend solely on prey species richness. More generally, one might expect that predator extinctions and coloniza-

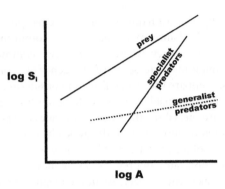

Figure 3.1 Species-area relationships as a function of trophic rank and trophic specialization. S_i is number of species of trophic rank i, and A is area.

tions could be directly influenced by island area, among islands with comparable prey species composition. For instance, large islands are larger targets for colonization, and can sustain larger population sizes. Incorporating direct area effects on predator colonization and extinction in the above formulation tends to increase the z-value of the predator assemblage (Holt et al. 1999, and unpublished results). If predators are much rarer than their prey (as is typically true of top endothermic predators), then these direct effects of area on predator-species richness may greatly outweigh the indirect effects of island area mediated through the prey trophic level.

Second, the above approach gives qualitative insight, but does not quantify the strength of the relationships. Some explicit food web models suggest rather weak dependencies with area (e.g., the cascade model; Cohen and Newman 1991). One practical complication in relating the model to field data is that food webs may not cleave neatly into distinct trophic levels (Polis 1994).

Third, we have ignored the potential for top-down impacts of predators on prey species richness (including apparent competition). As we saw in the sections on alternative stable states and apparent competition, top-down effects alter conditions for coexistence and could change extinction rates. For instance, Bengtsson and Ebert (1998) argue that parasites increase the extinction rates of *Daphnia* in rock pool metapopulations. Holt et al. (1999) suggest that top-down extinctions could weaken or even reverse the predicted relationship between trophic rank and z. A proper assessment of this suggestion will require the examination of more detailed models that make explicit assumptions about the web of interactions among predators and prey.

Finally, even if the suggested relationship exists, it may be obscured by other factors. For instance, although butterflies are typically somewhat restricted in larval host range (often monophagous or oligophagous in California, local butterfly species richness is not strongly correlated with host plant species richness, once one factors out the influence of environmental covariates such as temperature (Hawkins and Porter 2003). One obvious difference between consumers on islands in the Gulf of California (Holt et al. 1999) and parasitoids on the habitat patches studied by Kruess and Tscharntke (2000) is that the island populations of consumers on these islands are not solely dependent on island populations of prey, but can also be supplemented by allochthonous resources from outside the system (Polis and Hurd 1996). In many systems movements of organisms and materials across habitat boundaries have profound consequences for within-habitat trophic dynamics and species composition (Polis et al. 1997; McCann et al. 1998; Power and Rainey 2000; Polis, Power, et al. 2004). Allochthonous inputs can lead to systematic deviations in species-area relationships away from that expected from island biogeographic theory, particularly on small islands or habitat patches, where such inputs may greatly exceed local productivity (Anderson and Wait 2001). In chapter 20 of this volume, Holt et al. provide further discussion of spatial fluxes and landscape scale influences on food webs and metacommunity dynamics.

Linking Food Web Theory to Empirical Studies of Metacommunities

Empirical studies that examine entire, fully-articulated food webs in a metacommunity context have not yet been conducted. Many studies (e.g., the scale transition analyses of Melbourne et al., chapter 13) focus largely on dynamics within single trophic levels. However, several of the empirical contributions in this volume do consider communities with species at different trophic levels. Overall, a comparison of these studies suggests that different patterns will be observed in different ecosystems, with the relative strengths of the four major modes of metacommunity dynamics (patch dynamics, mass effects, species sorting, and neutrality) varying greatly among systems.

Cottenie and De Meester (chapter 8) in their analysis of zooplankton communities among ponds showed that species sorting along environmental gradients had large impacts, relative to mass effects. One of the gradients had to do with the presence / absence of a top predator (fish), comparing high fish predation with no fish predation, and the other was a habitat variable (macrophyte presence), which could indirectly influence the strength of predation. The rock pools assemblages examined by Kolasa et al. (chapter 9) also broadly fit a species sorting paradigm. An important implicit message in these results is that they suggest interspecific interactions are strong. If a species with density N and continuous population growth is rare and being excluded at rate f, but is at the same time being input at rate I from the regional species pool, the equilibrial standing crop is $N^* = I/|f|$, where $|f|$ is the absolute magnitude of the rate of exclusion (Holt 1993, 2004). Species that are being weakly excluded can thus be present in substantial abundance. If there is temporal variation in the rate of exclusion (e.g., due to fluctuations in the abundance of locally superior competing species), this, if anything, tends to increase the average abundance of the excluded species, particularly if exclusion is weak (Holt et al. 2003). The fact that Cottenie and De Meester and Kolasa et al. observed strong species sorting and weak mass effects, despite considerable opportunity for dispersal, suggests that interspecific interactions leading to exclusion is quite strong in these zooplankton and rock pool communities. It would be interesting to tie these experiments more explicitly to theory, so as to assess this prediction directly.

The metacommunities associated with butterflies described by Van Nouhuys and Hanski (chapter 4) closely match the modules approach discussed above. This correspondence arises because butterflies are often specific consumers on one to a few plant species, and many of their natural enemies, in particular parasitoids, are likewise host-specific. The Glanville fritillary (*Melitaea cinxia*) in south Finland utilizes just two plant species as hosts; it is attacked by two specialist parasitoids, which in turn are attacked by two hyperparasitoids. The relative simplicity of this food web permits close analysis of mechanisms at work influencing food web structure and dynamics. The authors conclude that both local, within-community, and large-scale, among-community processes contribute to observed patterns. For instance, the interaction of the two specialist parasitoids when they co-occur can be explained by local processes, largely independent of metacommunity dynamics. By contrast, the parasitoid with more limited mobility can only persist in patch networks with the highest metapopulation capacity, consistent with the metacommunity models of food chain length discussed earlier.

The inquiline communities in pitcher plant leaves discussed by Miller and Kneitel (chapter 5) also reveal the interplay of local interactions, such as predation, and metacommunity processes such as dispersal. In particular, the relationship of species richness to dispersal rate at intermediate trophic levels was unimodal in the absence of top predators, but flat in their presence (see also Kneitel and Miller 2003). In this system, however, the top predators (e.g., mosquito larvae) did not maintain separate populations in each pitcher, but instead were pieces of a population operating at a coarser spatial scale. The moss microcosms examined by Gonzalez (chapter 6) found that habitat fragmentation led to extinctions, particularly of those species that had low abundances in the original, prefragmentation communities. In one experiment, there was also a significant effect of fragmentation on the proportion of predators in the final community. The trophic island biogeographic theory we have presented suggests that the impact of fragmentation upon the proportion of predator species present should be sensitive to the degree of trophic specialization or generalization present in the predator guild. The detailed trophic information required to assess this hypothesis is not yet available for this microarthropod community.

Finally, Resetarits et al. (chapter 16) review empirical studies of habitat selection and tellingly observe that local trophic interactions can also strongly influence dispersal rates among communities, particularly when individuals can choose local habitats (e.g., to avoid predation). This is more likely for some components of food webs (e.g., large vertebrates) than for others (e.g., seed plants). Dispersal is a topic of great importance in behavioral ecology as well as metacommunity ecology, and explicitly drawing out these linkages is a theme that warrants much more empirical work.

Conclusions

There is enormous opportunity for further empirical and theoretical work on all the themes we have touched on in this chapter. We have shown that surprising effects may emerge, even in familiar modules, when considered in a metacommunity context. The food chain model with sequential colonization and interlinked extinctions revealed that landscapes may exist in alternative states, reflecting how local interactions influence extinction risk. Shared predation may lead to exclusion of prey species that are never found together in the same local community because of habitat specialization, due to predator spillover. Conversely, metacommunity dynamics may permit alternative prey to coexist, when such coexistence is not expected in a single local community closed to dispersal. All these predictions are amenable to experimental test (e.g., in microcosm studies). Moreover, the specific models we discussed considered transitions between qualitative states (e.g., food chain length, presence/absence of a prey species in a habitat patch). Such patch dynamic approaches need to be complemented with analyses that pay close attention to numerical dynamics in each habitat, and how such dynamics are modulated by flows of individuals among habitats (mass effects). As noted above, such flows or mass effects can either stabilize or destabilize local interactions (Huxel and McCann 1998; Holt 2002), depending on the detailed functional nature of the interactions (see also Holt et al., chapter 20). If species sorting turns out to be the norm in describing food web dynamics in heterogeneous landscapes, this has important implications for our understanding of the strength of local interspecific interactions as a force governing the structure of ecological communities, relative to dispersal.

An important task for future work will be to work systematically through other familiar modules in community ecology (e.g., intraguild predation, two consumers on two biotic resources, interactions involving mutualisms, competitive systems with ecosystem feedbacks through detrital pools), and explore the consequences of colonization-extinction dynamics and mass effects for species coexistence. In all these modules, as with apparent competition, permitting dispersal between communities is likely to open up additional avenues both for coexistence and exclusion.

Finally, it is important to embed these analyses of modules in analyses of full, complex food webs. Are there generalities that transcend the manifold complexity of food webs, or does the "curse of dimensionality" loom so large that it is difficult to discern generality in the face of the many idiosyncrasies of web structure? For instance, can signatures of metacommunity dynamics be discerned in properties of whole webs, such as connection or the stability-diversity relationship? Analyses of compartments in food webs suggest that habitat boundaries typically provide suture lines between compartments (e.g., Krause et al. 2003); this observation is consistent with the importance of species sorting along gradients as a major dimension of metacommunity structure. The recent literature has suggestive hints that broad scaling relationships may exist among food webs, in effect allometric relationship relating branching properties in the web to the size of the web (Garlaschelli et al. 2003). Relating such scaling relationships to spatial flows and dynamics could help sharpen our understanding of how metacommunity dynamics bears on food web structure.

These are important and difficult challenges—but the nettle must be grasped for food web ecology and metacommunity ecology to at least achieve a full and coherent integration.

Acknowledgments

This manuscript benefited from comments by Marcel Holyoak and the 2003 Metacommunities seminar at University of California Davis. This work was conducted as part of the Metacommunity Working Group at the National Center for Ecological Analysis and Synthesis (NCEAS), a center funded by the National Science Foundation (DEB-9421535), University of California Santa Barbara, and the state of California. MFH was supported by NSF grant DEB-9806635 and NIH grant R01 ES12067-01 to Cheryl J. Briggs. RDH was supported by NSF, NIH, and the University of Florida Foundation, and thanks to David Post for insightful conversations.

Literature Cited

Abrams, P. A. 1993. Effect of increased productivity on the abundance of trophic levels. American Naturalist 146:112–134.

Anderson, W. B. and D. A. Wait. 2001. Subsidized island biogeography hypothesis: Another new twist on an old theory. Ecology Letters 4:289–291.

Armstrong, R. A. 1989. Competition, seed predation, and species coexistence. Journal of Theoretical Biology 141:191–194.

Belyea, L. R. and J. Lancaster. 1999. Assembly rules within a contingent ecology. Oikos 86:402–416.

Bengtsson, J. and D. Ebert. 1998. Distributions and impacts of microparasites on *Daphnia* in a rock-pool metapopulation. Oecologia 115:213–221.

Bonsall, M. B., D. R. French, and M. P. Hassell. 2002. Metapopulation structures affect persistence of predator-prey interactions. Journal of Animal Ecology 71:1075–1084.

Bonsall, M. B. and M. P. Hassell. 1997. Apparent competition structures ecological assemblages. Nature 388:371–373.

———. 1998. The population dynamics of apparent competition in a host-parasitoid assemblage. Journal of Animal Ecology 67:919–929.

———. 2000. The effects of metapopulation structure on indirect interactions in host-parasitoid assemblages. Proceedings of the Royal Society of London, Series B 267:2207–2212.

Briggs, C. J. and M. F. Hoopes. 2004. Stabilizing effects in spatial parasitoid-host and predator-prey models: A review. Theoretical Population Biology 65:299–315.

Britton, N. F., G. P. Boswall, and N. R. Franks. 2001. Dispersal and conservation in heterogeneous landscapes. Pages 299–320 in I. P. Woiwod, D. R. Reynolds, and C. D. Thomas, eds. *Insect movement: Mechanisms and consequences.* CAB International, Wallingford, UK.

Caswell, H. 1978. Predator-mediated coexistence: A nonequilibrium model. American Naturalist 112:127–154.

Caswell, H. and J. E. Cohen. 1993. Local and regional regulation of species-area relations: A patch-occupancy model. Pages 99–107 in R. Ricklefs and D. Schluter, eds. *Species diversity in ecological communities.* University of Chicago Press, Chicago, IL.

Chaneton, E. J. and M. B. Bonsall. 2000. Enemy-mediated apparent competition: Empirical patterns and the evidence. Oikos 88:38–394.

Chase, J. D., P. A. Abrams, J. Grover, S. Diehl, R. D. Holt, S. Richards, T. Case, R. Nisbet and P. Chesson. 2002. The interaction between predation and competition: A review and synthesis. Ecology Letters 5:302–315.

Chase, J. M. and M. A. Leibold. 2003. *Ecological Niches.* University of Chicago Press, Chicago, IL.

Cohen, J. E., R. A. Beaver, S. H. Cousins, D. L. DeAngelis, L. Goldwasser, K. L. Heong, R. D. Holt et al. 1993. Improving food webs. Ecology 74:252–258.

Cohen, J. E., T. Jonsson and S. R. Carpenter. 2003. Ecological community description using the food web, species abundance, and body size. Proceedings of the National Academy of Sciences, USA 100:1781–1786.

Cohen, J. E. and C. M. Newman. 1991. Community area and food-chain length: Theoretical predictions. American Naturalist 138:1542–1554.

Cook, W. M., K. T. Lane, B. Foster and R. D. Holt. 2002. Island theory, matrix effects and species richness patterns in habitat fragments. Ecology Letters 5:619–623.

Crooks, K. R. and M. E. Soule. 1999. Mesopredator release and avifaunal extinctions in a fragmented system. Nature 400:563–566.

Davies, K. F., B. A. Melbourne and C. R. Margules. 2001. Effects of within- and between-patch processes on community dynamics in a fragmentation experiment. Ecology 2001:1830–1846.

Didham, R. K., J. H. Lawton, P. M. Hammond, and P. Eggleton. 1998. Trophic structure stability and extinction dynamics of beetles (Coleoptera) in tropical forest fragments. Philosophical Transactions of the Royal Society of London, series B: Biological Sciences 353:437–451.

Estes, J., K. Crooks and R. D. Holt. 2001. Predation and biodiversity. Pages 857–878 in S. Levin ed. *Encyclopedia of biodiversity.* Volume 4. Academic Press, New York, N.Y.

Evans, E. W. and S. England. 1997. Indirect interactions in biological control of insects: pests and natural enemies in alfalfa. Ecological Applications 6:920–930.

Garlaschelli, G. Caldarelli, and L. Pietronero. 2003. Universal scaling relations in food webs. Nature 423:165–168.

Grover, J. P. 1997. *Resource competition.* Chapman and Hall, London, UK.

Hanski, I. 1989. Fungivory: Fungi, insects and ecology. Pages 25–68 in N. Wilding, N. M. Collins, P. M. Hammand, and J. F. Webber, eds. *Insect-fungus interactions.* Academic Press, New York.

———. *Metapopulation ecology.* Oxford University Press, Oxford, UK.

Hassell, M. P., Comins, H. N. and R. M. May. 1994. Species coexistence and self-organising spatial dynamics. Nature 370:290–292.

Hawkins, B. A. and E. E. Porter. 2003. Does herbivore diversity depend on plant diversity? The case of California butterflies. American Naturalist 161:40–49.

Holt, R. D. 1977. Predation, apparent competition, and the structure of prey communities. Theoretical Population Biology 1:197–229.

———. 1984. Spatial heterogeneity, indirect interactions, and the coexistence of prey species. American Naturalist 124:377–406.

———. 1985. Population dynamics in two-patch environments: Some anomalous consequence of an optimal habitat distribution. Theoretical Population Biology 28:181–208.

———. 1993. Ecology at the mesoscale: The influence of regional processes on local communities. Pages 77–88 in R. E. Ricklefs, and D. Schluter, eds. *Species diversity in ecological communities: Historical and geographical perspectives.* University of Chicago Press, Chicago, IL.

———. 1996. Food webs in space: An island biogeographic perspective. Pages 313–323 in G. Polis, and K. Winemiller, eds. *Food webs: Integration of patterns and dynamics.* Chapman and Hall, New York.

———. 1997a. From metapopulation dynamics to community structure: some consequences of spatial heterogeneity. Pages 149–164 in I. P. Hanski, and M. E. Gilpin, eds. *Metapopulation dynamics: Ecology, genetics, and evolution.* Academic Press, New York.

———. 1997b. Community modules. Pages 333–349 in A. C. Gange and V. M. Brown, eds. *Multitrophic interactions in terrestrial ecosystems.* Blackwell Science, Oxford, UK.

———. 2002. Food webs in space: On the interplay of dynamic instability and spatial processes. Ecological Research 17:261–273.

———. 2004. Implications of system openness for local community structure and ecosystem function. Pages 96–114 in G. A. Polis, M. E. Power, and G. R. Huxel, eds. *Food Webs at the Landscape Scale.* University of Chicago Press, Chicago, IL.

Holt, R. D., M. Barfield, and A. Gonzalez. 2003. Impacts of environmental variability in open populations and communities: "Inflation" in sink environments. Theoretical Population Biology 64:315–330.

Holt, R. D. and J. H. Lawton. 1993. Apparent competition and enemy-free space in insect host-parasitoid communities. American Naturalist 142:623–645.

———. 1994. The ecological consequences of shared natural enemies. Annual Review of Ecology and Systematics 25:495–520.

Holt, R. D., J. H. Lawton, G. A. Polis and N. Martinez. 1999. Trophic rank and the species-area relationship. Ecology 80:1495–1504.

Holt, R. D. and D. Post. Manuscript. Spatial controls on food chain length: A review of mechanisms.

Holyoak, M. 2000. Habitat subdivision causes changes in food web structure. Ecology Letters 3:509–515.

Huxel, G. R. and K. McCann. 1998. Food web stability: The influence of trophic flows across habitats. American Naturalist 152:460–469.

Jansen, V. A. A. 1995. Effects of dispersal in a tri-trophic metapopulation model. Journal of Mathematical Biology 34:195–224.

Janzen, D. H. 1970. Herbivores and the number of tree species in tropical forests. American Naturalist 104:501–528.

Kareiva, P. 1990. Population dynamics in spatially complex environments: Theory and data. Philosophical Transactions of the Royal Society of London, series B: Biological Sciences 330:175–190.

Kneitel, J. M. and T. E. Miller. 2003. Dispersal rates affect species composition in metacommunities of Sarracenia purpurea inquilines. American Naturalist 162:165–171.

Komonen, A., R. Penttila, M. Lindgren and I. Hanski. 2000. Forest fragmentation truncates a food chain based on an old-growth forest bracket fungus. Oikos 90:119–126.

Krause, A. E., K. A. Frank, D. M. Mason, R. E. Ulanowicz, and W. W. Taylor. 2003. Compartments revealed in food-web structure. Nature 426:282–285.

Kruess, A. and T. Tscharntke. 1994. Habitat fragmentation, species loss, and biological control. Science 264:1581–1584.

———. 2000. Species richness and parasitism in a fragmented landscape: Experiments and field studies with insects on *Vicia sepium*. Oecologia 122:129–137.

Law, R., and R. D. Morton. 1996. Permanence and the assembly of ecological communities. Ecology 77:762–775.

Lawton, J. H. 1989. Food webs. Pages 43–78 *in* M. Cherrett, ed. *Ecological concepts*. Blackwell Scientific Publications, Oxford, UK.

Leibold, M. A. 1996. A graphical model of keystone predators in food webs: Trophic regulation of abundance, incidence, and diversity patterns in communities. American Naturalist 147:784–812.

Leibold, M. A. et al. The metacommunity concept: A framework for multi-scale community ecology. Ecology Letters 7:601–613.

Levins, R. and D. Culver. 1971. Regional coexistence of species and competition between rare species. Proceedings of the National Academy of Sciences, USA 68:1246–1248.

Lockwood, J. L., R. D. Powell, M. P. Nott and S. L. Pimm. 1997. Assembling ecological communities in time and space. Oikos 80:549–553.

Lomolino, M. V. 1984. Immigrant selection, predatory exclusion, and the distributions of *Microtus brevicauda* on islands. American Naturalist 125:310–316.

Luh, H.-K. and S. L. Pimm. 1993. The assembly of ecological communities: A minimalist approach. Journal of Animal Ecology 62:749–765.

MacArthur, R. H. and E. O. Wilson. 1967. The theory of island biogeography. Princeton University Press, Princeton, NJ.

May, R. M. 1973. Stability and complexity in model ecosystems. Princeton University Press, Princeton, NJ.

———. 1994. The effects of spatial scale on ecological questions and answers. Pages 1–17 *in* P. J. Edwards, R. M. May and N. R. Webb, eds. Large scale ecology and conservation biology. Oxford University Press, Oxford, UK.

McCann, K., A. Hastings and G. R. Huxel. 1998. Weak trophic interactions and the balance of nature. Nature 395:794–798.

Melian, C. J. and J. Bascompte. 2002. Food web structure and habitat loss. Ecology Letters 5:37–46.

Muller, C. B. and J. Brodeur. 2002. Intraguild predation in biological control and conservation biology. Biological Control 25:216–223.

Nisbet, R. M., C. J. Briggs, W. S. C. Gurney, W. W. Murdoch and A. Stewart-Oaten. 1993. Two-patch metapopulation dynamics. Pages 125–135 *in* S. A. Levin, T. M. Powell and J. H. Steele, eds. *Patch dynamics*. Springer-Verlag, Berlin, Germany.

Oksanen, L., S. D. Fretwell, J. Arruda and P. Niemala. 1981. Exploitation ecosystems in gradients of primary productivity. American Naturalist 118:240–261.

Oksanen, T., Oksanen, L. and M. Gyllenberg. 1992. Exploitation ecosystem in heterogeneous habitat complexes II: Impact of small-scale heterogeneity on predator-prey dynamics. Evolutionary Ecology 6:383–398.

Oksanen, T., M. Schneider, Ü. Rammul, P. Hambäck, M. Aunapuu. 1999. Population fluctuations of voles in North Fennoscandian tundra: contrasting dynamics in adjacent areas with different habitat composition. Oikos 86:463–478.

Pace, M. L., J. J. Cole, S. R. Carpenter, and J. F. Kitchell. 1999. Trophic cascades revealed in diverse ecosystems. Trends in Ecology and Evolution 14:483–488.

Persson, L. 1999. Trophic cascades: Abiding heterogeneity and the trophic level concept at the end of the road. Oikos 85:385–397.

Persson, A., L. A. Hansson, C. Bronmark, P. Lundberg, L. B. Pettersson, L. Greenberg, P. A. Nilsson, P. Nystrom, P. Romare, and L. Tranvik. 2001. Effects of enrichment on simple aquatic food webs. American Naturalist 157:654–669.

Pimm, S. L. 1982. *Food webs.* Chapman and Hall, London, UK.

Polis, G. A. 1994. Food webs, trophic cascades and community structure. Australian Journal of Ecology 19:121–136.

Polis, G. A., W. B. Anderson and R. D. Holt. 1997. Toward an integration of landscape ecology and food web ecology: The dynamics of spatially subsidized food webs. Annual Review of Ecology and Systematics 28:289–316.

Polis, G. A., R. D. Holt, B. A. Menge and K. Winemiller. 1996. Time, space and life history: Influence on food webs. Pages 435–460 *in* G. A. Polis, and K. O. Winemiller, eds. *Food webs: Integration of patterns and dynamics.* Chapman and Hall, London, UK.

Polis, G. A. and S. D. Hurd. 1996. Linking marine and terrestrial food webs: Allochthonous input from the ocean supports high secondary productivity in small islands and coastal land communities. American Naturalist 147:396–423.

Polis, G. A., M. Power, and G. R. Huxel. 2004. *Food webs at the landscape level.* University of Chicago Press, Chicago, IL.

Polis, G. A. and K. Winemiller. 1996. *Food webs: Integration of patterns and dynamics.* Chapman and Hall, London, UK.

Post, D. M. 2002. The long and short of food-chain length. Trends in Ecology and Evolution 17:269–277.

Post, D. M., M. L. Pace and N. G. Hairston, Jr. 2000. Ecosystem size determines food-chain length in lakes. Nature 405:1047–1049.

Power, M. E. and W. E. Rainey. 2000. Food webs and resource sheds: Towards spatially delimiting trophic interactions. Pages 291–314 *in* M. J. Hutchings, E. A. John, and A. J. A. Stewart, eds. *The ecological consequences of environmental heterogeneity.* Blackwell Science Limited, Oxford, UK.

Rey, J. R. and E. D. McCoy. 1979. Application of island biogeographic theory to pests of cultivated crops. Environmental Entomology 8:577–582.

Ritchie, M. E. 1999. Biodiversity and reduced extinction risks in spatially isolated rodent populations. Ecology Letters 2:11–13.

Rosenzweig, M. L. 1995. *Species diversity in space and time.* Cambridge University Press, Cambridge, UK.

Rosenzweig, M. L. and Y. Ziv. 1999. The echo pattern of species diversity: Pattern and processes. Ecography 22:614–628.

Ruxton, G. D. and P. Rohani. 1996. The consequences of stochasticity for self-organized spatial dynamics, persistence and coexistence in spatially extended host-parasitoid communities. Proceedings of the Royal Society of London, series B, 263:625–631.

Schoener, T. W. 1989. Food webs from the small to the large. Ecology 70:1559–1589.

Schoener, T. W., D. A. Spiller, and L. W. Morrison. 1996. Variation in the hymenopteran parasitoid fraction on Bahamian islands. Acta Oecologica 16:103–121.

Shurin, J. B. 2001. Interactive effects of predation and dispersal on zooplankton communities. Ecology 82:3404–3416.

Shurin, J. B. and E. G. Allen. 2001. Effects of competition, predation, and dispersal on species richness at local and regional scales. American Naturalist 158:624–637.

Shurin, J. B., P. Amarasekare, J. M. Chase, R. D. Holt, M. F. Hoopes and M. A. Leibold. 2004. Alternative stable states and regional community structure. Journal of Theoretical Biology 227:359–368.

Siemann, E. 1998. Experimental tests of effects of plant productivity and diversity on grassland arthropod diversity. Ecology 79:2057–2070.

Sinclair, A. R. E., C. J. Krebs, J. M. Fryxell, R. Turkington, S. Boutin, R. Boonstra, P. Seccombe-Hett, P. Lundberg, and L. Oksanen. 2000. Testing hypotheses of trophic level interactions: A boreal forest ecosystem. Oikos 89:313–328.

Spencer, M., L. Blaustein, and J. E. Cohen. 1999. Species richness and the proportion of predatory animal species in temporary pools: Relationships with habitat size and permanence. Ecology Letters 2:157–166.

Spencer, M. and P. H. Warren. 1996. The effects of habitat size and productivity on food web structure in small aquatic microcosms. Oikos 75:419–430.

Steffan-Dewenter, I. and T. Tscharntke. 2000. Butterfly community structure in fragmented habitats. Ecology Letters 3:449–456.

———. 2002. Insect communities and biotic interactions on fragmented calcareous grasslands—a mini review. Biological Conservation 104:275–284.

Swihart, R. K., Z. Feng, N. A. Slade, D. M. Mason and T. M. Gehring. 2001. Effects of habitat destruction and resource supplementation in a predator-prey metapopulation model. Journal of Theoretical Biology 210:287–303.

Terborgh, J., L. Lopez, V. P. Nunez et al. 2001. Ecological meltdown in predator-free forest fragments. Science 294:1923–1926.

Vander Zanden, M. J., B. J. Shuter, N. Lester and J. B. Rasmussen. 1999. Patterns of food chain length in lakes: a stable isotope study. American Naturalist 154:406–416.

Van Nouhuys, S. and I. Hanski. 2002. Multitrophic interactions in space: metacommunity dynamics in fragmented landscapes. Pages 124–147 in T. Tscharntke and B. A. Hawkins, eds. *Multitrophic interactions.* Cambridge University Press, Cambridge, UK.

Van Nouhuys, S. and W. T. Tay. 2001. Causes and consequences of small population size for a specialist parasitoid wasp. Oecologia 128:126–133.

Warren, P. H. 1994. Making connections in food webs. Trends in Ecology and Evolution 9:136–141.

Whittaker, R. H. 1975. Communities and ecosystems. MacMillan, New York.

Whittaker, R. J. 1992. Stochasticism and determinism in island ecology. Journal of Biogeography 19:587–591.

Wilson, D. S. 1992. Complex interactions in metacommunities, with implications for biodiversity and higher levels of selection. Ecology 73:1984–2000.

Wilson, H. B., R. D. Holt, and M. P. Hassell. 1998. Persistence and area effects in a stochastic tritrophic model. American Naturalist 151:587–596.

Wootton, J. T. 1998. Effects of disturbance on species diversity: a multitrophic perspective. American Naturalist 152:803–825.

Part 2

EMPIRICAL PERSPECTIVES

Marcel Holyoak and Robert D. Holt

Perhaps the greatest challenge in studying metacommunities is to link theoretical concepts to natural empirical systems. One problem is that many of the theoretical ideas recognized in this book have only recently been described (e.g., Chesson 1998; Hubbell 2001; Mouquet and Loreau 2002; Loreau et al. 2003; Leibold et al. 2004). Second, fully testing metacommunity ideas is demanding because it requires knowledge of both patterns and mechanisms involving multiple communities. Finally, testing the spatial dynamics of many interacting species could be challenging because it has been difficult to test theories involving the spatial dynamics of just two interacting species (e.g., Taylor 1990, Harrison and Taylor 1997). Nonetheless, there are ways of studying metacommunity ecology that are tractable and that go a long way toward testing the modern theories that dominate metacommunity ecology.

An obvious way to organize studies is to use the four conceptual metacommunity models Holyoak et al. have outlined in chapter 1. However, since addressing this framework was not the main aim of many chapters we instead organize our introduction around two of the greatest hurdles to applying theoretical definitions of metacommunities to empirical situations. These are that local communities do not always have discrete boundaries and that different species may respond to processes at different scales (Holyoak et al., chapter 1; Holt et al., chapter 20). Based on these problems, the examples presented in this volume might be placed into three categories.

Collections of Discrete Permanent Habitat Patches

If habitat patches are relatively permanent and large a species could persist by remaining within patches without dispersal. However, if populations within patches were small or species interactions caused extinctions, then additional persistence mechanisms would be required. Species could also evolve to persist through dormant propagules (the temporal storage effect, e.g., Chesson 2000), or through dispersal and metapopulation dynamics (reviewed by Hoopes et al., chapter 2).

Studies of permanent patches that cannot support viable populations are typified by many clusters of small oceanic islands, with oceans providing barriers to dispersal to varying degrees depending on the taxa considered (Mehranvar and Jackson 2001). Similarly, ponds and lakes often have biotas that are strongly bounded by terrestrial habitat, but the degree to which the intervening terrestrial habitat is a barrier varies between taxa. These systems also often contain considerable variation in physical or biotic conditions from patch to patch. Chapter 8 by Cottenie and De Meester describes an interconnected system of ponds, some with and some lacking fish predators. They show that local abiotic and biotic conditions are critical to the species composition and densities of assemblages of zooplankton. The system falls somewhere between the species-sorting and mass effects perspectives, depending on the data that are considered. The rock pool and zooplankton system presented by Kolasa and Romanuk (chapter 9) is another system in which patches are constant in their positions but vary considerably in their characteristics and may also dry out during the year. The authors suggest that physical conditions seem to be critical to organizing the communities and creating a hierarchy of scales, and argue that both species sorting and neutral dynamics could explain different aspects of dynamics.

Two chapters present empirical results from systems where patches are permanent but appear to be more uniform. In chapter 6, Gonzalez discusses experimental evidence from a system consisting of carpets of epilithic moss that contains a species-rich assemblage of microarthropods. These moss carpets represent readily manipulable microlandscapes. A series of experiments conducted in northern England demonstrated that altering landscape connectivity, and hence community isolation, influenced various community properties, such as local and regional diversity and secondary productivity. This example shows effects of dispersal on local and regional species diversity and the author suggests that the mass effects perspective is particularly relevant to dynamics in this system. An unmanipulated field system provides a second example of this kind. Van Nouhuys and Hanski (chapter 4), describe a system consisting of hundreds of patches containing a food web consisting of up to three plants, two butterfly species, five primary parasitoids and two hyperparasitoids. The study provides a convincing demonstration of the role of spatial dynamics and other factors in altering local species diversity, and does so in a food web context. It provides one of the best empirical examples of the patch dynamics perspective and presents interesting links with food web ideas (e.g., Holt and Hoopes, chapter 3).

Temporary Patches Distinct from a Background Habitat Matrix

Species in landscapes where patches vary in position both spatially and temporally may be strongly dependent on traits related to spatial dynamics such as dispersal and dormancy (Harrison and Taylor 1997). If patches are in constant positions across generations we are more likely to see selection for dormancy strategies than dispersal (McPeek and Kalisz 1998). However, if patches vary in location from generation to generation species will not necessarily be able to persist through becoming dormant. Pitcher plants form temporary patches of aquatic habitat, requiring dispersal of at least some of the inhabitants, which range from bacteria to insects (Miller and Kneitel, chapter 5). Miller and Kneitel show that their system shows elements of dispersal-limitation (indicating patch dynamics) and species sorting. They review both local and regional dynamics using a variety of experimental and observational evidence. The inhabitants of water-filled tree holes (Kitching 2000) and fungal-fruiting bodies (Worthen 1989) are other potential examples of this kind of community.

Permanent Habitats with Indistinct Boundaries

The final category is the most elusive, consisting of systems in which habitats are more permanent and boundaries are less distinct. In such systems it is not always clear that spatial dynamics are necessary for persistence; however a variety of pieces of empirical work suggest that they are important. Davies et al. (chapter 7), describe an experimentally manipulated landscape containing *Eucalyptus* forest that was fragmented by planting woodland consisting of a nonnative species of pine. They show that assemblages of ground dwelling beetles are characterized by different population dynamics within fragments than in more spatially continuous habitat. Mass effects, where species move from the pine habitat matrix to the *Eucalyptus* fragments offer one potential explanation for their findings. It is a fascinating study system that shows excellent opportunities for further analysis. In other study systems species have been shown to readily disperse across habitat boundaries (e.g., weeds invading roadside areas in serpentine chaparral habitats; Harrison 1999). In systems with fuzzy boundaries between habitats, the degree to which spatial dynamics are relevant is likely to vary depending on the degree of habitat specialization. Habitat specialization influences the organisms' perception of habitat size and isolation (Harrison 1997). The Davies et al. study provides some evidence that metacommunity dynamics based on discrete local communities can help us to understand these situations, in part by encouraging authors to think about different patterns within their datasets.

We encourage readers to question the nature of evidence for metacommunity dynamics and how inference might be further improved. The coda section at the end of the book summarizes some of the emerging patterns that the editors perceive. It is apparent from the chapters in part 2 of this book that an empirical-theoretical synthesis is well under way, and that continuing this work offers a rich world of possibilities to a broad range of kinds of ecologists.

Literature Cited

Chesson, P. 1998. Making sense of spatial models in ecology, Pages 151–166 *in* J. Bascompte, and R. V. Sole, eds. *Modeling spatiotemporal dynamics in ecology.* Springer Verlag, New York, NY.

———. 2000. Mechanisms of maintenance of species diversity. Annual Review of Ecology and Systematics 31:343–366.

Harrison, S. 1997. How natural habitat patchiness affects the distribution of diversity in Californian serpentine chaparral. Ecology 78:1898.

———. 1999. Local and regional diversity in a patchy landscape: Native, alien, and endemic herbs on serpentine. Ecology 80:70–80.

Harrison, S., and A. D. Taylor. 1997. Empirical evidence for metapopulation dynamics: a critical review. Pages 27–42 *in* I. Hanski, and M. E. Gilpin, eds. *Metapopulation dynamics: Ecology, genetics and evolution.* Academic Press, San Diego, CA.

Hubbell, S. P. 2001. The unified neutral theory of biodiversity and biogeography. Princeton University Press, Princeton, NJ.

Kitching, R. L. 2000, *Food webs and container habitats: The natural history and ecology of phytotelmata.* Cambridge University Press, New York, NY.

Leibold, M. A., M. Holyoak, N. Mouquet, P. Amarasekare, J. M. Chase, M. F. Hoopes, R. D. Holt, J. B. Shurin, R. Law, D. Tilman, M. Loreau, and A. Gonzalez, A.. 2004. The metacommunity concept: A framework for multi-scale community ecology. Ecology Letters 7:601–613.

Loreau, M., N. Mouquet, and R. D. Holt. 2003. Meta-ecosystems: A theoretical framework for a spatial ecosystem ecology. Ecology Letters 6:673–679.

McPeek, M. A., and S. Kalisz. 1998. On the joint evolution of dispersal and dormancy in metapopulations. Advances in Limnology 52:33–51.

Mehranvar, L., and D. A. Jackson. 2001. History and taxonomy: Their roles in the core-satellite hypothesis. Oecologia 127:131–142.

Mouquet, N., J. L. Moore, and M. Loreau. 2002. Plant species richness and community productivity: Why the mechanism that promotes coexistence matters. Ecology Letters 5:56–65.

Taylor, A. D. 1990. Metapopulations, dispersal, and predator-prey dynamics: An overview. Ecology 71:429–433.

Worthen, W. B. 1989. Effects of resource density on mycophagous fly dispersal and community structure. Oikos 54:145–153.

CHAPTER 4

Metacommunities of Butterflies, Their Host Plants, and Their Parasitoids

Saskya van Nouhuys and Ilkka Hanski

The great bulk of population biological research on butterflies is focused on single species, but in reality butterfly populations are dynamically coupled with other populations representing species at higher, lower, and equal trophic levels. Often these interactions take place in fragmented landscapes because the focal butterfly may use only one or a few larval host plant species, which may be habitat specialists with fragmented distributions. Butterfly eggs, caterpillars, and pupae are attacked by a range of more or less specialized parasitoids (Dempster 1983; Shaw and Fitton 1989; Van Nouhuys and Hanski 2004), which themselves host hyperparasitoids (parasitoids of the primary parasitoids). The behavior and population dynamics of the parasitoids are often influenced by the host plants of their host butterfly (Price et al. 1980; Vet and Dicke 1992; Hochberg and Ives 2000), so there are direct links between species on the first trophic level (plants) and species at the third trophic level (parasitoids). The patchy distribution of the host plants combined with the specificity of the butterflies and their natural enemies leads to metacommunity dynamics in the plant-butterfly-parasitoid assemblage, which often includes on the order of ten species. These species are typically interacting with yet other species, which may be less specific and whose populations are often less strongly spatially structured than the populations of the butterflies, their host plants and the specialist parasitoids. For instance, the host plants may support polyphagous herbivores in addition to the specialist butterflies, and all the insect species may be attacked by diseases, generalist arthropods, and vertebrate predators. Thus we view closely interacting species around the focal butterfly as embedded in a more comprehensive community made up of species that may have different spatial distributions than the focal plant-butterfly-parasitoid metacommunity. This may be a common situation in nature in general.

We present an overview of metacommunities associated with butterflies, with a particular focus on a well-studied species of checkerspot butterfly, the Glanville fritillary (Hanski 1999; Ehrlich and Hanski 2004). We illustrate several ecological processes taking place in metacommunities with examples drawn from the metacommunity of the Glanville fritillary and its host plants and parasitoids, including plant-butterfly interactions, the trade-off between competitive ability and dispersal rate / ability in competing species, tritrophic interactions involving the host plants, butterflies and parasitoids, and apparent competition. We then address two aspects of metacommunity structure. First, the length of food chains in relation to the spatial extent of the landscape occupied by the metacommunity. Second, we discuss to what extent different coexisting checkerspot community modules (Holt and Hoopes, chapter 3) are linked together via shared host plants and parasitoids to form more comprehensive metacommunities. Here we examine coexisting community modules around two checkerspot species in Finland, five checkerspot species in the steppe region in Buryatia, Russia, and five checkerspot species in the Montseny mountains in northern Spain. This analysis reinforces the view that community modules are often relatively independent. Therefore, the primary task is to understand how interspecific interactions within the relatively small community modules are affected by the spatial structure and dynamics of the constituent species, and vice versa, which sets a manageable challenge for research on metacommunities.

An Overview of Plant-Butterfly-Parasitoid Metacommunities

Butterflies and moths have been the subject of much study in population, meta-population, and community ecology, to a large extent because they are conspicuous and are frequently the targets of either conservation effort or pest control. Many butterfly species occur as metapopulations (Hanski and Kuussaari 1995; Thomas and Hanski 1997), and studies of butterflies have contributed greatly to the development of the concept (Harrison et al. 1988; Hanski et al. 1994; Hanski and Gaggiotti 2004) and theory (Hanski 1994; Hanski and Ovaskainen 2000) of metapopulations. Butterfly studies have yielded some of the most useful data to test the assumptions and predictions of metapopulation models (Boughton 2000; Thomas and Hanski 2004; Hanski et al. 2004), because butterflies possess characteristics that both facilitate their study and make them a good model group in which to study metapopulation biology (Murphy et al. 1990). Many habitats used by butterflies are naturally or anthropogenically fragmented, which leads to a fragmented structure of populations. Butterflies are small and therefore the number of individuals in even small habitat fragments may be large enough to constitute a local breeding population. Butterflies have a high rate of population growth, which means that following population establishment local populations may grow quickly to the local carrying capacity. Finally, butterflies have short generation times, and hence stochastic events are not buffered by great longevity of individuals, which would reduce the risk of population extinction.

Turning to community studies on butterflies and moths, they include, first of all, surveys of species diversity associated with different habitat types such as grassland and tropical forest; many recent studies demonstrate the impact of habitat fragmentation or degradation on butterfly diversity (Steffan-Dewenter and Tscharntke 2000; Kitahara and Sei 2001; Collinge et al. 2003; Horner-Devine et al. 2003). Other studies have examined the butterfly and moth communities associated with specific host plant species (Courtney and Chew 1987; Thomas et al. 1990; Harrison and Thomas 1991). Few if any of these large-scale butterfly community studies include the associated parasitoid communities, and in fact little is known about parasitoid communities associated with most butterflies (Shaw and Fitton 1989; Shaw 1994). There are however studies of insect communities that include Lepidoptera and parasitoids among other taxa (Memmott et al. 1994; Holyoak 2000; Kruess and Tscharntke 2000), such as the quantitative food web study by Lewis et al. (2002) describing the complex community of 93 leaf mining insect species (primarily moths, beetles, and flies) and their 84 parasitoid species in a 8500 m² area of tropical forest in Belize. There is also a handful of studies collating information for communities of Lepidoptera and their parasitoids across larger areas (Hawkins and Sheehan 1994; Hochberg and Ives 2000), such as the study of 60 tachinid parasitoids associated with 196 species of externally feeding Lepidoptera in a mesquite-oak savanna in Arizona (Stireman and Singer 2003), the parasitoids associated with tortricid moths (Mills 1993), and the rich invasive parasitoid community associated with forest moths in Hawaii (Henneman and Memmott 2001).

Literature-based host records have been used to generate hypotheses about parasitoid community structure and how it is affected by host characteristics such as feeding niche and abundance, affected by parasitoid characteristics such as host breadth and attributes of the host plants, and affected by habitat and the landscape (Askew and Shaw 1989; Hawkins et al. 1992; Hawkins 1994; Hawkins and Sheehan 1994; Sheehan 1994; Hawkins 2000; Holyoak 2000). These studies do not consider spatial population structure or dynamics in any detail, if at all, and the vast majority of studies are concerned with herbivores living in plant structures such as galls or leaf mines. In brief, though there is much knowledge about the spatial population structures and metapopulation dynamics of butterflies, and there is much information about community structure in Lepidoptera in terms of host plant and parasitoid associations, there is yet little knowledge of the combination of the two—that is, metacommunity structure and dynamics of butterflies and their associated species.

The Community Associated with the Glanville Fritillary Butterfly

The Glanville fritillary butterfly (*Melitaea cinxia*) is restricted, like many other checkerspot butterflies, to larval host plants that produce iridoid glycosides, supposedly in defense against generalist herbivores (Bernays and De Luca 1981; Bowers 1983; Wahlberg 2001). Throughout its range, *M. cinxia* feeds on plants in the genera *Plantago* and *Veronica*, in the family Plantaginaceae (Kuussaari et al. 2004). In the Åland Islands in southwestern Finland, the larval host plants are *Plantago lanceolata* and *Veronica spicata*, which grow in dry open meadows in a rural landscape (*P. lanceolata* is not limited to the meadows used by the butterfly, and it is found throughout the islands in open and disturbed habitats). The adult butterflies emerge, mate, and lay eggs in June. Eggs are laid in clusters of 100 to 200 on the underside of host plant leaves. Upon hatching in early July the larvae live gregariously in silken nests until they disperse to pupate in the leaf litter in the following May. The larvae winter in dense silken nests on or near the host plants (Kuussaari et al. 2004).

There are altogether around 4000 small meadows that are suitable for *M. cinxia* in the Åland Islands within an area of 50 by 70 km. In any one year 400 to 500 of these meadows are occupied by typically small local butterfly populations (figure 4.1; Nieminen et al. 2004). There is much population turnover, and the butterfly persists as a classic metapopulation in the highly fragmented landscape (Hanski et al. 1994; Ehrlich and Hanski 2004). A local population of *M. cinxia* consists of individuals on a single meadow. A local community includes the species that interact directly or indirectly within the same meadow, whereas a metacommunity is a set of local communities connected by dispersal of the constituent species.

Melitaea cinxia experiences little direct interspecific competition because few other insect herbivores feed predominantly on *P. lanceolata* and *V. spicata*. However, there is probably more indirect competition, mediated by the host plants, with a powdery mildew, thrips, aphids, agromyzid flies, several moth species, and seed-head feeding weevils (Nieminen et al. 2004). Powdery mildew is present in only some of the *M. cinxia* populations, whereas thrips and aphids are abundant only during some years. We have observed no predation of *M. cinxia* by vertebrates, but eggs, larvae, and occasionally adults are consumed by generalist invertebrate predators such as lacewing, ladybird beetle larvae, and pentatomid bugs. The red ant *Myrmica rubra* consumes *M. cinxia* eggs, and spiders, and dragonflies have been observed to catch adult butterflies (Van Nouhuys and Hanski 2004).

The parasitoid community associated with *M. cinxia* in the Åland Islands is relatively simple (figure 4.2). The larvae are attacked by two parasitoid wasps, *Cotesia melitaearum* (Wilkinson; Braconidae: Microgastrinae), which is a gregarious endoparasitoid that has two to three generations during each host generation, and *Hyposoter horticola* (Gravenhorst; Ichneumonidae: Campoplaginae), a solitary endoparasitoid with one generation during each host generation (Lei et al. 1997; Van Nouhuys and Hanski 2004). Neither parasitoid has other hosts in Åland. Though restricted to the same host species, the two parasitoids differ greatly in natural history, behavior, spatial dynamics, and impact on host population dynamics (Lei and Hanski 1997, 1998; Lei and Camara 1999; Van Nouhuys and Hanski 2002a; Van Nouhuys and Ehrnsten 2004; Van Nouhuys and Lei 2004). Additionally, there are several generalist pupal parasitoids (Lei et al. 1997), but no egg parasitoids, larval-pupal parasitoids, nor tachinid fly parasitoids.

Each larval parasitoid has an abundant secondary parasitoid. *Cotesia melitaearum* is parasitized by the generalist pseudohyperparasitoid *Gelis agilis* (Fabricius) (Ichneumonidae: Cryptinae), which is a solitary ectoparasitoid that lays

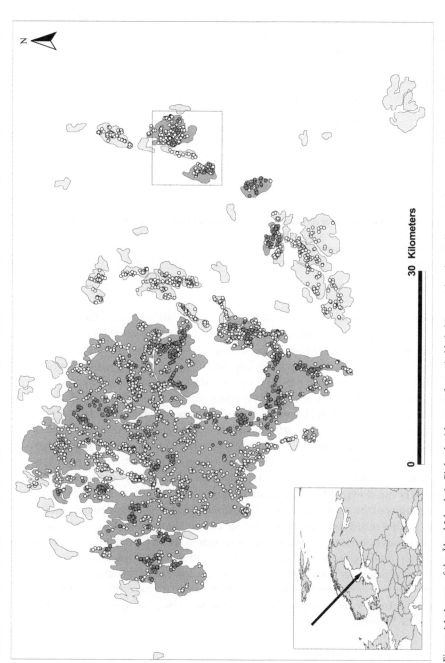

Figure 4.1 A map of the Åland Islands in Finland with empty but suitable habitat patches for *Melitaea cinxia* represented by open circles, and those occupied by the butterfly in 2002 by filled circles. The square indicates the habitat patch networks on the islands of Kumlinge and Seglinge. Dark background indicates regions occupied by the butterfly, and the light background areas not occupied.

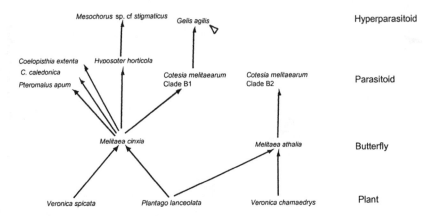

Figure 4.2 The food web of the checkerspot butterfly community in the Åland Islands. Alternate hosts of the generalist hyperparasitoid *Gelis agilis,* such as *Cotesia glomerata* (a parasitoid of Pierid butterflies), are present in surrounding habitats. Note: *Coelopisthia extenda, C. caledonica, and Pteromalus apum* may use other melitaeine and nonmelitaeine hosts. *Gelis agilis* may use other *Cotesia* and non-*Cotesia* hosts.

eggs on host parasitoid pupae in silken cocoons after they have left the butterfly larva (hence the term pseudohyperparasitoid). *Hyposoter horticola* is parasitized by the solitary hyperparasitoid *Mesochorus* sp. cf. *stigmaticus* (Brischke; Ichneumonidae: Mesochorinae), which lays eggs in *H. horticola* larvae within the butterfly host larvae. The taxonomy of the genus *Mesochorus* is not well worked out, but probably in Åland *M. stigmaticus* is restricted to parasitoids using *M. cinxia* larvae, and principally to *H. horticola* (Van Nouhuys and Hanski 2004).

The Glanville fritillary butterfly, its two host plant species, two specialist parasitoids, and two common hyperparasitoids comprise a group of seven species that interact closely with one another. These species represent a relatively independent module (Holt and Hoopes, chapter 3) in the large community of species found in open dry meadows with *V. spicata* and *P. lanceolata.* We have studied the metacommunity of the seven species over large spatial and temporal scales. Because we work with a relatively small number of species we do not address here conventional issues of (meta)community ecology, such as species diversity, community assembly, food web topology, species sorting, and so forth. Instead, we focus on the mechanisms, such as dispersal, extinction-colonization dynamics, host specialization and competition, which contribute to metacommunity structure and dynamics from local to regional scales in the assemblage of species around the Glanville fritillary.

Processes Operating in Butterfly-Parasitoid Metacommunities

In this section we review the key processes in the dynamics of the metacommunity associated with *M. cinxia* with reference to other butterfly-parasitoid metacommunities where relevant. To count as a metacommunity process in this context, we require that the interaction involves at least three species and that the interaction is critically influenced by the spatial structure of the (meta)populations of particular species. For comprehensive accounts of the biologies of the species involved see chapters in Ehrlich and Hanski (2004).

Host Plant–Butterfly Interactions

Many butterflies are specialized to use only one or a few host plant species in a particular region, though they may use a wider range of host species across their geographical range. This is a common situation in checkerspot butterflies including *M. cinxia,* and is especially well documented for the North American *Euphydryas editha* (Singer and Thomas 1987; Singer et al. 1992; Kuussaari et al. 2004; Singer 2004). In fragmented landscapes, it is likely that there is substantial variation in host plant composition from one habitat patch to another within butterfly metapopulations. Results for *M. cinxia* demonstrate that such variation may have profound consequences both for the ecological and the evolutionary dynamics of the host plant-butterfly interaction.

The two host plants used by *M. cinxia* in Åland are not distributed uniformly across the main island: *V. spicata* is absent from the east and is increasingly abundant toward the west, whereas *P. lanceolata* is present throughout the area but varies in density among habitat patches (Kuussaari et al. 2004). Genetically determined butterfly oviposition preference shows broadly matching geographic variation, shifting from preference for *P. lanceolata* in the east, where only *P. lanceolata* is available, to preference for *V. spicata* in the west, where both host plants are present (Kuussaari et al. 2000). Host plant use by larvae varies spatially because of spatial variation both in host plant occurrence and female oviposition preference. Unexpectedly, variation in adult oviposition preference appears to be unrelated to variation in larval performance or host plant suitability. Thus *P. lanceolata* and *V. spicata* are equally suitable for larval development where they are used by the butterfly and where one or the other is not used, and larvae are not locally adapted, as assessed by larval growth and survival, to the locally used host plant (Van Nouhuys et al. 2003).

Though butterfly oviposition preference does not appear to be correlated with larval survival, oviposition preference does affect regional population dynamics. Hanski and Singer (2001) found that the establishment of new populations by female butterflies was strongly influenced by the match between the host species composition of an empty habitat patch and the relative host use by larvae in the surrounding patches during previous years. For instance, a patch with only or mostly *P. lanceolata* remained frequently uncolonized in a region where most eggs were laid on *V. spicata*. A comparable patch was significantly more likely to become colonized if located in a region where *P. lanceolata* is widely used. Hanski and Singer (2001) investigated several possible mechanisms that could lead to such biased colonizations. They concluded that the most likely mechanism is female movement behavior—emigration from and immigration to patches being influenced by the correspondence between their oviposition preference and the plant species composition in the habitat patches.

Hanski and Heino (2003) demonstrated that biased colonizations due to host plant preference (Hanski and Singer 2001) may influence the evolution of oviposition preference in fragmented landscapes. The idea is simple: assuming that butterflies evolve to prefer the more common host plant species, the realized commonness of the host plants is influenced by how frequently butterflies actually encounter them, which is influenced, among other things, by the colonization rate of habitat patches with dissimilar host plant composition. One challenge for empirical research is to understand why the butterflies tend to evolve to prefer the more common host plant species, given that there is no evidence for preference-performance correlation in this case (Van Nouhuys et al. 2003). One possibility is that specialization is selected for by decreased cost of host searching, which the females do using visual cues (host plant morphology).

We have described here how butterfly metapopulation structure is in part the result of the behavioral response of females to the spatial distribution of the two host plant species, and how metacommunity dynamics may be critically influenced by the behavioral responses of the constituent species (see Resetarits et al., chapter 16, "Interactive Habitat Selection"). This interaction between the butterfly and the host plants influences both directly and indirectly the dynamics of other community members as discussed below in the section on tritrophic interactions between the host plants, the butterfly, and its parasitoids.

Contrasting Spatial Dynamics and Competition among Parasitoids

If two or more species are specialized to use the same host species, our first expectation might be that their biologies are broadly similar due to similar shared selection pressures. In the Åland Islands, *C. melitaearum* and *H. horticola* parasitize only *M. cinxia*, which has a highly fragmented and dynamic metapopulation structure. We might expect that, for example, the movement behaviors of the two parasitoids are similar. But in reality, just the opposite is the case. *Hyposoter horticola* is very mobile, and can disperse up to at least 5 to 8 km in one generation, which is more than the host dispersal range of up to 3 to 4 km (Van Nouhuys and Hanski 2002a). In contrast, *C. melitaearum* has a clearly shorter dispersal range than the host, up to 1 km (Lei and Hanski 1998; Lei and Camara 1999; Van Nouhuys and Hanski 2002b). As a consequence, *C. melitaearum* has a classic metapopulation structure in Åland, like the host butterfly, whereas *H. horticola* experiences the host metapopulation more like a single patchily distributed population. In other words, at any one time only 10% of the host populations can be reached by *C. melitaearum* (because of dispersal limitation), whereas practically all host populations are accessible to *H. horticola.*

Two species sharing the same resource must partition the resource to persist. The two parasitoids of *M. cinxia* compete directly for the same host individuals where they co-occur (Lei and Hanski 1998; Van Nouhuys and Hanski 2002a). One possible mechanism of coexistence in a fragmented landscape is based on a trade-off between competitive and dispersal abilities: the species that is an inferior competitor locally is a superior disperser (Levins and Culver 1971; Hastings 1980; Nee and May 1992; Nee et al. 1997). *Hyposoter horticola* is extremely dispersive while *C. melitaearum* is quite sedentary, sometimes attending individual larval groups in a host population for days. Therefore, the two might coexist if *C. melitaearum* is a stronger competitor locally, while *H. horticola* would largely avoid the adverse effects of competition by moving to host populations currently unoccupied by *C. melitaearum.* Lei and Hanski (1998) interpreted observational data in support of the competition-dispersal trade-off, but more recent research has shown that the competitive interaction between the two parasitoids is more complex. There are two and sometimes three generations of *C. melitaearum* during each host and *H. horticola* generation, hence there are two to three different stages in host development during which the two immature wasps compete. Surprisingly, it is only during the second (overwintering) generation that *C. melitaearum* has a competitive advantage. During the two other generations, *C. melitaearum* eggs or larvae in hosts previously parasitized by *H. horticola* die (Van Nouhuys and Hanski 2004; Punju 2002). These results provide little or no support for the notion that *C. melitaearum* is able to coexist with *H. horticola* because it is a superior competitor. Instead, it now appears that the persistence of *C. melitaearum* is enhanced by *H. horticola* typically leaving a fraction of the larvae in each host larval group unparasitized (for reasons discussed by Van Nouhuys and Ehrnsten 2004). The previous observation that parasitism by *H. horticola* is reduced in the presence of *C. melitaearum* (Lei and Hanski 1998) is probably explained by many of the doubly-parasitized host larvae dying (Punju 2002; Van Nouhuys and Tay 2001).

The idea that two competing species in a metacommunity coexist because one is a good local competitor and the other one is a good disperser is compelling, especially in situations like the present one where the two competing species differ greatly in host searching behavior and dispersal ability. However, our detailed studies of the mechanisms of host finding and parasitism behavior, and of the competitive interactions inside the host larvae, have shown that resource partitioning takes place primarily at the level of host larval groups rather than among populations. The few other studies critically testing the idea of competition-dispersal trade-off in parasitoids have similarly failed to support this mechanism

(Hopper 1984; Amarasekare 2000). The competition-dispersal trade-off has been suggested to operate in several nonparasitoid systems, but convincing examples are very few. Perhaps the best one is Pajunen's work on two species of corixid water bugs living in a network of rock pools (1979, 1982). In this case there is direct experimental evidence both for asymmetric competitive and dispersal abilities. Two other likely examples involve fungi in a laboratory system (Armstrong 1976) and mosses colonizing dung pats in the field (Marino 1991).

Tritrophic Interactions among Plants, Butterflies, and Parasitoids

Interactions between species at two trophic levels, such as between a parasitoid and its host butterfly, or the butterfly and its host plant, may be influenced directly and indirectly by other trophic levels (Price et al. 1980; Tscharntke 1992; Roininen et al. 1996; Abrahamson and Weis 1997; Sullivan and Völk 1999; Tscharntke and Hawkins 2002; Dyer and Letourneau 2003). Traditionally multitrophic interactions are studied within local communities or under laboratory conditions, but such studies can be profitably extended to include dispersal of the species among multiple local communities. The indirect effects of lower and higher trophic levels may increase the stability of populations by moderating the use of a potentially limiting resource, or decrease the stability of local populations by facilitating the consumption of the resource until it has gone locally extinct (Van Nouhuys and Hanski 2002b). The extent to which these local multitrophic interactions contribute to metacommunity dynamics then depends on the movement rate of the interacting species among local communities (see Holt and Hoopes, chapter 3). Conversely, movement among communities (species sorting, assembly or mass effects) can influence greatly which local interactions actually occur.

One much-studied tritrophic interaction involves a parasitoid using a host insect that feeds on two or more host plant species (Price et al. 1980; Vet and Dicke 1992; Hochberg and Ives 2000; Tscharntke and Hawkins 2002). The pattern of host plant use by the insect host may influence the rate and success of parasitism both directly through differential host performance and indirectly through host plant signaling and apparency. Spatial variation in the attributes of host plant individuals and species may then contribute to the dynamics of parasitoids and ultimately to the dynamics of the entire metacommunity. For example, we have found that local populations of *C. melitaearum* persist for a longer time, and that the probability of establishment of new populations is greater, where the host larvae use *V. spicata* rather than *P. lanceolata* as a host plant (Van Nouhuys and Hanski 1999). There are several possible mechanisms that could lead to these effects in natural populations. Adult parasitoids may be more attracted to caterpillar-infested *V. spicata* than *P. lanceolata*; caterpillars eating *V. spicata* may be more vulnerable to parasitism; or immature parasitoid development may be retarded in *P. lanceolata*-feeding larvae. In laboratory experiments *C. melitaearum* adults appear equally willing to parasitize larvae reared on the two host plant species, and there is no difference in the number and sizes of progeny produced. However, female *C. melitaearum* are more attracted to hosts feeding on *V. spicata* than to those feeding on *P. lanceolata* (Anton 2001; Van Nouhuys and Hanski 2004). Therefore, *C. melitaearum* populations may perform better in host populations using *V. spicata* because foraging parasitoids find hosts more easily on *V. spicata* than on *P. lanceolata*.

Apparent Competition

The indirect interaction between two species that share a natural enemy is termed apparent competition, because the two species appear to have a negative impact on one another that is similar to the impact of direct competition (Holt 1977; Holt and Hoopes, chapter 3). Apparent competition may occur wherever herbivores share a pathogen, predator or parasitoid (Bonsall and Hassell 1997; Müller and Godfray 1997; Bonsall and Hassell 1998; Pope et al. 2002), and at higher trophic levels when predators or parasitoids share hyperparasitoids or their own predators. At the metacommunity level, (hyper)parasitoid-mediated apparent competition is likely to be stabilizing if host species occur in different patches and the (hyper)parasitoids aggregate, but destabilizing if host species co-occur (Hanski 1981; Bonsall and Hassell 1999; Holt and Hoopes, chapter 3).

We have used the species assemblage consisting of *M. cinxia*, the primary parasitoids *C. melitaearum* and *C. glomerata*, and the shared hyperparasitoid *G. agilis* to study apparent competition in the field (Van Nouhuys and Hanski 2000). The hyperparasitoid *G. agilis* is a very common, wingless generalist ectoparasitoid of many kinds of insects that build silken cocoons. *Gelis agilis* is found in virtually all *C. melitaearum* populations in Åland (Van Nouhuys and Tay 2001), and it aggregates locally where the host density is high (Lei and Camara 1999). *Cotesia glomerata* parasitizes *Pieris* butterflies in agricultural areas in Åland. Though it does not naturally occur in the meadows with *C. melitaearum*, the hyperparasitoid *G. agilis* uses *C. glomerata* as well as *C. melitaearum*. *Cotesia glomerata* is therefore a potential member of the food web if the movement of *G. agilis* spans both meadows and agricultural areas. We hypothesized that the addition of *C. glomerata* in *C. melitaearum* populations, close to the existing *C. melitaearum* aggregates and larval nests of *M. cinxia*, would reduce the population size of *C. melitaearum* by causing *G. agilis* to aggregate in response to high total host density, resulting in a high rate of hyperparasitism of *C. melitaearum* in the same and subsequent generations.

In the spring of the first season of the experiment cocoon clusters of *C. glomerata* were added to a randomly selected population in each of three pairs of *M. cinxia* populations occupied by *C. melitaearum*. Parasitism by *G. agilis* was later confirmed in each population by finding *G. agilis* exit holes in the cocoons of both *Cotesia* species. Following the one-time augmentation of hosts for *G. agilis* all three treatment populations of *C. melitaearum* declined, two of them to extinc-

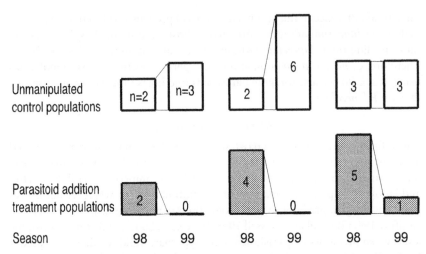

Figure 4.3 Change in *Cotesia melitaearum* population sizes, measured as the number of host larval groups parasitized in the spring between 1998, before the treated populations were augmented with *Cotesia glomerata* cocoons, and 1999, one butterfly generation and two to three parasitoid generations after the augmentation (redrawn with permission, Van Nouhuys and Hanski 2000).

tion, while the untreated control populations remained the same or increased in size (figure 4.3). We did not detect an increase in butterfly population size following the decline in parasitoid populations, but this is not surprising considering the many other factors that contribute to changes in *M. cinxia* population size (Nieminen et al. 2004).

In this experiment the cocoons of the second parasitoid were added to *C. melitaearum* populations next to the host butterfly larval groups, which likely caused the hyperparasitoid to aggregate in the patch close to the sites where *C. melitaearum* would be found subsequently. The dispersal behavior of the hyperparasitoid and the spatial distribution of potential shared host parasitoids, within and among host populations, are expected to influence the extent to which apparent competition links species within communities and metacommunities.

Metacommunity Structures

Food Chain Length and the Size of the Habitat Patch Network

As the amount of habitat available for a community decreases, the number of species decreases—this is the familiar species-area relationship (Rosenzweig 1995). Since species at higher trophic levels tend to have smaller and more extinction-prone populations than species at lower trophic levels, habitat loss and fragmentation should lead to disproportionate loss of species at higher trophic levels (Lawton and May 1995; Holt 1997, 2002). This tendency should increase with decreasing resource breadth or host range. Empirical examples include the insect community associated with bracket fungi in old-growth forests (Komonen et al. 2000) and the insect community associated with bush vetch (*Vicia sepium*) seed pods (Kruess and Tscharntke 2000).

In the case of a metacommunity living in a fragmented landscape, the expectation is that species at higher trophic levels would be absent from small, sparse, or low-quality host population networks, because the probability of a natural enemy persisting as a metapopulation in a host metapopulation decreases with decreasing size of the host metapopulation (Hanski 1999; Holt 2002; Van Nouhuys and Hanski 2002a, 2002b; Holt and Hoopes, chapter 3). The suitable habitat patches for *M. cinxia* in the Åland Islands are clustered into patch networks, between which butterfly movement is restricted by distance or barriers such as forest or sea. Therefore different networks have relatively independent butterfly metapopulations (Hanski et al. 1996; Hanski 1999). These networks, of which there are more than 100 in the Åland Islands, differ in terms of the number of patches (1 to 192, mean = 34), host plant species composition, and other attributes.

The presence of *M. cinxia* in the patch networks shows clear evidence for an extinction threshold, such that the least favorable networks as judged by their metapopulation capacities (Hanski and Ovaskainen 2000) do not have butterfly metapopulations (figure 4.4a; the metapopulation capacity is determined by the number of habitat patches in the network, and by their areas and spatial connectivities). Considering the occurrence of the parasitoid *C. melitaearum* in the same networks, only those butterfly metapopulations that occur in networks with the highest metapopulation capacities have been occupied by the parasitoid (Van Nouhuys and Hanski 2002a; figure 4.4b). On the other hand, the occurrence of the second primary parasitoid, *H. horticola*, is not similarly restricted by the properties of the patch networks or of the host butterfly metapopulations. The two parasitoids show contrasting patterns because *H. horticola* is more mobile than its host and hence little affected by habitat fragmentation, whereas *C. melitaearum* disperses shorter distances than the host and is consequently restricted to the subset of the most favorable patch networks (Van Nouhuys and Hanski 2002a).

Moving up to the fourth trophic level, the generalist hyperparasitoid *G. agilis* has a spatial population structure that is largely independent of the butterfly and the host parasitoid *C. melitaearum* at large spatial scales. However, because *G. agilis* aggregates locally in response to small-scale variation in host density, it is likely to have a great impact on *C. melitaearum* where the density of the latter is high (Lei and Camara 1999; Van Nouhuys and Tay 2001). The specialist hyperparasitoid *Mesochorus* sp. cf. *stigmaticus* parasitizes *H. horticola* larvae inside the host butterfly larvae. We expect that a fourth trophic level with a restricted host range should be absent from small isolated patch networks. Our rearing records support this prediction. The hyperparasitoid is present in most host populations in the main Åland Islands, and there is no evidence that it is limited by host population

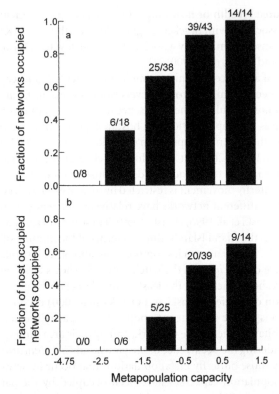

Figure 4.4 The fraction of habitat patch networks that have been occupied by *Melitaea cinxia* (a) and the butterfly-occupied networks that have been occupied by the parasitoid *Cotesia melitaearum* (b) as a function of the metapopulation capacity of the network (modified from Van Nouhuys and Hanski 2002a).

isolation at this spatial scale (Lei et al. 1997; Van Nouhuys and Hanski 2002a). On the other hand, the hyperparasitoid is absent from the butterfly metapopulations inhabiting relatively small and isolated patch networks on the islands of Kumlinge and Seglinge, at a distance of 30 km from the main islands (figure 4.1).

To summarize, the parasitoid community associated with *M. cinxia* in Åland illustrates the effects of network size, species' dispersal range, and host range on the spatial occurrence of the parasitoids. In particular, this system provides support for the notion that higher trophic levels are more limited by isolation and network size than species at lower trophic levels. Distributional patterns for two species, the primary parasitoid *C. melitaearum* and the hyperparasitoid *M.* sp. cf. *stigmaticus,* fit well the expectations based on network size. The other two species have less restricted spatial distributions, in one case (the hyperparasitoid *G. agilis*) because the species is a generalist and hence not solely dependent on the focal host

(*C. melitaearum*), and in the second case because the species (*H. horticola*) is extremely mobile and experiences the local host population (*M. cinxia*) as a single patchily distributed population rather than a fragmented metapopulation.

Linkages between Different Checkerspot Community Modules

The section on apparent competition above illustrates how population dynamics of two or more species may be dynamically linked via shared enemies. More broadly, butterfly species living in different habitat types, or even in overlapping habitats, may be linked by shared host plants and parasitoids. Considering the species assemblages around each butterfly species as a community module, these linkages potentially create more extensive metacommunities, in which direct and indirect interactions may be locally and regionally stabilizing or destabilizing, depending in part on the dispersal behavior of the species involved.

Many checkerspot species occupy overlapping habitat types, some of them even sharing the same host plant species (Wahlberg 2001; Ehrlich and Hanski 2004). They also share parasitoids, many of which parasitize only checkerspots (Van Nouhuys and Hanski 2004; Kankare and Shaw 2004; Kankare, Van Nouhuys, et al. 2005, Kankare, Stefanescu, et al. 2005). These features, along with what we have learned about the population and metapopulation ecology of the checkerspot butterflies and their host plant use (Ehrlich and Hanski 2004), make this group of butterflies an informative model system in which to address spatial aspects of food web ecology. Here we will discuss the situation in the Åland Islands in Finland with two butterfly species (Kankare, Van Nouhuys, et al. 2005), as well as in the Russian Republic of Buryatia in Siberia (Wahlberg et al. 2001) and in the Montseny mountains in Catalonia, Spain (Kankare, Stefanescu, et al. 2005), both with five butterfly species.

The Glanville fritillary *M. cinxia* and the heath fritillary *Melitaea athalia* are both relatively common in the Åland Islands. *Melitaea cinxia* is restricted to open dry meadows and it feeds on *P. lanceolata* and *V. spicata* (figure 4.2). *Melitaea athalia* uses a much more widespread habitat, forest edges, and it feeds primarily on *Veronica chamaedrys*. Both butterfly species are similar in their movement behavior and distances (Wahlberg et al. 2002), but because its habitat is more continuous, *M. athalia* has a widespread distribution in the Åland Islands whereas *M. cinxia* is restricted to the sparse network of dry meadows. The distributions of the two butterflies overlap in some *M. cinxia* habitat patches bordering forests, and occasionally larvae of the two species are even found together. Both butterflies are parasitized by *C. melitaearum*. However, phylogenetic analyses of the six recognized species of *Cotesia* parasitizing melitaeine butterflies using mitochondrial DNA (mtDNA) and microsatellite markers have revealed that each butterfly species is parasitized by a genetically distinct form of *C. melitaearum* (figure 4.2; Kankare and Shaw 2004). In behavioral experiments adult female parasitoids were only willing to parasitize the host species from which they had been reared

(Kankare, Van Nouhuys, et al. 2005). Therefore, and contrary to what traditional taxonomy would suggest, the dynamics of the two butterfly species are not linked by a shared *Cotesia* parasitoid. Much is known about the metapopulation structure of the *Cotesia* parasitizing *M. cinxia*, as we have described in the previous sections. This form is constrained by the distribution and dynamics of its host, and its occurrence is effectively limited to tightly clustered host population networks. If this wasp could use both host species it would experience a much less fragmented habitat, and could occupy a much larger faction of host populations, perhaps even influencing host metapopulation dynamics to a greater extent than it does at present. Unfortunately, nothing is known of the population structure of the *C. melitaearum* form using *M. athalia* in Åland, as this parasitoid is very difficult to sample. The two distinct forms of *C. melitaearum* must be linked by the hyperparasitoid *G. agilis* where the butterfly habitats overlap, but for the most part the respective habitats are separate, and because the wingless *G. agilis* likely disperses slowly, this coupling should be weak. There are several generalist pupal parasitoids of *M. cinxia* in Åland that are likely to also parasitize *M. athalia*, though there have been no studies of this.

In the steppe region of Buryatia in Russia, larvae of five species of checkerspot butterflies feeding on three different host plants in one habitat were collected and reared to adults by Wahlberg et al. (2001). Two species of *Cotesia* were collected from four of the butterfly species. Two butterfly species using different host plants in the same habitat were parasitized by the same genetic form of *C. melitaearum* (figure 4.5a); hence here the host butterflies are linked by a shared parasitoid. The two other butterfly species that were parasitized by *Cotesia* were each host to a distinct form or cryptic species of *Cotesia acuminata* (clades A2 and A4 in the phylogeny of Kankare and Shaw 2004; figure 4.5a). Interestingly, the three butterfly species that shared the same host plant (*Veronica incana*) did not share even closely related *Cotesia* (though not all the host-parasitoid combinations may have been sampled). Four non-*Cotesia* parasitoid species were collected from three of the butterfly species (Wahlberg et al. 2001; figure 4.5a), each of which may in fact use the other checkerspot butterflies as hosts (Van Nouhuys and Hanski 2004) but were not present in the sample.

The third metacommunity, at El Puig in the Montseny mountains in northern Spain, is similar in structure to the previous communities (figure 4.5b). There are five co-occurring species of checkerspot butterflies, three of which are parasitized by one recognized *Cotesia* species (*melitaearum*) and another one by another recognized species (*acuminata*). However, extensive sampling, genetic analyses and behavioral observations revealed again that the four host species were each parasitized by a member of different a clade in the *Cotesia* phylogeny (Kankare, Stefanescu, et al. 2005; Kankare and Shaw 2004), each possibly representing a cryptic species. In this case three of the butterfly species are linked by the shared host pant *P. lanceolata* (figure 4.5b).

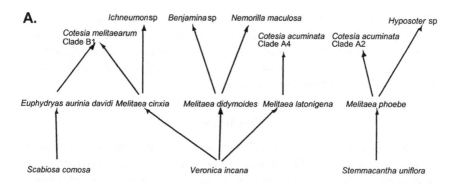

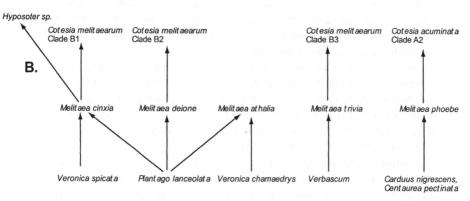

Figure 4.5 Food webs of checkerspot butterfly communities in (a) the steppe habitat in Buryatia, Russia (based on data from Wahlberg et al. 2001; Kankare and Shaw 2004), and (b) meadow habitats in El Puig in the Montseny mountains of northern Spain (based on data from Kankare and Shaw 2004; Kankare, Stefanescu, et al 2005). Note: In figure 4.5a, *Ichneumon* sp., *Benjamina* sp., and *Hyposoter* sp. may use other melitaeine hosts; *Nemorilla maculosa*, first on record on a melitaeine, may use non-melitaeine hosts; and for *Cotesia acuminata*, there is no molecular data for this parasitoid, but it most probably falls in clade A2, which includes *C. acuminata* collected from *M. phoebe* in Spain and France.

Surprisingly, *Cotesia* parasitoids appear to link species of host butterflies in only one of the three metacommunities. These checkerspot-associated *Cotesia* present a strikingly clear example of parasitoids evolving closely with their hosts, and leading to more compartmentalized ecological interactions than previously suspected. In this case evolution tends to isolate the dynamics of each community module, which have, however, remained to some extent coupled due to shared host plant species and other primary parasitoids, and between the different cryptic species of *Cotesia* and their hyperparasitoids (such as *Gelis*). Other parasitoids in these communities have genuinely broader host ranges, and they serve to link the population dynamics of their host species. For instance, the chalcid pupal parasitoids of *M. cinxia* in Åland (figure 4.2) parasitize many other host species

(Lei et al. 1997), as do tachinid flies in many checkerspot communities (Van Nouhuys and Hanski 2004). It is however noteworthy that the interaction between the host butterflies and *Cotesia* parasitoids is the potentially strongest host-parasitoid interaction in these communities, occasionally even driving local host populations to extinction (Lei and Hanski 1997).

Conclusions

We have described in this chapter the operation of several processes in butterfly metacommunities using *M. cinxia* and its host plants and parasitoids in Finland as an example. One important result to emerge from these studies is the coupling of behavioral, ecological, and evolutionary dynamics in the metacommunity context. A prime example is the contribution of variation in host plant species composition and female butterfly oviposition preference throughout the Åland Islands to metacommunity dynamics.

Another point that we have illustrated is that both local-scale processes operating within communities and large-scale processes operating among relatively independent communities may contribute to metacommunity structure and dynamics (as in Miller and Kneitel, chapter 5). Different scales may dominate in particular interactions. A case in point is the two parasitoids *C. melitaearum* and *H. horticola* sharing the same host in the same landscape. In this case the among-community processes appear to be immaterial for the interaction of the two competing parasitoids, though among-community processes explain the dissimilar distributions and dynamics of the host and the two parasitoids. The large-scale population dynamics of the butterfly, largely driven by the spatial configuration of the habitat and spatially correlated weather (Hanski 1999), affects the dynamics of the two primary parasitoids in a very different manner because of their contrasting dispersal behaviors.

We have compared the small food webs or community modules in the Åland Islands with similar modules in other checkerspot communities to explore the degree to which host species can be linked into larger metacommunities by shared resources and enemies. In all of the communities butterflies are linked by shared host plants and probably also by shared generalist enemies. Many of the butterflies appear superficially to be linked by shared parasitoids in the genus *Cotesia*. However, detailed molecular analyses and studies of host acceptance behavior indicate that most of these wasps are actually even more host specific than previously thought, and the recognized species appear to include several cryptic species, each specialized to use a single host species. Our current understanding of the structures of these butterfly-parasitoid metacommunities makes it clear that we cannot assume that identifying species by name is necessarily sufficient for understanding the ecological links among them.

A still largely open question is the population dynamic significance of the less obvious indirect interactions in which the focal butterfly species are engaged. One example is the plant pathogenic fungus that shares the host plant *P. lanceolata* with *M. cinxia* and may interfere with larval feeding, and influences butterfly and plant metapopulation dynamics (Laine 2004a, 2004b). Another example is seed-feeding beetles, which may have a substantial impact on the recruitment of the host plant (Nieminen et al. 2004). Another area where more research is needed, and which is central to our interests in this volume, is the role of habitat fragmentation in molding the dynamics of entire metacommunities. Conducting empirical research on many species at a large spatial scale poses obvious problems; replicating such studies across landscapes with dissimilar structures means even more challenging work. Such research is however badly needed to answer some of the fundamental questions in the interface between the biology of metacommunities and changing landscapes structures.

Literature Cited

Abrahamson, W. G., and A. E. Weis. 1997. *Evolutionary ecology across three trophic levels goldenrods, gallmakers, and natural enemies*. Princeton University Press, Princeton, NJ.

Amarasekare, P. 2000. Coexistence of competing parasitoids on a patchily distributed host: Local vs. Spatial mechanisms. Ecology 81:1286–1296.

Anton, C. 2001. Host selection and performance of the parasitoid *Cotesia melitaearum* in tritrophic context. Diploma thesis. Institute for Ecology, Friedrich-Schiller University, Jena, Germany.

Armstrong, R. A. 1976. Fugitive species: Experiments with fungi and some theoretical considerations. Ecology 57:953–963.

Askew, R. R., and M. R. Shaw. 1989. Parasitoid communities: Their size, structure and development. Pages 225–26 *in* J. Waage and D. Greathead, eds. *Insect parasitoids*. Academic Press, Cambridge, UK.

Bernays, E. A., and C. De Luca. 1981. Insect anti-feedant properties of an iridoid glycoside: Ipolamiide. Experimentia 37:1289–1290.

Bonsall, M. B., and M. P. Hassell. 1997. Apparent competition structures ecological assemblages. Nature 388:371–378.

———. 1998. Population dynamics of apparent competition in a host-parasitoid assemblage. Journal of Animal Ecology 67:918–929.

———. 1999. Parasitoid-mediated effects: Apparent competition and the persistence of host-parasitoid assemblages. Researches in Population Ecology 41:59–68.

Boughton, D. A. 2000. The dispersal of a butterfly: A test of source-sink theory suggests the intermediate-scale hypothesis. American Naturalist 156:131–144.

Bowers, M. D. 1983. The role of iridoid glycosides in host-plant specificity of checkerspot butterflies. Journal of Chemical Ecology 9:475–493.

Collinge, S. K., K. I. Prudic and J. Oliver. 2003. Effect of local habitat characteristics and landscape context on grassland butterfly diversity. Conservation Biology 17:178–187.

Courtney, S. P., and F. S. Chew. 1987. Coexistence and host use by a large community of Pierid butterflies: Habitat is the template. Oecologia 71:210–220.

Dempster, J. P. 1983. The natural control of populations of butterflies and moths. Biological Review 58:461–481.

Dyer, L. A., and D. K. Letourneau. 2003. Top-down and bottom-up diversity cascades in detrital vs. living food webs. Ecology Letters 6:60–68.

Ehrlich, P. R., and I. Hanski. 2004. *On the wings of checkerspots: A model system for population biology.* Oxford University Press, Oxford, UK.

Hanski, I. 1981. Coexistence of competitors in patchy environment with and without predation. Oikos 37:306–312.

———. 1994. A practical model of metapopulation dynamics. Journal of Animal Ecology 63:151–162.

———. 1999. *Metapopulation ecology.* Oxford University Press, Oxford, UK.

Hanski, I. and Gaggiotti, O. E. 2004. *Ecology, genetics and evolution of metapopulations.* Elsevier Academic Press, London, UK.

Hanski, I. and M. Heino. 2003. Metapopulation-level adaptation of insect host plant preference and extinction-colonization dynamics in heterogeneous landscapes. Theoretical Population Biology 64:281–290

Hanski, I., J. J. Hellmann, C. L. Boggs, and J. F. McLaughlin. 2004. Checkerspots as a model system in population biology. Pages 245–263 *in* P. R. Ehrlich, and I. Hanski, eds. *On the wings of checkerspots: A model system for population biology.* Oxford University Press, Oxford, UK.

Hanski, I., and M. Kuussaari. 1995. Butterfly metapopulation dynamics. Pages 149–171 *in* P. Price and N. Cappuccino, eds. *Population dynamics: New approaches and synthesis.* Academic Press, San Diego, CA.

Hanski, I., M. Kuussaari, and M. Nieminen. 1994. Metapopulation structure and migration in the butterfly *Melitaea cinxia.* Ecology 75:747–762.

Hanski, I., A. Moilanen, T. Pakkala and M. Kuussaari. 1996. The quantitative incidence function model and persistence of an endangered butterfly metapopulation. Conservation Biology 10:578–590.

Hanski, I. and O. Ovaskainen. 2000. Metapopulation capacity of a fragmented landscape. Nature 404:755–758.

Hanski, I., and M. C. Singer. 2001. Extinction-colonization dynamics and host-plant choice in butterfly metapopulations. American Naturalist 158:341–353.

Harrison, S., D. D. Murphy, and P. R. Ehrlich. 1988. Distribution of the bay checkerspot butterfly, *Euphydryas editha bayensis:* Evidence for a metapopulation model. American Naturalist 132:360–382.

Harrison, S. and C. D. Thomas. 1991. Patchiness and spatial pattern in the insect community on ragwort *Senecio jacobaea.* Oikos 62:5–12.

Hastings, A. 1980. Disturbance, coexistence, history and the competition for space. Theoretical Population Biology 18:363–373

Hawkins, B. A. 1994. *Pattern and process in host-parasitoid interactions.* Cambridge University Press, Cambridge, UK.

Hawkins, B. A. 2000. Species coexistence in parasitoid communities: Does competition matter? Pages 198–214 *in* M. E. Hochberg and A. R. Ives, eds. *Parasitoid population biology.* Princeton University Press, Princeton, NJ.

Hawkins, B. A., M. R. Shaw, and R. R. Askew. 1992. Relations among assemblage size, host specialization, and climatic variability in North American parasitoid communities. American Naturalist 139:58–79.

Hawkins, B. A., and W. Sheehan. 1994. Parasitoid community ecology. Oxford University Press, Oxford, UK.

Henneman, M. L., and J. Memmott. 2001. Infiltration of a Hawaiian community by introduced biological control agents. Science 293:1314–1316.

Hochberg, M. E., and A. R. Ives. 2000. *Parasitoid Population Biology.* Page 366. Princeton University Press, Princeton, NJ.

Holt, R. D. 1977. Predation, apparent competition and the structure of prey communities. Theoretical Population Ecology 12:197–229.

————. 1997. From Metapopulation dynamics to community structure. Pages 149–165 *in* I. Hanski and M. E. Gilpin, eds. Metapopulation biology. Academic Press, San Diego. CA.

————. 2002. Food webs in space: On the interplay of dynamic instability and spatial processes. Ecological Research 17:261–273.

Holyoak, M. 2000. Comparing parasitoid-dominated food webs with other food webs: Problems and future promises. Pages 184–197 *in* M. E. Hochberg and A. R. Ives, eds. *Parasitoid population biology.* Princeton University Press, Princeton, NJ.

Hopper, K. 1984. The effects of host-finding and colonization rates on abundance of parasitoids of a gall midge. Ecology 65:20–27.

Horner-Devine, C. M., G. C. Daily, P. R. Ehrlich, and C. L. Boggs. 2003. Countryside biogeography of tropical butterflies. Conservation Biology 17:168–177.

Kankare, M., and M. R. Shaw. 2004. Molecular phylogeny of *Cotesia* Cameron, 1891 (Insecta: Hymenoptera: Braconidae: Microgastrinae) parasitoids associated with Melitaeini butterflies (Insecta: Lepidoptera: Nymphalidae: Melitaeini). Molecular Phylogenetics and Evolution 32:207–220.

Kankare, M., C. Stefanescu, S. van Nouhuys, and M. R. Shaw. 2005. Host specialization by *Cotesia* wasps (Hymenoptera: Braconidae) parasitizing species rich Melitaen (Lepidoptera: Melitaeine) communities in northeastern Spain. Biological Journal of the Linnean Society, in press.

Kankare, M., S. van Nouhuys, I. Hanski. 2005. Genetic divergence among host specific cryptic species in Cotesia Melitaearum agg. a parasitoid of checkerspot butterflies. Annals of the Entomological Society of America, in press.

Kitahara, M., and K. Sei. 2001. A comparison of the diversity and structure of butterfly communities in semi-natural and human-modified habitats at the foot of Mt. Fuji, central Japan. Biodiversity and Conservation 10:331–351.

Komonen, A., R. Penttilä, M. Lindgren, and I. Hanski. 2000. Forest fragmentation truncates a food chain based on an old-growth forest bracket fungus. Oikos 90:119–126.

Kruess, A., and T. Tscharntke. 2000. Species richness and parasitism in a fragmented landscape: experiments and field studies with insects on *Vicia sepium*. Oecologia 122:129–137.

Kuussaari, M., M. C. Singer, and I. Hanski. 2000. Local specialization and landscape-level influence on host use in a herbivorous insect. Ecology 81:2177–2187.

Kuussaari, M., S. van Nouhuys, J. Hellmann, and M. C. Singer. 2004. Larval biology of checkerspot butterflies. Pages 138–160 *in* P. R. Ehrlich, and I. Hanski, eds. *On the wings of checkerspots: A model system for population biology*. Oxford University Press, Oxford, UK.

Laine, A.-L. 2004a. A powdery mildew infection on a shared host plant affects the dynamics of the Glanville fritillary butterfly populations. Oikos 107 (2):329–337.

————. 2004b. Resistance variation within and among host populations in a plant pathogen metapopulation—implications for regional pathogen dynamics. Journal of Ecology, in press.

Lawton, J. H., and R. M. May. 1995. *Extinction rates.* Oxford University Press, Oxford, UK.

Lei, G. C. and M. D. Camara. 1999. Behaviour of a specialist parasitoid, *Cotesia melitaearum:* From individual behavior to metapopulation processes. Ecological Entomology 24:59–72.

Lei, G. C., and I. Hanski. 1997. Metapopulation structure of *Cotesia melitaearum*, a specialist parasitoid of the butterfly *Melitaea cinxia.* Oikos 78:91–100.

————. 1998. Spatial dynamics of two competing specialist parasitoids in a host metapopulation. Journal of Animal Ecology 67:422–433.

Lei, G. C., V. Vikberg, M. Nieminen, and M. Kuussaari. 1997. The parasitoid complex attacking the Finnish populations of Glanville fritillary *Melitaea cinxia* (Lep: Nymphalidae), an endangered butterfly. Journal of Natural History 31:635–648.

Levins, R., and D. Culver. 1971. Regional coexistence of species and competition between rare species. Proceedings of the National Academy of Sciences, U.S.A. 68:1246–1248.

Lewis, O. T., J. Memmott, J. Lasalle, C. H. C. Lyal, C. Whitefoord, and H. C. J. Godfray. 2002. Structure of a diverse tropical forest insect-parasitoid community. Journal of Animal Ecology 71:855–873.

Marino, P. C. 1991. Dispersal and coexistence of mosses (Splachnaceae) in patchy habitats. Journal of Ecology 79:1047–1060.

Memmott, J., H. C. J. Godfray, and I. D. Gauld. 1994. The structure of a tropical host-parasitoid community. Journal of Animal Ecology 63:521–540.

Mills, N. J. 1993. Species richness and structure in the parasitoid complexes of tortricid hosts. Journal of Animal Ecology 62:45–58.

Müller, C. B. and H. C. J. Godfray. 1997. Apparent competition between two aphid species. Journal of Animal Ecology 66:57–64.

Murphy, D. D., K. E. Freas, and S. B. Weiss. 1990. An environment-metapopulation approach to population viability analysis for a threatened invertebrate. Conservation Biology 4:41–51.

Nee, S., and R. M. May. 1992. Dynamics of metapopulations: Habitat destruction and competitive coexistence. Journal of Animal Ecology 61:37–40.

Nee, S., R. M. May, and M. P. Hassell. 1997. Two Species Metapopulation Models. Pages 123–148 in I. Hanski and M. E. Gilpin, eds. Metapopulation biology. Academic Press, San Diego, CA.

Nieminen, M., M. Siljander, and I. Hanski. 2004. Structure and dynamics of Melitaea cinxia metapopulations. Pages 63–91 in P. R. Ehrlich, and I. Hanski, eds. On the wings of checkerspots: A model system for population biology. Oxford University Press, Oxford, UK.

Pajunen, V. I. 1979. Competition between rock pool corixids. Annales Zoologici Fennici 16:138–143.

————. 1982. Replacement analysis of non-equilibrium competition between rock pool Corixids (Hemiptera: Corixidae). Oecologia 52:153–155.

Pope, T., E. Croxson, J. K. Pell, and H. C. J. Godfray. 2002. Apparent competition between two species of aphid via the fungal pathogen Erynia neoaphidis and its interaction with the aphid parasitoid Aphidious ervi. Ecological Entomology 27:196–203.

Punju, E. 2002. Larval competition between two specialist parasitoids Cotesia melitaearum and Hyposoter horticola inside the hose larva. Diploma thesis. Department of Ecology and Systematics, University of Helsinki, Finland.

Price, P. W., C. E. Bouton, P. Gross, B. A. McPheron, J. N. Thompson, and A. E. Weis. 1980. Interaction among three trophic levels: Influence of plants on interactions between insect herbivores and natural enemies. Annual Review of Ecology and Systematics 11:41–65.

Roininen, A., P. W. Price, and J. Tahvanainen. 1996. Bottom-up and top-down influences in the trophic system of a willow, a galling sawfly, parasitoids and inquilines. Oikos 77:44–50.

Rosenzweig, M. L. 1995. Species Diversity in Space and Time. Cambridge University Press, Cambridge, UK.

Shaw, M. R. 1994. Parasitoid host ranges. Pages 111–144 in B. A. Hawkins and W. Sheehan, eds. Parasitoid community ecology. Oxford University Press, Oxford, UK.

Shaw, M. R., and M. G. Fitton. 1989. Survey of parasitoids of British butterflies. Entomologist's Record 101:69–71.

Sheehan, W. 1994. Parasitoid community structure: Effects of host abundance, phylogeny, and ecology. Pages 90–110 in B. A. Hawkins and W. Sheehan, eds. Parasitoid community ecology. Oxford University Press, Oxford, UK.

Singer, M. C. and Hanski, I. 2004. Oviposition preference: Its measurement, its correlates and its importance in the life of checkerspots. Pages 181–198 in P. R. Ehrlich, and I. Hanski, eds. On the wings of checkerspots: A model system for population biology. Oxford University Press, Oxford, UK.

Singer, M. C., D. Ng, D. Vasco, and C. D. Thomas. 1992. Rapidly evolving associations among oviposition preferences fail to constrain evolution of insect diet. American Naturalist 139:9–20.

Singer, M. C., and C. D. Thomas. 1987. Variations in host preference affects movement patterns within a butterfly population. Ecology 68:1262–1267.

Steffan-Dewenter, I., and T. Tscharntke 2000. Butterfly community structure in fragmented habitats. Ecology Letters 3:449–456.

Stireman III, J. O., and M. S. Singer. 2003. Determinants of parasitoid-host associations: insights from a natural Tachinid-Lepidoptera community. Ecology 84:296–310.

Sullivan, D. J., and W. Völk. 1999. Hyperparasitism: Multitrophic ecology and behaviour. Annual Review of Entomology 44:291–315.

Thomas, C. D., and I. Hanski. 1997. Butterfly metapopulations. Pages 359–386 in I. Hanski and M. E. Gilpin, eds. Metapopulation biology. Academic Press, San Diego, CA.

———. 2004. Metapopulation dynamics in changing environments: Butterfly responses to habitat and climate change. Pages 489–514 in I. Hanski and O. Gaggiotti, eds. Ecology, genetics and evolution of metapopulations. Academic Press, San Diego.

Thomas, C. D., D. Vasco, M. C. Singer and D. Ng. 1990. Diet divergence in two sympatric congeneric butterflies: Community or species level phenomenon? Evolutionary Ecology 4:62–74.

Tscharntke, T. 1992. Coexistence, tritrophic interactions and density dependence in a species-rich parasitoid community. Journal of Animal Ecology 61:59–67.

Tscharntke, T., and B. A. Hawkins. 2002. Multitrophic Level Interactions. Cambridge University Press, Cambridge, UK.

Van Nouhuys, S., and J. Ehrnsten. 2004. Wasp behavior that leads to uniform parasitism of a host available only a few hours per year. Behavioral Ecology 15:661–665.

Van Nouhuys, S., and I. Hanski. 1999. Host diet affects extinctions and colonizations in a parasitoid metapopulation. Journal of Animal Ecology 68:1–12.

———. 2000. Apparent competition between parasitoids mediated by a shared hyperparasitoid. Ecology Letters 3:82–84.

———. 2002a. Colonisation rates and distances of a host butterfly and two specific parasitoids in a fragmented landscape. Journal of Animal Ecology 71:630–650.

———. 2002b. Multitrophic interactions in space: Metacommunity dynamics in fragmented landscapes. Pages 124–147 in T. Tscharntke, and B. A. Hawkins, eds. Multitrophic level interactions. Cambridge University Press, Cambridge, UK.

———. 2004. Natural enemies of checkerspot butterflies. Pages 161–180 in P. R. Ehrlich, and I. Hanski, eds. On the wings of checkerspots: A model system for population biology. Oxford University Press, Oxford, UK.

Van Nouhuys, S., and G. C. Lei. 2004. Parasitoid and host metapopulation dynamics: The influences of temperature mediated phenological asynchrony. Journal of Animal Ecology 73:526–535.

Van Nouhuys, S., M. C. Singer, and M. Nieminen. 2003. Spatial and temporal patterns of caterpillar performance and the suitability of two host plant species. Ecological Entomology 28:193–202.

Van Nouhuys, S., and W. T. Tay. 2001. Causes and consequences of mortality in small populations of a parasitoid wasp in a fragmented landscape. Oecologia 128:126–133.

Vet, L. E. M., and M. Dicke. 1992. Ecology of infochemicals used by natural enemies in a tritrophic context. Annual Review of Entomology 37:141–172.

Wahlberg, N. 2001. The phylogenetics and biochemistry of host plant specialization in melitaeine butterflies (Lepidoptera: Nymphalidae). Evolution 55:522–537.

Wahlberg, N., T. Klemetti, V. Selonen, and I. Hanski. 2002. Metapopulation structure and movements in five species of checkerspot butterflies. Oecologia 130:33–43.

Wahlberg, N., J. Kullberg, and I. Hanski. 2001. Natural history of some Siberian melitaeine butterfly species (Melitaeini: Nymphalidae) and their parasitoids. Entomologica Fennica 12:72–77.

Chapter 5

Inquiline Communities in Pitcher Plants as a Prototypical Metacommunity

Thomas E. Miller and Jamie M. Kneitel

Introduction

It has been argued that ecology is "awash in all manner of untested (and often untestable) models" with theory being developed too far ahead of its empirical support (Simberloff 1981). While this is an extreme view, it does appear that metacommunity theory has now matured sufficiently to require evaluation by means of experimental work, as Holyoak et al. and Law and Amarasekare argue in this volume. Experiments can test basic assumptions and predictions of metacommunity theory, while providing direction for future theory. For example, current metacommunity theory assumes that an understanding of some communities requires understanding the relative contributions of local and regional forces. However, if either force dominates to a large degree, simpler models are adequate to predict community dynamics and the metacommunity view is unnecessary. Such ties between theory and experimental work are critical to nurture the metacommunity concept as it develops.

The community found in water-filled leaves of the pitcher plant *Sarracenia purpurea* is an ideal system for testing assumptions and predictions associated with the metacommunity framework (see, e.g., Kneitel and Miller 2002, 2003). Local communities of invertebrates and bacteria in each leaf are discrete and easily manipulated. Species that inhabit pitcher-plant leaves have relatively short generation times (two hours to two weeks), allowing relatively short experiments to capture the dynamics of this system. The discrete nature of individual leaves, leaves on plants, and different plants in different bogs allows us to examine community variation across a variety of scales and the interactions of different processes at these scales.

We have used the inquiline system to begin to address a number of important questions about metacommunities. What constitutes a local community? What defines the boundaries of the larger metacommunity? What are the relative effects of species interactions and migration within and among communities? Are the rates of migration among discrete local communities appropriate for the metacommunity concept to be useful? Further, metacommunities are defined by the interaction between habitat and species, but different species, guilds, and trophic levels may interact with the environment at different spatial scales (Holt 1996).

Ultimately, how do these interactions affect the value of the metacommunity concept?

Our work in the pitcher-plant inquiline community has examined processes that occur at a variety of spatial scales: within an individual leaf, among leaves within an individual plant, among plants within a population, and among plant populations across the full geographic range of the species. This chapter presents an overview of such work by us and others, framed within the important components of metacommunity theory. After a description of the natural history of the community as a whole, we discuss spatial scales at which this community is appropriately studied, present natural patterns of variation and experiments conducted at each scale, and address the appropriateness of the metacommunity framework to each scale.

Natural History of Inquiline Communities

The purple pitcher plant, *Sarracenia purpurea*, is generally restricted to low-pH bogs from southern Canada to Florida (figure 5.1). The systematics of the species and, in fact, the entire genus are not well understood (Bayer et al. 1996). Two subspecies are recognized; *S. purpurea* ssp. *purpurea* is found north of Maryland and *S. purpurea* ssp. *venosa* from Maryland south at least through South Carolina (Schnell 1979; but see also Reveal 1993). Naczi et al. (1999) have recently proposed that the southernmost, somewhat disjunct, populations in northern Florida, southern Alabama, and southern Mississippi be considered a separate species, *Sarracenia rosea* (this group was sometimes formerly described as *S. purpurea* ssp. *venosa* var. *burkii*). As no differences in the inquiline communities associated with these subspecies are known (Buckley et al. 2003), we refer to the whole group as *S. purpurea* in the present chapter.

New leaves are produced by plants during much of the year, particularly just after flowering in the summer through the fall. The leaves are generally cup-shaped, with taller, narrower leaves produced in warmer climates (Ellison et al. 2004). Each leaf can contain from 0 to over 50 ml of rainwater and, in north Florida, survive for a year or longer. Leaves capture a large variety of invertebrate prey that may be attracted to nectar or by leaf coloration and drown in the water. The capture rate of insect visitors is apparently quite low (more than 99% of the visitors to each leaf escape; Newell and Nastase 1998); the successfully captured prey is composed primarily of ants (Heard 1998; Miller et al. 1994) but includes dipterans, collembolans (Cresswell 1991), and spiders (TEM, pers. observ.). The leaves of *S. purpurea* may produce digestive enzymes that aid in breaking down captured prey, especially in the first ten days after the leaf opens (Gallie and Chang 1997). Leaves vary in their ability to capture prey; larger pitchers may attract and consume more prey (Wolfe 1981; Cresswell 1993; but see Newell and Nastase 1998).

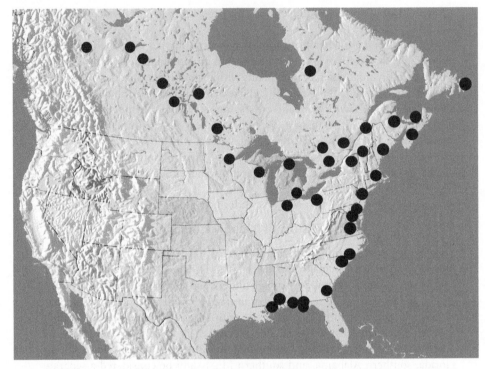

Figure 5.1 The distribution of *Sarracenia purpurea* across North America. Circles indicate populations that were sampled by Buckley et al. (2003), chosen to cover the range of the plant's distribution.

The individual leaves also act as hosts for an active community of nonprey invertebrates (inquilines), which are the subject of the present chapter (figure 5.2). Virtually all components of the community can appear within days after the leaf opens (TEM, pers. observ.; Judd 1959; but see Fish and Hall 1978). Inquiline species include several dipterans that occupy pitchers during their larval development: the pitcher-plant mosquito (*Wyeomyia smithii* [Coq.]), midge (*Metriocnemus knabi* [Coq.]), and sarcophagid (*Fletcherimyia fletcheri* [Aldrich]). These species are univoltine in the northern distribution of *S. purpurea* but may go through several generations per year in the southeastern United States (Bradshaw and Lounibos 1972). The dipterans, along with the pitcher plant mite (*Sarraceniopus gibsoni* Nesbitt; Fashing and O'Conner 1984) and a bdelloid rotifer (*Habrotrocha rosa* Donner; Bateman 1987; Petersen et al. 1997; Bledzki and Ellison 1998), are found almost exclusively in *Sarracenia purpurea*. A number of other species are also commonly found in pitchers, including many species of protozoans (Addicott 1974; Cochran-Stafira and von Ende 1998), copepods, cladocerans (Miller et al. 1994; Hamilton et al. 2000), and bacteria (Addicott 1974; Prankevicius and Cameron 1989, 1991; Cochran-Stafira and von Ende 1998).

Each leaf must undergo succession after it is formed, so one might expect community (or leaf) age to explain much of the variation in community composition. In a greenhouse study, Fish and Hall (1978) found that individual leaves underwent a very regular progression of assembly of bacteria, protozoa, and dipterans. They suggested that, after 100 days, the community essentially dies when all available resources are consumed and the univoltine dipteran larvae pupate and leave the leaf. In habitats with short growing seasons, such patterns may occur (but see Judd 1959), but we have found little evidence of such strong successional patterns in pitchers from north Florida. Our preliminary data suggest that leaves attract large numbers of ants and mosquitoes in the first few weeks after opening but that they continue to attract some prey and mosquitoes until the leaf senesces or is so damaged that it cannot hold water (figure 5.3). Average leaf age is over 180 days, and many leaves survive more than a year. We suggest that, as pitchers get older, the effects of successional age on the inquiline community become relatively minimal compared to the effects of stochastic prey capture, mosquito deposition, and climate.

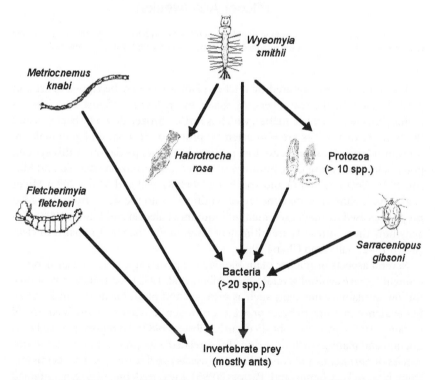

Figure 5.2 General food web of the inquilines found in *Sarracenia purpurea*. The species shown are numerically dominant both within and among leaves in populations *S. purpurea;* other species, such as copepods, cladocerans, and nematodes, are generally rare. The plant's invertebrate prey (rather than primary productivity) serves as the base of the web, and mosquito larvae are at the top.

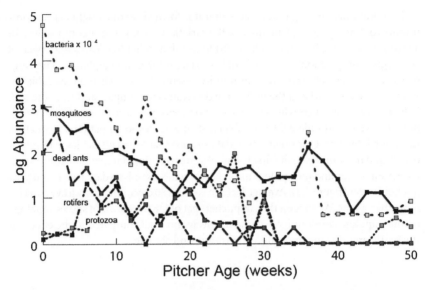

Figure 5.3 The pattern of succession of inquiline communities in *Sarracenia purpurea* leaves in the Crystal Swamp, Florida. Leaves were marked before opening and sampled at weekly intervals.

Invertebrate prey captured by pitcher-plant leaves may be ripped apart and partially eaten by the scavengers, *M. knabi* and *F. fletcheri* (figure 5.2). The remains provide particulate matter, which is further broken down by bacteria, and the bacteria themselves are also eaten by protozoa, *H. rosa*, and zooplankton. Mosquito larva, *W. smithii*, are top, perhaps keystone, predators in this system, potentially consuming large numbers of rotifers and protozoa (Bledzki and Ellison 1998; Cochran-Stafira and von Ende 1998; Kneitel and Miller 2002; Miller et al. 2002). Nutrients are either released directly into the water from degraded prey or released through excrement of inquilines (Bledzki and Ellison 1998). Exactly how these nutrients are ultimately taken up by the plant remains unclear (Lloyd 1942; Gallie and Chang 1997).

Several aspects of pitcher-plant biology are poorly understood. Carnivory is thought to have evolved several times (Albert et al. 1992) as an alternative source for low-availability nutrients such as nitrogen and phosphorus (Givnish 1989), but evidence that invertebrate prey are a nitrogen source for plants is equivocal (Adamec 1997). In an unpublished work, Gibson (1983) did report that seed production and plant growth were directly correlated with prey-capture rates in two species of *Sarracenia* and concluded that insects provide some nutrients missing from bog soils. Chapin and Pastor (1995) measured the nitrogen obtained through prey capture in *S. purpurea* and found that it constituted as little as 10% of the total amount required. Similarly, both Eleuterius and Jones (1969) and

Christensen (1976) found that prey capture had little beneficial effect on plant growth. Although insects are generally thought to be the major carbon source for inquiline communities, at least two studies have mentioned the presence of minor amounts of filamentous green algae in some pitchers (Heard 1994b; Miller et al. 1994), and the bacterial assemblage may include a small number of photosynthetic species (Bradshaw and Creelman 1984; Cochran-Stafira and von Ende 1998) and nitrogen-fixing flavobacteria (Prankevicius and Cameron 1989; R. H. Reeves, pers. comm.). Clearly, the full relationship between the plant and the inquiline community found in its leaves has yet to be fully explored.

Within Pitcher: Local Dynamics

Ultimately, all community dynamics within an individual leaf are limited by productivity or the energy that comes into the system (Power 1992). Inquiline communities are distinctive in that they seldom include primary productivity; instead, they depend on detritus from the prey captured by the pitcher-plant leaf. Several investigators have studied bottom-up effects in inquiline communities by adding dead insects and have generally obtained very similar results. Increasing prey abundance in natural inquiline communities leads to increased standing abundances of bacteria (cells/ml; see, e.g., figure 5.4; Kneitel and Miller 2002; Miller et al. 2002). This increase appears in turn to affect the next trophic level; both protozoa and rotifer abundances generally increase with prey addition. Increased prey availability also leads to increased bacterial and protozoan diversity, but whether the mechanism is decreased competition or other factors such as increased prey heterogeneity is unclear. Nutrient-addition experiments suggest that ants and other prey captured by pitcher plants are providing limiting carbon, rather than nitrogen or phosphorus, for the bacteria (Gray et al. in preparation).

Resource effects may be passed up to the highest trophic level, larvae of the mosquito *W. smithii*. Mosquito growth rates have been shown to increase with food availability (Istock et al. 1975; Farkas and Brust 1985) and to decrease with intraspecific density (Istock et al. 1975; Bradshaw and Holzapfel 1986, 1992; Miller et al. 1994; Broberg and Bradshaw 1995). Resource effects may vary with latitude; *Wyeomyia smithii* from southern populations have been shown to grow faster at high densities than those from more northern populations (Bradshaw and Holzapfel 1989). Laboratory experiments have shown that increasing resources increases larval size and decreases time to pupation (Istock et al. 1975; Kneitel, submitted). However, in two field experiments, increasing prey levels significantly increased bacterial and protozoan abundances without affecting mosquito size (Miller et al. 2002; Gray et al. in prep.). The sarcophagid fly *Fletcherimyia fletcheri* may also be influenced by resource levels and intraspecific interactions (Forsyth and Robertson 1975), but the densities of their potentially cannibalistic larva are generally very low (less than one per leaf) unless supplemental food is provided (Rango 1999). Overall, the bottom-up control of dipterans by basal resources is generally indirect and diffuse and may often be relatively minor.

The relationship between the mosquito *W. smithii* and the midge *M. knabi* is complex. The presence of *M. knabi* has been shown to increase growth rates in *W. smithii*, apparently either directly by increasing the amount of particulate food in

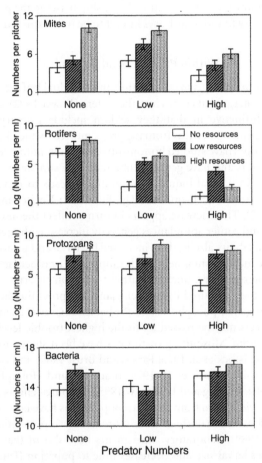

Figure 5.4 Effects of resources and predators on the abundances of several inquiline species in *Sarracenia purpurea* found in the Crystal Bog, Florida. We controlled resources by maintaining 0 (none), 5 (low), or 25 (high) ant carcasses per leaf; we controlled predator levels by maintaining 0 (none), 2 (low), or 10 (high) *Wyeomyia smithii* larvae per leaf. Data shown are means and standard errors after twenty-three days (from Kneitel and Miller 2002). Mite and protozoan populations increased with increasing resource availability, but other groups were affected by complex combinations of indirect effects.

the water or indirectly by increasing bacterial numbers (Heard 1994a). *Metrioc-nemus knabi* may also feed selectively on larvae of other mosquito species, eliminating potential competitors of *W. smithii* (Petersen et al. 2000).

The composition of local communities is also affected by predation at several trophic levels. Experiments have demonstrated that a reduction in the abundance of the top predator, larvae of *W. smithii*, generally leads to increased abundance of their preferred prey, the rotifer *H. rosa*, as well as increased abundances of other prey, notably protozoa (figure 5.4; Addicott 1974; Cochran-Stafira and von Ende 1998; Kneitel and Miller 2002; Miller et al. 2002). Bledzki and Ellison (1998) documented that rotifer consumption by mosquitoes and sarcophagid flies was in direct proportion to rotifer density. In some cases, the effects of mosquitoes have been shown to cascade down trophic levels; increasing abundance of mosquito larvae has also been associated with increases in bacterial abundance (Miller et al. 2002) and richness (Kneitel and Miller 2002; Miller et al. 2002), apparently through decreases in the abundance of consumers in the middle trophic level.

Clearly, both bottom-up (resource-availability) and top-down (predation) forces are important in this community, as is probably true in many communities (Power 1992; Osenberg and Mittelbach 1996; Stiling and Rossi 1997). Resource availability and consumption by top predators could interact in a variety of ways through nonadditive effects on intermediate trophic levels. However, we have found that, in general, resources and predation have additive effects on abundances of bacteria, protozoa, and rotifers in inquiline communities (Kneitel and Miller 2002; Miller et al. 2002). An interaction between top-down and bottom-up forces was found only for protozoan species richness; predation decreased richness only at very low resource levels (Kneitel and Miller 2002).

Metacommunity Context: Our current view of metacommunities requires both local and regional processes and, in particular, that local forces such as competition and predation contribute to community patterns at local scales. Clearly, species interactions are important in inquiline communities; both predation and resource limitation operate to determine species abundances and richness. The question is therefore whether interactions at larger scales, such as migration and habitat destruction, are also important and require the integrative metacommunity perspective.

Among-Community Dynamics—Regional Forces

The communities found in different leaves in the same area vary widely in the presence and abundances of inquiline species (Harvey and Miller 1996). Some of this variation may be due to community age (figure 5.3), although age of the leaf is not always a good predictor of community composition (Miller et al. 1994; Buckley et al., 2004). More likely, variation among the communities found in leaves in the same area is due to aspects of the leaf that influence prey capture (e.g., leaf height; Cresswell 1991, 1993) or inquiline migration, especially of predators (Heard 1994a). Some of the variation may also arise from the inherently stochastic nature of prey capture and migration.

Migration can have several different effects on local communities in a metacommunity (Leibold and Miller 2004; Holyoak et al., chapter 1). First, low rates of migration can overcome migration limitation and allow species to disperse to new habitats. Second, migration among habitats with different conditions may allow species sorting, in which the conditions in each local community favor the growth and dominance of different species (Chesson 2000). Third, if rates are sufficiently high, migration may augment local population growth, allowing mass (Mouquet and Loreau 2002) or rescue (Brown and Kodric-Brown 1977) effects. Finally, relatively high rates of migration can have a homogenizing effect, allowing a smaller group of species to dominate regionally (Mouquet and Loreau 2002).

In pitcher plants, we think that at least the first two effects of migration, dispersal to open habitats and species sorting, are important. We know that initial colonists reach new leaves within days, if not hours, after they fill with water (figure 5.3; see also Judd 1959), but we know almost nothing about mechanisms of natural dispersal and the origins of dispersers. We can make several conjectures, however, based on the known natural history of these species. Some colonists, especially bacteria and protozoa, are probably generalists that may occur in soils or open waters in the area and disperse as propagules carried in the air. Others are obligate inquilines of *S. purpurea* (figure 5.2) that probably colonize from surrounding leaves. The dipterans (*W. smithii, M. knabi,* and *F. fletcheri*) can all disperse as adults. *Wyeomyia smithii* are apparently quite poor fliers but have been observed to lay single eggs in numerous leaves, preferring larger leaves with high numbers of either conspecific larvae or *M. knabi* (Heard 1994a). Further, *M. knabi* and *F. fletcheri* larvae have been observed crawling on wet surfaces on the outsides of leaves and may actively migrate from pitcher to pitcher. The component species in inquiline communities are therefore likely to have very different dispersal rates and distances.

Although migration limitation can be very important for metacommunity dynamics, very few experimental studies have demonstrated such limitation in any natural community (but see Shurin 2001). We used experimental introductions of several species of protozoa and the rotifer *H. rosa* into leaves to test for migration limitation in communities in which we manipulated resource and predator abundances (Miller et al. 2002). Three of the six protozoan species successfully became established after introduction; establishment rates were increased by high resource availability and the presence of predators (figure 5.5). The other three protozoan species and the rotifer failed to colonize any new communities; although all four are naturally found in these communities, some factor other than migration seems to limit their distribution. These results show that some

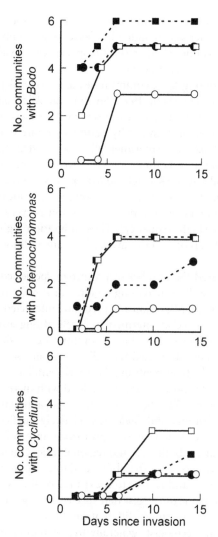

Figure 5.5 Numbers of communities successfully invaded over time by three protozoans (*Bodo* sp., *Poterioochromonas* [a chrysomonad], and *Cyclidium* sp.) in manipulated pitcher-plant leaves in the Pleaphase Savanna near Sumatra, Florida. Solid squares denotes mosquito predators removed and prey added; solid circles denotes mosquito predators removed, no prey added; open squares denotes mosquitoes present, prey added; open circles denotes controls; mosquitoes present, no prey added (from Miller et al. 2002). Resource and predator levels affected invasion by different protozoans to different degrees.

inquiline species are migration limited and suggest that species sorting may also occur, whereby characteristics of individual leaves (e.g., number of mosquitoes or prey captured) determine whether certain species can invade the community (Holt 1993; Tilman and Pacala 1993).

Dispersal rates among local communities are a central mechanism for the metacommunity concept (Holt 1993; Shurin and Allen 2001; Mouquet and Loreau 2002). Experimental demonstrations of the effects of varying dispersal rates on local and regional community patterns are rare (Gonzalez et al. 1998; Forbes and Chase 2002). We experimentally manipulated the rate of migration among regions of inquiline communities that were composed of five local communities (individual leaves) (Kneitel and Miller 2003). Three different migration levels were implemented: no migration among the five local communities and small amounts of fluid moved among the communities every two weeks (low frequency) and weekly (high frequency). These migration regimes were superimposed on different levels of resource availability (ant addition) and predator (*W. smithii*) abundance.

Migration increased the number of protozoan species co-occurring in each local community, as well as increasing regional protozoan diversity (figure 5.6; Kneitel and Miller 2003). Migration and predation interacted, such that the highest local diversity occurred at intermediate levels of migration when predators were absent. Dispersal also allowed local communities to share more species as the variance in diversity decreased with dispersal rate. Several species showed specific responses to the treatments in this experiment. For example, abundances of the protists *Colpoda* and *Colpidium* increased with increasing migration and food availability (figure 5.7), whereas chrysomonad abundances decreased with increasing food availability and *Bodo* was relatively unaffected by both treatments. Trade-offs were evident among protozoan and rotifer species and may allow species to coexist at both local and regional scales (Kneitel 2002). In particular, trade-offs between competitive ability and tolerance of predation appear to allow protozoan species and *H. rosa* to coexist regionally, even if few species can coexist locally. Such patterns suggest that species sorting is quite important and that each species grows best under different conditions.

Finally, community ecologists generally treat evolutionary change as something that happened a long time ago rather than as an active process. Although this static view may be a good assumption in some cases, many ecological processes occur over periods longer than a single generation of the component species, so understanding community patterns may require understanding evolutionary processes as well (McPeek and Miller 1996; Miller and Travis 1996). Particularly in a metacommunity context, migration provides gene flow among different communities, which can both provide genetic variation upon which selection can act and potentially swamp local adaptation. Preliminary work with bacteria in *S. purpurea* suggests that such local adaptation can occur and may influence species interactions. Characteristics of bacteria isolated from one-week-old, early successional communities and six-week-old communities were compared in the laboratory (Ellison et al. 2003). Bacteria from older, established communities were frequently poorer competitors, perhaps because of a trade-off in the abilities to tolerate low resource levels and predation.

Metacommunity Context: The key factor that gives value to a metacommunity perspective is that migration among local communities influences local commu-

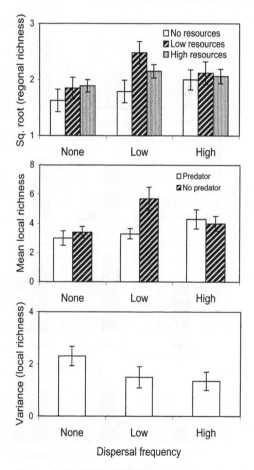

Figure 5.6 The effects of migration on within-leaf and among-leaf species richness and variance in species richness of protozoa in *Sarracenia purpurea* (from Kneitel and Miller 2003). Small amounts of fluid were moved among suites of five pitchers at either low (biweekly) or high (weekly) rates; the suites of pitchers were maintained either with or without mosquitoes present. Resource levels were as described in figure 5.4. The effect of migration on increasing local protozoan richness depends on the level of predation.

nity dynamics and patterns. In inquiline communities, some species have been shown to be migration limited, so even low rates of migration may allow the species to spread throughout a region. Further, migration has been shown to increase both local and regional species diversity and apparently interacts with local predation to determine the occurrence and abundance of species in the community. Understanding of the inquiline community thus requires a metacommunity perspective: local processes, regional migration, and regional heterogeneity all influence species distributions and abundances.

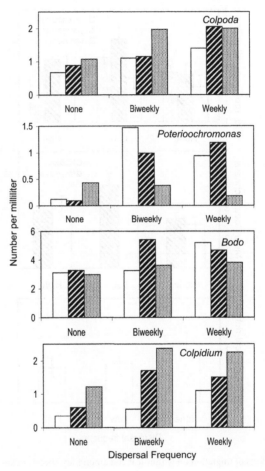

Figure 5.7 Effects of migration and resources on the abundances of several protozoan species in the inquiline communities of *Sarracenia purpurea* (from Kneitel and Miller 2003). Migration took place in small amounts of fluid moved among suites of five leaves at three levels: none, biweekly, and weekly. Resources were varied as described in figure 5.4. Abundances of *Colpoda* and *Colpidium* increased with increasing migration and food availability, whereas chrysomonad (*Poterioochromonas*) abundances decreased with increasing food availability, and *Bodo* was relatively unaffected by both treatments.

Larger-scale Dynamics

Although the metacommunity perspective adds a layer of complexity to our understanding of community dynamics, it still presents a gross oversimplification of the spatial structure of real communities. For most communities, the local and, especially, regional scales are vague constructs that probably represent points along a continuum rather than a true dichotomy. The local scale can be defined as where species interactions occur or in terms of some disjunction in the habitat that limits the rates of individual movement. The limits to the regional scale are more difficult to define and are related to dispersal distances and rates in which species in the same community can vary widely. In discussing metapopulations, Hanski and Gilpin (1991) define three scales: local, metapopulation, and geographic; the geographic scale encompasses distances greater than those over which an individual moves over its lifetime. Scale distinctions are therefore likely to be very specific to particular systems and must be arbitrarily defined by the researcher.

The inquiline communities in pitcher plants are an interesting system for studying questions of scale, because several scales are well defined by the natural history of the host plant, *Sarracenia purpurea*. Here, we have defined the local scale as that of the individual leaf. This choice seems well justified, as individuals can mix freely within the leaf, and we know that competition and predation within a leaf are important in determining species abundances. The regional scale is more difficult to define, however: each plant can consist of up to twelve water-filled leaves, the plants are often patchily distributed within populations, bogs themselves are distributed nonrandomly within some larger locality, and *S. purpurea* populations occur in bogs across North America. We have not been very specific about the regional scale but have described dispersal among leaves in the same area, intentionally leaving "area" vague. Dispersal could be among leaves on the same plant, leaves in the same population of *S. purpurea*, or even leaves from different populations.

Some indication of the importance of different spatial scales can be gained from a comparison of patterns of species abundance and community similarity at different scales. Communities that are more similar can be assumed to be affected by similar forces, including resources and predation, or may be linked by strong migration. We have conducted three different studies that compare patterns at different scales: among leaves and plants in the same population, among local populations, and among populations across North America.

We investigated spatial relationships among inquiline communities within a 10 m by 20 m area of the Crystal Bog in north Florida (Buckley et al. 2004). The locations of all 25 plants with a total of 141 water-filled leaves were mapped. For each leaf, we also determined length, aperture size, volume, and a rough estimate of relative age based on leaf position on the stem (Fish and Hall 1978). We partitioned the variance in species richness and abundance patterns into among-leaf and among-plant components. Communities from leaves on the same plant were significantly more similar to one another, in terms of both diversity and species relative abundances, than leaves from different plants. The among-plant component explained 42% of the variance in invertebrate species richness and 34% of the variance in protozoan species richness, but the spatial positions of plants themselves had no significant effect on the community composition. In general, diversities within different functional groups were positively correlated; at both the among-leaf and among-plant scales, the species richness of bacteria and protozoa was significantly correlated with invertebrate richness, which was itself positively correlated with leaf length. Contrary to patterns of species richness, species abundances were not generally correlated with environmental variables such as leaf size, relative leaf age, number of leaves per plant, light availability, or soil moisture.

This study demonstrates that one important scale for inquiline communities is the plant itself but that, otherwise, the spatial position of leaves within the larger population was unimportant. Communities in leaves on the same plant may be influenced by similar habitat factors that we did not measure, migration among leaves on the same plant, or both.

Small distances can separate populations of S. purpurea within the same geographic area. Inquiline communities in these plant populations are likely to be subjected to similar environmental conditions and exchange migrants of at least some of their component species; both processes could result in greater community similarity between neighboring populations. Harvey and Miller (1996) quantified species patterns within leaves on the same plant and plants in the same area of the population in each of three different populations of S. purpurea in north Florida. Like Buckley et al. (2004), they found that leaves on the same plant were more similar than leaves on different plants, especially for bacterial communities, but different species demonstrated different spatial patterns, possibly as a result of different dispersal abilities. For example, neither plants in the same field nor populations of S. purpurea differed significantly in mosquito abundance, but populations differed in abundance of M. knabi, and areas within the same population of S. purpurea differed in cladoceran abundance.

At the largest scale, we can consider the distribution of the entire community type. The S. purpurea inquiline community is unusual in that the component species are thought to co-occur over the plant's entire range. This large range, up the eastern portion of North America from northern Florida to Labrador and across Canada to British Columbia and the Northwest Territories (figure 5.1), encompasses a wide range of climates, soil types, and hydrology. Further, because of glacial events, we can consider aspects of historical biogeography; S. purpurea is thought to have spread northward from its southern populations over the last 10,000 years, creating the habitat for the inquiline community as it spread (Ellison and Parker 2002).

The mosquito *W. smithii* has been especially well studied over the range of *S. purpurea* by Bradshaw, Holzapfel, and their colleagues (Bradshaw and Creelman 1984). They have demonstrated significant geographic variation in life-history traits of *W. smithii*, especially in developmental responses to day length (Bradshaw and Holzapfel 1989). *Wyeomyia smithii* also demonstrates significant changes in genetic structure; allozyme heterozygosity is greater in populations south of the maximum extent of the Laurentide ice sheet (Armbruster et al. 1998).

Less is known about how other inquiline species or whole community patterns vary over the range of the host plant. Buckley et al. (2003) visited thirty-nine sites across North America (figure 5.1), measuring both plant and inquiline community traits from twenty randomly chosen plants at each site. To minimize seasonal differences among the communities, they visited each site during the period of maximum leaf production, three to four weeks after the flowering of *S. purpurea*. Their study confirmed that, while leaves within a site may contain very different communities, among-site variation in the inquiline community is remarkably small across this large range; all the major species are found at virtually all sites, and only 4% of the total variation among 780 leaves investigated is explained by site (partial canonical correspondence analyses). Bacterial and protozoan species richness at among-leaf and among-site scales do increase with latitude (figure 5.8); numbers of species were significantly higher in the Canadian communities. Per-site invertebrate richness is not affected by latitude, and *W. smithii* abundances demonstrate a small but significant decline.

An interesting result is that this community is more variable in species composition at the within-site than at the among-site scale (Buckley et al. 2003). This pattern is unusual because at large scales we would expect to find that regional or biogeographic processes, such as dispersal, speciation, and landscape change over geological time, were more important than within-site variation in resource and predator abundances. Instead, the structure of this community at the among-site scale seems remarkably robust, despite a strong gradient in length of active season (eleven to twelve months in northern Florida to three months in the Northwest Territories), climate, soil conditions, and even life-history of dominant species (e.g., *W. smithii* and other dipterans are multivoltine in northern Florida but univoltine in northern populations). We do not know of any other community type that appears this stable or consistent over a range of this size or under growth conditions that vary to this degree, but few similar studies have been conducted.

Metacommunity Context: While the local scale appears to be well defined as that of an individual water-filled leaf, larger, regional scales are less discrete, and have different roles in a metacommunity framework. If we assume that dispersal rates decrease with distance between communities, then we can speculate that dispersal among leaves on the same plant or population of *S. purpurea* is relatively high and could be important for species sorting and mass effects (Leibold and Miller 2004). Migration among populations of *S. purpurea* in the same geo-

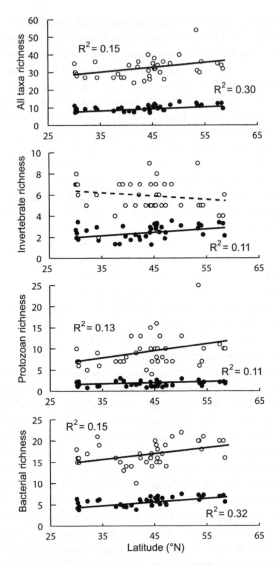

Figure 5.8 Relationships between latitude and species richness of inquiline species found in *Sarracenia purpurea* populations across North America for all species, invertebrates, protozoa, and bacteria at the among-site scale (open circles) and the pitcher scale (filled circles). Bacterial and protozoan species richness increase with latitude at both scales, but per-site invertebrate richness is not affected by latitude.

graphic area may be important for species sorting but is probably most important for overcoming dispersal limitation and allowing recolonization of some sites after disturbances such as drought or fire. Ultimately, however, this community type appears to have spread from the southeastern United States to Canada; migration among sites appears to have been sufficient to overcome migration limitation in entire populations of *S. purpurea*, leading to the remarkable uniformity of this community type across North America. We suggest that simple local versus regional partitioning of scale may be inappropriate for some, especially biogeographic, questions in community ecology.

Discussion: Are These Metacommunities? How Does One Deal with Scale Issues?

We began this chapter by suggesting that metacommunity theory has advanced sufficiently such that evaluations of its assumptions and predictions are needed from the field. We conclude that experiments have demonstrated that inquiline communities found in the water-filled leaves of the purple pitcher plant, *Sarracenia purpurea*, do indeed act as metacommunities with important interactions at different spatial scales.

At the local scale, competition for resources affects the abundances and richness of bacteria, protozoa (Cochran-Stafira and von Ende 1998; Kneitel and Miller 2002; Miller et al. 2002), and mosquitoes (Bradshaw and Holzapfel 1989; Miller et al. 1994). Further, predation by *W. smithii* affects the abundances and richness of its direct prey and may have cascading effects on bacterial abundance and richness (Addicott 1974; Cochran-Stafira and von Ende 1998; Kneitel and Miller 2002; Miller et al. 2002). Other, more complicated species interactions can occur at this local scale, such as the processing-chain mutualism (Heard 1994b). Interactions at the scale of the individual leaf habitat are clearly important in structuring these communities.

The structural habitat (individual leaves) itself provides a dynamic backdrop for this community. The birth-and-death dynamic of leaves creates a constant source of empty patches that drives one whole scale of interactions. Some inquiline species have now been shown to be migration limited (Miller at al. 2002). Migration has been shown to influence species richness and composition (Kneitel and Miller 2003). Further, the importance of migration can depend on the current occupants of the communities and their interactions with new migrants (Miller et al. 2002).

The metacommunity view of inquiline communities has not translated perfectly to the field and this may suggest areas for further theoretical work. Although the local community appears to be as well defined as that contained by an individual leaf, the regional scale is more difficult to determine and is almost certainly less discrete. We have not actually been able to follow migration from source to destination, so actual dispersal distances are still unknown. We do know that colonization can occur relatively quickly and bacteria seem not to be dispersal limited (Kneitel and Miller, unpublished data). Observational evidence suggests that, for some species, migration occurs primarily among leaves within a plant, but for other species, migration may span much longer distances (especially for the adult dipterans). The two alternative views of migration in this system are (1) that migration is simply a declining function of distance between leaves, and (2) that migration is best represented as a lottery in which species colonize pitchers from a large pool (air).

Even if we had a good knowledge of dispersal distances and rates in this community, it is likely that species have different migration functions, making a specific regional scale difficult to define. Species may fall into dispersal groups based on similarity in the mechanisms of dispersal or dispersal distance, but differences between groups must still be taken into account when the regional scale is defined. Further, the definition of any regional scale may depend on the questions of interest. For example, in a study of protozoan diversity, plants or local neighborhoods of plants may be the important regional scale, but entire populations of S. purpurea can be subject to occasional droughts and fires, so differences among populations may also be relevant. Local adaptation of inquiline species in different populations may even occur and could affect invasion success in other populations. Large-scale or long-term questions may simply require larger, biogeographic, scales of study.

Are the communities associated with Sarracenia purpurea too unusual, even obscure, to yield results applicable to other community types? Very similar systems are found in other carnivorous plants. Naeem (1988), in a related system, found competitive interactions between midges (M. edwardsi) and mites (Sarraceniopus darlingtoniae) in the California pitcher plant, Darlingtonia californica. A variety of similar experiments have been conducted by Kitching (1987; Clarke and Kitching 1993) and others (Beaver 1985; Mogi and Yong 1992) in old-world pitcher plants (Nepenthes spp.) in Indonesia. The results from pitcher-plant experiments are also similar to studies in other phytotelmata, including tree holes (Srivastava and Lawton 1998) and the water-filled axils of bromeliads and other plants (review in Kitching 2000, 2001). Work to date on local interactions suggests that other container communities resemble our inquiline communities in harboring significant species interactions determined by both resources and predators; we know of no work on the regional processes in these systems. All the above studies are small aquatic communities whose energy input is allocthonous rather than arising through photosynthesis. The energy source may be less important than the discrete nature of the local community. The dynamics of other communities, both aquatic and terrestrial, with relatively discrete boundaries appear similar to those we find in pitcher-plant communities (e.g., Gonzalez, chapter 6; Cottenie and De Meester, chapter 8). Ultimately, we argue that the inquiline communities associated with S. purpurea are fairly typical, and represent a broad variety of natural communities.

Conclusions

In most communities, especially aquatic ones, processes such as competition and predation occur locally, driving the local abundances of most resident species (see, e.g., Forbes and Chase 2002). At some larger regional scale, however, most communities are subject to disturbance and habitat variation; colonization and migration may then be critical to persistence of some species in the region. An understanding of community composition and dynamics will require an understanding of processes important at both local and regional scales.

Work with the inquiline communities associated with *S. purpurea* demonstrates that species interact and patterns of species association occur at several spatial scales. Local scales may be the most easily defined and manipulated to demonstrate that habitat characteristics and species interactions affect local community composition. Larger, regional scales are also important but are more difficult to define because different trophic levels (and species within trophic levels) respond differently to processes on these scales.

Current metacommunity theory focuses on small-scale migration and ecological interactions as the important processes at the regional scale. We suggest that future theory should also consider two related lines of research: First, a variety of regional processes occurring at different scales must be recognized and investigated (Ricklefs 1987). These include migration among local communities, as well as very different factors such as species' historical spread and biogeography (Holt and Gomulkiewicz 1997). Researchers may have to abandon the simple local versus regional dichotomy in favor of something like multiple, nested, spatial scales (sensu Loreau and Mouquet 1999) and develop an understanding of the feedback dynamics between local community interactions and regional-scale movement patterns. Second, evolutionary processes should be integrated into our understanding of metacommunities. For example, local adaptation may change species' responses to resources and predation, affecting patterns of species interactions and coexistence (Kneitel 2002, Kneitel and Chase 2004) and ultimately affecting community dynamics in different *S. purpurea* populations (Ellison et al. 2003).

Acknowledgments

We are grateful to a number of colleagues for sharing ideas, information, and techniques related to working with pitcher plant communities, including L. Bledzki, W. Bradshaw, H. Buckley, J. Burns, L. Cochran-Stafira, A. Ellison, N. Gotelli, S. Gray, S. Heard, C. Holzapfel, N. Mouquet, R. Naczi, J. Reeves, R. Reeves, and K. Trzcinski. We also thank K. Cottenie and J. Shurin for thoughtful reviews of this chapter.

Literature Cited

Adamec, L. 1997. Mineral nutrition of carnivorous plants: A review. Botanical Review 63:273–299.

Addicott, J. F. 1974. Predation and prey community structure: An experimental study of the effect of mosquito larvae on the protozoan communities of pitcher plants. Ecology 55:475–492.

Albert, V. A., S. E. Williams, and M. W. Chase. 1992. Carnivorous plants: Phylogeny and structural evolution. Science 257:1491–1495.

Armbruster, P. A., W. E. Bradshaw, and C. M. Holzapfel. 1998. Effects of postglacial range expansion on allozyme and quantitative genetic variation in the pitcher-plant mosquito, *Wyeomyia smithii*. Evolution 52:1697–1704.

Bateman, L. E. 1987. A bdelloid rotifer living as an inquiline in leaves of the pitcher plant, *Sarracenia purpurea*. Hydrobiologia 147:129–133.

Bayer, R. J., L. Hufford, and D. E. Soltis. 1996. Phylogenetic relationships in the Sarraceniaceae based on rbcL and ITS sequences. Systematic Botany 21:121–134.

Beaver, R. A. 1985. Geographical variation in food web structure in *Nepenthes* pitcher plants. Ecological Entomology 10:241–248.

Bledzki, L. A., and A. M. Ellison. 1998. Population grown and production of *Habrotrocha rosa* Donner (Rotifera: Bdelloidea) and its contribution to the nutrient supply of its host, the northern pitcher plant, *Sarracenia purpurea* L. (Sarraceniaceae). Hydrobiologia 385:193–200.

Bradshaw, W. E., and R. A. Creelman. 1984. Mutualism between the carnivorous purple pitcher plant and its inhabitants. American Midland Naturalist 112:294–303.

Bradshaw, W. E., and C. M. Holzapfel. 1986. Geography of density-dependent selection in pitcher-plant mosquitoes. Pages 48–65 *in* F. Taylor and R. Karban, eds. *The evolution of insect life cycles.* Springer-Verlag, New York.

————. 1989. Life-historical consequences of density-dependent selection in the pitcher-plant mosquito, *Wyeomyia smithii*. American Naturalist 133:869–887.

————. 1992. Reproductive consequences of density-dependent size variation in the pitcher-plant mosquito, *Wyeomyia smithii* (Diptera: Culicidae). Annals of the Entomological Society of America 85:274–281.

Bradshaw, W. E., and L. P. Lounibos. 1972. Photoperiodic control of development in the pitcher-plant mosquito, *Wyeomyia smithii*. Canadian Journal of Zoology 50:713–719.

Broberg, L., and W. E. Bradshaw. 1995. Density-dependent development in *Wyeomyia smithii* (Diptera: Culicidae): Intraspecific competition is not the result of interference. Annals of the Entomological Society of America 88:465–470.

Brown, J. H., and A. Kodric-Brown. 1977. Turnover rates in insular biogeography: Effects of immigration on extinction. Ecology 58:445–449.

Buckley, H. L., J. H. Burns, J. M. Kneitel, E. L. Walters, P. Munguía, and T. E. Miller. 2004. Small-scale patterns in community structure of *Sarracenia purpurea* inquiline communities. Community Ecology 5:181–188.

Buckley, H. L., T. E. Miller, A. M. Ellison, and N. J. Gotelli. 2003. Reverse latitudinal trends in species richness of pitcher-plant food webs. Ecology Letters 6:825–829.

Chapin, C. T., and J. Pastor. 1995. Nutrient limitations in the northern pitcher plant *Sarracenia purpurea*. Canadian Journal of Botany 73:728–734.

Chesson, P. 2000. General theory of competitive coexistence in spatially-varying environments. Theoretical Popululation Biology 58:211–237.

Christensen, N. L. 1976. The role of carnivory in *Sarracenia flava* L. with regard to specific nutrient deficiencies. Journal of the Elisha Mitchell Scientific Society 92:144–147.

Clarke, C. M., and R. L. Kitching. 1993. The metazoan food webs from six Bornean *Nepenthes* species. Ecological Entomology 18:7–16.

Cochran-Stafira, D. L., and C. N. von Ende. 1998. Integrating bacteria into food webs: Studies with *Sarracenia purpurea* inquilines. Ecology 79:880–898.

Cresswell, J. E. 1991. Capture rates and composition of insect prey of the pitcher plant *Sarracenia purpurea*. American Midland Naturalist 125:1–9.

———. 1993. The morphological correlates of prey capture and resource parasitism in pitchers of the carnivorous plant *Sarracenia purpurea*. American Midland Naturalist 129:35–41.

Eleuterius, L. N., and S. B. Jones, Jr. 1969. A floristic and ecological study of pitcher plant bogs in south Mississippi. Rhodora 71:29–34.

Ellison, A. M., H. L. Buckley, T. E. Miller, and N. J. Gotelli. 2004. Morphological variation in *Sarracenia purpurea* (Sarraceniaceae): Geographic, environmental, and taxonomic correlates. American Journal of Botany 91:1930–1935.

Ellison, A. M., N. J. Gotelli, J. S. Brewer, L. Cochran-Stafira, J. Kneitel, T. E. Miller, A. S. Worley, and R. Zamora. 2003. The evolutionary ecology of carnivorous plants. Advances in Ecological Research 33:1–74.

Ellison, A. M., and J. N. Parker. 2002. Seed dispersal and seedling establishment of *Sarracenia purpurea* (Sarraceniaceae). American Journal of Botany 89: 1024–1026.

Farkas, M. J., and R. A. Brust. 1985. The effect of a larval diet supplement on development in the mosquito *Wyeomyia smithii* (Coq.) under field conditions. Canadian Journal of Zoology 63:2110–2113.

Fashing, N. J., and B. M. O'Conner. 1984. *Sarraceniopus*—a new genus for histiostomatid mites inhabiting the pitchers of Sarraceniaceae (Astigmata: Histiostomatidae). International Journal of Acarology 10:217–227.

Fish, D., and D. W. Hall. 1978. Succession and stratification of aquatic insects inhabiting the leaves of the insectivorous pitcher plant *Sarracenia purpurea*. American Midland Naturalist 99:172–183.

Forbes, A. E., and J. N. Chase. 2002. The role of habitat connectivity and landscape geometry in experimental zooplankton metacommunities. Oikos 96:433–440.

Forsyth, A. B., and R. J. Robertson. 1975. K reproductive strategy and larval behavior of pitcher plant sarcophagid fly, *Blaesoxipha fletcheri*. Canadian Journal of Zoology 53:174–179.

Gallie, D. R., and S. Chang. 1997. Signal transduction in the carnivorous plant *Sarracenia purpurea*. Plant Physiology 115:1461–1471.

Gibson, T. C. 1983. Competition, disturbance and the carnivorous plant community in the southeastern United States. Ph.D. dissertation, University of Utah, Salt Lake City.

Givnish, T. J. 1989. Ecology and evolution of carnivorous plants. Pages 234–290 in W. G. Abrahamson, ed. Plant animal interactions. McGraw-Hill, New York.

Gonzalez, A., J. H. Lawton, F. S. Gilbert, T. M. Blackburn, and I. Evans-Freke. 1998. Metapopulation dynamics, abundance, and distribution in a microecosystem. Science 281:2045–2047.

Gray, S., T. E. Miller, N. Mouquet, and T. Daufresne. In preparation. Nutrient limitation in a detritus-based container community.

Hamilton, R., J. W. Reid, and R. M. Duffield. 2000. Rare copepod, *Paracyclops canadensis* (Willey), common in leaves of *Sarracenia purpurea* L. Northeastern Naturalist 7:17–24.

Hanski, I., and M. Gilpin. 1991. Metapopulation dynamics—brief history and conceptual domain. Biological Journal of the Linnean Society 42:3–16.

Harvey, E., and T. E. Miller. 1996. Variance in composition in inquiline communities in leaves of *Sarracenia purpurea* L. on multiple spatial scales. Oecologia 108:562–566.

Heard, S. B. 1994a. Imperfect oviposition decisions by the pitcher plant mosquito, *Wyeomyia smithii*. Evolutionary Ecology 8:493–502.

———. 1994b. Pitcher-plant midges and mosquitoes: A processing chain commensalism. Ecology 75: 1647–1660.

———. 1998. Capture rates of invertebrate prey by the pitcher plant, *Sarracenia purpurea* L. American Midland Naturalist 139:79–89.

Holt, R. D. 1993. Ecology at the mesoscale: The influence of regional processes on local communities. Pages 77–88 *in* R. E. Ricklefs and D. Schluter, eds. *Species diversity in ecological communities: Historical and geographical perspectives.* University of Chicago Press, Chicago.

———. 1996. Temporal and spatial aspects of food web structure and dynamics. Pages 255–257 *in* G. A. Polis and K. O. Winemiller, eds. *Food webs: Integration of patterns and dynamics.* Chapman and Hall, New York.

Holt, R. D., and R. Gomulkiewicz. 1997. How does immigration influence local adaptation? A reexamination of a familiar paradigm. American Naturalist 149:563–572.

Istock, C. A., S. S. Wasserman, and H. Zimmer. 1975. Ecology and evolution of the pitcher-plant mosquito: 1. Population dynamics and laboratory responses to food and population density. Evolution 29:296–312.

Judd, W. W. 1959. Studies of the Byron Bog in southwestern Ontario X. Inquilines and victims of the pitcher-plant, *Sarracenia purpurea* L. Canadian Entomologist 91:171–180.

Kitching, R. L. 1987. Spatial and temporal variation in food webs in water-filled treeholes. Oikos 48: 280–288.

———. 2000. *Food webs and container habitats: The natural history and ecology of phytotelmata.* Cambridge University Press, Cambridge, UK.

———. 2001. Food webs in phytotelmata: "Bottom-up" and "top-down" explanations for community structure. Annual Review of Entomology 46:729–760.

Kneitel, J. M. Submitted. Intermediate-consumer identity and resources alter an omnivory food web. Oikos.

———. 2002. Species diversity and trade-offs in *Sarracenia purpurea* inquiline communities. Ph.D. dissertation. Florida State University.

Kneitel, J. M. and J. M. Chase. 2004. Trade-offs in community ecology: Linking spatial scales and species coexistence. Ecology Letters 7:69–80.

Kneitel, J. M., and T. E. Miller. 2002. The effects of resource and top-predator addition to the inquiline community of the pitcher plant *Sarracenia purpurea.* Ecology 83:680–688.

———. 2003. Dispersal rates affect community composition in metacommunities of *Sarracenia purpurea* inquilines. American Naturalist 162:165–171.

Leibold, M. A., and T. E. Miller. 2004. From metapopulations to metacommunities. Pages 133–150 *in* I. Hanski and O. Gaggiotti, eds. *Ecology, genetics and evolution of metapopulations.* Academic Press, San Diego.

Lloyd, F. E. 1942. *The carnivorous plants.* Ronald Press, New York.

Loreau, M., and N. Mouquet. 1999. Immigration and the maintenance of local species diversity. American Naturalist 154:427–440.

McPeek, M. A., and T. E. Miller. 1996. Evolutionary biology and community ecology. Ecology 77: 1319–1320.

Miller, T., D. Cassill, C. Johnson, C. Kindell, J. Leips, D. McInnes, T. Bevis, D. Mehlman, and B. Richard. 1994. Intraspecific and interspecific competition of *Wyeomyia smithii* (Coq.) (Culicidae) in pitcher plant communities. American Midland Naturalist 131:136–145.

Miller, T., J. M. Kneitel, and J. Burns. 2002. Effects of community structure on invasion success and rate. Ecology 83:898–905.

Miller, T. E., and J. Travis. 1996. The evolutionary role of indirect effects in communities. Ecology 77:1329–1335.

Mogi, M., and H. S. Yong. 1992. Aquatic arthropod communities in *Nepenthes* pitchers—the role of niche differentiation, aggregation, predation, and competition in community organization. Oecologia 90:172–184.

Mouquet, N., and M. Loreau. 2002. Coexistence in metacommunities: The regional similarity hypothesis. American Naturalist 159:420–426.

Naczi, R. E., E. M. Soper, F. W. Case, and R. B. Case. 1999. *Sarracenia rosea* (Sarraceniaceae), a new species of pitcher plant from the southeastern United States. Sida 18:1183–1206.

Naeem, S. 1988. Resource heterogeneity fosters coexistence of a mite and midge in pitcher plants. Ecological Monographs 58:215–227.

Newell, S. J., and A. J. Nastase. 1998. Efficiency of insect capture by *Sarracenia purpurea* (Sarraceniaceae), the northern pitcher plant. American Journal of Botany 85:88–91.

Osenberg, C. W., and G. G. Mittelbach. 1996. The relative importance of resource limitation and predator limitation in food chains. Pages 134–148 *in* G. A. Polis and K. O. Winemiller, eds. *Food webs: Integration of patterns and dynamics.* Chapman and Hall, New York.

Petersen, R. L., A. Faust, J. Nagawa, C. Thomas, and A. Vilmenay. 2000. Foreign mosquito survivorship in the pitcher plant *Sarracenia purpurea*—the role of the pitcher-plant midge *Metriocnemus knabi.* Hydrobiologia 439:13–19.

Petersen, R. L., L. Hanley, E. Walsh, H. Hunt, and R. M. Duffield. 1997. Occurrence of the rotifer, *Habrotrocha* cf. *rosa* Donner, in the purple pitcher plant, *Sarracenia purpurea* L., (Sarraceniaceae) along the eastern seaboard of North America. Hydrobiologia 354:63–66.

Power, M. E. 1992. Top down and bottom up forces in food webs: Do plants have primacy? Ecology 73:733–746.

Prankevicius, A. B., and D. M. Cameron. 1989. Free-living dinitrogen-fixing bacteria in the leaf of the northern pitcher plant (*Sarracenia purpurea* L.). Naturaliste Canadien 116:245–249.

———. 1991. Bacterial dinitrogen fixation in the leaf of the northern pitcher plant (*Sarracenia purpurea*). Canadian Journal of Botany 69:2296–2298.

Rango, J. J. 1999. Resource dependent larviposition behavior of a pitcher plant flesh fly, *Fletcherimyia fletcheri* (Aldrich) (Diptera: Sarcophagidae). Journal of the New York Entomological Society 107:82–86.

Reveal, J. L. 1993. A list of validly published, automatically typified, ordinal names of vascular plants. Taxon 42:825–844.

Ricklefs, R. E. 1987. Community diversity—relative roles of local and regional processes. Science 235:167–171.

Schnell, D. E. 1979. Critical review of published variants of *Sarracenia purpurea* L. Castanea 44:47–59.

Shurin, J. B. 2001. Interactive effects of predation and dispersal on zooplankton communities. Ecology 82:3404–3416.

Shurin, J. B., and E. G. Allen. 2001. Effects of competition, predation, and dispersal on species richness at local and regional scales. American Naturalist 158:624–637.

Simberloff, D. 1981. A succession of paradigms in ecology: essentialism to materialism to probabilism. Synthese 43: 3–39.

Srivastava, D. S., and J. H. Lawton. 1998. Why more productive sites have more species: an experimental test of theory using tree-hole communities. American Naturalist 152:510–529.

Stiling, P., and A. M. Rossi. 1997. Experimental manipulations of top-down and bottom-up factors in a tri-trophic system. Ecology 78:1602–1606.

Tilman, D., and S. Pacala. 1993. The maintenance of species richness in plant communities. Pages 13–25 *in* R. Ricklefs and D. Schluter, eds. *Species diversity in ecological communities.* University of Chicago Press, Chicago.

Wolfe, L. M. 1981. Feeding behavior of a plant: Differential prey capture in old and new leaves of the pitcher plant (*Sarracenia purpurea*). American Midland Naturalist 106:352–359.

CHAPTER 6

Local and Regional Community Dynamics in Fragmented Landscapes

Insights from a Bryophyte-based Natural Microcosm

Andrew Gonzalez

Introduction

Habitat destruction is now so prevalent that it is thought to be the dominant cause of anthropogenic extinction (Vitousek 1994; Kareiva and Wennergren 1995; Riitters et al. 2002; Brooke et al. 2003). Furthermore, destruction has increased the frequency of local extinctions to such an extent that taxonomic extinction is presently several orders of magnitude greater than the background rate (May et al. 1995; Pimm et al. 1995). In response ecologists have published a large body of literature on the effect habitat destruction, and a substantial research effort is now focused on understanding how habitat fragmentation alters community structure and function (e.g., Andrén 1994; Debinski and Holt 1999; Laurance et al. 2002; Mc-Garigal and Cushman 2002). To date this research has rested heavily on the formal foundations of island biogeography and metapopulation theory (Hanski and Simberloff 1997); both of these theories focus on extinction and immigration, which are the phenomena perceived to dominate ecological dynamics in fragmented landscapes. Although these theories have greatly improved our understanding of the basic patterns that we see in fragmented landscapes it remains a challenge to predict the dynamic distribution of species in fragmented landscapes. It is likely, therefore, that a deeper understanding of the effects of habitat fragmentation will require a more inclusive theoretical framework. Such a framework should acknowledge that habitat fragmentation not only alters the fluxes of individuals, nutrients, and energy, but also alters patterns of disturbance across the landscape (Turner 1989; Wu and Loucks 1995). Metacommunity theory involves the study of the patterns and processes derived from intercommunity connection across a range of spatial scales. It promises a fuller understanding of the effects of habitat fragmentation through the integration of the process oriented vision of landscape ecology with the species oriented vision of community ecology.

Current (meta)population and (meta)community theory makes a number of predictions concerning the consequences of disrupting community coupling, which derive from a rather disjointed ensemble of models. Below I give a brief review of the central ideas in this literature. However, I do not attempt to be exhaustive, or to review the large and highly relevant landscape ecology literature (Turner 1989; Turner and Gardner 1991; Pickett and Cadenasso 1995; Goodwin 2003). This is followed by a description and synthesis of results from a series of microcosm experiments conducted to examine the effects of habitat fragmentation on community pattern and process, both at the local (within community) and regional (among community) scale. The results of these experiments are discussed in light of the theoretical expectations reviewed below.

A Brief Review of Theoretical Expectations
HABITAT FRAGMENTATION AS PATTERN AND PROCESS

Habitat destruction is rarely a random or uniform process (Seabloom et al. 2002) and in many cases it results in the transformation of a landscape into a mosaic of habitat fragments differing in quantity and quality through time (Riitters et al. 2002). Fortunately this transformation is not without some identifiable pattern. It is often dominated by two components as original habitat is lost: reduction in fragment size and increasing fragment isolation. How do these factors interact to alter community pattern and process?

In the initial stages of habitat destruction species richness and population abundances in the remaining often large and interconnected fragments are expected to reflect a random sample of the original habitat. Here fragment species richness will follow the species-area relation of the prefragmented landscape (Conner and McCoy 1979). However, as more habitat is lost, the reduced fragment sizes and connectivity will lead to levels of species loss and population declines over and above those expected from the random sample hypothesis (Andrén 1994; Bascompte and Solé 1996; With et al. 1997; Bender et al. 1998; Fahrig 2002). The greater loss of species from the smaller fragments creates a steeper species-area relation. This process of community relaxation is time dependent, such that there will be a period immediately after the attainment of a given level of fragmentation when changes in species richness and abundance have yet to take effect, and the number of species present in the landscape is greater than its ultimate steady-state value (Diamond 1984). Between the start and end of this relaxation process there is an extinction "debt" equal to the difference between the initial and final steady state (Stone 1995; Tilman et al. 1997). To date the notion of an extinction debt has been restricted to species loss, although similar delayed dynamics may be expected for other more functional community attributes such as productivity or nutrient cycling (Kareiva and Wennergren 1995). In principal, community relaxation can be observed at both the local and regional scale, although the patterns of region-wide changes in species richness have received relatively little attention (Harrison 1999). Various explanations for community relaxation have been proposed.

LOCAL COMMUNITY DISASSEMBLY

A widespread pattern in fragmented landscapes is the existence of nested subsets of species; the species comprising smaller local assemblages constitute a subset of the species richer ones (Patterson and Brown 1991). Nested patterns of species composition suggest relatively predictable sequences of species loss (Patterson 1990), but on their own they can tell us little of the possible causal explanations. In fragmented landscapes community disassembly will be brought about by two principal effects: (1) reduced movement and dispersal, and (2) modified habitat quality, for instance, lower productivity, or higher disturbance rates (Saunders et al. 1991; Fahrig 1992). Theory makes contrasting predictions regarding the identity of species that will be lost initially because of these effects. Rare species will suffer extinction because of the altered patterns of environmental and demographic stochasticity (Lawton 1995). However, relaxation may be characterized by declines and changes in dominant, widely distributed, taxa because their persistence may be compromised by low dispersal ability (Nee and May 1992; Tilman et al. 1997; Didham, Hammond, et al. 1998).

Food web theory suggests that dynamic constraints (Pimm and Lawton 1977), in addition to enhanced disturbance rates in fragments, should result in shorter food chains (Pimm and Lawton 1977; Patterson 1984; Schoener 1989; Holt et al. 1999; Holt 2002; Melián and Bascompte 2002). More generally those species that are the least resilient (here defined as the capacity of a species' population to tolerate disturbance; Carpenter et al. 2001) will tend to suffer extinction sooner in fragmented landscapes. Likely candidates are large-bodied species, trophic specialists, and species of high trophic rank requiring large home ranges, all of which are correlates of rarity (Lawton 1995). Thus we might predict that food webs will collapse from the top down, and hence for food chain length to be correlated with fragment area (Cohen and Newman 1991; Holt et al. 1999), and some evidence exists for this prediction (Jenkins et al. 1992; Kruess and Tscharntke 1994; Didham, Lawton, et al. 1998; Holyoak 2000; Hoyle 2004; Komonen et al. 2000; Laurance et al. 2002). In contrast, Mikkelson (1993) detected no differences in the extinction susceptibilities of different trophic levels in a variety of data sets from "natural fragments," although the theoretical basis for this empirical observation is weak.

The order and identity of species loss during community relaxation is of fundamental importance because it may have substantial consequences for local community functioning (Lawton 1994; Sala et al. 1996; Grime 1998). In particular, nonlinear losses of community function over time may be generated by variation in the identity of species going extinct, and the timing of their extinctions and abundance declines (Kareiva and Wennergren 1995). Species may vary in factors like rarity or dominance (Sala et al. 1996; Grime 1998). For example, recent metacommunity models (based on source-sink dynamics) predict that disruption of interfragment dispersal will lead to local extinction and lower community productivity (Loreau et al. 2003; Loreau et al., chapter 18). Similarly, higher rates of species turnover observed in remnant fragments are expected to have far reaching consequences for community stability (Boulinier et al. 1998; Laurance et al. 2002). This temporal aspect of community disassembly has been missing from biodiversity-functioning studies, although it will be critical to understanding the effects of fragmentation on ecosystem functioning (Gonzalez and Chaneton 2002).

The retention of habitat corridors between remnant fragments has been identified as a potentially effective means of maintaining movement and dispersal, and hence ameliorating the deleterious effects of fragmentation. Conceptually the idea of habitat corridors follows from island biogeography and metapopulation theory; the suggestion being that by enhancing movement corridors will allow increased effective population sizes, fragment recolonization following local extinction, and sustained local species richness (Wilson and Willis 1975; Saunders and Hobbs 1991; Harris and Scheck 1991; Noss 1994; Beier and Noss 1998). In general corridors are not thought to compensate for deterioration in fragment quality, and may in fact create a source-sink system by permitting the movement of individuals from high to low quality habitat (Donahue et al. 2003); although such sink habitats may make significant contributions to regional abundance if they varying in quality through time (Gonzalez and Holt 2003). Corridors may have other negative effects, for example dispersal may eliminate the possibility of recolonization if it acts to synchronize metapopulation fluctuations (Earn et al. 2000), or enhance the spread of disease (Simberloff et al. 1992). For the most part, theory (e.g., Brooker et al. 1999; Hugens and Haddad 2003) and a great deal of field data suggest that corridors do work to promote movement and migration between fragments (e.g., Haddad et al. 2003). A growing number of relatively short-term experimental studies have also established that corridor-induced dispersal slows species loss and maintains community structure and functioning (Gonzalez et al. 1998; Golden and Crist 1999; Haddad and Baun 1999; Collinge 2000; Gonzalez and Chaneton 2002; Tewksbury et al. 2002).

REGIONAL COMMUNITY DISASSEMBLY

The effects of habitat fragmentation on regional species richness will depend on the principal processes determining coexistence in the landscape (Harrison 1999). In contrast to island biogeography theory, which assumes an external regional species pool, metacommunity theory seeks to explain regional diversity by explicitly considering dispersal and local interactions. If source-sink dynamics, or mass effects, dominate then declines in landscape connectivity will hinder rescue effects, and ultimately result in reduced levels of regional diversity. However, recent metacommunity models (e.g., Mouquet and Loreau 2002; Mouquet et al., chapter 10) suggest that dispersal can have nonintuitive, nonmonotonic effects.

Regional diversity may be low in highly connected systems because dispersal homogenizes the metacommunity, and one species (the best competitor at the regional scale) excludes all others (even good local competitors); Forbes and Chase (2002) provide some experimental evidence for this prediction. Regional richness is also expected to increase as interfragment dispersal declines because habitat differences among fragments will support as many habitat specialists as there are fragment types. A similar conclusion was arrived at by Leibold (1998) based on a model of species-sorting in a heterogeneous landscape.

Habitat fragmentation is predicted to affect regional scale patterns of distribution and abundance. For example, there are good theoretical grounds for believing that the commonly reported positive relation between local abundance and regional distribution (Gaston et al. 2000) will be strongly influenced by habitat fragmentation (Hanski 1994). This relationship has been explained using a number of ecological processes, including niche breadth and metapopulation dynamics (Gaston et al. 1997; Kolasa and Drake 1998). The best empirically supported of these are two theories based on metapopulation dynamics, which both predict a positive abundance-occupancy relationship. First, the carrying capacity hypothesis assumes that interspecific variation in carrying capacity (local abundance) causes some species to have lower extinction rates and higher colonization rates than others (Nee et al. 1991). Second, the rescue effect hypothesis (Hanski 1991) emphasizes the importance of dispersal in preventing local extinction (the rescue effect); here, a positive abundance-occupancy relation arises because of the positive relation between immigration rate (causing rescue) and fragment occupancy (Hanski et al. 1993). Metapopulation models predict the loss of the positive abundance-occupancy relationship over time because fragmentation directly influences dispersal, fragment quality, and disturbance rate. In other words, species should slide-down the relation becoming increasingly less abundant and more spatially restricted. It is worth noting that these predictions derive from metapopulation theory in which interspecific interactions are either assumed to be nonexistent or weak. Recent metacommunity theory has not addressed the very general relation between abundance and patch occupancy (but see Bell 2001), nor has it addressed the expected relation between these two variables when interspecific interactions are explicitly incorporated (Holt et al. 2002).

This brief review of theory has highlighted a range of testable predictions regarding the effects of habitat fragmentation on natural communities. Unfortunately, the large temporal and spatial scales involved in the empirical study of habitat fragmentation have slowed the iterative cycle between theory and experiments, and limited our understanding of its effects (but see Debinski and Holt 1999; McGarigal and Cushman 2002). Natural microcosm experiments, unconstrained by the difficult problems of conducting experiments at large spatial scales, offer a more immediate means of testing and developing new theory (Srivastava et al. 2004). The microcosm experiments I describe below were conceived as a general experimental test of some of the theory described above. Here I shall use the results to address the following predictions:

LOCAL SCALE

1) Fragmentation will induce a community disassembly process characterized by declines in species richness (over and above that of an equivalent-sized random sample from continuous tract of habitat), abundance, and secondary biomass.

2) Species loss will be characterized by the preferential extinction of rare species (and traits correlated with rarity), as opposed to abundant and numerically dominant species.

3) By ensuring dispersal, habitat corridors will buffer fragments from the effects of fragmentation. In particular, they will delay or slow declines in species richness, abundance and biomass.

REGIONAL SCALE

4) Fragmentation will induce declines in regional species richness, an affect ameliorated by the presence of habitat corridors.

5) Fragmentation induced disruption of dispersal will prompt widespread declines in species' distribution and abundance.

All of the results I present here have been reported elsewhere (Gilbert et al. 1998; Gonzalez et al. 1998; Gonzalez 2000; Gonzalez and Chaneton 2002) where further statistical and methodological details can be found.

The Experimental System

Temperate moss ecosystems harbor a high density, species-rich microfauna of protists, rotifers, tardigrades, nematodes, collembolans, and mites (Gerson 1969; Davis 1981; Kinchin 1992). Epilithic mosses often form continuous carpets that may be considered as miniature landscapes that are large compared to the size and dispersal capacity of the organisms that inhabit them. The size, biotic complexity, and short generation time of many of the component taxa, makes this an ideal model system for confronting the practical problems posed by the large spatiotemporal scales associated with community responses to habitat fragmentation (Debinski and Holt 1999). The results reported here focus on the effect of habitat loss on the mites (Arachnida; Acarina), a diverse group of organisms including predatory and herbivorous species (feeding on fungi and decaying matter; Evans et al. 1961; Krantz 1978; Kinchin 1992) that account for more than 70% of the total arthropod density in this system.

The Experiments

Three experiments were conducted in the Derbyshire Peak District, northern England (OS map ref. SK264566). For each experiment, we selected large, flat limestone boulders covered by a mixture of the moss species *Hypnum cupressiforme* (Hedw.), *Thuidium tamariscinum* (Hedw.), and *Tortella tortuosa* (Hedw.). I used three distinct experimental designs to manipulate habitat connectivity (figure 6.1), two of which used continuous and broken habitat corridors to manipulate dispersal.

Experiment 1 examined the effects of fragmentation over time and did not manipulate habitat connectivity; as such it directly addresses predictions 1, 2, and 5. In March 1995 I established two treatments, (i) *control* and (ii) *fragmented,* in a randomized block design using eight moss-covered boulders. Each replicate boulder contained twelve randomly distributed circular moss fragments, measuring six 20 cm² and six 200 cm² (the slower dynamics of which are not considered here, but see Gonzalez 2000), and an adjacent continuous moss carpet acting as an undisturbed control of minimum area of 50 × 50 cm (figure 6.1a). The fragmented treatment was created using a template to ensure constancy in frag-

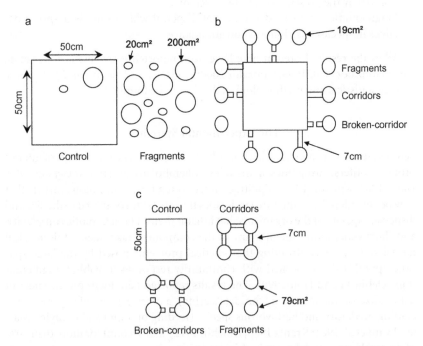

Figure 6.1 Schematic representation (not to scale) of a single replicate of the treatment design: (a) experiment 1, (b) experiment 2, and (c) experiment 3.

ment area and distance ($\geq$ 15 cm) between adjacent fragments. Habitat fragments were created on one half of the boulder by scraping and removing the moss cover; the remaining moss fragments were left surrounded by a sea of bare rock—a habitat considered inhospitable for most microarthropod taxa. Community responses to fragmentation were monitored over a twelve-month period that encompassed several generations for the larger predatory mites (and tens of generations for many of their nematode prey species). Every two months, one moss fragment was randomly chosen and removed from each block. Moss samples of equal area were also removed from the control treatment on each sampling date. This control allowed for seasonal changes in species abundance and diversity (Davis 1981; Schenker and Block 1989). Microarthropod fauna were extracted and enumerated in the laboratory (see below).

Experiment 2 tested the effects of habitat corridors and therefore addresses predictions 3 and 4. It was conducted between November 1996 and May 1997. We chose seven moss-covered boulders and set up four treatments in a randomized block design. Each block comprised a large continuous carpet of moss (50 $\times$ 50 cm) surrounded by twelve circular fragments of 19 cm^2 each. These fragments were randomly assigned to each of three treatments differing in degree of connectivity (figure 6.1b): (i) a *corridor* treatment, which connected fragments to the *mainland* with a 7 $\times$ 2 cm strip of live moss; (ii) a *broken corridor* treatment, which split corridors by a 2–3 cm gap of bare rock, and (iii) an *insular* treatment, which had no corridor to the mainland. The broken corridor treatment was created to control for the increase in habitat area represented by the presence of corridors and permitted the separation of habitat quantity and quality effects from the immigration (rescue) effect expected because of the presence of habitat corridors. In this design, the fragments may not be independent because they share a common mainland block of moss. The risk of a Type 1 error therefore depends upon whether there is an interdependence amongst the fragments that affects their spatial dynamics (Scheiner 2001). This should be kept in mind when interpreting the effects presented below for this experiment. The microarthropod fauna was sampled three and six months after establishment. At each date, and from each block, we removed one randomly chosen fragment from each treatment, and one equivalently sized sample from the control mainland. Two of the four replicate fragments were therefore not used. Only the results of the last date are presented here.

Experiment 3 was a second corridor experiment that adopted a different design and was intended to verify predictions 1, 3, and 4. It ran for six months between November 1995 and May 1996. This design also consisted of four experimental treatments, each consisting of four circular islands 79 cm^2. Islands centers were placed at the corners of a square with a side of length 17cm; treatments were at least 10 cm apart on each rock. The fragmentation treatments (figure 6.1c) were (i) *mainland*—four circular samples taken from the adjacent matrix of continuous moss; (ii) *corridor*—four fragments connected along the sides of the square by corridors 7 cm long by 1 cm wide; (iii) *broken corridor*—like the corridor treatment, but corridors split in the middle and separated by a gap of 5 cm; and (iv) *insular*—no corridors present. There were thus four independent, replicate fragments per treatment, with all four treatments replicated on six different rocks in a randomized block design. All replicates were destructively sampled once after six months whereon the experiment was terminated.

The microarthropods were extracted from the moss samples using a heat and light gradient generated by a Tullgren funnel. The moss samples were treated for seventy-two hours; emerging specimens were collected in an alcohol/glycerol/water (7 : 2 : 1) mix and sorted into morphospecies (Krantz 1978). Mites with distinct sets of morphological characters were considered the operational taxonomic entities. Thus, the term species diversity (or richness) refers to different microarthropod morphotypes. A species was considered locally extinct from a replicate landscape when it no longer appeared in the suite of organisms extracted with the Tullgren funnel. This is necessarily a working definition. Although the extraction efficiency of this system is high, it was impossible to ascertain whether a species did not occur in a given fragment. Most important for this definition of "extinction" was the identical extraction regimes experienced by the control and fragmented treatments. Mean population abundance and standing biomass were assessed for all individual morphospecies on each sampling date. I calculated microarthropod species biomass in each moss sample of experiments 1 and 2 using the following empirical relationship between individual weight and body size (length): weight $= -1.15$ length$^{3.07}$ (Gonzalez and Chaneton 2002). Biomasses (mg) of individual species were summed to estimate total community biomass within a moss fragment. The principal response variables were local species richness, regional richness (only for experiment 3, calculated as the cumulative richness of the four fragments), population abundance, patch occupancy (proportion of fragments occupied), and total community abundance and biomass.

Results
Local Scale
PREDICTION 1: COMMUNITY DISASSEMBLY

Fragmentation of the moss-habitat had a strong and temporally delayed effect on mean local richness in remnant fragments. Figure 6.2a shows that a significant separation of the control and treatment trajectories became apparent at six months and increased over time. By the end of the experiment, fragmentation reduced species richness by 40%. These delayed extinction dynamics (an extinction debt) depended on fragment surface area (Gonzalez 2000).

Fragmentation also induced declines in community abundance and biomass (figure 6.2b, c), although for both variables there were only small, nonsignificant

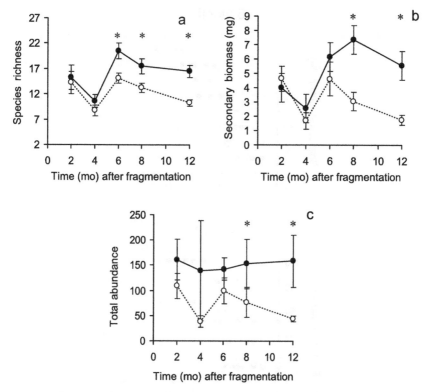

Figure 6.2 Changes in mean species diversity: (a) species richness, (b) community biomass, and (c) total abundance of microarthropods in 20 cm² moss patches extracted from continuous (solid circles) and fragmented (open circles) moss carpets. Vertical bars show ± 1 standard error; asterisks indicate significant (P < 0.01) treatment effects with ANOVA following unplanned pairwise comparisons using a Bonferroni *t*-test within dates (modified from Gonzalez and Chaneton 2002).

(P > 0.05), effects over the first six months. The delayed effect of isolation on heterotrophic biomass emerged after eight months, by which time microarthropod biomass in the control was twice that in the remnant fragments. The relation between biomass, abundance, and species richness, in particular the relative contribution of the latter two variables in statistically explaining the change in community biomass, was examined by multiple regression. This analysis indicated that community abundance (slope = 0.63, SE = 0.11, P < 0.001) and species richness (slope = 0.02, SE = 0.01, P < 0.008) were both positively and linearly related with community biomass; together they explained 65% of its variation ($F_{2,35}$ = 35.5, P < 0.0001). However, species richness explained only a small fraction (6.7%) of the change in community biomass when the effect of community abundance was accounted for.

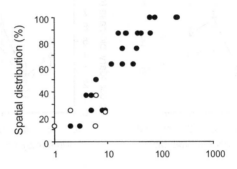

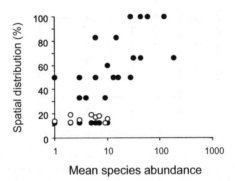

Figure 6.3 Figure depicts the patterns of microarthropod extinction in response to fragmentation relative to individual species abundance (mean density per fragment, log scale) and distribution (percentage of occupied fragments) in continuous moss landscapes. Local species extinctions in the fragments are denoted by the empty symbols (modified from Gonzalez and Chaneton 2002).

PREDICTION 2: BIASED EXTINCTION

Logistic regression revealed that species suffering extinction in fragments had lower population abundance prior to extinction ($\chi^2 = 20.2$, d.f. $= 1$, $P < 0.001$), and occupied fewer fragments prior to extinction ($\chi^2 = 16.4$, d.f. $= 1$, $P < 0.001$) than the extant species (figures 6.3a, b). Only in experiment 3 did we establish a significant effect of fragmentation on the proportion of predators in the local communities; we found a lower proportion of predators in the broken corridor treatment than in the corridor treatment (ANOVA [analysis of variance], $F_{3,15} = 11.6$, $P < 0.001$; Gilbert et al. 1998).

PREDICTION 3: HABITAT CORRIDORS WILL SLOW OR DELAY THE EFFECTS
OF FRAGMENTATION

The presence of corridors ameliorated the negative effects of habitat loss and isolation on arthropod community structure and biomass. In experiment 2, patch isolation decreased mean local species richness after six months (figure 6.4a). Species richness did not differ between mainland and corridor treatments, or between the broken-corridor and insular ones. When pooled, the insular and broken corridor treatments contained an average of 34% fewer species than the corridor and mainland treatments. After six months, mean community biomass was significantly higher in connected than in disconnected moss patches. Insular and broken corridor patches experienced, on average, a 62% reduction in community biomass relative to connected patches (figure 6.4b). There was no difference in mean community biomass between the mainland and corridor treatments. Overall microarthropod abundance was 60–80% greater in patches of mainland and corridor treatments than in broken-corridor and insular ones (figure 6.4c). No significant difference in arthropod abundance was found between connected or between disconnected treatments.

Experiment 3 confirmed the positive effects of corridors on local species richness. After six months local richness was 30% lower in the broken corridor treatments than the corridor treatment, and total community abundance in the corridor treatments was 60% greater than the broken corridor treatments.

Regional Scale

PREDICTION 4: FRAGMENTATION WILL REDUCE REGIONAL SPECIES
RICHNESS—AN AFFECT AMELIORATED BY THE PRESENCE OF
HABITAT CORRIDORS

In experiment 3 we found a significant negative effect of fragmentation on the cumulative species richness of all four fragments ($F_{3,23} = 28.1$, $P < 0.001$). Regional richness was 27% lower in the broken corridor treatment than in the corridor treatment. Analysis of the total number of individuals of all four fragments (regional abundance) also indicated a strong effect of fragmentation ($F_{3,23} = 20.3$, $P < 0.001$). Regional abundance was 52% lower in the broken corridor than in the corridor treatment.

PREDICTION 5: FRAGMENTATION-INDUCED DISRUPTION OF DISPERSAL
WILL PROMPT WIDESPREAD DECLINES IN SPECIES' DISTRIBUTION
AND ABUNDANCE

Figure 6.5a confirms the existence of the positive relation between and abundance and patch occupancy for the microarthropods in this system (data from experiment 1). Comparison of the continuous and broken corridor treatments at the end

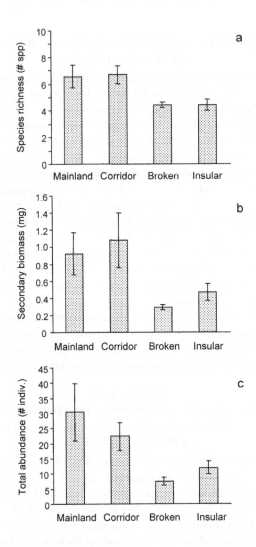

Figure 6.4 Experiment 2: Effect of isolation and habitat corridors after six months on (a) species diversity, (b) standing biomass, and (c) total abundance of moss-living microarthropods. Bars represent means ± 1 standard error of the mean. ANOVA models that included the appropriate orthogonal contrasts were used to test for significant differences in arthropod community structure between connected (mainland and corridor) and disconnected (fragment and broken corridor) fragments. For each of the three response variables studied there were significant differences ($P \ll 0.05$) between the connected and disconnected treatments. No significant differences were found between the connected treatments, or between the disconnected treatments (modified from Gonzalez and Chaneton 2002).

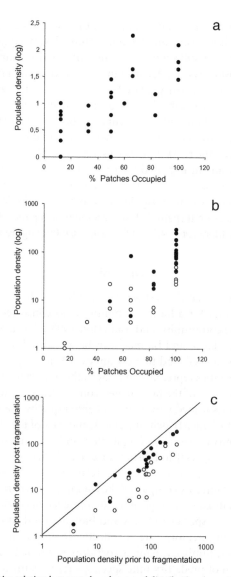

Figure 6.5 (a) The positive relation between abundance and distribution (percentage of fragments occupied) for the microarthropods inhabiting the control moss fragments at twelve months; (b) the relation between abundance and distribution for those species surviving fragmentation in the broken corridors (open circles), and for the same species from the control sample communities (filled circles); (c) the difference in abundance (log scale) of each species in the corridor (filled circles) and broken corridor (open circles) communities before and after fragmentation. Points below the 1:1 lines show declines due to fragmentation (modified from Gonzalez et al. 1998).

of experiment 3 allowed us to directly assess the role dispersal plays in maintaining the positive abundance-occupancy relation. Although surviving species in the broken corridor treatment still showed a significant positive abundance-occupancy relationship, almost all species declined in abundance (paired t test: $t = 4.98$, d.f. $= 20$, $P < 0.001$) and occupancy (paired t test: $t = 4.50$, d.f. $= 20$, $P < 0.001$) relative to the controls (figure 6.5b). Species surviving in the broken corridor and fragmented treatments did not differ in occupancy, but the abundances were significantly lower in the broken corridor treatment (paired t tests: $t = -2.46$, d.f. $= 20$, $P < 0.001$). Although the remaining species in the corridor treatment also showed declines in both abundance (paired t tests: $t = 4.92$, d.f. $= 20$, $P < 0.001$) and occupancy (paired t tests: $t = 3.19$, d.f. $= 20$, $P < 0.001$) relative to the continuous controls, these declines were of significantly lower magnitude than those in the broken corridor treatment (paired t tests: fragment occupancy, $t = 2.90$, d.f. $= 20$, $P < 0.01$; $\log_{10}$ abundance, $t = 4.31$, d.f. $= 20$, $P < 0.001$; figure 6.5c).

Discussion

It has been suggested that the best way to establish the short-term consequences of regional dynamics for a local community is to construct a dispersal-proof barrier around the community, and wait (Holt 1993). Any large changes in species richness, abundance and biomass can be attributed to the breakdown of intercommunity dispersal, and to a certain extent the response of these microarthropod communities represents an assay of the importance of dispersal in determining the structure of the nonisolated communities. However, habitat fragmentation is also associated with reduced habitat quality (Saunders et al. 1991) and increased disturbance (Cochrane 2001), and part of the challenge of understanding community dynamics in fragmented landscapes is establishing the relative importance of dispersal and disturbance (e.g., Fahrig 1992, Warren 1996).

In the experiments presented here the relative roles of dispersal and disturbance can be ascertained by comparing the community dynamics in corridor and broken-corridor landscapes. In general, the presence of corridors helped maintain microarthropod species richness and biomass production (figure 6.4). In contrast, the broken corridors were not effective in preventing the disassembly of these communities. Similarly, habitat corridors permitted the persistence of a greater proportion of predatory mite species; a result confirming the top-down community disassembly predicted by theory (Cohen and Newman 1991; Holt 2002). Given that the only difference between the two corridor treatments was a small break in the moss cover, it is unlikely that the corridor effects can be explained by differences in microclimate conditions between the two treatments. Together these results indicate that dispersal plays an important role in these microarthropod communities (see also Cottenie and De Meester, chapter 8, and Miller and Kneitel, chapter 5).

The Creation of Sink Conditions

The observation that the corridor communities in experiments 2 and 3 continued to show declines in species richness and biomass (figure 6.5b, c), albeit at a substantially slower rate than the broken corridor communities, suggests that dispersal cannot entirely compensate for the negative effects of fragmentation. This is probably because habitat fragmentation generally reduces habitat quality and increases disturbance rates (Saunders et al. 1991; Fahrig 1992; Harrison 1994; Didham et al. 1998a; Laurance et al. 2002). In this system the increased edge to area ratio of the fragments increased the frequency of the wetting and drying cycle of the moss substrate (pers. observ.); this is the factor most likely to alter the long-term abundance and diversity of microarthropods within isolated moss patches (Berthet 1964; Block 1966; Davis 1981; Schenker and Block 1986; Siepel 1996). The decreased abundance and richness of microarthropods within corridor treatments presumably reflected this change in moss-substrate quality (Vannier 1972; Davis 1981), and suggests that dispersal could not entirely compensate for these changes. Thus the declines in population abundances indicate that, for a large proportion of the microarthropod fauna, fragmentation transformed the moss fragments into habitat sinks (with long-term mortality greater than long-term reproduction; Holt 1993).

Spatial Aspects of Community Structure and Function

The delayed separation in the levels of species richness in the control and fragmented treatments of experiment 1 (figure 6.2a) provides evidence for the existence of an extinction debt in this system (Tilman et al. 1997). The declines in secondary biomass concurrent with declines in species richness suggest that habitat fragmentation also generated a *functioning debt:* a delayed alteration in ecosystem attributes driven by the decline of species persisting in remnant patches. This result also demonstrates that although declines in both abundance and species richness were concurrent with postfragmentation losses in community biomass, the initial declines in species richness had a relatively small impact on community functioning. For example, the first significant reduction of species richness (25% relative to the control patches after six months) was associated with rare species and did not correspond to significant changes in secondary biomass (figure 6.2b). At eight months the mean proportional species loss remained the same, whereas the mean heterotrophic biomass in the remnant fragments had decreased to 59% of control levels. These species abundance dynamics (figure 6.2c), as opposed to those of species loss, dominated post-fragmentation change in heterotrophic biomass (the losses of rare species also provided early signs of altered community functioning). This result suggests that analyses of diversity-function relations in recently fragmented communities should track the transient dynamics of remnant populations.

The beneficial effect of dispersal for local heterotrophic biomass production suggests the importance of considering diversity dynamics at the metacommunity scale for a more complete understanding of community function (Loreau et al., chapter 18; Loreau et al. 2003; Lundberg and Moberg 2003). Our experimental results indicate that the persistence of rare species (including predators), and the abundance of common species, may strongly depend on immigration from nearby patches. By preserving a flow of immigrants from less disturbed habitat, corridors prevented the short-term changes in moss substrate quality from being translated into reduced secondary biomass. These results thus support the broader hypothesis that disruption of habitat connectivity will induce widespread local community disfunctioning (Bond and Chase 2002; Gonzalez and Chaneton 2002; Loreau et al. 2003; Lundberg and Moberg 2003).

Response at the Regional Scale

Regional diversity declined in the absence of dispersal corridors, suggesting that dispersal was involved in the maintenance of larger scale diversity in these communities. The regional diversity (the species richness of the fragments in experiment 3) is interesting in the context of whether many small fragments can support more species than an equivalent sized continuous area (Higgs 1981; Boecklen 1997). The presence of corridors might affect this because movement could increase the homogeneity of species composition across connected islands. The idea predicts that the local per-island species richness may be greater than the equivalent disconnected system, but the cumulative species richness for the whole connected system might be lower (see also Chase et al., chapter 14). However, the data presented here (Gilbert et al. 1998) indicate that the regional diversity of connected systems is greater than that of disconnected ones; there is no evidence of reduced diversity caused by homogenizing movement. These results contrast with those of Forbes and Chase (2002), who found the opposite effect of dispersal in experimental zooplankton communities; dispersal had a homogenizing effect that reduced regional diversity. Further experiments directly manipulating dispersal intensity are required if we are to better understand the relation between dispersal and regional diversity in this and other systems (i.e., unimodal as predicted by patch dynamic models, or linear and negative as predicted by metacommunity models, Leibold et al. 2004).

The results of experiment 3 show that, at the regional scale, habitat fragmentation resulted in a near universal decline in the distribution and abundance of microarthropod species. This is exactly as theory predicts. These declines are in part a consequence of reducing or eliminating immigration between patches in the fragmented landscape (Hanski et al. 1993). The presence of corridors, but not broken corridors, ameliorated the effects of fragmentation. The differences between the continuous and broken corridor treatments count against explanations based on changes in microclimate; the small structural differences between the broken and continuous corridor treatments were unlikely to have generated the large differences in abundance and patch occupancy observed between the two. Thus, the corridors in this experiment facilitated intercommunity dispersal and helped maintain the positive relation between distribution and abundance, presumably through the rescue effect.

Dispersal may not be the ultimate cause of the positive abundance-distribution relationship, but the necessity of dispersal for its maintenance suggests that it is at the very least important in its establishment and persistence (recent neutral metacommunity models suggest that niche differences between species are not required to generate this relation; Bell 2001). Interestingly both the rescue effect hypothesis (Hanski 1991) and the carrying capacity hypothesis (Nee et al. 1991) derive from models that assume that interspecific interactions are weak or absent; in other words the landscape is made up of an ensemble of metapopulations. This begs the question, to what extent do we need a metacommunity theory to explain the positive relation between abundance and patch occupancy? Current metacommunity theory has not addressed this pattern (but see Bell 2001), although current source-sink metacommunity models do not predict a general positive relation between range and abundance (M. Loreau, pers. comm.). Whether species interactions are an important part of the theoretical explanation for the positive relation between abundance and patch occupancy remains to be seen. However, Holt et al. (2002) present results from a laboratory experiment using protozoan communities that tested the effect of species interactions in a multipatch landscape. They demonstrate "better defined" abundance-occupancy relations in interacting metapopulations (metacommunities) than in noninteracting metapopulations (although it seems that in their system coexistence was not dependent on mass effects). This result suggests that our understanding of widespread ecological patterns such as the relation between distribution and abundance may have much to gain from the development of metacommunity theory.

Conclusions

A series of experiments, using a bryophyte-based microecosystem, were conducted to examine the manifold effects of habitat fragmentation. I have documented the temporal declines in microarthropod species abundance, biomass and species richness during community disassembly. Extinctions were delayed and biased toward rare species, and thus occurred initially without any significant changes in total community abundance and biomass. Eventual reductions in abundance and biomass lagged behind the observed declines in species richness. The presence of moss-habitat corridors coupling fragments to each other, or to a large "mainland" area of moss, greatly slowed the declines in microarthropod richness, abundance, and biomass at both local and regional scales. Some evidence was found indicating a greater susceptibility of predators to fragmentation.

Regional diversity, the cumulative species richness of four fragments, was greater in connected than fragmented landscapes. Declines in abundance and patch occupancy in response to fragmentation were widespread, emphasizing the importance of dispersal for the maintenance of this mesoscale pattern.

By illustrating the fundamental importance of dispersal for community dynamics at both local and regional scales these results appeal strongly to metacommunity theory (i.e., study of the patterns and processes deriving from community coupling across space). In particular, because fragmentation alters habitat quality and landscape connectivity these results would seem to be best explained within a mass effects paradigm (see Holyoak et al., chapter 1). Future experiments should seek to establish the relative importance of changes in habitat quality (altered disturbance rates, edge effects, etc.) and connectivity as explanations for the effects of fragmentation in this and other systems. Microcosm experiments of this sort, coupled to appropriate metacommunity models, represent an obvious avenue for the direct application of metacommunity theory to issues of important conservation concern. Indeed, explication of the multifarious effects of habitat fragmentation on community pattern and process will represent an important challenge for the future application of metacommunity theory.

Acknowledgments

I would like to thank the editors for inviting me to make this contribution. I would also like to thank John Lawton for his support during early stages of this work. Francis Gilbert was a constant source of encouragement throughout this research. Thanks also to all the people who have coauthored papers with me on this project—for all the help they have provided, and the many thought provoking discussions we've had on these topics. Karl Cottenie, Marcel Holyoak, and an anonymous reviewer provided many very helpful comments that much improved the text.

Literature Cited

Andrén, H. 1994. Effects of habitat fragmentation on birds and mammals in landscapes with different proportions of suitable habitat: A review. Oikos 71:355–366.

Bascompte, J., and R. V. Solé. 1996. Habitat fragmentation and extinction thresholds in spatially explicit models. Journal of Animal Ecology 65:465–473.

Beier, P., and R. F. Noss. 1998. Do habitat corridors really provide connectivity? Conservation Biology 12:1241–1252.

Bell, G. 2001. Neutral macroecology. Science 293:2413–2418.

Bender, J., T. A. Contreras, and L. Fahrig. 1998. Habitat loss and population decline: A meta-analysis of the patch size effect. Ecology 79:517–533.

Berthet, P. L. 1964. A field study of the mobility of Oribatei (Acari) using radioactive tagging. Journal of Animal Ecology 33:443–449.

Block, W. 1966. Seasonal fluctuations and distribution of mite populations in moorland soils, with a note on biomass. Journal of Animal Ecology 35:487–503.

Boecklen, A. J. 1997. Nestedness, biogeographic theory, and the design of nature reserves. Oecologia 112:123–142.

Bond, E. M., and J. M. Chase. 2002. Biodiversity and ecosystem functioning at local and regional scales. Ecology Letters 5:467–470.

Boulinier, T., J. D. Nichols, J. E. Hines, J. R. Sauer, C. H. Flather, and K. H. Pollock. 1998. Higher temporal variability of forest breeding bird communities in fragmented landscapes. Proceedings of the National Academy of Sciences, USA 95:7497–7501.

Brooke, B. W., N. S. Sodhi, and P. K. L. Ng. 2003. Catastrophic extinctions follow deforestation in Singapore. Nature 424:420–423.

Brooker, L., M. Brooker, and P. Cale. 1999. Animal dispersal in fragmented habitat: Measuring habitat connectivity, corridor use, and dispersal mortality. Conservation Ecology 3:4.

Carpenter, S., B. Walker, M. J. Anderies, and N. Abel. 2001. From metaphor to measurement: Resilience of what to what? Ecosystems 4:765–781.

Cochrane, M. A. 2001. Synergistic interactions between habitat fragmentation and fire in evergreen tropical forests. Conservation Biology 15:1515–1521.

Cohen, J., and C. M. Newman. 1991. Community area and food-chain length—theoretical predictions. American Naturalist 138:1542–1554.

Collinge, S. K. 2000. Effects of grassland fragmentation on insect species loss, colonization, and movement patterns. Ecology 81:2211–2226.

Conner, E. F., and E. D. McCoy. 1979. The statistics and biology of the species-area relationship. American Naturalist 113:791–833.

Davis, R. C. 1981. Structure and function of two Antarctic terrestrial moss communities. Ecological Monographs 5:125–143.

Debinski, D. M., and R. D. Holt. 1999. A survey and overview of habitat fragmentation experiments. Conservation Biology 14:342–355.

Diamond, J. M. 1984. "Normal" extinctions of isolated populations. Pages 191–246 in M. Nitecki. ed. Extinctions. University of Chicago Press, Chicago, IL.

Didham, R. K., P. M. Hammond, J. H. Lawton, P. Eggleton, and N. E. Stork. 1998. Beetle species responses to tropical forest fragmentation. Ecological Monographs 68:295–323.

Didham, R. K., J. H. Lawton, P. M. Hammond, and P. Eggleton. 1998. Trophic structure stability and extinction dynamics of beetles (Coleoptera) in tropical forest fragments. Philosophical Transactions of the Royal Society of London, B 353:437–451.

Donahue, M. J., M. Holyoak, and C. Feng. 2003. Patterns of dispersal and dynamics among habitat patches varying in quality. American Naturalist 162:302–317.

Earn, D. J. D., S. A. Levin, and P. Rohani. 2000. Coherence and conservation. Science 290:1360–1364.

Evans, G. O., J. G. Sheals, and D. McFarlane. 1961. The terrestrial Acari of the British Isles. Bartholomew Press, London, UK.

Fahrig, L. 1992. Relative importance of spatial and temporal scales in a patchy environment. Theoretical Population Biology 42:300–314.

———. 2002. Effect of habitat fragmentation on the extinction threshold: A synthesis. Ecology 12: 346–353.

Forbes, A. E., and J. M. Chase. 2002. The role of habitat connectivity and landscape geometry in experimental zooplankton metacommunities. Oikos 96:433–440.

Gaston, K. J., T. M. Blackburn, J. J. D. Greenwood, R. D. Gregory, R. M. Quinn, and J. H. Lawton. 2000. Abundance-occupancy relationships. Journal of Applied Ecology 37:39–59.

Gaston, K. J., T. M. Blackburn, and J. H. Lawton. 1997. Interspecific abundance-range relationships: an appraisal of mechanisms. Journal of Animal Ecology 66:579–601.

Gerson, U. 1969. Moss-arthropod associations. Bryologist 72:495–500.

Gilbert, F. S., A. Gonzalez, and I. Evans-Freke. 1998. Corridors maintain species richness in the fragmented landscapes of a microecosystem. Proceedings of the Royal Society of London, series B, Biological Sciences 265:577–582.

Golden, D. M., and T. O. Crist. 1999. Experimental effects of habitat fragmentation on old-field canopy insects: Community, guild and species responses. Oecologia 118:371–380.

Gonzalez, A. 2000. Community relaxation in fragmented landscapes: The relation between species, area and age. Ecology Letters 3:441–448.

Gonzalez, A., and E. Chaneton. 2002. Heterotroph species extinction, abundance and biomass dynamics in an experimentally fragmented microecosystem. Journal of Animal Ecology 71:594–602.

Gonzalez, A., and R. D. Holt. 2002. The inflationary effects of environmental fluctuations in source-sink systems. Proceedings of the National Academy of Sciences, USA 99:14872–14877.

Gonzalez, A., J. H. Lawton, F. S. Gilbert, T. M. Blackburn, and I. Evans-Freke. 1998. Metapopulation dynamics, abundance, and distribution in a microecosystem. Science 281:2045–2047.

Goodwin, B. J. 2003. Is landscape connectivity a dependent or independent variable? Landscape Ecology 18:687–699.

Grime, J. P. 1998. Benefits of plant diversity to ecosystems: Immediate, filter and founder effects. Journal of Ecology 86:902–910.

Haddad, N. M., and K. A. Baun. 1999. An experimental test of corridor effects on butterfly densities. Ecological Applications 9:623–633.

Haddad, N. M., D. R. Browne, A. Cunningham, B. J. Danielson, D. J. Levey, S. Sargent, and T. Spira. 2003. Corridor use by diverse taxa. Ecology 84:609–615.

Hanski, I. 1991. Reply to Nee, Gregory and May. Oikos 62:88–89.

———. 1994. Patch-occupancy dynamics in fragmented landscapes. Trends in Ecology and Evolution 9:131–135.

Hanski, I., J. Kouki, and A. Halkka. 1993. Three explanations of the positive relationship between distribution and abundance of species. Pages 108–116 in R. Ricklefs, and D. Schluter, eds. Species diversity in ecological communities: Historical and geographical perspectives. University of Chicago Press, Chicago, IL.

Hanski, I., and D. Simberloff. 1997. The metapopulation approach, its history, conceptual domain, and application to conservation. Pages 5–26 in I. A. Hanski and, M. E. Gilpin eds. Metapopulation Biology: Ecology, genetics and evolution. Academic Press, London, UK.

Harris, L. D., and J. Scheck. 1991. From implications to applications: the dispersal corridor principal applied to the conservation of biological diversity. Pages 189–220 in D. A. Saunders, and R. J. Hobbs eds. Nature conservation 2: The role of corridors. Surrey Beatty and Sons, Chipping Norton, UK.

Harrison, S. 1994. Metapopulations and conservation. Pages 111–128 in P. J. Edwards ed. Large scale ecology and conservation biology. Blackwell Science, Oxford, UK.

Harrison, S. 1999. Local and regional diversity in a patchy landscape: native, alien, and endemic herbs on serpentine. Ecology 80:70–80.

Higgs, A. J.1981. Island biogeography theory and nature reserve design. Journal of Biogeography 8: 117–124.

Holt, A. R., P. H. Warren, and K. J. Gaston. 2002. The importance of biotic interactions in abundance-occupancy relationships. Journal of Animal Ecology 71:846–854.

Holt, R. D. 1993. Ecology at the mesoscale: The influence of regional processes on local communities. Pages 108–116 in R. Ricklefs, and D. Schluter eds. Species diversity in ecological communities: Historical and geographical perspectives. University of Chicago Press, Chicago, IL.

Holt, R. D. 2002. Food webs in space: On the interplay between dynamical instability and spatial processes. Ecological Research 17:261–273.

Holt R. D., J. H. Lawton, G. A. Polis, and N. D. Martinez. 1999. Trophic rank and the species-area relationship. Ecology 80:1495–1504.

Holyoak, M. 2000. Habitat subdivision causes changes in food web structure. Ecology Letters 3:509–515.

Hoyle, M. 2004. Causes of the species-area relationship by trophic level in a field-based microecosystem. Proceedings of the Royal Society, series B, Biological Sciences 271:1159–1164.

Hugens, B. R., and N. M. Haddad. 2003. Predicting which species will benefit from corridors in fragmented landscapes from population growth models. American Naturalist 161:808–820.

Jenkins, B., R. L. Kitching, and S. L. Pimm. 1992. Productivity, disturbance and food web structure at a local spatial scale in experimental container habitats. Oikos 65:242–255.

Kareiva, P., and U. Wennergren. 1995. Connecting landscape patterns to ecosystem and population processes. Nature 373:299–302.

Kinchin, I. M. 1992. An introduction to the invertebrate microfauna associated with mosses and lichens, with observations from maritime lichens on the west coast of the British Isles. Microscopy 36:721–731.

Kolasa, J., and J. A. Drake. 1998. Abundance and range relationship in a fragmented landscape: connections and contrasts between competing models. Coenoses 13:79–88.

Komonen, A., R. Penttila, M. Lindgren, and I. Hanski. 2000. Forest fragmentation truncates a food chain based on an old-growth forest bracket fungus. Oikos 90:119–126.

Krantz, G. W. 1978. A manual of acarology. Oregon State University, Corvallis, OR.

Kruess, A., and T. Tscharntke. 1994. Habitat fragmentation, species loss, and biological control. Science 264:1581–1584.

Laurance, W. F., T. E. Lovejoy, H. L. Vasconcelos, E. M. Bruna, R. K. Didham, P. C. Stouffer, C. Gascon, R. O. Bierregaard, S. G. Laurance, and E. Sampaio. 2002. Ecosystem decay of Amazonian forest fragments: A twenty-two-year investigation. Conservation Biology 16:605–618.

Lawton, J. H. 1994. What do species do in ecosystems? Oikos 71:367–374.

———. 1995. Population dynamic principles. Pages 146–163 in Lawton, J. H. and, R. M. May eds. Extinction rates. Oxford University Press, Oxford, UK.

Leibold, M. A. 1998. Similarity and coexistence of species in regional biotas. Evolutionary Ecology 12:95–110.

Leibold, M. A., M. Holyoak, N. Mouquet, P. Amarasekare, J. Chase, M. Hoopes, R. Holt, J. Shurin, R. Law, D. Tilman, M. Loreau, and A. Gonzalez. 2004. The metacommunity concept: A framework for multi-scale community ecology. Ecology Letters 7:601–613.

Loreau, M., N. Mouquet, and A. Gonzalez. 2003. Biodiversity as spatial insurance in heterogeneous landscapes. Proceedings of the National Academy of Sciences, USA 100:12765–12770.

Lundberg, J., and F. Moberg. 2003. Mobile link organisms and ecosystem functioning: Implications for ecosystem resilience and management. Ecosystems 6:87–98.

May, R. M., Lawton, J. H., Stork, N. E. 1995. Assessing Extinction Risk. Pages 1–25 in Lawton, J. H. and, R. M. May, eds. Extinction rates. Oxford University Press, Oxford, UK.

McGarigal, K., and S. A. Cushman. 2002. Comparative evaluation of experimental approaches to the study of habitat fragmentation effects. Ecological Applications 12:335–345.

Melián, C. J., and J. Bascompte. 2002. Food web structure and habitat loss. Ecology Letters 5:37–46.

Mikkelson, G. M. 1993. How do food webs fall apart - a study of changes in trophic structure during relaxation on habitat fragments. Oikos 67:539–547.

Mouquet, N., and M. Loreau. 2002. Coexistence in metacommunities: The regional similarity hypothesis. American Naturalist 159:420–426.

Nee, S., R. D. Gregory, and R. M. May. 1991. Core and satellite species: Theory and artefacts. Oikos 62:83–87.

Nee, S., and R. M. May. 1992. Dynamics of metapopulations: Habitat destruction and competitive coexistence. Journal of Animal Ecology 61:37–40.

Noss, R. F. 1994. Creating regional reserve networks. Pages 289–290 in G. K. Meffe, and C. R. Carroll eds. Principles of conservation biology. Sinauer, Sunderland, MA.

Patterson, B. D. 1984. Mammalian extinction and biogeography in the Southern Rocky mountains. Pages 247–293 *in* M. Nitecki, ed. *Extinctions*. University of Chicago Press, Chicago, IL.

———. 1990. On the temporal development of nest subsets patterns of species composition. Oikos 59:330–342.

Patterson, B. D. and J. H. Brown. 1991. Regionally nested patterns of species composition in granivorous rodent assemblages. Journal of Biogeography 18:395–402.

Pickett, S. T. A., and M. L. Cadenasso. 1995 Landscape ecology: Spatial heterogeneity in ecological systems. Science 269:331–334.

Pimm, S. L., and J. H. Lawton. 1977. The number of trophic levels in ecological communities. Nature 268:329–331.

Pimm, S. L., G. J. Russell, J. L. Gittleman, and T. M. Brooks. 1995. The future of biodiversity. Science 269:347–250.

Riitters, K. H., J. D. Wickham, R. V. O'Neill, K. B. Jones, E. R. Smith, J. W. Coulston, T. G. Wade, and J. H. Smith. 2002. Fragmentation of continental United States forests. Ecosystems 5:815–822.

Sala, O. E., W. K. Lauenroth, S. J. McNaughton, G. Rusch, and X. Zhang. 1996. Biodiversity and ecosystem function in grasslands. Pages 129–149 *in* H. A. Mooney, J. H. Cushman, E. Medina, O. E. Sala, and E. D. Schulze, eds. *Functional roles of biodiversity: A global perspective.* John Wiley and Sons, London, UK.

Saunders, D., R. J. Hobbs, and C. R. Margules. 1991. Biological consequences of ecosystem fragmentation: A review. Conservation Biology 5:18–32.

Scheiner, S. M. 2001.Theories, hypotheses and statistics. Pages 3–14 *in* S. M. Scheiner, and J. Gurevitch eds. *Design and analysis of ecological experiments.* Oxford University Press, Oxford, UK.

Schenker, R., and W. Block. 1986. Micro-arthropod activity in three contrasting terrestrial habitats on Signy Island, Maritime Antarctic. British Antarctic Survey Bulletin 71:36–43.

Schoener, T. W. 1989. Food webs from small to large. Ecology 70:1559–1589.

Seabloom, E. W., A. P. Dobson, and Stoms, D. M. 2002. Extinction rates under nonrandom patterns of habitat loss. Proceedings of the National Academy of Sciences, USA 99:11229–11234.

Siepel, H. 1996. The importance of unpredictable and short-term environmental extremes for biodiversity in oribatid mites. Biodiversity Letters 3:36–34.

Simberloff, D., J. A. Farr, J. Cox, and D. W. Mehlman.1992. Movement corridors: Conservation bargains or poor investments? Conservation Biology 1:63–71.

Srivastava, D. S., J. Kolasa, J. Bengtsson, A. Gonzalez, S. P. Lawler, T. E. Miller, P. Munguia, T. Romanuk, D. C. Schneider, and M. K. Trzcinski. 2004. Are natural microcosms useful model systems for ecology. Trends in Ecology and Evolution 19:379–384.

Stone, L. 1995. Biodiversity and habitat destruction: A comparative study of model forest and coral reef ecosystems. Proceedings of the Royal Society of London, series B, Biological Sciences 261:381–388.

Tewksbury, J. J., D. J. Levey, N. M. Haddad, S. Sargent, J. L. Orrock, A. Weldon, B. J. Danielson, J. Brinkerhoff, E. I. Damschen, and P. Townsend. 2002. Corridors affect plants, animals and their interactions in fragmented landscapes. Proceedings of the National Academy of Sciences, USA 99: 12923–12926.

Tilman, D., C. L. Lehman, and C. J. Yin. 1997. Habitat destruction, dispersal, and deterministic extinction in competitive communities. American Naturalist 149:407–435.

Turner, M. G. 1989. Landscape Ecology: The effect of pattern on process. Annual Review of Ecology and Systematics 20:171–197.

Turner, M. G., and R. H. Gardner. 1991. *Quantitative methods in landscape ecology: The analysis and interpretation of landscape heterogeneity.* Springer-Verlag, New York, NY.

Vannier, G. 1972. Modèle expérimental pour analyser le comportement des Microarthropodes terrestres soumis à différents régimes hydriques dans leur biotope. Comptes Rendus de l'Académie des Sciences, Paris 274:1942–1945.

Vitousek, P. M. 1994. Beyond global warming: Ecology and global change. Ecology 75:1861–1876.

Warren, P. H. 1996. The effects of between-habitat dispersal rate on protist communities and meta-communities in microcosms at two spatial scales. Oecologia 105:132–140.

Wilson, E. O., and E. Willis 1975. Applied biogeography. Pages 522–534 *in* M. L. Cody, and J. M. Diamond, eds. *Ecology and evolution of communities.* Harvard University Press, Cambridge, MA.

With, K. A., R. H. Gardner, and M. G. Turner. 1997. Landscape connectivity and population distributions in heterogeneous environments. Oikos 78:151–169.

Wu, J. and O. L. Loucks. 1995. From balance of nature to hierarchical patch dynamics: A paradigm shift in ecology. Quarterly Review of Biology 70:439–466.

CHAPTER 7

Metacommunity Structure Influences the Stability of Local Beetle Communities

Kendi F. Davies, Brett A. Melbourne, Chris R. Margules,
and John F. Lawrence

Introduction

How useful is the metacommunity concept to empiricists? For example, can it help us understand how a real-world beetle community works? To answer these questions, we attempt to interpret the results from a habitat fragmentation experiment using metacommunity concepts.

We contrast two beetle metacommunities in habitats that had different spatial structures and where we would have expected different levels of dispersal. In one, which we call the low-dispersal treatment, communities occur in fragments of native *Eucalyptus* forest embedded in an alien pine matrix. In the other, which we call the high-dispersal treatment, communities occur in patches equivalent to the *Eucalyptus* fragments that are, in fact, demarcated by invisible boundaries in continuous native *Eucalyptus* forest (figure 7.1). In the latter, the similarity of the habitat patches and matrix between is expected to promote interpatch movement. Specific aspects of community structure and dynamics in this system have been reported in Davies et al. (2004), Davies, Melbourne, et al. (2001), Davies and Margules (1998, 2001), and Davies et al. (2000).

Ultimately, we have emphasized metacommunity concepts to reexamine evidence from a fragmentation experiment. We believe that habitat fragmentation experiments may provide some of the best systems in which to explore metacommunity ideas. First, they are usually set up to compare the properties of systems of patches that are connected, with systems of patches that are not connected (Debinski and Holt 2000). Consequently, they provide the opportunity to contrast metacommunities with different spatial structure and level of dispersal. Second, they are generally real, complex, speciose systems, and thus provide a balance to the micro and mesocosm style systems in which metacommunity ideas have usually been explored. Therefore, we believe that fragmentation studies have the potential to offer unique insights.

The way we precede in this chapter is as follows. In the first section, "The System," we describe (1) the study system and (2) the measures of community dynamics and community structure that we used to detect differences between low-

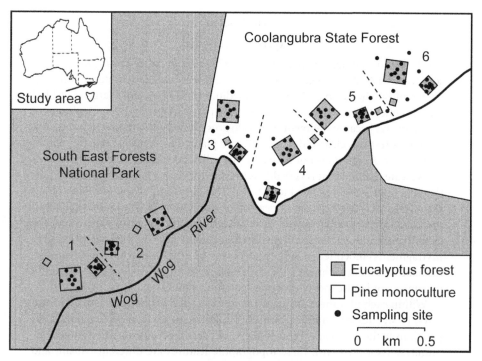

Figure 7.1 A map of the experimental site showing *Eucalyptus* forest fragments and unfragmented plots in continuous forest. The numbers refer to replicates. Dots represent the location of a pair of monitoring sites (a slope site and a drainage-line site) that were established in the pine matrix between the fragments after fragmentation. There are eight monitoring sites within each fragment. At each monitoring site, pitfall traps were opened in each season for two years before fragmentation and for five years after fragmentation. The fragment sizes are 0.25 ha, 0.875 ha, and 3.062 ha. Fragments are separated by at least 50 m.

and high-dispersal metacommunities. In the second section, "Evidence for Meta-community Dynamics," we organize our empirical evidence around a framework for using the metacommunity concept. From an empiricist's perspective, a meta-community has three elements: (1) local communities, (2) dispersal between those local communities, and (3) dynamics. Our first aim was to investigate whether the metacommunity concept could help us to understand how the bee-tle community works in a complex, real-world system. The third element, dy-namics, was the focus of our second and more specific aim, which was to discover how local community dynamics differed between the two metacommunities. We note that in fragmentation studies, it is the dynamics of *local* communities that is of interest because the focus is the persistence of communities in fragments.

The System

Experimental Design

The Wog Wog habitat fragmentation experiment is in native *Eucalyptus* forest in southeastern Australia (figure 7.1). The experiment has six replicates (figure 7.1). Each replicate has a small (0.25 ha), a medium (0.875 ha), and a large (3.062 ha) plot (fragment). This gives a total of eighteen plots (fragments). Four replicates of each plot size became fragments of *Eucalyptus* habitat when the surrounding *Eucalyptus* forest was cleared during 1987 and planted to *Pinus radiata*, for plantation timber. *P. radiata* is not native to Australia and when planted in monocultures creates habitat that is very different from the native *Eucalyptus* forest. Thus, the pine matrix was intended to act as a "sea" surrounding the *Eucalyptus* fragment "islands." Two replicates of each plot size remain in continuous habitat of uncleared *Eucalyptus* forest and serve as unfragmented control plots. Thus, the experimental design contrasts two different landscapes: one with native *Eucalyptus* fragments embedded in a pine matrix and the other with continuous *Eucalyptus* habitat. The experiment provides an opportunity to study metacommunity dynamics, for landscapes of different spatial structure.

Within each plot there are eight monitoring sites, which are stratified in two ways. First, sites are stratified into slope habitat and drainage line habitat (depressions in the landscape where water would drain) because the vegetation communities associated with these topographic features are different (Austin and Nicholls 1988). Second, sites are stratified by proximity to the fragment edge (edge or interior). There are two monitoring sites in each of the four strata (slope edge, slope interior, drainage-line edge, drainage-line interior), totaling 8 sites within each plot and a total of 144 sites over the 18 plots (fragments). Following clearing, an additional 44 sites were established in the *P. radiata* plantation between the habitat fragments. Two permanent pitfall traps are located at each of the 188 sites to catch beetles and other invertebrates. Traps were opened four times a year, once during each season, for seven days. Monitoring started in February 1985. Thus, two years of data were collected before the fragmentation treatment for all *Eucalyptus* plots.

Records of beetle species were processed up until 1991 (five years postfragmentation) over which time 655 beetle species were captured. More than a third of these species were trapped only one or two times, while six species were trapped over 1000 times each. The incidental captures may represent species that are either rare, are not habitually ground dwelling, are "tourists" (just passing through), or that move little and are therefore unlikely to fall into pitfall traps. Thus in this study, we considered only those species that were caught three or more times (325 species).

Comparing Low- and High-Dispersal Habitats

We expected the pine matrix surrounding fragments to be uninhabitable for most beetle species. We also expected that most beetle species would find the pine matrix a greater barrier to dispersal between fragments than the *Eucalyptus* forest between equivalent sized plots in the continuous forest. Thus, our expectation was that this fragmentation experiment would allow us to contrast metacommunity dynamics in two different kinds of metacommunities, one with low dispersal between local communities (eucalypt fragments in a pine matrix) and one with high dispersal between local communities (eucalypt patches in continuous forest). To contrast dynamics, we measured differences in the stability of local communities between these two different metacommunities. Stability was measured as the turnover of species in local communities. We then used two other types of evidence to help understand changes in turnover. (1) We measured differences in community structure between local communities in the low- and high-dispersal metacommunities. (2) We looked at which species were driving these changes by observing how the occurrences of individual species differed between local communities in low- and high-dispersal metacommunities, and how these differences in occurrence related to traits of species. Below, we describe how we measured differences in local community dynamics using a turnover metric, differences in local community structure, and differences in the occurrences of individual species.

Measuring Differences in Local Community Dynamics

We compared the stability of community dynamics in low- and high-dispersal metacommunities, by measuring the rate of change in species composition, or species turnover. We calculated percent turnover in species composition from year to year as

$$\frac{C_{obs} + E_{obs}}{S_i + S_{i+1}} \times 100 \tag{7.1}$$

where S_i is the number of species observed at a site in year i, C_{obs} is the number of species present in year $i + 1$ but not in year i (i.e., the number of species observed to colonize) and E_{obs} is the number of species absent in year $i + 1$ but present in year i (i.e. the number of species observed to become extinct). We compared the turnover of local communities in habitat fragments and continuous forest using ANOVA (analysis of variance). We also tested for the effects of fragment size, edges and year on turnover. Turnover between the two years before fragmentation was used as a covariate to control for spatial variation in turnover across the landscape before fragmentation (details are in Davies, Melbourne, et al. 2001).

Measuring Differences in Local Community Structure

We looked for differences in the local community structure of low- (fragmented) and high-dispersal metacommunities (unfragmented) by contrasting the species richness, species composition, and relative abundance of species in local communities. Here we give only brief descriptions; details appear in Davies, Melbourne, et al. (2001). Because we established that a local community was structured in space at about the monitoring site scale (see below in "Defining the Local Community"), all of our analyses were focused at this scale. That is, we made all of our comparisons of community structure at the local-community scale.

SPECIES RICHNESS

We used Poisson regression to compare the annual beetle species richness over five years postfragmentation in habitat fragments and plots in continuous forest. Poisson regression is a generalized linear model (like logistic regression) for count data. We also looked at the effects of fragment size, and edge effects, and how richness changed through time. The natural logarithm of the average richness in the two years before fragmentation was included as a covariate in this analysis to control for spatial variability in richness across the landscape before fragmentation.

SPECIES COMPOSITION AND RELATIVE ABUNDANCE

To examine the effects of fragmentation on species composition and relative abundance, we used an approach based on dissimilarity measures. We asked two questions about dissimilarities: (1) did communities in the fragments become more dissimilar from communities in continuous forest after fragmentation, and (2) did they become less dissimilar (more similar) to communities in the pine matrix after fragmentation? This told us two things. (1) Were communities in the fragments diverging in structure from those in the continuous forest? (2) If so, were communities in the fragments being influenced by communities in the pine matrix?

Two dissimilarity measures were used. We used the Bray-Curtis measure (Bray and Curtis 1957) to measure the dissimilarity in relative abundance. The Bray-Curtis measure is widely used in community studies and has been shown to provide a robust estimate of the difference in structure between communities (Faith et al. 1987). It is most sensitive to differences in the relative abundance of species between communities, but it is also affected by species richness and species composition. We used the Sorensen-Czekanowski measure (Czekanowski 1913; Digby and Kempton 1987) to measure the dissimilarity in species composition (i.e., the presence-absence pattern). The Sorensen-Czekanowski measure is most sensitive to species composition but is also affected by species richness. It is equivalent to the Bray-Curtis measure, only for presence-absence data rather than relative abundance data. Also, it is equivalent to turnover (equation 7.1). For the

Sorensen-Czekanowski measure, i in equation 7.1 indexes sites instead of years and C_{obs} and E_{obs} contrast sites instead of years.

We used Mantel tests to determine whether communities in the fragments became more dissimilar from the continuous forest after fragmentation, for each of species composition and relative abundance (Legendre and Legendre 1998). We conducted separate tests for each combination of size, edge, and year, as well as tests for an overall effect of fragmentation (i.e., averaged over site and edge). We used the same approach to determine whether beetle communities in the fragments became more similar to beetle communities in the pine matrix than the continuous forest patches were to the pine matrix (Davies, Melbourne, et al. 2001).

Differences in the Occurrence of Individual Species by Traits

We analyzed the difference in occurrence between low- and high-dispersal metacommunities for each of the 325 species. This difference measures the response of the species to the habitat fragmentation treatment. We then looked at how a single trait of species, degree of isolation or habitat specialization, correlated with this response.

In the absence of dispersal information, we calculated an isolation index for each species as the ratio of capture rate in the pines to capture rate in the fragments. Beetles fell into two categories: (1) Never trapped in the pine matrix (isolated); and (2) Trapped in the pine matrix (not isolated). For a species to be isolated two conditions are necessary: the species does not occur in the matrix, and it therefore is a habitat specialist of eucalypt habitat; and the species does not disperse between fragments. Given that roughly one quarter of the trapping effort was in the pine matrix, we are confident that if a species was not caught in the matrix in five years of trapping, it did not occur there. Similarly, we are confident that these species did not disperse along the ground through the matrix, or rarely did so. Unfortunately, we can say little about other modes of dispersal, such as flight. Thus, it is possible that some species categorized as isolated met the first condition but were able to disperse, undetected between fragments. Species classified as not isolated either had populations established in the pine matrix (were habitat generalists), or dispersed through the matrix between fragments. Given these caveats, we believe these isolation categories are useful as a first approximation, particularly given that the contrasts that we make are relative (habitat fragments versus continuous *Eucalyptus* forest) rather than absolute.

A correlation between degree of isolation and the response of species to the different types of landscape structure can help to identify important processes influencing the dynamics of the two metacommunity types. For example, if isolation correlates with reduced occurrence in fragments, then this would suggest that dispersal is an important process influencing individual species dynamics and by implication metacommunity dynamics.

Evidence for Metacommunity Dynamics

Defining the Local Community

In metapopulation theory, local populations refer to the spatial scale over which individuals are well mixed so that they can be considered to share the same location and environment, and thus interact with other individuals (Hanski and Simberloff 1997). By analogy, we define a local community as the scale at which individuals are well mixed, and have the potential to interact with individuals of their own and other species. The term *metapopulation* refers to a collection of local populations connected by dispersal (Hanski and Simberloff 1997); by analogy a metacommunity is a collection of local communities connected by dispersal. In metapopulation theory, too much dispersal between local populations means that a set of local populations acts more like one large population than a metapopulation, because individuals in different locations frequently interact. By analogy, too much dispersal between local communities also means that a set of local communities acts more like one large community than a metacommunity. We might therefore take the scale of the local metacommunity to be some scale at which there is not too much dispersal. However, one problem in the metacommunity case is that species operate on different scales. Some species disperse more than others (see Holyoak et al., chapter 1 for further discussion). For the beetle community, we consider a pragmatic working definition of a local community to be the spatial scale of the local population for most of the species in the assemblage.

We combined two pieces of evidence to help determine the most common scale of a local beetle population, and hence the scale of the local community: (1) the spatial scale of variation in abundance for each species, and (2) dispersal studies of some species. We determined the spatial scale at which the abundance of each of the 325 species varied in the landscape using variance components analyses (Underwood and Chapman 1996). For each habitat type (slopes and drainage lines), we determined variance components for three spatial scales, each corresponding to the scales of the experiment (figure 7.1): between replicates (~500 m), between plots/fragments (~100 m), and between sites (~25 m). A fourth scale, samples within sites, removed sampling error. For nearly all beetle species, the largest variance component of abundance was at the between-site scale (~25 m, figure 7.1), the smallest spatial scale. Variance components tended to be small or negligible between plots (~100 m) and between replicates (~500 m). In other words, abundances were most variable at the between-site scale, suggesting that individuals were not well mixed between sites with similar habitat. We also discovered, at least for some of the larger beetle species, that an individual beetle had a range of roughly 5–10 m and rarely moved further (Davies and Margules, in review). This is discussed more in the next section on dispersal. Thus, we can interpret the scale of most local beetle populations, and therefore of the local beetle community, to be an area of roughly 10–25 m in diameter. Thus, in what follows, we focus our analyses on changes in the dynamics and structure of local communities at the site scale. Remember that there are eight sites within each fragment/plot so this definition means that fragments of *Eucalyptus* habitat and experimental plots in the continuous *Eucalyptus* forest contain multiple local communities.

It might seem more intuitive to use obvious boundaries to define local communities, for example, a lake, an island, a serpentine patch in a matrix of nonserpentine soil, or, as in this study, a forest fragment (e.g., Holyoak et al., chapter 1). However, it is prudent to ask whether the scale of the local community matches with these apparent patch boundaries, because this can have implications for the dynamics of the metacommunity. Using metapopulations as an example, at one extreme, the scale of a local population could be larger than a single apparent patch. In this case, individual patches support only parts of a local population but not an entire local population, that is, a "patchy population" (Harrison 1991, Diffendorfer et al. 1995, Andreassen et al. 1998). At the other extreme, like in this example for beetles, if the scale of a local population is smaller than that of a fragment, then patches may be internally heterogeneous, supporting many populations (i.e., a metapopulation within a fragment). In those cases, recolonization of local populations from within patches increases persistence at both fragment and between-fragment scales (Holt 1992). The case where apparent patches (fragments) also define local populations, and by extension local communities, is an intermediate case and is perhaps special.

Finally, our definition of a local community stipulates that the species making up a local community have the potential to interact. We use the word *potential* because metacommunity theory does not make it clear whether interactions between species are necessary for an assemblage of species to be considered a local community. However, we wonder whether species that share a location but do not interact should be considered a collection of metapopulations rather than a metacommunity (see also Holyoak et al., chapter 1). In this system, we did not study interactions between beetle species, but we presume that species interact via competition and predation.

Dispersal

We did not directly measure dispersal between fragments or between plots. However, two pieces of evidence helped us to understand how populations and local communities of beetle species were connected in the two metacommunities. First, we have data about which species were present in the pine matrix between fragments and in what abundances. Species that were trapped in the matrix either were able to colonize the habitat there, or were able to disperse through the matrix between fragments. About half of the species were trapped in the matrix. Our second piece of evidence came from a mark and recapture study of eight numerically dominant carabid species (Davies and Margules, in review). We marked and recaptured beetles over two summers in a grid of 13 × 13 pitfall traps spaced 5 m apart in continuous forest adjacent to the experiment. Recaptures of marked beetles suggested that individual beetles had a mean range roughly 5–10 m in diameter. Further, the distance that an individual moved was independent of the time between mark-release and recapture. Since plots / fragments were separated by at least 50 m, these short movement distances suggest that individuals trapped in the pine matrix probably had populations that were established there.

Metacommunity Dynamics

We compared the stability of community dynamics (as turnover) in low- and high-dispersal metacommunities. Taking a standard view of fragmented landscapes for noninteracting species, we would predict that local communities in the low-dispersal metacommunity would be less stable than in the high-dispersal meta-community. The reasoning is as follows. In the high dispersal metacommunity of the continuous forest, extinction rates should be low because the size of local populations is large. Also, the colonization rate should be higher because dispersal is higher: species can easily colonize from nearby populations. Then, the number of species observed to go extinct (as opposed to the extinction rate) and the number observed to colonize (as opposed to the colonization rate) is relatively low. This occurs because we will not observe a colonization unless the species is first absented from the community by extinction and we will not observe an extinction if the colonization rate is high, so as to rescue the species from extinction. Stability is thus enhanced and turnover is diminished. In contrast, in the low-dispersal metacommunity of the fragmented landscape, we would predict a higher extinction rate by the standard argument of reduced population size on habitat fragments and the factors that contribute to extinction in small populations (Caughley 1994; Harrison and Taylor 1997). We would also predict a lower colonization rate by the standard argument that fragments are now isolated, requiring species to disperse through the alien *P. radiata* plantation. Then, we should observe more species to drop in and out of the community, increasing turnover.

Intriguingly, and contrary to these predictions of standard fragmentation theory, there was less temporal variability in species composition in the low-dispersal metacommunity than in the high-dispersal metacommunity (figure 7.2). In other words, local communities in the low-dispersal metacommunity of the fragmented landscape were more stable. We can think of five hypotheses that might contribute to this result, including two hypotheses that reframe the standard arguments in terms of metacommunity dynamics.

HYPOTHESIS ONE

The first explanation invokes mass effects, such that the pine matrix acted as a source for a subset of the beetle community. A mass effect is where patches with a

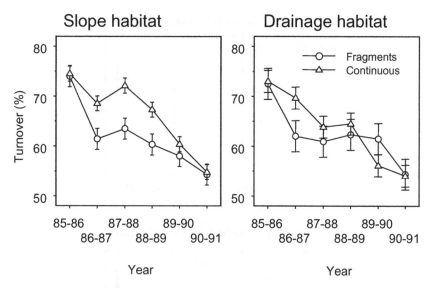

Figure 7.2 Percent change in species composition (turnover) from one year to the next for sites in fragments and continuous forest. Fragmentation occurred during 1987. In slope habitat, there was a significant difference in turnover between fragments and continuous forest averaged over all years postfragmentation (P = 0.05), and for *fragmentation × year*, P = 0.066. In drainage habitat, there was not a significant difference between fragments and continuous forest for turnover averaged over all years postfragmentation (P = 0.72), but for *fragmentation × year*, P = 0.066. Means and standard errors are from ANOVA.

high output of individuals have a large influence on the dynamics of nearby patches through dispersal (Shmida and Wilson 1985). Thus, this subset of species was much less likely to go extinct in fragments than the rest of the local community. Then, because there was a subset of species whose composition changed little from year to year, the temporal variability in the species composition of entire local communities was reduced.

This means that the role of the pine matrix between fragments turned out to be quite different from what was originally intended. The pine matrix was intended to be habitat that was inhospitable to beetles so that communities on fragments were isolated. As it turned out, the pine matrix habitat was preferred by some species to the extent that they were trapped there in greater abundances than in the eucalypt habitat of fragments (figure 7.3b). In the habitat fragmentation literature, the matrix is the habitat between fragments that replaces the original habitat. Fragments are usually composed of the original habitat. Thus, our experiment illustrates how real experiments force us to think about complications that arise in the real world compared to the simplifications made in micro / mesocosm studies and in theory.

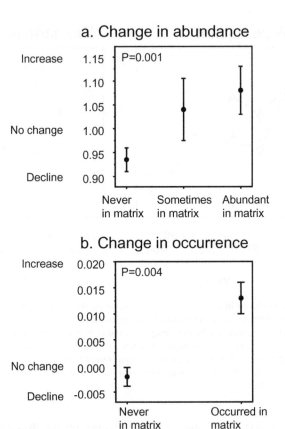

Figure 7.3 (a) Change in abundance of individual species in fragments compared to continuous forest, regressed against degree of isolation on fragments. Species were either (1) never trapped in the pine matrix between fragments in the five years after fragmentation, (2) trapped in the pine matrix but in lower abundances than in fragments, or (3) trapped in the pine matrix in higher abundances than fragments. A species' response to fragmentation was calculated as $\log_e$ (mean-abundance fragments/mean-abundance controls). Here values have been back transformed so that 1.15 represents an increase of 1.15 times in abundance in fragments compared to continuous forest, and 0.85 represents a decline of 15% in populations in fragments compared to continuous forest. The y-axis is on a natural log scale. Error bars represent standard errors. (b) Change in occurrence of individual species in fragments compared to continuous forest, regressed against degree of isolation. Species were either (1) never trapped in the pine matrix between fragments in the five years after fragmentation, or (2) they did occur in the pine matrix. A change in occurrence of 0.02, for example, means that species occurred at 2% more sites in the fragments than in the continuous forest, after fragmentation.

There are four pieces of evidence that suggest that the pine matrix might have acted as a source for a subset of species. The first two pieces of evidence are about differences in the distribution of individual species. (1) Species that occurred in the pine matrix occurred at significantly more sites in the low-dispersal metacommunity than in the high dispersal metacommunity (figure 7.3a; Davies, Melbourne, et al. 2001). (2) Species that were more abundant in the pine matrix than in the eucalypt fragments were significantly more abundant in the low-dispersal metacommunity than in the high dispersal metacommunity (figure 7.3b, Davies et al. 2000). These correlative observations suggest that the presence and abundance of a species in the matrix could have altered its presence and abundance in fragments compared to equivalent plots in continuous forest. Thus, for a subset of the community, and contrary to the standard view, a net flow of individuals from the pine matrix into the eucalypt fragments might have reduced the probability of extinction in the local communities of fragments.

Changes in the occurrences and abundances of this subset of species were also detected as differences in the species richness, composition, and relative abundance of local communities in low- and high-dispersal metacommunities, which provide our third and fourth pieces of evidence. These changes were greatest at fragment edges. (3) Species richness was about 10% higher (significantly so) at fragment edges than in fragment interiors or in plots in continuous forest. (4) Finally, species composition and the relative abundance of species at fragment edges tended to be most like the composition and relative abundance of the pine matrix, while the composition and relative abundance of fragment interiors was like that of continuous forest (Davies, Melbourne, et al. 2001).

Altogether, this evidence suggests that a mass effect of matrix communities on fragment communities could have occurred, resulting in greater temporal stability in local communities in fragments. A stronger test of this mechanism would be to look at whether the subset of species that are abundant in the matrix have lower temporal turnover in fragments compared to species that are isolated on fragments. This test will be performed in future work.

Finally, an alternative is that the continuous forest acted as a source for fragment communities. This hypothesis would require that individual beetles moved long distances and were more likely to colonize fragment interiors than edges, as it was the composition and relative abundance of the interiors of fragments (large fragments in particular) that were most like continuous forest.

HYPOTHESIS TWO

An alternate explanation for the stabilizing effect of the matrix on the low-dispersal metacommunity was that the pine matrix provided a homogenizing force because the habitat there, the pine monoculture, was more homogeneous than the natural eucalypt forest. This explanation also involves mass effects. The pine matrix is a monoculture of *Pinus radiata* of even age. *Pinus radiata* is not native to Australia and when grown as a plantation timber, it forms stands with little or no herbaceous layer or understorey. Also, the litter of the forest floor is much simpler in structure than that of the native eucalypt forest that it replaces. By contrast, the vegetation in the eucalypt forest is heterogeneous. Slopes are characterized by a grassy understorey and scattered shrubs that lie below open eucalypt forest. Drainage lines are dominated by Ti-tree, *Leptospermum phylicoides* (Myrtaceae), which is a small shrubby tree that forms dense stands. Slopes tend to be dry and open, whereas drainage lines tend to be dark and moist. We hypothesized that, as a result of the lack of heterogeneity of the plant community in the pines, the pine beetle community might be correspondingly homogeneous in space and time. Then, if the beetle community in the matrix was dispersing into eucalypt fragments, it could provide a stabilizing influence. However, the spatial variability in species composition (turnover) was actually higher in the pine matrix than in fragments (figure 7.4; Davies, Melbourne, et al. 2001). Thus, this hypothesis is unlikely to explain the greater stability of fragment communities.

HYPOTHESES THREE AND FOUR

The third hypothesis is that the environment of fragments was more homogeneous in time and space than the plots in continuous forest. Although we think this is plausible, we have no data with which to examine this hypothesis. However, we do have environmental data for 1997–98. Contrary to this hypothesis, these data showed that there was more habitat heterogeneity in fragments than in continuous forest. For example, counts of fallen logs were significantly higher at fragment edges than fragment interiors or plots in continuous forest, and more solar radiation reached the forest floor in fragments than in continuous forest plots but only for slope habitat (Davies and Margules, in review). A final possibility is the fourth hypothesis: that the edges of fragments acted as reflective boundaries so that on reaching the boundary, rather than dispersing out of fragments, beetles tended to turn around. This could result in reduced turnover in local communities, since colonists would be retained within fragments. However, it does not explain why the composition of fragment edges was most like the composition of the pine matrix (Davies, Melbourne, et al. 2001b).

HYPOTHESES FIVE

Local communities in the high-dispersal metacommunity (continuous forest) could appear less stable because there might be more species that occur only infrequently because of the higher connectivity there. In fragments, these species may be lost if fragments were more isolated. If this hypothesis is correct, then we would expect continuous forest to have more species that turned over at relatively few sites (say three or less) than forest fragments. However, in fact, in forest fragments more species turned over at three or fewer sites (eighty-four) than in continuous forest (twenty species), suggesting that this mechanism was not respon-

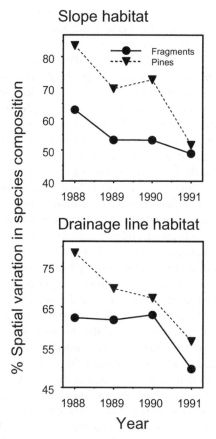

Figure 7.4 Comparison of percentage spatial variation in species composition for sites in fragments and sites in the pine matrix for both slope and drainage line habitat. Two dissimilarities were computed: fragment versus fragment and pine matrix versus pine matrix. The spatial variation in species composition was greater at pine matrix sites than at fragment sites.

sible for the more stable local community dynamics in fragments compared to continuous forest.

AN IMPORTANT ROLE FOR THE MATRIX

As our example illustrates, the matrix in which patches are embedded has the potential to influence patch communities by providing alternative habitat for some patch-community species. This idea has emerged as a generality from the empirical habitat fragmentation literature (Davies, Gascon, et al. 2001; see also examples in Lindenmayer and Franklin 2002), and has been explored theoretically for metapopulations (Vandermeer and Carvajal 2001) but it also is likely to apply to metacommunities. For example, consider pitcher plant metacommunities (Kneitel and Miller 2002; Miller et al. 2002; Miller and Kneitel, chapter 5). In this system, discrete local communities of bacteria, mites, rotifers, protozoans, and mosquito larvae occur inside the water-filled leaves of the pitcher plant *Sarracenia purpurea*. The pitchers contain simple food webs, and are connected by dispersal to form metacommunities. It is likely that at least some species in these communities persist and reproduce in other places (alternatives to the pitcher plants) in the bog matrix in which the pitcher plants are embedded. The lesson from metacommunity studies in fragmented landscapes is that populations and communities in the matrix should not be overlooked, since they can have important consequences for the embedded communities.

Three other habitat fragmentation examples are illustrative. First, the dynamics of populations of an understorey herb (*Heliconia acuminata*) in tropical forest fragments in Brazil were more stable in both space and time in ten-hectare fragments than in continuous forest, and populations were even more stable in one-hectare fragments (Bruna, pers. comm.). The dynamics of populations in the matrix, which is a mix of pasture and secondary forest, were not measured but the plant is known to occur there. Thus, a possible explanation for the reduction in variability in fragments is an influence of populations in the matrix. Second, and also in tropical-forest fragments in Brazil, birds, small mammals, and frog species that were able to tolerate or exploit the matrix were less vulnerable to fragmentation, while species that avoided the matrix tended to decline or disappear (Gascon et al. 1999). Third, in successional old-field patches in a closely mowed matrix in Kansas, species that were established in the matrix influenced patterns of species richness in patches. Roughly one quarter of species that occurred in patches also occurred in the matrix. The effects of distance from the "mainland" and patch size on the species richness of patches (as predicted by the theory of island biogeography; MacArthur and Wilson 1967) was only detected once matrix species were removed from the analyses (Cook et al. 2002).

OTHER STUDIES ON DIFFERENCES IN METACOMMUNITY DYNAMICS

In contrast to our findings, three studies recorded increased community variability as the result of habitat fragmentation. First, in a twenty-two-year study of forest breeding birds in the eastern United States, forest fragmentation was associated with increased temporal variability in species richness and reduced richness (Boulinier et al. 1998). Second, tree turnover was significantly higher at the edges of eighteen-year-old tropical rainforest fragments in Brazil as the result of increased mortality (Laurance et al. 1998). Third, revisiting the successional old-field patches in Kansas over six years following fragmentation showed that spatial heterogeneity in vegetation, measured as local community dissimilarity, had increased in all patches but less in large than small patches. However, even though some clonal species disappeared from the smallest patches, there was not a significant difference in local species turnover in time between small and large patches (Holt et al. 1995).

Conclusions

This study had two aims, one general and one specific. At a general level, we investigated whether the metacommunity concept helped us to understand how a beetle community works in a complex, real-world system. Specifically, we examined how local community dynamics were affected by being part of metacommunities with either low dispersal (forest fragments) or high dispersal (plots in continuous forest).

At a general level, we used the metacommunity concept as an organizing framework around which to arrange empirical evidence about our beetle communities. The framework that we used had three elements: determine the spatial scale of local communities, investigate dispersal between local communities, and measure how dynamics changed because of differences in the spatial structure of metacommunities.

Our findings were as follows: (1) The spatial scale of a local community was smaller than that of a fragment, so plots and fragments contained multiple local communities. (2) Most beetle species probably did not regularly disperse large enough distances to travel between fragments, which means that those species that were trapped in the pine matrix in which fragments were embedded probably had populations established there. (3) We compared the temporal variability in species composition (turnover), a measure of stability. Contrary to standard expectations of fragmented systems, which are based on the dynamics of small and isolated populations, local communities in the low-dispersal metacommunity were more stable than those in the high-dispersal metacommunity. The most likely explanation was that mass effects of matrix communities on fragment communities occurred, resulting in greater temporal stability in fragment communities. Thus, an important point to emerge from our study was that populations and communities in the matrix in which local communities are embedded should not be overlooked, since they can have important consequences for the embedded communities.

Our study contributes to the metacommunity theme by illustrating how the metacommunity concept can be used to understand complex real-world communities. The metacommunity concept provided a framework from which to decide which questions to ask and how to interpret the evidence. We used metacommunity theory to explain our findings and to generate new hypotheses. This may be how metacommunity theory is best used in empirical studies like this one, with many species and real-world complexity.

We used just one measure of community dynamics—the spatial variability of species composition, measured as turnover. An important future direction is to develop other measures of community dynamics that can help to quantify meta-community processes. We can think of at least two other useful measures. The first is a simple extension of turnover: the rate of change of the relative abundance of species. This measure contains the turnover signal but also provides informa-tion about how the abundance of species change in relation to one another. A sec-ond measure is the degree of spatial synchrony in community dynamics. Quanti-fying synchrony could be particularly useful for determining the extent and pattern of spatial coupling between local communities, thus quantifying an im-portant spatial characteristic of the metacommunity. As in single species popula-tions (Lande et al. 1999), the spatial scale of synchrony could influence the regional dynamics of the metacommunity, and determine properties such as the regional species richness. We intend to investigate the community dynamics of beetles using these measures in the future.

Acknowledgments

We thank Susan Harrison, Martha Hoopes, and Marcel Holyoak for their com-ments, which greatly improved the manuscript. We also thank George Milkovits who ran the experiment.

Literature Cited

Andreassen, H. P., K. Hertzberg, and R. A. Ims. 1998. Space use responses to habitat fragmentation and connectivity in the root vole. *Microtus oeconomus*. Ecology 79:1223–1235.

Austin, M. P., and A. O. Nicholls. 1988. Species associations within herbaceous vegetation in an Aus-tralian eucalypt forest. Pages 95–114 *in* H. J. During, M. J. A. Werger, and J. H. Willems, eds. *Diver-sity and pattern in plant communities*. SPB Academic Publishing, The Hague, The Netherlands.

Boulinier, T., J. D. Nichols, J. E. Hines, J. R. Sauer, C. H. Flather, and K. H. Pollock. 1998. Higher tem-poral variability of forest breeding bird communities in fragmented landscapes. Proceedings of the National Academy of Sciences, U.S.A. 95:7497–7501.

Bray, J. R., and J. T. Curtis. 1957. An ordination of the upland forest communities of Southern Wis-consin. Ecological Monographs 27:325–349.

Caughley, G. 1994. Directions in conservation biology. Journal of Animal Ecology 63:215–244.

Cook, W. M., K. T. Lane, B. L. Foster, and R. D. Holt. 2002. Island theory, matrix effects and species richness patterns in habitat fragments. Ecology Letters 5:619–623.

Czekanowski, J. 1913. Zarys metod statystycznych do zastosowaniu do anthropolgii. Warszawa, Naklad Towaraystwa Naukowego Warszawskiego.

Davies, K. F., C. Gascon, and C. R. Margules. 2001. Habitat Fragmentation: Consequences, manage-ment and future research priorities. Pages 81–97 *in* M. E. Soulé, and G. H. Orians, eds. Conserva-tion Biology: Research Priorities for the Next Decade. Washington, Island Press.

Davies, K. F., and C. R. Margules. 1998. Effects of habitat fragmentation on carabid beetles: Experi-mental evidence. Journal of Animal Ecology 67:460–471.

———. 2001. The beetles at Wog Wog: A contribution of Coleoptera systematics to an ecological field experiment. Invertebrate Taxonomy 14:953–956.

———. In review. Spatial population dynamics and habitat modification drive six species' responses to experimental fragmentation.

Davies, K. F., C. R. Margules, and J. F. Lawrence. 2000. Which traits of species predict population declines in experimental forest fragments? Ecology 81:1450–1461.

———. 2004. A synergistic effect puts rare, specialized species at greater risk of extinction. Ecology 85:265–271.

Davies, K. F., B. A. Melbourne, and C. R. Margules. 2001. Effects of within- and between-patch processes on community dynamics in a fragmentation experiment. Ecology 82:1830–1846.

Debinski, D. M., and R. D. Holt. 2000. A survey and overview of habitat fragmentation experiments. Conservation Biology 14:342–355.

Diffendorfer, J. E., M. S. Gaines, and R. D. Holt. 1995. Habitat fragmentation and movements of three small mammals (*Sigmodon, Microtus,* and *Peromyscus*). Ecology 76:827–839.

Digby, P. G. N., and R. A. Kempton. 1987. *Multivariate analysis of ecological communities.* Chapman and Hall, London, U.K.

Faith, D. P., P. R. Minchin, and L. Belbin. 1987. Compositional dissimilarity as a robust measure of ecological distance. Vegetatio 69:57–68.

Gascon, C., T. E. Lovejoy, R. O. Bierregaard, J. R. Malcolm, P. C. Stouffer, H. L. Vasconcelos, W. F. Laurance, et al. 1999. Matrix habitat and species richness in tropical forest remnants. Biological Conservation 91:223–229.

Hanski, I., and D. Simberloff. 1997. The metapopulation approach, its history, conceptual domain, and application to conservation. Pages 5–26 *in* I. Hanski, and M. E. Gilpin, eds. *Metapopulation biology: Ecology, genetics, and evolution.* Academic Press, San Diego, CA.

Harrison, S. 1991. Local extinction in a metapopulation context: An empirical evaluation. Biological Journal of the Linnean Society 42:73–88.

Harrison, S., and A. D. Taylor. 1997. Empirical evidence for metapopulation dynamics. Pages 27–42 *in* I. Hanski and M. E. Gilpin, eds. *Metapopulation biology: Ecology, genetics, and evolution.* Academic Press, San Diego, CA.

Holt, R. D. 1992. A neglected facet of island biogeography: The role of internal spatial dynamics in area effects. Theoretical Population Biology 41:354–371.

Holt, R. D., G. R. Robinson, and M. S. Gaines. 1995. Vegetation dynamics in an experimentally fragmented landscape. Ecology 76:1610–1624.

Kneitel, J. M., and T. E. Miller. 2002. Resource and top-predator regulation in the pitcher plant (*Sarracenia purpurea*) inquiline community. Ecology 83:680–688.

Lande, R., S. Engen, and B. E. Saether. 1999. Spatial scale of population synchrony: Environmental correlation versus dispersal and density regulation. American Naturalist 154:271–281.

Laurance, W. F., L. V. Ferreira, J. M. Rankin-De Merona, and S. G. Laurance. 1998. Rain forest fragmentation and the dynamics of Amazonian tree communities. Ecology 79:2032–2040.

Legendre, P., and L. Legendre. 1998. *Numerical ecology.* Elsevier Science, Ltd., Amsterdam, The Netherlands.

Lindenmayer, D. B., and J. F. Franklin. 2002. *Conserving forest biodiversity: A comprehensive multiscaled approach.* Island Press, Washington, D.C.

MacArthur, R. H., and E. O. Wilson. 1967. *The theory of island biogeography.* Princeton University Press, Princeton, N.J.

Miller, T. E., J. M. Kneitel, and J. H. Burns. 2002. Effect of community structure on invasion success and rate. Ecology 83:898–905.

Shmida, A., and M. V. Wilson. 1985. Biological determinants of species diversity. Journal of Biogeography 12:1–20.

Underwood, A. J., and M. G. Chapman. 1996. Scales of spatial patterns of distribution of intertidal invertebrates. Oecologia 107:212–224.

Vandermeer, J., and R. Carvajal. 2001. Metapopulation dynamics and the quality of the matrix. American Naturalist 158:211–220.

Wilson, D. S. 1992. Complex interactions in metacommunities, with implications for biodiversity and higher levels of selection. Ecology 73:1984–2000.

CHAPTER 8

Local Interactions and Local Dispersal in a Zooplankton Metacommunity

Karl Cottenie and Luc De Meester

Introduction

Metacommunities might be described by four perspectives, as outlined by Leibold et al. (2004) and by Holyoak et al. in chapter 1 of this book. In the present chapter we consider freshwater pond systems that have large differences in physical and biotic conditions between ponds. Species also show strong responses to local pond conditions (Cottenie et al. 2001, 2003; Cottenie and De Meester 2004), which is likely to make the patch dynamics and neutral perspectives less relevant for explaining metacommunity structure. The species sorting and mass effects perspectives are therefore obvious starting points for describing metacommunity structure. These perspectives could be differentiated by the degree to which dispersal perturbs local communities away from their equilibrium conditions of density and species composition. If local dynamics prevail, species sorting will occur, whereas if dispersal is overwhelming, mass effects and source-sink dynamics will occur. Moreover, stronger local environmental variation corresponds to stronger local population dynamics. Both the mass effects and species sorting perspectives also assume that species are able to reach local communities, so that species and communities are not "dispersal-limited." Mass effects imply that there is an effect of dispersal on local dynamics (Holyoak et al., chapter 1), whereas source-sink dynamics imply that there are sink habitat patches in which the population growth rates at low densities and in the absence of immigration are negative, and source patches in which the growth rates at low density are positive (Pulliam 1988).

Given these different perspectives, the principal axes to study metacommunity dynamics are environmental heterogeneity between patches and the amount of interpatch dispersal. In this chapter, we study these two important axes in a field metacommunity. We studied the characteristics of a system with between patch heterogeneity to determine the relative importance of local dynamics (species sorting) versus regional factors (mass transport of individuals).

The zooplankton in the shallow ponds in De Maten (Limburg province, Belgium) represents an interesting study system in which to address how species in a site combine to form a community. First, in this very small area a diverse set of ponds is available, with considerable variation in environmental characteristics in

189

the different ponds (table 8.1). A large body of literature is available that illustrates the importance of these factors on zooplankton community structure in general and on zooplankton communities in shallow ponds in particular (Carpenter and Kitchell 1993; Jeppesen et al. 1997; Scheffer 1998). This implies that the ponds harbor potentially very different zooplankton communities. Second, due to the small geographical extent of the area, it can be assumed that the ponds all have the same geographical species pool. Third, the ponds are also connected to one-another through a very particular system of overflows and rivulets that act as dispersal pathways. Moreover, the ponds are very old (they were dug in the Middle Ages), and since zooplankton are known to create an extensive egg bank (De Stasio 1989; Hairston 1996; Brendonck and De Meester 2003), it is likely that most species are present in every pond, either in the active or in the passive species pool. Effective dispersal and the presence of a resting egg bank may result in every species present in the pond complex (the metacommunity) being available in every local pond. This is very different from other systems studied so far in which dispersal among local communities is often considered to be very low because it depends on long range and low probability events (transport with wind or animal vectors, e.g., Rodriguez et al. 1993; Keller and Conlon 1994; Pinel-Alloul et al. 1995; Jeppesen et al. 2000). Thus in the ponds of De Maten, two opposing forces potentially contribute to local zooplankton community structure: different local environmental factors lead to divergence of the zooplankton communities along environmental gradients, while high dispersal rates tend to homogenize the zooplankton communities in different ponds; this is likely to occur even if species are entering habitat patches where they show negative population growth rates.

We first describe the study system and then outline the research approach used, which combines observations and experiments. We then describe the methods and findings for three specific questions: (1) What is the relative importance of local (environmental) and regional (spatial) variables on zooplankton community structure? We used relationships of local community attributes with local environmental factors to indicate species sorting, and relationships of community attributes with spatial structure of the habitat to indicate mass effects (because dispersal is expected to depend on spatial structure). (2) Does the connectivity pattern result in temporal coherence (synchrony) in seasonal succession between the different ponds? If the pattern of connection of ponds did influence the degree of coherence this would indicate that dispersal and mass effects are structuring communities. (3) Do local processes and dispersal have a deterministic effect on zooplankton community structure in this system? This question was addressed by using a manipulative experiment to see whether movement of individuals between pond conditions altered local community structure, which would support the mass effects perspective. An absence of an effect of movement of individuals on local community structure would indicate that species sorting acts rapidly and deterministically. Finally we synthesize our understanding of the processes underlying the observed metacommunity structure.

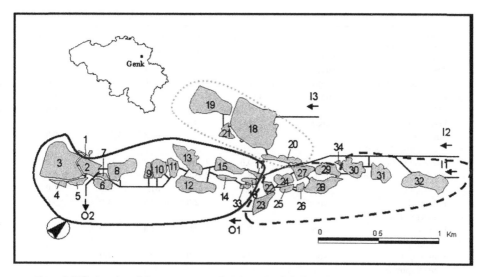

Figure 8.1 The location of the nature reserve De Maten, Genk (Belgium), the spatial configuration of the different ponds, and the connecting elements (overflows and rivulets). There are two different input sources (I1 and I2 come from the Stiemerbeek, and I3 from the Heiweyerbeek), and two output streams (O1 and O2). Also indicated are the three hydrological pond clusters: the upstream ponds (cluster 1: ponds 32–22), the downstream ponds (cluster 2: ponds 17–2), and the ponds located northwest of the remainder of the complex (cluster 3: ponds 18–21).

Part of the data and analyses used here are taken from published papers (Cottenie, Nuytten, et al. 2001; Cottenie, Michels, et al. 2003; Cottenie and De Meester 2003, 2004), while some data and analyses are new (see table 8.2).

The Setting: The Interconnected Ponds of De Maten

The pond complex is part of the De Maten nature reserve (50°57′ N, 5°27′ E) situated in the territory of Genk and Diepenbeek (Limburg province, Belgium; figure 8.1). The area has been classified as a protected nature reserve since 1976 and now covers a surface of nearly 310 ha, of which 217 ha are managed by Natuurpunt vzw. The pond complex is a typical example of an ancient chain of fishing ponds. The ponds were created in the Middle Ages through peat digging and the construction of dykes. Fish farming continued until 1990 (Daniels 1998; W. Peumans, pers. comm.), but fish are still present in the ponds. All ponds studied are shallow (mean depth approximately 0.5 m) and eutrophied (except pond 1). We measured several environmental variables that are generally considered to be important for understanding the ecology of shallow lakes and zooplankton community structure: turbidity, chlorophyll-a concentration, total fish biomass, macroinvertebrate diversity, submerged macrophytes, lake area, nutrient concentrations, pH, conductivity, and oxygen concentration (Cottenie, Nuytten, et al. 2001; Cottenie, Michels, et al. 2003). Table 8.1 gives an overview of the environ-

Table 8.1 An overview of environmental heterogeneity in the different ponds of De Maten in 1996.

Variable	Units	Minimum	25%	Mean	75%	Maximum
Depth	(cm)	12.0	40.0	54.5	70.0	100.0
Area	(ha)	0.1	0.4	1.7	1.8	9.5
O_2-concentration	(mg l^{-1})	2.6	5.8	6.6	7.7	9.6
Conductivity	(μS cm^{-1})	114.0	340.0	397.3	484.0	731.0
pH		4.1	7.2	7.3	7.7	8.6
Fe-concentration	(mg l^{-1})	0.0	0.3	1.5	1.9	6.0
Turbidity (Secchi disk depth)	(cm)	20.0	37.0	66.4	100.0	100.0
N-concentration	(μg l^{-1})	150.0	173.0	264.2	333.0	550.0
Total P-concentration	(μg l^{-1})	198.0	235.0	308.5	339.0	506.0
Chlorophyll-a	(μg l^{-1})	8.0	19.0	61.8	58.0	301.0
Emergent macrophytes	(score*)	1.0	2.0	3.0	4.0	5.0
Floating macrophytes	(score*)	0.0	0.0	1.4	3.0	6.0
Submerged macrophytes	(score*)	0.0	0.0	0.4	1.0	2.0
Macroinvertebrate density	(CPUE*)	1.0	13.0	45.4	88.0	110.0
Fish density	(CPUE*)	0.2	36.6	231.9	373.5	690.0

Note: For every variable, the measurement units are shown, together with minimum, 25th percentile (25%), mean, 75th percentile (75%), and maximum values. Information on the data from each pond can be found in Cottenie et al. (2001).

*The scoring system for emergent macrophytes consists of 6 scores ranging from 0 (absent), to 5 (well developed littoral zone with high abundances and high diversity); for floating macrophytes 7 scores ranging from 0 (absent) to 6 (50–100% coverage), and for submerged macrophytes 3 scores: 0 (absent), 1 (not abundant), and 2 (abundant). The catch per unit effort (CPUE) for macroinvertabrates is the total number of individuals caught during twelve minutes, and for fish the total number of individuals caught per fyke per day.

mental conditions, as observed in 1996. Most of these characteristics show considerable variation. For instance, the chlorophyll-a concentrations and Secchi disk depths (a measure of water clarity) cover the spectrum of mesotrophic to eutrophic habitats.

The pond complex is 3 km long and 1.5 km wide (figure 8.1). There is a unidirectional flow of water from the upstream pond in the northeast corner of the nature reserve (pond 32) at 55 m above sea level to the ponds in the southern end of the system at approximately 40 m above sea level. Almost half of the water in the system comes from two rivulets, and the remainder comes from ground water. The main rivulet, the Stiemerbeek, feeds pond 32 and all downstream ponds. The second rivulet, the Heiweyerbeek, feeds a subset of ponds located in the northwest corner of the area and flows directly into pond 18 (figure 8.1). At the southwest end of the pond complex, the water is diverted back into the Stiemerbeek, which flows into the Demer River. Since the ponds are directly connected to each other it is more appropriate to use landscape connectivity rather than Euclidean distances to estimate dispersal; an alternative would have been to assume that there is dispersal of resting stages through air, for example by birds (Proctor 1964; Proctor et al. 1967), and to calculate Euclidian distances. Landscape connectivity can be defined as the effect of landscape structure on species' use, on their ability to move and risk mortality in the various landscape elements, and on the movement rate among habitat patches in the landscape (Tischendorf and Fahrig 2000). We mapped the connecting elements (landscape features) and estimated both water and species flow rates in the different connecting elements. Specifically, Michels et al. (2001a) measured dispersal rate in connecting elements to be about 7000 zooplankton individuals per hour, which is equivalent to up to about 1% of the local zooplankton density per twenty-four hours. The probability of dispersal limitation in our system is thus very low. These data (connecting elements, flow, and dispersal rates) were used as inputs to model the effective geographic distances within a GIS (Geographical Information Systems) environment (Michels et al. 2001b). The model divides the ponds into three hydrological clusters, with ponds within a cluster sharing the same water flow, and no direct connections between the different clusters (see figure 8.1).

The Research Approach

We integrated observations and experiments to obtain insight into the relative importance of species sorting and mass effects in determining local community structure. The observational component consists of data on summer zooplankton densities in the different ponds of De Maten during three consecutive years, and a time series of zooplankton densities during one growing season in a subset of the ponds. The experimental data are from an *in situ* enclosure study, manipulating both dispersal and environmental variables.

We consider three scales of connectivity: directly connected ponds, ponds in a hydrological cluster, and ponds of De Maten as a whole (figure 8.1). The effect of exchange of individuals through dispersal might be limited to directly connected ponds, but might also indirectly influence the other ponds in a hydrological cluster, and even the whole pond system. Krebs (2001) described four traditional characteristics of community structure: (1) biodiversity, the number of species in a community; (2) the densities of the different species; (3) the functional types of species within a community; and (4) trophic structure—the feeding relations that determine the flow of energy and materials. We used both biodiversity and density as the main descriptive variables of community structure. However, our results allowed us to determine functional types for zooplankton taxa within shallow lakes. Since we studied the community structure within a trophic level, we did not determine trophic structure.

We give an overview of the analyses done, including the sources of the data, in table 8.2. We also related the patterns found in our experimental treatments to the observational data.

Table 8.2 Overview of the analyses

	Direct connection			Cluster		Regional	
	LR	SY	EX	LR	SY	LR	SY
Abundance	1	N	3	N(1)	N	ns	N
Diversity	2	N	3	N(2)	N	2	N

Notes: This analysis uses three the scales of three scales of connectivity (the directly connected ponds, ponds in the different hydrological clusters, and the pond system as a whole), and two characteristics of community structure (relative abundances of the different species and species diversity, measured as cladoceran species richness). The three research questions are (A) what is the relative importance of local and regional processes (LR), (B) do the ponds show synchronicity (SY), and (C) is there experimental evidence for local and regional processes (EX).

1: data and analyses from Cottenie et al. (2003).
2: data and analyses from Cottenie and De Meester (2003).
3: data and analyses from Cottenie and De Meester (2004).
N: data and analyses are new for this chapter.
N(x), where x equals 1 or 2: data were used in the respective paper, but the analyses are new.
ns: that the pattern was not studied because of lack of comparative data on other pond systems.

Question 1: What Is the Relative Importance of Local and Regional Variables?

Methods

Quantitative zooplankton samples were collected from the pelagic zone of each pond on three days in June–July of 1996, 1997, and 1998; samples were collected for cladocerans, copepods, and large rotifers using a 12 L Schindler-Patalas sampler equipped with a 64 µm mesh, and four of these samples were pooled per pond and sampling occasion. Different levels of identification were used: (1) cladocerans were identified to species level, (2) copepods were divided into cyclopoid, calanoid, harpacticoid copepods, and nauplii, and (3) large rotifers were identified to genera. Zooplankton densities for these taxa were determined as the number of individuals per taxon per liter. We determined biodiversity only for cladocerans (number of species per sample), since this is the only taxonomic group containing a uniform level of taxonomic resolution. We also determined important environmental characteristics, as described above in "The Setting."

To determine the independent contributions of both spatial and environmental variables to zooplankton density and cladoceran species richness, we used a partial regression approach. For the multivariate zooplankton density data, we used a (partial) Redundancy Analysis (RDA; Cottenie et al. 2003), and for the cladoceran species richness data, a (partial) multiple regression approach was used (Cottenie and De Meester 2003). We refer readers to these previous works for a detailed description of the sampling procedure as well as construction of the different spatial and environmental data sets. For the analyses with pond clusters (figure 8.1), we used two binary variables as spatial variables. The presence of a pond in cluster 1 was indicated by a 1 for the first spatial variable, and a 0 for the second spatial variable; a pond in cluster 2 by a 0 for the first spatial variable and a 1 for the second; and a pond in cluster 3 by a 0 for both spatial variables.

Results and Discussion

The results of the analyses are presented in figure 8.2. The pure spatial component contributed significantly to explaining the variation in zooplankton densities for the three years; this was independent of the variation explained by the environ-

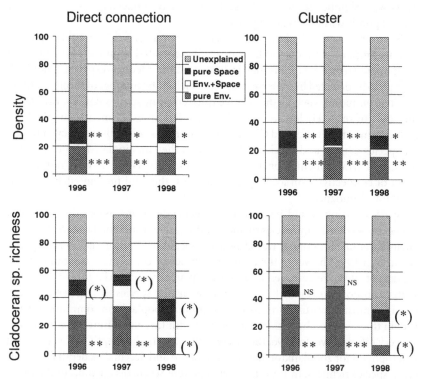

Figure 8.2 Variance partitioning of the zooplankton data matrix in the three different years of study. The analyses were done for two levels of community structure and two scales of connectivity (direct connections among ponds, pond clusters). Four different components are given: pure environmental variation, pure spatial variation, the spatial structuring in the species data that is shared by the environmental data, and the unexplained variation. The significance of the pure components are obtained with 999 new permutations under the reduced model (values of P noted below). If the pure environmental or spatial component is not significant, but the corresponding total variation is significant, the significance level is indicated between parentheses. If the pure environmental or spatial variation and the corresponding total variation are not significant, it is indicated by NS (data from Cottenie and De Meester 2003, Cottenie et al. 2003).

* = P < 0.05
** = P < 0.01
*** = P < 0.001

mental variables. The spatial component, however, is confounded within the environmental component for the cladoceran species richness data. The significance of the pure spatial components for zooplankton density suggests that the ponds have a metacommunity structure following the mass effect perspective. Whereas in the species sorting perspective we would expect that environmental variables would be able to explain variation in density, and that pure spatial variation would not be significant (providing that the effects of environmental variables on density are adequately described). We found that pure spatial components were significant both at the scale of directly connected ponds and at the scale of ponds within a hydrological cluster.

By contrast to zooplankton density, cladoceran species richness did not have a spatial component that could be statistically separated from the environmental component (figure 8.2). This may indicate that the effect of dispersal is not high enough to be detected by a coarse endpoint, such as the number of species. A more sensitive endpoint, such as the relative differences in species density, can detect the effect of relatively low levels of dispersal. For an experimental zooplankton metacommunity, Forbes and Chase (2002) found a similar discrepancy in results for species density and species richness when they manipulated dispersal rates, the spatial arrangement of enclosures, and nutrients. While they did not find an effect of connectivity on local species richness, they did find an effect on zooplankton abundances.

Species sorting would be expected to generate strong associations between either species density or species richness and local environmental factors. The pure environmental component was always significant, except for cladoceran species richness in 1998 (however, the total environmental component is significant for the data of 1998). Percentages of variation explained were also large relative to pure spatial components (figure 8.1), illustrating that the species sorting perspective is often the most important process in determining zooplankton community structure in this system. The variables associated with this environmental component are very informative. In 1996, they were depth, Secchi disk depth, and submerged macrophytes for zooplankton density, and depth and Secchi disk depth for cladoceran species richness. In 1997, the variables were Secchi disk depth, conductivity, and diversity of macroinvertebrates for density, and Secchi disk depth and oxygen concentration for richness. In 1998, they were pH, nitrate concentration, and submerged macrophytes for density, and pH and diversity of macroinvertebrates for richness. Given that similar variables were selected as important for both zooplankton density and cladoceran species richness in each year, it is likely that similar processes determined community structure at different levels.

Interestingly, almost all of the factors selected are related to the occurrence of alternative equilibria in shallow lakes. Scheffer et al. (1993) developed a model explaining the occurrence of two alternative stable states in shallow lakes. One equilibrium state, which predominates at low nutrient concentrations, is characterized by abundant macrophytes and clear water (i.e., large Secchi disk depths).

This state is stabilized by high zooplankton grazing rates, low planktivorous and benthivorous fish abundances, and high piscivorous fish abundances. The other state is characterized by abundant phytoplankton and turbid water, and is most common at relatively high nutrient concentrations. The turbid state is stabilized by light limitation due to algal blooms. At intermediate nutrient levels, both alternative stable states can occur. Several of the selected variables, such as Secchi disk depth, submerged macrophytes, and nitrate concentration, are key variables in this model. Moreover, ponds in De Maten occur in both turbid and clear-water states (Cottenie et al. 2001). Clear-water ponds in De Maten are characterized by high densities of *Daphnia pulex, Polyphemus pediculus, Simocephalus vetulus,* and low abundances of *Bosmina longirostris, Daphnia ambigua,* cyclopoid copepods, and rotifers. Ponds in the turbid state are mainly characterized by a high abundance of rotifers and cyclopoid copepods. An intermediate group of ponds was also observed and characterized by the presence of cyclopoid copepods, *Chydorus sphaericus, Ceriodaphnia pulchella, Daphnia ambigua, Daphnia galeata,* and *Bosmina longirostris* (Cottenie et al. 2001). Two alternative explanations are available for this last group: it could either be a true transition phase between the two stable states or be the result of continuous dispersal between clear and turbid ponds. Cottenie et al. (2003) showed that this intermediate state did not result from continuous dispersal from upstream ponds. We will provide evidence for an alternative interpretation below.

The previous data show that the metacommunity structure of the ponds of De Maten influences the local community structure. However, an interesting question is whether metacommunity structure influences the regional species pool (see Hoopes et al., chapter 2, and Shurin and Srivastava, chapter 17)? For the system-wide scale (all of the ponds), we did not attempt to answer this question for relative densities because it would require data on ponds situated close to De Maten that were unconnected and had similar environmental conditions and history. We believe that for a derived measure, such as species diversity, these requirements can be relaxed somewhat. Previous studies showed that the local communities in the ponds of De Maten are richer in species than ponds with a similar surface area that are reported in the literature (Cottenie and De Meester 2003). We hypothesized that the high connectivity within this system may increase local diversity by reducing the impact of local extinctions, thus also illustrating the impact of mass effects.

Question 2: Do the Ponds Show Temporal Coherence?

Methods

Temporal coherence is the phenomenon of synchronous fluctuations in one or more parameters among locations within a geographic region (Magnuson et al. 1990). Leibold and Mikkelson (2002) propose that coherence along an environmental gradient is a useful feature to study in metacommunities. Temporal coherence is affected by both the spatial correlation in the environment and dispersal (see Hoopes et al., chapter 2). At the level of all ponds or at the level of the pond cluster, temporal coherence can indicate the importance of both factors. Conversely, more temporal coherence between directly connected ponds than in ponds lacking direct connections would indicate a stronger relationship between dispersal and temporal coherence, which would indicate the importance of mass effects in the metacommunity. The previous paragraph addressed the question of habitat differentiation along a spatial gradient. With the second question, we extend the idea of coherence to a temporal gradient.

A subset of the pond complex (thirteen ponds, eight from the upstream subgroup, and five from the downstream subgroup, with no direct water connection between the two subgroups) was sampled on seven different dates between April 13 and July 17, 1999, following an experimental manipulation of the different ponds (stocking with pike fingerlings and drying treatments). The results concerning this biomanipulation will be detailed elsewhere (Cottenie and De Meester, in preparation). We will use the data here to test whether directly connected ponds show more temporal coherence than ponds lacking direct connections.

The sampling procedure and data manipulation were identical to the June–July samples. We selected taxa with more than thirty individuals per liter summed over the seven time points (nine cladoceran species, three copepod taxa, and eight rotifer taxa). We concurrently measured phytoplankton abundances as chlorophyll-a biomass (Cottenie, Nuytten, et al. 2001; Cottenie, Michels, et al. 2003). For the analyses at the scales of the pond clusters and all ponds, we followed the procedure outlined by Rusak et al. (1999). We computed the intraclass correlation coefficient to estimate temporal coherence from a two-way ANOVA without replication. This measure can simultaneously estimate the relationship between more than two sets of time series data. To determine whether directly connected ponds show more synchrony than unconnected ponds, we computed the intraclass correlation coefficients for all pairs of ponds, and compared directly connected ponds with ponds lacking direct connections. For the zooplankton abundance analysis, we computed the principal components of all of the data together (for all of the ponds using all seven time points). The first principal component axis offers the best approximation to the variation present in the original species matrix. It can thus be considered to be a "composite species," incorporating the information from the whole species matrix in one variable. The second principal component extracts the second largest amount of information from the original species matrix, but is independent from the variation extracted by the first principal component.

Results and Discussion

The first principal component of density showed significant temporal coherence (synchrony) at the regional level (figure 8.3a). At the level of pond clusters, the results differed, with more evidence for synchrony in the upper pond cluster than the

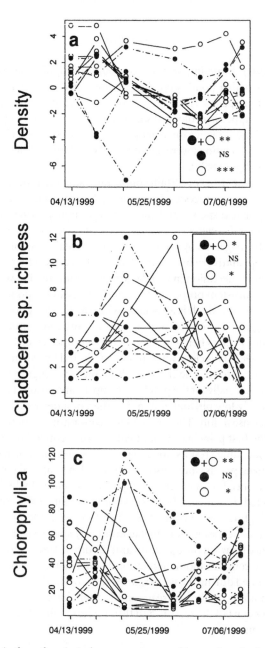

Figure 8.3 Time series for *a*: the principal component scores of the ponds on the first axis of the zooplankton densities; *b*: cladoceran species richness, and *c*: chlorophyll-a concentrations. One time series is one pond. The solid circles with dashed lines indicate ponds from the downstream pond cluster; open circles with full lines indicate ponds from the upstream pond cluster. Parallel lines show temporal coherence, and the intraclass correlation coefficient computes the overall "parallelness" of all the lines, together with its significance. For every graph, three intraclass correlation coefficients and associated significances were computed, which correspond to two levels of spatial scale: one for the region (all ponds together) ($\bullet + \bigcirc$); one for downstream ($\bullet$), and one for upstream ($\bigcirc$) ponds. $P < 0.05 =$ *, $P < 0.01 =$ **, $P < 0.001 =$ ***.

lower pond cluster. More importantly, we found no significant relation between synchrony and whether ponds were directly connected (results are not shown).

For cladoceran species richness, synchrony was significant for the regional level and for the upper hydrological subgroup, while the lower hydrological subgroup did not show evidence for synchrony in species richness (figure 8.3b). The same intraclass correlations were significant for chlorophyll-a as for cladoceran species richness (figure 8.3c).

The regional synchrony of certain community and environmental patterns indicated that there were processes determining dynamics on a regional scale. This does not imply that dispersal and the metacommunity structure of the ponds caused this temporal coherence. The response of the ecosystems in the different ponds is probably triggered by the same environmental cues; for instance, timing of increased zooplanktivory by young-of-year fish is caused by regional temperature cues. Seasonal succession in zooplankton communities is a well-studied phenomenon. One consistent property of seasonal succession in ponds is the presence of a clear-water phase, which is associated with low turbidity, low phytoplankton biomass, high grazing pressure, and a zooplankton community dominated by cladocerans, with rotifers being less abundant (Sommer et al. 1986; Luecke et al. 1990; Scheffer 1998). This clear-water phase is followed by a switch to turbid water with the opposite patterns. The chlorophyll-a time series (figure 8.3c) showed this general pattern, especially for the ponds in the upstream cluster. The ponds showed either continuously low phytoplankton densities, or a decrease in May–June, followed by an increase in July. The zooplankton community structure mirrored this pattern, with the first principal component axis (determined by rotifer densities) being low in May–June (results not shown). However, this normal seasonal succession in unlinked ponds would also have resulted in temporal synchrony. We did not observe evidence for an effect on temporal coherence of whether ponds were directly connected. This may be because even ponds that are not connected to each other showed high temporal coherence. Therefore, we could not disentangle the effects of seasonal succession from metacommunity spatial structure.

The difference between the upstream and the downstream subgroups might be caused by a difference in biomanipulation history: all but two of the upstream ponds were drained during the winter, while none of the downstream ponds were drained. The draining in the upstream ponds might have reduced environmental variability at the start of the season, and thus increased the likelihood of detecting temporal coherence.

Question 3: Do Local Processes and Dispersal Have a Deterministic Effect on Zooplankton Community Structure in This System?

Methods

To further understand the relative effects of local and regional processes, we conducted a factorial field experiment using communities from two of the De Maten ponds. Zooplankton communities in pond 13 were typical of a clear-water pond, and those in pond 12 were typical of a turbid pond. The experiment was carried out in the downstream, turbid water pond. Enclosures (bags sealed from the pond water) were constructed and filled with filtered pond water from this turbid pond. Filtration eliminated the large zooplankton community from the bags. We manipulated fish predation pressure and macrophyte cover in a factorial way, and inoculated the enclosures with a mix of species from the two different ponds. Fish predation was controlled by adding 0 or 2 individuals of *Lepomis gibbosus*, and macrophyte cover was manipulated by adding 0 or 5 strands of *Polygonum amphibium*. We also established an additional treatment in which the enclosure was inoculated with only the community from the resident pond without fish predation and macrophytes. We compared this treatment with the No-fish/No-macrophyte enclosures with a mixed inoculum to check for an effect of dispersal on the zooplankton community (Cottenie and De Meester 2004). In twelve of the fifteen enclosures, a mix of zooplankton communities from the downstream (turbid) and upstream (clear) pond, with a very different species composition, was added. In the other three enclosures, only the community from pond 12 was added. In the three enclosures without a mix of pond 12 and 13 communities, neither fish nor macrophytes were added. This resulted in three replicates of each of five treatments: resident community (RC); and mixed communities with fish/macrophytes (FM), fish/no-macrophytes (FN), no-fish/macrophytes (NM), and no-fish/no-macrophytes (NN). The resident community treatments are similar to the no-fish/no-macrophytes treatment, except that the latter received an additional inoculum from pond 13.

Results and Discussion
DOES THE LOCAL ENVIRONMENT INFLUENCE COMMUNITY STRUCTURE?

After twenty-five days, the zooplankton densities in the enclosures were similar to the densities found in the upstream and downstream pond. The experimental manipulation had a significant effect on zooplankton density (figure 8.4) and cladoceran species richness (figure 8.5). The community that developed in the fish/no-macrophyte enclosures was similar to the resident community of the pond in which the experiment was done (in figure 8.4 see the *a posteriori* addition of the samples taken from pond 12). This reflects the ecological conditions of a turbid water state, high fish predation pressure and no macrophytes in this pond (Cottenie et al. 2001). The community from the upstream pond 13 (from which the immigrating community came) takes an intermediate position between the communities in the fish/macrophyte and no-fish/macrophyte enclosures. Pond 13 is indeed in the clear-water state, and is characterized by abundant submerged vegetation and intermediate levels of fish predation (Cottenie et al. 2001). The results of this experiment therefore strongly indicate that fish predation and macrophyte abundance are important factors structuring zooplankton communities in shallow ponds (Brooks and Dodson 1964; Carpenter and Kitchell 1993; Scheffer et al. 1993).

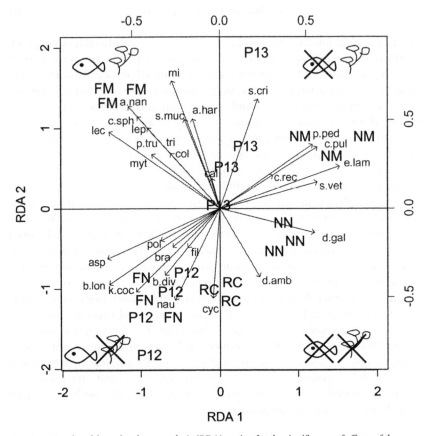

Figure 8.4 Results of the redundancy analysis (RDA) testing for the significance of effects of the presence of fish and macrophytes on zooplankton community structure in *in situ* enclosures. The different enclosures are represented by symbols, indicating the treatment they received: FM = fish / macrophyte, NM = no-fish / macrophyte, NN = no-fish / no-macrophyte, FN = fish / no-macrophyte, RC = resident community only. The samples from the two original ponds are added as supplementary variables: P13 = pond 13, P12 = pond 12. The relative species composition in the enclosures can be determined from the relation between the position of the enclosures and the species arrows: orthogonal projection of an enclosure on a species arrow determines the relative abundance of this species in the enclosure. Thus the fish / macrophyte enclosures are characterized by small, macrophyte-associated species such as lec (*Lecane*), c.sph (*Chydorus sphaericus*), a.nan (*Alonella nana*), lep (*Lepadella*), s.muc (*Scapholeberis mucronata*), mi (macroinvertebrates), a.har (*Acroperus harpae*); the no-fish / macrophyte by large, macrophyte-associated species such as s.cri (*Sida cristallina*), p.ped (*Polyphemus pediculus*) c.pul (*Ceriodaphnia pulchella*), e.lam (*Eurycercus lamellatus*), s.vet (*Simocephalus vetulus*); the no-fish / no-macrophyte by large pelagic species such as d.gal (*Daphnia galeata*), d.amb (*Daphnia ambigua*); the fish / no-macrophyte enclosures by small pelagic species such as nau (nauplii), k.coc (*Keratella cochlearis*), b.lon (*Bosmina longirostris*), asp (*Asplanchna*), cyc (cyclopoid copepods), b.div (*Brachionus diversicornis*). The species with small arrow lengths are myt (*Mytillina*), p.tru (*Pleuroxus truncatus*), tri (*Trigonellus*), col (*Colurella*), cal (calanoid copepods), c.rec (*Camptocercus rectirostris*), fil (*Filinia*), bra (*Brachionus angularis*), pol (*Polyarthra*). Figure taken from Cottenie and De Meester (2004).

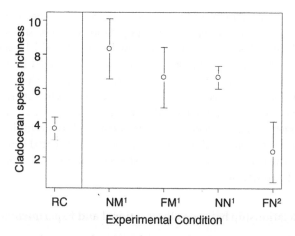

Figure 8.5 Mean cladoceran species richness (circles) in the different experimental conditions plus or minus two times the standard error (bars): RC = resident community (species of only pond 12 without fish and macrophytes), NM = no-fish/macrophyte, FM = fish/macrophyte, NN = no-fish/no-macrophyte, FN = fish/no-macrophyte. Post-hoc multiple comparisons using the Scheffé method on the experimental conditions that received a mix of species from pond 12 and pond 13 (for treatments on the right side of the vertical line), showed that FN was significantly different from the other three treatments (indicated by a different superscript), which were not different from each other. Figure taken from Cottenie and De Meester (2004).

In figure 8.4 the first axis discriminates between the enclosures with and without fish, resulting in enclosures with small and large species. By contrast, the second axis discriminates between enclosures with and without macrophytes, resulting in enclosures with macrophyte-associated and pelagic species (Cottenie and De Meester 2004).

Using the data on cladoceran species densities to compute cladoceran diversity in the different experimental treatments also gives evidence for the importance of local factors (figure 8.5). The fish/no-macrophytes enclosures had significantly lower cladoceran species richness than the other three treatments with a mixed inoculum.

DOES DISPERSAL INFLUENCE COMMUNITY STRUCTURE?

Although the results of the transplant experiment underscore the importance of local environmental conditions, the effect of dispersal is also evident in this experiment. The effect of dispersal can be seen by comparing the resident community (RC) treatment and the no-fish/no-macrophyte (NN) treatment, which differ only in whether they received an inoculum from pond 13. Figure 8.5 shows that the enclosures with the RC treatment had approximately three cladoceran species less than the NN community. There were indeed three cladoceran species not present in the RC enclosures and in the samples from pond 12 but present in the NN enclosures and in the samples of pond 13 (*Simocephalus vetulus, Eurycercus lamellatus,* and *Sida cristallina*). Since populations of these species in the NN enclosures displayed positive population growths after the initial dispersal events (data not shown), they are not sink populations maintained by the dispersal event. This indicates that dispersal of taxa is responsible for the higher diversity in the NN enclosures compared to the RC enclosures. This is in line with studies by Shurin (2000, 2001) and Amezcua and Holyoak (2000) that show that local dispersal rates may strongly influence the capacity of local communities to adjust to changing local environmental conditions. The results presented here also suggest that dispersal is necessary to deliver the species that are best adapted to local environmental conditions.

The Relationship between Observational and Experimental Data

The four types of communities found in the transplant experiment, consisting of combinations of small versus large species and pelagic versus macrophyte species, can be considered the four prototype communities in shallow lakes at the local scale. They define the range of possible local communities. Real communities can be thought of as being a mix of these four types, depending on the amount of macrophyte cover and fish predation. We have already shown where the samples from ponds 12 and 13 fit into this set of prototype communities. Below, we show two other examples of how integrating experimental and observational data might explain our field system.

The first example concerns differences in the relationships between zooplankton relative densities or cladoceran species richness and local environmental variables (see question 1) in the three years studied. We found similar relations for 1996 and 1997, but not for 1998. In 1998, the total variance explained was lower for species richness, and the pure environmental variation was not significant (figure 8.2). Also, for both 1996 and 1997, Secchi disk depth was selected, while for 1998 this was not the case. In Figure 8.6a, we fitted all communities observed in the three years to our four prototype communities. In 1996, the zooplankton communities covered the whole spectrum of different communities. In 1997, there was a shift to communities more typical of ponds without macrophytes, while in 1998 communities were mainly typical of turbid ponds. This might well explain the observed differences between the three years: the communities sampled in 1998 were not yet in the clear-water stage, and consequently no relation could be found between zooplankton community structure or species richness and local environmental variables associated with the alternative stable states in shallow lakes.

In the second example, the use of prototype communities as observed in the experiment not only explained the differences between years, but can also explain the differences observed during 1996. In 1996, we identified three types of

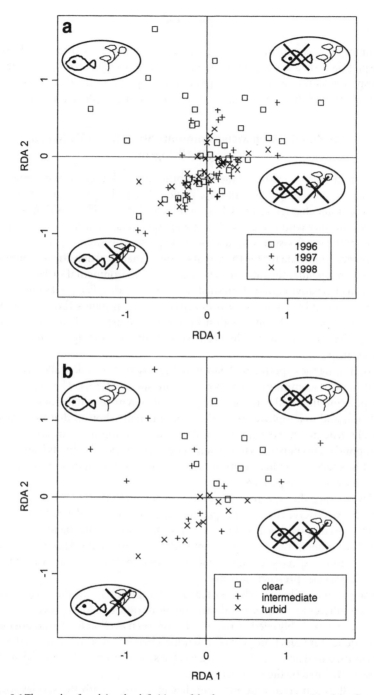

Figure 8.6 The results of applying the definitions of the four prototype communities from figure 8.4 to a: the observational data from the three different years, and b: the observational data of 1996 only. Both panels indicate the different types of zooplankton communities delineated by Cottenie et al. (2000). The placement of the four prototype communities is identical to figure 8.4.

communities (clear, turbid, and intermediate; see above and Cottenie et al. 2001). If we plot these data in our prototype community plot (figure 8.6b), the clear type corresponds to the no-fish/macrophyte prototype, the turbid to the fish/no-macrophyte, and the intermediate state to the fish/macrophyte prototype.

The Zooplankton Metacommunity Structure of De Maten: Putting It All Together

Two defining characteristics of metacommunities are the degree of heterogeneity between the habitat patches and the dispersal rates between the patches (see Holyoak et al., chapter 1, and Chase et al., chapter 14). The position of a field metacommunity with respect to these two parameters will influence how appropriate are the species-sorting and mass effect perspectives.

Observational data for the three years showed that local densities of zooplankton were independently influenced both by local environmental variables associated with alternative equilibria in shallow lakes (turbidity, fish predation, macrophyte cover) and the pattern of connectivity between ponds (both for directly connected ponds and ponds within hydrological clusters). Thus, zooplankton community structure was influenced by both species sorting and mass effects (mass transport).

Combining the experimental data with the observational data allows us to describe the metacommunity structure in more specific terms. The experimental manipulation imposed extremes along separate environmental gradients of fish predation and macrophyte cover. We experimentally showed that these factors strongly structure the communities and that these communities are also observed in the ponds. This illustrates that heterogeneity among local habitats is strong and that this is reflected in local community structure, implying that mass effects do not overwhelm these patterns.

If we reduce the information content in the analyses by looking at species diversity, we observe that the independent effect of the spatial component in the variance decomposition becomes insignificant, and the independent environmental component remains significant with the density data. This indicates that local dynamics (species sorting) are more important than mass transport in determining species presence/absence.

The presence of high levels of synchrony, irrespective of the direct connection between different ponds, also indicates the importance of local processes. These processes are probably cued by regional factors such as climate. We did not detect any evidence for mass transport of individuals between directly connected ponds contributing to synchrony. If there is a level of synchrony based on local dispersal, it is overwhelmed by the seasonal signal.

Overall, our metacommunity complies more often with the species sorting perspective than with the mass effect perspective. This is true despite it being characterized by strong connectivity. Other evidence also indicates that dispersal is important but not overwhelming in this metacommunity. For instance, we did not find evidence for sink populations of species in our transplant experiment. If, in the transplant experiment, we define sink populations as species that have no individuals in the samples from the downstream pond, have high population densities in the upstream pond, and show up in the enclosures of the resident community treatment. Then we did not find any species that met these requirements. This indicates that if sink populations are present in the downstream pond, the densities are so low that even after twenty-five days they are still not detectable. Time itself was not truly limiting in this experiment because twenty-five days was enough to establish population densities of up to twenty-five individuals per liter of a species (*Daphnia galeata*), which was only present in one sample of the upstream pond. This observation actually illustrates that local dynamics can be overwhelmingly important in this type of community. Also, there were three cladoceran species that we did not detect in the downstream pond or in the resident community enclosures, but that we did detect in both the no-fish/no-macrophyte enclosures and the upstream pond. These species thus were "dispersal limited" in the context of the experiment. Dispersal is necessary to deliver the species best adapted to environmental conditions within a pond and thus to allow effective species sorting. Since the communities of the experimental treatments correspond to the communities in the field, however, our results also indicate that most species are present in the habitats with the "correct" environmental conditions. Summarizing, our observation of the absence of sink populations and the absence of dispersal limitation in most species both underscore that the metacommunity studied is characterized by intermediate dispersal rates relative to the environmental heterogeneity.

Comparing our pond system to other published aquatic metacommunity systems, we see an interesting gradient of metacommunity types, typified by an increasing importance of dispersal. One extreme is exemplified by inquiline communities in pitcher plants (Miller and Kneitel, chapter 5). The middle of the gradient is covered by the large body of literature on ponds and lakes that are not directly connected (e.g., Pinel-Alloul et al. 1995). This chapter covers the other extreme—highly connected ponds in a small area. All three study systems show conclusive evidence for the importance of the species sorting view. Analysis of observational data using the variance decomposition method leads all three studies to find an important and significant component of the community structure associated with environmental variables. This is backed up by controlled experiments manipulating local environment variables (the study by Pinel-Alloul et al. does not include this, but a large literature on experimental manipulation of zooplankton community structure in lakes is available). The effect of dispersal on these three types of communities is, however, different. In the inquiline metacommunity, the spatial configuration of the communities explained only a small and nonsignificant portion of community structure. Moreover, several inquiline species showed dispersal limitation. In the lake system studied by Pinel-Alloul et al. (1995), the pure spatial component also explained only a small and non-significant portion of community structure. Conversely, in the connected ponds of De Maten, the pure spatial component explains a significant portion of variation, giving some evidence for effects of mass transport on community structure.

Given our results from a system of strongly interconnected ponds, we conclude that large mass effects are expected to be less frequent than species sorting in zooplankton communities of standing waters. Furthermore, the zooplankton metacommunity in connected pond and lake systems will often correspond to a species sorting model in which the response to environmental heterogeneity may be large and rapid. Zooplankton metacommunities in ponds that are not connected may also show species sorting, but in a less efficient way because of dispersal limitation. The impact of disturbance events is expected to differ in these different kinds of metacommunities. For instance, the ponds in De Maten exhibit frequent disturbance events due to drying out of some of the ponds. Compared to the same disturbances in a closed system, the response of the zooplankton community to the refilling with water and to the changed environmental conditions (increased macrophyte cover and decreased planktivorous fish) will be faster and will include better adapted species.

Conclusions

We believe we can safely state that the ponds of De Maten show a high degree of connectivity compared to other pond systems, both because of the small spatial scale and the presence of direct connections through overflows and rivulets with measurable dispersal rates. Despite these high dispersal rates, we found little evidence for a metacommunity conforming to the mass effect perspective. Most of the evidence indicates that the species occur in the environment to which they are best adapted. However, we did find some evidence for mass transport effects on zooplankton community structure, acting alongside species sorting.

Although there are obvious differences, the parallel with evolutionary responses and local adaptation in natural populations is striking (see also Leibold et al., chapter 19). In natural populations, local factors drive evolutionarily divergent selection. In this process, dispersal both homogenizes populations and increases genetic variation within populations, and therefore increases the ability to adapt. In this respect, it is an interesting notion that evolutionary change within species may be more important in more closed systems. This is because a reduced rate of species sorting in relatively closed systems gives resident populations more time to adapt to changed environments before they are wiped out by the arrival of preadapted immigrants. In the metacommunity system of De Maten, dispersal is a necessary condition to deliver the potential building blocks of local communities. The nature and diversity of these building blocks are determined by the regional species pool (Shurin and Srivastava, chapter 17). Biotic interactions then select the species best adapted to the different local environmental conditions. Although strong dispersal may homogenize communities, our results indicate that local species sorting is powerful in the face of very substantial dispersal rates.

The next step in the study of this metacommunity study system could go in several directions. It would be interesting to further explore the effect of different rates of dispersal: (1) By experimentally manipulating dispersal rates between enclosures with different communities and environmental conditions, we could determine the rate at which mass effects alter community structure. (2) A modeling study of this system could determine theoretical dispersal rates necessary to achieve mass effects. (3) These rates could then be compared to the actual dispersal rates between different ponds in the system. It may also be revealing to compare the results on zooplankton metacommunity structure with results based on other components of the aquatic ecosystem: bacteria, phytoplankton, periphyton, macroinvertebrates, amphibians, and fish. Since these components have very different population densities, dispersal modes, and rates, comparing the patterns obtained across taxa may greatly increase our insight into metacommunity structure. Finally, we believe that comparing our results on zooplankton metacommunity structure with results on the genetic metapopulation structure of several key zooplankton species (e.g., Michels et al. 2001b) may contribute to the synergy and integration of community ecology and evolutionary biology.

Acknowledgments

We would like to thank the editors for inviting us to contribute to this exciting book on metacommunity dynamics. Much of the data and concepts used in this chapter were collected and contributed by numerous colleagues, including Erik Michels, Steven Declerck, Nele Nuytten, Lies Neys, Yolente Delaunoy, Frank Van De Meuter, to mention a few. The valuable comments of Jon Shurin, Marcel Holyoak, and two anonymous reviewers greatly improved the clarity and coherence of the text. Funding has been provided by the National Fund of Scientific Research—Flanders, grant G.0358.01, and by EU project BIOMAN, EVK2-CT-1999-00046. KC has conducted this work as a research assistant of the Fund for Scientific Research—Flanders (Belgium) (F.W.O), and as a postdoctoral associate at the National Center for Ecological Analysis and Synthesis, a Center funded by NSF (Grant #DEB-0072909), the University of California, and the Santa Barbara Campus.

Literature Cited

Amezcua, A. B., and M. Holyoak. 2000. Empirical evidence for predator-prey source-sink dynamics. Ecology 81:3087–3098.
Brendonck, L., and L. De Meester. 2003. Egg banks in freshwater zooplankton: Evolutionary and ecological archives in the sediment. Hydrobiologia 491: 65–84.

Brooks, J. L., and S. I. Dodson. 1965. Predation, body size, and composition of zooplankton. Science 150:28–35.

Carpenter, S. R., and J. F. Kitchell. 1993. *The trophic cascade in lakes*. Cambridge University Press, Cambridge, UK.

Cottenie, K., and L. De Meester. 2003. Connectivity and cladoceran species richness in a small interconnected pond system. Freshwater Biology 48:823–832.

———. 2004. Metacommunity structure: Synergy of biotic interactions as selective agents and dispersal as fuel. Ecology 85:114–115.

———. Biomanipulation in interconnected ponds. In preparation.

Cottenie, K., E. Michels, N. Nuytten, and L. De Meester. 2003. Zooplankton metacommunity structure: Regional versus local processes in highly interconnected ponds. Ecology 84:991–1000.

Cottenie, K., N. Nuytten, E. Michels, and L. De Meester. 2001. Zooplankton community structure and environmental conditions in a set of interconnected ponds. Hydrobiologia 442:339–350.

Daniels, L. 1998. Kansen voor natuurbehoud en -herstel. Natuurreservaten 10:4–7.

De Stasio, B. T. 1989. The seed bank of a fresh-water crustacean—copepodology for the plant ecologist. Ecology 70:1377–1389.

Forbes, A. E., and J. M. Chase. 2002. The role of habitat connectivity and landscape geometry in experimental zooplankton metacommunities. Oikos 96:433–440.

Hairston, N. G., Jr. 1996. Zooplankton egg banks as biotic reservoirs in changing environments. Limnology and Oceanography 41:1087–1092.

Jeppesen, E., J. P. Jensen, M. Sondergaard, T. Lauridsen, and F. Landkildehus. 2000. Trophic structure, species richness and biodiversity in Danish lakes: Changes along a nutrient gradient. Freshwater Biology 45:201–218.

Jeppesen, E., M. Sondergaard, M. Sondergaard, and K. Christoffersen. 1997. *The structuring role of submerged macrophytes in lakes*. Springer Verlag, Berlin, Germany.

Keller, W., and M. Conlon. 1994. Crustacean zooplankton communities and lake morphometry in precambrian shield lakes. Canadian Journal of Fisheries and Aquatic Sciences 51:2424–2434.

Krebs, C. J. 2001. *Ecology: The experimental analysis of distribution and abundance*. Benjamin Cummings, San Francisco, CA.

Leibold, M. A., M. Holyoak, N. Mouquet, P. Amarasekare, J. M. Chase, M. F. Hoopes, R. D. Holt, J. B. Shurin, R. Law, D. Tilman, M. Loreau, A. Gonzalez. 2004. The metacommunity concept: A framework for multi-scale community ecology. Ecology Letters 7:601–613.

Leibold, M. A., and G. M. Mikkelson. 2002. Coherence, species turnover, and boundary clumping: Elements of meta-community structure. Oikos 97:237–250.

Luecke, C., M. J. Vanni, J. J. Magnuson, J. F. Kitchell, and P. T. Jacobson. 1990. Seasonal regulation of *Daphnia* populations by planktivorous fish: Implications for the spring clear-water phase. Limnology and Oceanography 35:1718–1733.

Magnuson, J. J., B. J. Benson, and T. K. Kratz. 1990. Temporal coherence in the limnology of a suite of lakes in Wisconsin, U.S.A. Freshwater Biology 23:145–159.

Michels, E., K. Cottenie, L. Neys, and L. De Meester. 2001a. Zooplankton on the move: First results on the quantification of dispersal in a set of interconnected ponds. Hydrobiologia 442:117–126.

Michels, E., K. Cottenie, L. Neys, K. De Gelas, P. Coppin, and L. De Meester. 2001b. Geographical and genetic distances among zooplankton populations in a set of interconnected ponds: a plea for using GIS modelling of the effective geographical distance. Molecular Ecology 10:1929–1938.

Pinel-Alloul, B., T. Niyonsenga, and P. Legendre. 1995. Spatial and environmental components of freshwater zooplankton structure. Ecoscience 2:1–19.

Proctor, V. W. 1964. Viability of crustacean eggs recovered from ducks. Ecology 45:656–658.

Proctor, V. W., C. R. Malone, and V. L. DeVlaming. 1967. Dispersal of aquatic organisms: Viability of disseminules recovered from the intestinal tract of captive killdeer. Ecology 48:672–676.

Pulliam, H. R. 1988. Sources, sinks, and population regulation. American Naturalist 132:652–661.

Rodriguez, M. A., P. Magnan, and S. Lacasse. 1993. Fish species composition and lake abiotic variables in relation to the abundance and size structure of cladoceran zooplankton. Canadian Journal of Fisheries and Aquatic Sciences 50:638–647.

Rusak, J. A., N. D. Yan, K. M. Somers, and D. J. McQueen. 1999. The temporal coherence of zooplankton population abundances in neighboring north-temperate lakes. American Naturalist 153:46–58.

Scheffer, M. 1998. *Ecology of shallow lakes*. Chapman and Hall, London, U.K.

Scheffer, M., S. H. Hosper, M.-L. Meijer, B. Moss, and E. Jeppesen. 1993. Alternative equilibria in shallow lakes. Trends in Ecology and Evolution 8:275–279.

Shurin, J. B. 2000. Dispersal limitation, invasion resistance and the structure of pond zooplankton communities. Ecology 81:3074–3086.

———. 2001. Interactive effects of predation and dispersal on zooplankton communities. Ecology 82:3404–3416.

Sommer, U., Z. M. Gliwicz, W. Lampert, and A. Duncan. 1986. The PEG-model of seasonal succession of planktonic events in fresh waters. Archiv für Hydrobiologie 106:433–471.

Tischendorf, L., and L. Fahrig. 2000. On the usage and measurement of landscape connectivity. Oikos 90:7–19.

CHAPTER 9

Assembly of Unequals in the Unequal World of a Rock Pool Metacommunity

Jurek Kolasa and Tamara N. Romanuk

Introduction

Scale is an important component of an integrated metacommunity theory (Leibold et al. 2004; Holyoak et al., chapter 1), and incorporating various elements of scale is likely to continue to be a major force shaping the metacommunity framework. A number of specific insights have proved useful. Mouquet and Loreau (2002) considered two spatial scales—local and regional—and demonstrated potential interactions between these scales. Hanski and Gyllenberg (1997) showed that local abundance, geographical range, and species-area relationships could have a common conceptual core and represent differently scaled aspects of metapopulations. Some aspects of scale, particularly spatial ones, are formalized in metacommunity theory, such as patch size and distance among patches (Holyoak et al., chapter 1). By contrast, theoretical or experimental investigations of multiple spatial scales are rare (cf. Cadotte and Fukami 2004), and scales other than those discussed above may be important in developing an integrated metacommunity perspective. For example, scale may include dimensions such as gradients in species specialization or gradients in habitat heterogeneity and habitat structure.

In this chapter we consider the importance of different spatial and organizational scales on ecological processes in a metacommunity of rock pools inhabited by a diverse assembly of aquatic invertebrates using a hierarchical view of habitat structure. We also attempt to relate the empirical patterns seen in the rock pool metacommunity to the four metacommunity models articulated by Leibold et al. (2004) and Holyoak et al. (chapter 1). The primary structural feature of the rock pool metacommunity is the nested hierarchy of spatial habitat units, from individual pools to the entire regional array of pools. This structural hierarchy suggests that there is no simple local-regional scale dichotomy in the metacommunity (sensu Mouquet and Loreau 2002; Mouquet et al., chapter 10). Instead, habitat is portrayed as a nested hierarchy of habitat units, with each unit reflecting a fraction of multidimensional habitat space and with the whole structure potentially acting as a species sorting mechanism (Kolasa 1989).

First, we will outline and explain a hierarchical model of habitat, its predictions, and relationship to species and metacommunities. Subsequently, we will introduce the study system and illustrate the hierarchical nature of its spatial configuration and the possible constraints the hierarchy may impose on the movement and interactions among species. After describing the system in which metacommunity processes take place, we proceed to identify patterns in support of the hierarchical perspective and patterns that relate our empirical observations to various metacommunity models. In that section we highlight the existence of discontinuities in the observed metacommunity descriptors, positive links between density of individual species, occupancy of sites and degree of habitat specialization, the relationship between local and regional density, and the dependence of local species richness on scale of habitat variability.

A Hierarchical Model of Metacommunities

The habitat-based model (HBM) uses assumptions about (1) species attributes, (2) habitat attributes, and (3) the relation between species and habitat. The model is scale independent and can be applied at any scale of habitat structure.

Species Attributes

With respect to species we assume that: (1) they differ in their ecological needs such that there is a specialization gradient from species with narrow habitat requirements (habitat specialists) to species with broad habitat requirements (habitat generalists), and (2) any habitat may host a set of species representing the full range of habitat specializations. We contrast habitat generalists and specialists, but when referring to these two categories we recognize a gradient of habitat specialization. The differences potentially go far beyond core and satellite dynamics (Hanski 1982) because species interact with multiscale habitat heterogeneity. Different habitat requirements include tolerance limits, resource needs, and conditions for successful reproduction and dispersal such as nursery needs or presence of dispersal conduits. The terms specialist and generalist have a variety of meanings in ecology. The meaning here is general but precise: a specialist is a species restricted to a small volume of the multidimensional habitat space. In contrast, a generalist is a species that is not as restricted (Kolasa and Waltho 1998). Note that a species is evaluated by its relation to the habitat, and it is defined as a specialist or generalist relative to other species in the community, which jointly define the multidimensional habitat space.

Habitat Assumptions

We assume that habitat is a nested hierarchy of multidimensional volumes (for detailed discussion see Kolasa and Waltho 1998). Because any dimension except time can be mapped onto space, we interpret the habitat as (1) a hierarchical mosaic of patches with (2) lower level patches nesting within higher level patches, and (3) with each level and patch at any level involving a distinct set of attributes. For the sake of illustration, we use two dimensions only, which give the model a

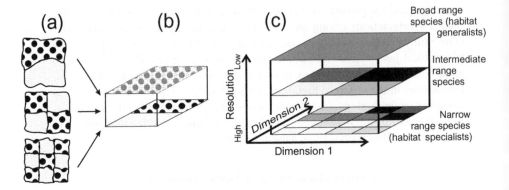

Figure 9.1 A simplified view of habitat hierarchy showing species categories defined by their use of various levels of resolution: (a) each of two types of habitat can take continuous or fragmented forms as illustrated by three combinations; (b) these types are represented in the model as a single two-level habitat hierarchy that can be used by two specialists at the lower level and one generalist at the higher level; (c) a hypothetical habitat structure able to accommodate twenty-one species. Note that habitat grain, size, and diversity changes with resolution. From species' perspective, the habitat becomes more fragmented at higher levels of resolution (i.e., lower levels of structure). While types of habitat patch are shown here as regular subdivisions of the total habitat unit, real structures are much more variable (figure 9.2; for further explanation see Kolasa 1989).

spatial appearance even though this appearance represents a considerable simplification because it does not purport to convey any specific configuration of actual habitats in space. The model only identifies the total amount of space a particular habitat unit occupies on average relative to a higher level unit. Each unit can take various configurations in space as a patch that is either contiguous or fragmented to varying degrees (figure 9.1a). Regardless of spatial configuration, two habitat types are represented in the model as two subunits (figure 9.1b). Extending this approach permits representation of the whole habitat as a nested structure of units emerging at increasing levels of resolution (figure 9.1c).

Relationship between Species and Habitat Structure

The relationship between the species and the habitat assumes the following: (1) habitat generalists use the largest habitat units, thus increasing habitat specialization results in the use of progressively smaller habitat units nested within the larger ones; and (2) the density of a species varies positively with the habitat unit size and negatively with the degree of unit isolation from other similar units. Thus, according to the model, the generalists' density will be proportional to the continuous area of the top habitat defined by the two dimensions in figure 9.1c. The situation is different for the specialists at the two lower levels. Although their densities are proportional to the area of the squares used, the expected density is different because their habitats appear as single patches in the generalist's habitat.

In a habitat unit even larger than one shown in figure 9.1c, these habitats would multiply but remain separated from each other by other, unusable, patches. The more specialized the species, the greater the geometrical and ecological distance (e.g., barriers of hostile habitats) between suitable units of the environment. Fragmentation of a habitat unit itself (cf. figures 9.1a and 9.2) can exacerbate this barrier effect. In a sense, resources available to a specialist can be viewed as being diluted in the patchwork of other habitat units. It is not unreasonable to assume that a specialist species faces higher energy and population costs when it uses patchily distributed habitats. By contrast, generalists in the model experience a more continuous distribution of habitat. Theoretically there may be several components to the costs experienced by specialists, such as energetic costs of travel, mortality during dispersal, decreased efficiency through failure to find patchily-distributed resources, or finding and settling in low-quality, sink patches. From now on, the sum of these costs will be assumed to reduce a species' density in proportion to the degree that the habitat is fragmented at the specialist level.

Predictions of the HBM

Consequences of the HBM are that (a) there are few species of habitat generalists and many specialists in species rich communities, (b) specialists have lower local density and habitat fragmentation plays a progressively larger role in determina-

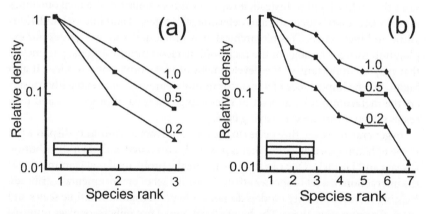

Figure 9.2 Effect of spatial fragmentation on relative density of species predicted by the HBM model: (a) the relative density of species using habitat unit in figure 9.1b as a function of fragmentation depicted in figure 9.1a; (b) the relative density of species using a habitat unit with one more level of resolution (inset shows a habitat able to accommodate 7 species: 1 + 2 + 4), with spatial fragmentation introduced for habitats units at the intermediate level of resolution only. Fragmentation in both panels is similar to that shown in figure 9.1a, that is, each habitat type is split into 1, 2, or 4–5 patches, with corresponding labels of 1, 0.5, and 0.2 to reflect the resulting patch size. Note that the fragmentation of the intermediate level (figure 9.2b) affects habitat specialists at the lower level of structure (ranks 4–7) even though no specific fragmentation was assumed for this level.

tion of their density relative to other factors, (c) community structure is discontinuous, with groups of species having similar degrees of habitat specialization and local density, and (d) local and regional density should converge for generalists.

Furthermore, the model implies that most local communities may transition gradually into metacommunities as spatial scale is increased. A thought experiment illustrates this observation. If the graphical representation of habitat structure in figure 9.1c grew in space by copying itself by 25%, it would mean no change for the top generalist because its habitat would remain continuous. However, two of the four intermediate species and four of the specialists would acquire additional subunits and thus enter the realm of metacommunity dynamics. A further expansion in the same direction by 5% would replicate patches suitable for four more habitat specialists. However, neither the top generalist nor additional intermediate species would experience a different habitat structure in the second step. A complete duplication of the structure would create a complete two-patch metacommunity, except for the top species (unless the duplication was discontinuous in space). Consequently, depending on the actual structure of the habitat, or the structure included in the study, a community of interest would require different degrees of metacommunity level insights.

This model is general in that it does not rely on physical properties, competition, predation or any other specific mechanism acting among the many dimensions that define habitat. Instead, it captures and summarizes the final outcomes of many processes, stochastic and deterministic, external and internal, individualistic and interactive. This hierarchical view of habitat structure guides our exploration into the ecology of the rock pool metacommunity and the parameters that may be important in the development of metacommunity models. It also highlights the importance of a metacommunity paradigm that explicitly considers the metacommunity as a set of interacting continua of properties and processes staged on an uneven playing field.

In general, however, the species in the rock pool metacommunity display a full range of habitat specialization that is positively correlated with their distribution (Kolasa and Li 2003). One can thus formulate the problem of metacommunity as the interaction of a set of species attributes with a set of environmental attributes. Existing metacommunity models do not explicitly recognize multiple scales and species' responses to them. The hierarchical model not only recognizes multiple scales, but also makes predictions about the patterns of densities and degrees of habitat specialization that should result from them.

Habitat-based Model and Other Metacommunity Models

Both the species sorting and mass effects models include a habitat gradient along which species sort. The hierarchical model (HBM) is most like the species sorting model but it explicitly recognizes the hierarchical structure of habitat and species responses to the habitat. Unlike the mass effects model the hierarchical model assumes that a species will be present in habitats that meet two conditions: they do have required resources (a niche based explanation) and are not excluded from it via biotic interactions such as direct or indirect effects of competition, parasitism, or predation, or by intolerable physical conditions. The HBM does not assume or include explicit mass effects or source-sink dynamics (e.g., Pulliam 2000). However, in empirical tests of the model (e.g., Kolasa 1989) expected density was calculated from the occupancy patterns. Because such patterns do not differentiate between values of the finite rate of increase λ, smaller or larger than 1, both effects may have been subsumed in the relationship between the distribution and density. Indeed, mass effects could promote species occupying habitats that are unsuitable for them and so affect occupancy and distribution patterns. Unlike the neutral and patch dynamics models, in the hierarchical model species are assumed to be found in particular kinds of habitat that are suitable for them, although the model is not restrictive as to how suitable habitats must be. At small spatial and temporal scales, a suitable habitat may simply mean habitat sufficient for survival of immigrants ($\lambda < 1$), while at a larger scale this implies a persistent population ($\lambda \geq 1$). Hence, in the neutral and patch dynamics models, species compositions of particular patches are expected to drift randomly through time and to show no particular habitat associations (see "Differences between Habitat Specialists and Generalists" for a test).

Ritchie and Olff (1999) have proposed an approach to regional diversity that also relies on differences among species in their perception of habitat patches. This model has an advanced formal structure built on a set of specific assumptions concerning body size, habitat fractal structure, and home range of individuals that do not apply to the rock pool metacommunity. The major difference between the HBM and Ritchie and Olff's model is that species' perceptions are analyzed at the individual level in the latter while they are a trait of populations in HBM. Consequently, home range of an individual and its effect on access to the amount and quality of resources forms the foundation of the Ritchie and Olff's model. In HBM, though, the perceptions of a single copepod or midge do not matter. What matters is the ability of a copepod subpopulation to persist in a habitat. Consequently, the home range does not matter in the HBM; the access to resources is determined by the ability of a population to use the habitat. Finally, while in both models habitat is seen as a nested mosaic of patches (which can be more than three-dimensional in the HBM), the HBM does not require the patch shape and arrangement to approach fractal characteristics. Nestedness is a universal habitat feature and thus it is always present. However no evidence exists that the self-affinity of habitat structure assumed by Ritchie and Olff (1999) is a common condition for most species assemblages. Finally, for aquatic invertebrates the body size does not seem to offer any biological insights to their ability to move among patches such as natural pools. Adaptations such as airborne stages, eggs or ephippia, or ability to withstand transport by water and wind are known to play crucial role (Bohonak and Jenkins 2003).

The Rock Pool Metacommunity
Site

The rock pool metacommunity is located around a small cove near the grounds of the Discovery Bay Marine Laboratory along the northern coast of Jamaica. The system involves a set of 49 coastal rock pools that are formed primarily by rain erosion of fossil reefs (Kolasa, Drake, et al. 1996; Kolasa, Hewitt, et al. 1998; Romanuk and Kolasa 2002). Within the study area shown in figure 9.3a, an additional 180 pools have been identified and analyzed for their community structure. The pools are on average ~ 60 × 30 cm, have a mean depth of 13 cm and mean volume of 15 liters. On average the rock pools are located within 1 m of the nearest neighbor and none is separated by more than 5m from the next nearest rock pool. Their elevation above sea level is 1–235 cm (mean ~ 80 cm) at high tide, with the tide rarely exceeding a 30 cm range. Seven of the 49 rock pools are tidal, although tidal flooding is not daily. Most pools are maintained by atmospheric precipitation and occasional wave splash water.

The strengths of the analyses stem from the nature of the study system. The rock pool metacommunity is highly diverse, both biotically and abiotically. The pools have similar history, climate, greatly reduced seasonality, and the system is small enough to allow any species to reach any habitat. The rock pool system shares several general advantages with other aquatic habitat landscapes listed by Resetarits et al. (chapter 16). Furthermore, it can be sampled for community composition and abundance within hours of measuring physical attributes. The main weakness of the approach to date stems from the fact that we infer the community and metacommunity dynamics from temporal snapshots instead of population trajectories. Nevertheless, we believe that such snapshots are informative, especially because they often cover tens or hundreds of generations.

Biotic Composition

We have collected invertebrate rock pool fauna and made physical measurements of the rock pools at least once a year, since the winter of 1989–90 (Kolasa et al. 1998; Therriault and Kolasa 1999, 2000; Romanuk and Kolasa 2001; Therriault 2002). Over seventy species have been identified and counted totaling more than 300,000 individuals from all samples. The majority of species are small benthic animals ranging from 60 μm to 0.5 mm but some are plankton-like (*Orthocyclops modestus, Ceriodaphnia* sp.) and swim in the water column. The metacommunity has unusual origin and composition. It is a mix of marine, brackish water, and freshwater organisms, some with poor dispersal abilities and some with non-aquatic dispersive stages (e.g., insects, crabs). The full list of the taxa identified to date includes: Turbellaria (7 species), Nematoda (1), Polychaeta (5), Oligochaeta (2), Ostracoda (21), Copepoda (8), Cladocera (4), Decapoda (crab) larvae (1), Decapoda (shrimps) (3), Amphipoda (1), Isopoda (1), and Insecta (18). The pool communities experience high desiccation, especially through the summer months when most of the shallow pools may become completely dry (Therriault and Kolasa 2001).

Coupled with the short generation times of the organisms (between less than a week and three months) the annual samples are relatively independent, that is, the current community structure is only partially determined by the community state a year prior to the sampling due to partial reassembly from a species-pool of short-lived organisms and cumulative stochastic effects of environmental and biotic variability (Romanuk and Kolasa 2002). Thus, the sampling regime misses detailed population dynamics but provides a long-term data set consisting of statistically independent snapshots, each separated by many generations from the previous one. Metacommunity patterns are therefore expected to be either at or approaching equilibrium with the environmental template.

The system of rock pool communities exhibits at least five distinct spatial scales (figure 9.3) although additional intermediate but less defined scales may also exist. This multiscale aspect appears to be an inherent feature of natural metacommunities (cf. Miller and Kneitel, chapter 5). Dispersal is likely to be modified by all of the five scales: regional (figure 9.3a), groups of rock terraces (figure 9.3b), pools on a single rock (figure 9.3c), pools lying in a single watershed (figure 9.3d), and individual pools (figure 9.3e). Rock pool species, which include taxa with diverse life histories such as turbellarians and mosquitoes, vary strongly in their dispersal abilities, and perceive individual patches on a gradient from either totally isolated to fully accessible or connected (Kolasa and Drake 1998). This does not differ from the general patterns established for inland aquatic invertebrates (Bohonak and Jenkins 2003). Consequently, a local pool community is likely to result from a blend of variably scaled immigration rates and the interaction of processes that are akin to population dynamics, metapopulation dynamics, and island biogeography, depending on which of the five or more scales one chooses to look at (figure 9.3).

As spatial scale changes, we would expect different biotic and abiotic dispersal processes to become relevant. The intermediate scales include subregions of the regional scale that share similar patterns of wind and precipitation or portions of a rock terrace where large invertebrate territories may affect several but not all watersheds. In particular, crabs that move actively in and out of pools on a daily basis can act as vectors for dispersal of rock pool invertebrates by carrying propagules in their digestive system or gill chambers. A different set of dispersal mechanisms operates at each scale (cf. Bohonak and Jenkins 2003). For example, adjacent pools that remain unconnected by water channels may exchange propagules or organisms via rain drop splash and overland transport by am-

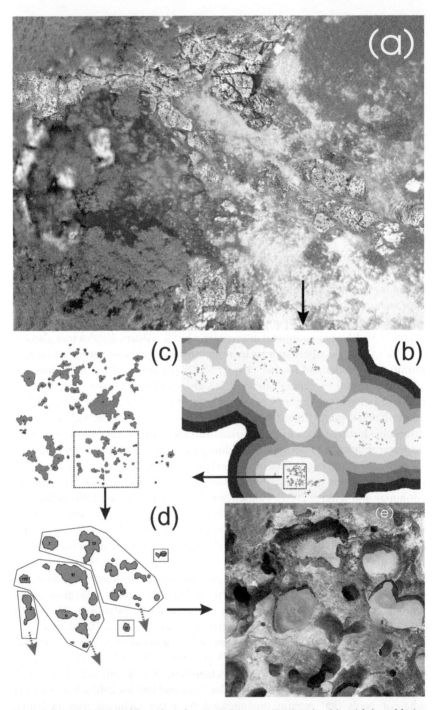

Figure 9.3 Physical layout of the rock pool system in Discovery Bay, Jamaica: (a) aerial view of the inlet with rock terraces and outlines of rock pools, ~60 m across; (b) distance based groups of pools; each group may encompass several rock terraces; (c) pools of one rock terrace; (d) five natural watersheds (bounded by polygons) on one rock terrace (courtesy of D. G. Jenkins, University of Central Florida); (e) individual pools in one of the watersheds. Shaded arrows indicate watershed outflow.

phibious animals as evidenced by specimens caught after rain in beaker traps set nearby. Pools in a watershed experience directional transfer of animals in addition to the previous mechanisms (Jenkins, pers. comm.). Pools of different watersheds receive animals as a cumulative function of pool-to-pool exchanges and a directional transfer within a watershed. As the spatial scale increases, the role of wind, storm, and large animal mediated dispersal increases; the relative contribution of these factors changes depending on the rock terrace elevation and relative position on a land-ocean gradient. The pools, watersheds, rock terraces, groups of rocks, and the region form a distinct spatial nested hierarchy, with a number of important consequences for modeling organism exchanges and their dynamics.

Recolonization experiments have shown that dispersal is high enough to allow reconstitution of local communities within six to twelve months, irrespective of their degree of isolation (A. Biggers, pers. comm.). However, the hierarchical structure of the habitat and habitat specialization of species mediates colonization patterns (Kotliar and Wiens 1990; Kolasa and Rollo 1991). If habitats of similar qualities are within the same spatial subunit (such as on a rock terrace), then predictions based on regional (entire metacommunity) dispersal rates will fail quantitatively because they assume a mean degree of interpatch isolation greater than the effective isolation of patches within the rock terrace.

Background Analyses and Review of Results
Specialization and Occupancy

Rock pool species show a strong link between distribution measured as occupancy, p, and habitat specialization. This is true for a variety of measures of habitat specialization but one method illustrates the point. Numerous other measures of specialization can be used and most produce similar results (unpublished). We have used a method of calculating ecological range that is largely immune to occasional occurrence of individuals outside conditions preferred and tolerated by a species. This immunity is achieved by weighing exposure to environmental variables by local density. We obtained tolerance ranges for each species using CANOCO's (software package) Canonical Correspondence Analysis module from four significant CCA axes. These were averaged and regressed on the occupancy (used in this chapter) expressed as the cumulative number of pools a species occupied over nine years ($r = 0.818$, $p < 0.0001$, $F = 145$; or $r = 0.896$, $F = 292$; if $y = ax^b$ is fitted). The regression indicates that distribution (occupancy) and specialization to physical conditions have a significant degree of equivalency in the system of interest.

Because occupancy reflects both the habitat suitability and the stochastic effects of immigration and extinction, the unexplained variance could represent a coarse indicator of the importance of patch dynamics. At the same time, occupancy can be used as an empirical substitute for specialization but not, indeed, in direct tests of models involving occupancy independently of specialization. This need for caution is supported by the fact that dispersal and hence distribution does affect measures of habitat specialization (Pulliam 2000), with more mobile species sampling a greater range of habitat conditions. We believe, however, that the methods employed here to evaluate the ecological specialization of species reduces any potential bias to an acceptable minimum.

Hierarchical Nature of the Rock Pool Metacommunity

Space is the most obvious dimension in the habitat hierarchy, however the spatial map is not a strict organizational tool in the hierarchy, nor do the habitat attributes necessarily match the spatial map. In fact, for the rock pool metacommunity the match is very weak because pools that cluster according to their physicochemical properties may be scattered in different locations of the spatial hierarchy. To add to this complexity, the degree of match is scale-dependent, with the two hierarchies—one spatial and one of habitat properties—intersecting at some scales but not at others. For example, groups of rock terraces at higher elevations share pools that are more similar to each other due to the dominant effect of rain and leaf litter on their water quality. A GIS (Geographic Information Systems) analysis of the rock pool metacommunity linking species patterns and physicochemical properties (Romanuk 1999) showed that all physical variables exhibited some spatial autocorrelation. However, the clustering scale differed among variables. The net result is that clusters of physicochemical properties may be scattered in different locations of the spatial hierarchy. For example, elevation and salinity showed subregional clustering while depth, temperature, oxygen, and pH all showed smaller scales of clustering.

The hierarchical structure of the metacommunity can also be seen in patterns of species habitat use. Species view the habitat at different levels in the habitat hierarchy ranging along a gradient from habitat specialists to generalists. However, this gradient is not necessarily continuous, since scale breaks or discontinuities are expected consequences of the hierarchical structure of habitat (cf. Kolasa 1989; Allen and Holling 2002). Such discontinuities are registered in assemblage attributes. Data from the rock pool metacommunity provide a good example of this phenomenon (figure 9.4) and can be seen in the graph combining degree of specialization and local density for ostracod species. Ostracods group into three distinct scales of habitat use; a group of highly specialized species with low local densities, a group of generalists with high local densities, and an intermediate group (figure 9.4). In addition to discontinuities ostracods illustrate a trend in the distribution-density relationship. The scale discontinuities in how species perceive their habitat can be used to guide decisions regarding the number of scales that would be important for modeling. For ostracods, any model using three scales might do a satisfactory job.

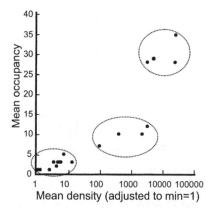

Figure 9.4 Groups of ostracod species identified by similarity of occupancy and mean local density based on one 0.5 liter water samples taken over twelve annual snapshot censuses of forty-nine rock pool communities. Density is scaled up to the minimum of one and has no relationship to observed values per liter. Arbitrary ellipses help to delineate the groups.

Habitat Specialization

Two primary observations on the ecology of metacommunities are integral to an understanding of rock pool metacommunity processes. We have already discussed the importance of viewing the habitat as a nested hierarchy of patches that exist in multidimensional space. In this section we present the results of analyses that suggest that species perceive habitat fragments differently, depending on their dispersal abilities and habitat tolerances.

Differences between Habitat Specialists and Generalists

Several observations are relevant to applying the HBM to the rock pool metacommunity system. Species vary in their local density over both space and time. In metacommunities, relative variability in density will have strong effects on the probability of whether a species will be present or not in a patch. This occurs because species with low population numbers are more likely to become locally extinct than species with high population numbers. In rock pools, specialists vary more than generalists in their relative variability (coefficients of variation) but tend to vary less in absolute terms (Kolasa and Li 2003). Mean local densities of species (represented as ranks) also relate strongly to occupancy (figure 9.5a), which can be interpreted in favor of the link between local presence (and thus regional occupancy) and habitat suitability. The density and occupancy relationship could also arise under special circumstances of uniform habitat space and random distribution of individuals, with the relationship being described by the Poisson probability distribution (Wright 1991). These circumstances do not

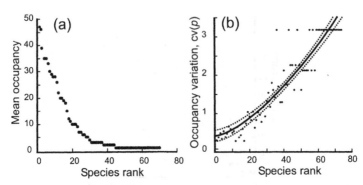

Figure 9.5 Species occupancy (a) and (b) its variability (coefficient of variation, cv) as a function of mean species rank in abundance. Occupancy and cv of occupancy are based on ten annual censuses.

appear to apply to the rock pool metacommunity for which abundant evidence of nonrandom, environmentally determined distribution (Therriault and Kolasa 1999) exists. Wright (1991) proposed that a set of species conforms to the Poisson null model if the slope of a regression (b) between mean abundance and $\ln(1 - p)$ is -1.0. For the rock pool metacommunity this slope is markedly different ($b = -0.126$, $r^2 = 0.289$).

The density-occupancy relationship however results in a triple-edged sword for specialists in the rock pool community where specialists are locally and regionally rare, and their local population densities as well their occupancy are relatively more variable (figure 9.5b). These species' attributes suggest that the population dynamics of specialists are fundamentally different than the dynamics of generalists in the metacommunity. The importance however of specialists to the dynamics of the local communities in the metacommunity is unknown. Statistical analyses of the rock pool metacommunity described earlier indicate that the population dynamics of generalists dominate the dynamics of local communities (Therriault and Kolasa 2000). These effects were positive or negative depending on the identity of the generalist and specifically affected long-term changes in evenness. In this system, generalists may represent large-scale integrators of the metacommunity (like in Figure 9.1c), a role analogous to hyperparasitoids in butterfly metacommunities (Van Nouhuys and Hanski, chapter 4), or common coral reef species (Waltho and Kolasa 1994). If habitat generalists vary over time, then rare species should also vary in response to habitat generalists, thus increasing community variability (Therriault and Kolasa 2000).

The hierarchical model (HBM) postulates a general abundance pattern as an outcome of the interaction between the habitat structure and a gradient of species attributes. Unlike the regional similarity model (Mouquet and Loreau 2002), the HBM implies poor regional performers will also be, on average, poor local performers in terms of their density. This further implies that mean and regional density of individual species should be strongly correlated, which is the case in the rock pool metacommunity (figure 9.6). This relationship captures that regionally rare species fail to attain high densities at sites where they manage to persist; without exception, local rarity implied regional rarity (figure 9.6). Furthermore, there is an accelerated convergence between local and regional densities as predicted by the HBM for abundant and thus broadly distributed species.

One other consequence of the assumptions of the HBM is a relative constancy of local composition. This consequence is shared with the species sorting model as the HBM relies on species sorting at multiple levels of resolution. The patch dynamics model and the neutral model imply a continuous drift of community composition. In the HBM, however, as long as the environment retains its attributes over time, local communities should retain not only its composition but also the density structure. Furthermore, habitat specialists should show less site tenacity than habitat generalists. Various analyses support this postulate. The community similarity to itself declined dramatically with the number of desiccation events and the pattern was evident whether just composition (Jaccard's index), rank structure (Kendall's W), or composition and density were considered (Therriault and Kolasa 2001). Where the mean richness was reduced, its decline was predominantly due to the differential loss of habitat specialists (Kolasa et al. 1998).

Habitat generalists and specialists also differ in variation of their distribution measured by occupancy (p), with specialists showing greater variability in the number of pools occupied (figure 9.5a, b). Recall that we define occupancy as the number of sites a species is found in and specialization as a species position in the multidimensional hierarchy of habitat attributes. However, this hierarchy imposes only a partial relationship between occupancy and specialization due to

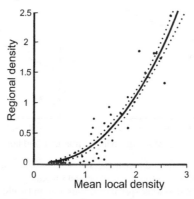

Figure 9.6 Regional density and local density ($\log_{10}$-transformed) of species are strongly related ($y = 0.215x^{2.368}$, $r^2 = 0.902$, $p = 0.0000$). Local density of a species is the mean of that species of all the pools it was present on a single day, averaged over ten annual censuses. The regional density of a species is the total population of the species in a region divided by the volume of water sampled (region is treated here as a single large pool) and averaged over ten censuses.

differences in habitat attributes. Furthermore, rock pool species with high local densities have the greatest occupancy and the lowest temporal variability in density. These descriptive trends may both result from and affect metacommunity dynamics.

Other Effects of Scale

We postulated earlier that the metacommunity patterns appear to differ qualitatively depending on scale. The effect of environmental variability on species richness and abundance is clearly scale dependent (figure 9.7; Romanuk and Kolasa 2002). We found that species richness is lower in pools with higher temporal variability in abiotic conditions, while species richness is higher in pools with greater

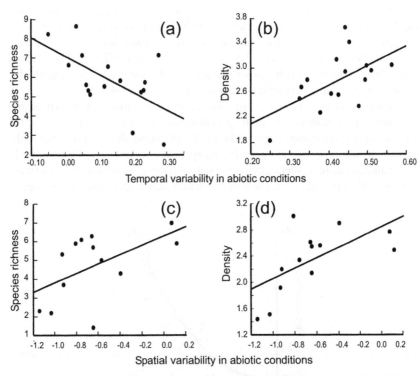

Figure 9.7 Effect of heterogeneity at different scales on species richness and combined pool density for different groups of rock pools (Reid 2003). Temporal variability of the whole pool among dates represents a larger scale (a and b) while spatial variability at 5 cm lag distances and 10 cm depth increments (c and d) represents a smaller scale. Panel (a) shows richness declines in freshwater pools that vary more in time; (b) density increases in saline pools; (c) richness increases in brackish water pools that are internally more heterogeneous; and (d) density also increases in such pools. The variability at either scale is a geometric mean of the standard deviations (SD) of the four abiotic variables (temperature, oxygen, pH, and salinity).

spatial variability in abiotic conditions (figure 9.7). In contrast, mean community density is higher in pools with greater spatial and greater temporal variability in abiotic conditions. The scale-dependent effect of environmental variability is most likely due to different processes arising at different scales. Spatial variability in abiotic conditions may increase species coexistence through increases in the number of suitable habitats (cf. Holyoak 2000), while temporal variability in environmental conditions may decrease species coexistence through intermittent reductions of suitable habitat that may drive some populations to local extinction. Frequent desiccation is an extreme example of such a process.

Applicability of the Four Metacommunity Perspectives to the Rock Pool System

Neutral Model

Fuller et al. (2004) have calibrated and tested the neutral model (Hubbell 2001) of community assembly using data from the rock pool metacommunity. On the whole the model performed fairly well, however it underestimated the density of species that are rare as defined by their rank. This was likely due to high predator density in some pools where predators improved the survival of rare species thus elevating the representation of rare species in the metacommunity. This result underscores the potential of cross-trophic interactions to influence community structure and the proportional abundance of species in the metacommunity (Holt 2002). These results also affirm the importance of habitat structure, inclusive of its biotic component, argued throughout this chapter. In addition, the hierarchical structure of the habitat and the species attributes might also contribute to the departure from predictions of the neutral model (Kotliar and Wiens 1990; Kolasa and Rollo 1991). If habitats of similar qualities occur within the same spatial subunit, then predictions based on mean regional dispersal rates (cf. Resetarits at al., chapter 16) will fail quantitatively. This is because predictions based on mean dispersal rates assume a mean degree of interpatch isolation that is greater than the effective isolation. Consequently, habitat specialists should perform somewhat better when the habitat patches they need are clustered. A GIS analysis supports this argument by revealing a nontrivial degree of spatial autocorrelation among physical properties of rock pools (Romanuk 1999).

Furthermore, using the hierarchical framework produces an intriguing pattern that may be relevant to testing the neutral model. When relative density predicted by the HBM is plotted against species ranks, an S-shaped curve emerges (figure 9.2b). This S-shaped curve is also predicted by the neutral model and is reminiscent of those seen in natural communities.

Species Sorting

Various aspects of this model of local community assembly and regional persistence gain support in the analyses to date. Experiments in which we emptied and cleaned rock pools and allowed their recolonization showed that dispersal was effective enough to allow most if not all the species in the pool to reach any rock pool within days or weeks. Furthermore, we found that habitat specialists were differentially absent from physically variable pools, which led to impoverished and yet more predictable composition (Kolasa et al. 1998). Ample evidence exists for linking individual species, species richness, and community metrics to individual physical variables, composite gradients expressed as principal components, or their variability in rock pools (Therriault and Kolasa 2000, 2001; Romanuk and Kolasa 2002; Kolasa and Li 2003). All of these links are strongly consistent with the idea that most species are restricted to specific subsets of pools defined by the interaction between the physical attributes of the pools and the tolerance ranges of species.

Species sorting processes have important but previously unrecognized consequences for metacommunities. One is the asymmetry of interactions that arise from nested distributions. Habitat generalists interact with more species than habitat specialists, have larger populations, and are regionally buffered against abiotic and biotic variability. This results in different population dynamics at all scales. Furthermore, habitat generalists are members of both species-rich and species-poor local communities where they influence local dynamics differently as a result of their different contribution to the local abundance structure. Finally, habitat generalists are likely to dominate community patterns of variability without being themselves affected by local competitors or predators. Habitat generalists also dominate the community patterns of specialists (Therriault and Kolasa 2000).

These general constraints are likely to interact with species' life histories, the presence of predators, and the availability of resources. Such effects have been demonstrated to have strong effects in other aquatic microcosms (Miller et al. 2002), but have yet to be examined in the rock pool system.

Mass Effect Model

Preliminary work on the rock pool communities (D. G. Jenkins, pers. comm.) shows that dispersal may occur in large pulses and may be directional. However, the intensive dispersal that takes place after major rain events is uneven and largely restricted to the watersheds that encompass several pools. The frequency of dispersal reflects the stochastic external forcing (storms) and exhibits strong directionality due to downstream movement of water. Upstream flow and dispersal can also occur when ocean waves and high tides combine to flood some rock terraces, which was observed twice between 1989 and 2003. The two kinds of flow have different consequences. The rain-generated flows move only a fraction, albeit substantial, of a local community to pools downstream from the focal pool, thus adding species and individuals to an already established community. The ocean influence has an entirely different impact—it removes most individuals of the original community and homogenizes species composition across all affected patches. In sum, given the frequency of rains relative to the pace of population dynamics of constituent populations, the mass dynamics model is likely to apply at the watershed level in the rock pool system while other models may be more appropriate at different scales.

Patch Dynamics Model

The examination of the natural rock pool metacommunity has provided strong evidence that both habitat, at multiple scales, and species are highly differentiated. This evidence appears to violate the basic assumptions of the patch dynamics model. We do not see how it could apply to the complex, hierarchical structure of the rock pool metacommunity. Earlier, we suggested (Kolasa and Drake 1998) that the explanatory power of the patch dynamics model as applied to metapopulations declines with the increasing differences among quality of habitat patches and their isolation. We believe that this conclusion applies to metacommunities as well.

All Four Models

Some observations pertain to all models. For example, the fact that regional and mean local densities of all species are strongly correlated does not flow directly from any of the four perspectives on metacommunity. According to the Mouquet and Loreau (2002) model, a species should compensate for poor regional performance in at least some patches (regional similarity; the mass effect paradigm). The species sorting model (Leibold 1998) appears more likely to accommodate this pattern because it explicitly allows for a gradient of performance along a gradient of environmental constraints (e.g., resources). Our data suggest that such constraints are indeed very strong. A further analysis of perturbation and physical variability patterns, which also differ substantially among individual rock pools, might lead to tightening of this line of inference. The increasing variability of occupancy along the gradient of species specialization has a biologically relevant consequence. As species diversity increases during assembly or in reflection of rich habitat structure, the number of specialists in a community also increases but the predictability of their density declines. Thus, the determinacy of community composition declines with increasing species diversity. It is unclear if any of the models are able to reproduce this effect; however a promising approach taken by Holt (2002) yields results that converge with the above idea.

The most recent extension of the source-sink model (Mouquet and Loreau 2003) led those authors to quantitative predictions concerning the abundance structure of local communities and the whole metacommunity. Some of these predictions appear to converge with the hierarchical model. For instance, high dispersal homogenizes species distribution (cf. habitat generalists in the HBM). Other predictions, such as emergence of the logarithmic series distribution of relative densities in highly isolated or highly invasible communities do not. Future empirical tests and field experiments could discriminate between such alternatives.

Desirable Additions to Theory

The hierarchical view of habitat structure implies a more complex approach to calculation of the dispersal rates. Dispersal is a function of both organism vagility and spatial resistance imposed by an emergence of dispersal boundaries within habitat hierarchy. Thus, an average (mean field) dispersal rate would not take care of the problem, even though it might sometimes provide a reasonable approximation or serve as a useful theoretical tool (e.g., Holt 2002). An appropriate model should make dispersal rate dependent on the availability of *accessible* patches. Formally, this could be obtained by weighting the dispersal success (cf. Mouquet and Loreau 2002) or dispersal distance (Hanski 1998) by the number of hierarchical levels separating suitable patches (cf. figure 9.1, where the lowest-level specialist would have to cross, on average, three hostile patches and two other [potentially hostile] habitats to reach another suitable patch). Such weighting could employ one of the hierarchical tree distance algorithms already used in phylogenetic studies (cf. Mitchell 1996). Furthermore, differential performance of species in different patches translates into different effective patch density and spatial configuration for each species. The dispersal should thus be further modified to reflect this differential effect.

A similar challenge emerges with respect to incorporation of scale discontinuities into the conceptualization of the relationship between species and their habitat. Hierarchy of structure, whether spatial, temporal, or organizational, implies discontinuities in response variables such as occupancy, densities, or patterns of their variability (Kolasa 1989). Empirical work on this and other systems shows strong discontinuities in various attributes. A question emerges as to what additional components or modifiers of metacommunity models could reproduce such discontinuities.

Conclusions

In this paper we review the ecology of a rock pool metacommunity using a hierarchical view of habitat structure and present several aspects of the ecology of metacommunities that may prove valuable in the construction of metacommunity models. We suggest that three observations are integral to an understanding of rock pool metacommunity processes: (1) the habitat is arrayed in a nested hierarchy of discrete patches that exist in a multidimensional habitat space; (2) species perceive habitat fragments differently, depending on their dispersal abilities and habitat tolerances, along a specialist to generalist gradient; and (3) interactions between the hierarchical structure of habitat and species attributes produces discontinuities in species' response patterns that may be a common feature of multispecies systems. The hierarchical model views the metacommunity in multidimensional space as a set of interacting continua of properties and processes staged on an unequal hierarchical playing field. Species at different levels of the hierarchy exist and interact asymmetrically with their environment and each other. Of the four metacommunity models, we found a stronger correspondence between patterns and processes in the rock pool system and the species sorting model than the patch dynamic, mass effect, or neutral metacommunity models.

Acknowledgments

JK's students, Lesley Reid and Andy Biggers, provided some unpublished results. The research has been funded by the Natural Sciences and Engineering Research Council of Canada. Many other individuals assisted in collecting, processing and analyzing data from the rock pool system. The incomplete list includes Susan Marsh, Thomas Therriault, Carol Chapman, undergraduate thesis students, and Discovery Bay Marine Lab staff. The chapter benefited greatly from the substantive and editorial suggestions of Marcel Holyoak and an anonymous reviewer.

Literature Cited

Allen, C. R. and C. S. Holling. 2002. Cross-scale structure and scale breaks in ecosystems and other complex systems. Ecosystems 5:315–318.

Bohonak, A. J. and D. G. Jenkins. 2003. Ecological and evolutionary significance of dispersal by freshwater invertebrates. Ecology Letters 6:783–793.

Cadotte, M., and T. Fukami. 2004. Community dispersal, scale and the partitioning of diversity in a hierarchically-structured landscape. Ecology Letters.

Fuller, M. M., T. N. Romanuk, and J. Kolasa. 2004. Community structure and metacommunity dynamics of aquatic invertebrates: A test of the neutral theory. Electronic archive: http://arxiv.org/pdf/q-bio.PE/0406023.

Hanski, I. 1982. Dynamics of regional distribution: The core and satellite species hypothesis. Oikos 38:210–221.

———. 1998. Metapopulation dynamics. Nature 396:41–49.

Hanski, I. and M. Gyllenberg. 1997. Uniting two general patterns in the distribution of species. Science 275:397–400.

Holt, R. D. 2002. Food webs in space: On the interplay of dynamic instability and spatial processes. Ecological Research 17:261–273.

Holyoak, M. 2000. Habitat subdivision causes changes in food web structure. Ecology Letters 3:509–515.

Hubbell, S. P. 2001. *The unified neutral theory of biodiversity and biogeography.* Princeton University Press, Princeton.

Kolasa, J. 1989. Ecological systems in hierarchical perspective: Breaks in the community structure and other consequences. Ecology 70:36–47.

Kolasa, J. and J. A. Drake. 1998. Abundance and range relationship in a fragmented landscape: Connections and contrasts between competing models. Coenoses 13:79–88.

Kolasa, J., J. A. Drake, G. R. Huxel, and C. L. Hewitt. 1996. Hierarchy underlies patterns of variability in species inhabiting natural microcosms. Oikos 77:259–266.

Kolasa, J., C. L. Hewitt, and J. A. Drake. 1998. The Rapoport's rule: An explanation or a byproduct of the latitudinal gradient in species richness? Biodiversity and Conservation 7:1447–1455.

Kolasa, J. and B.-L. Li. 2003. Removing the confounding effect of habitat specialization reveals stabilizing contribution of diversity to species variability. Proceedings of the Royal Academy of Sciences of London, series B, Biological Sciences (suppl.) 270:198–201.

Kolasa, J. and D. C. Rollo. 1991. Heterogeneity of heterogeneity: A glossary. Pages 1–23 in J. Kolasa and S. T. A. Pickett, eds. Ecological heterogeneity. Springer-Verlag, New York.

Kolasa, J. and N. Waltho. 1998. A hierarchical view of habitat and its relation to species abundance. Pages 55–76 in D. Peterson and V. T. Parker, eds. Ecological scale: Theory and applications. Columbia University Press, New York.

Kotliar, N. B. and J. A. Wiens. 1990. Multiple scales of patchiness and patch structure: A hierarchical framework for the study of heterogeneity. Oikos 59:253–260.

Leibold, M. A. 1998. Similarity and local co-existence of species in regional biotas. Evolutionary Ecology 12:95–110.

Leibold, M. A., M. Holyoak, N. Mouquet, P. Amarasekare, J. M. Chase, M. Hoopes, R. D. Holt, J. B. Shurin, R. Law, D. Tilman, M. Loreau, and A. Gonzales. 2004. The metacommunity concept: A framework for multi-scale community ecology. Ecology Letters 7:601–613.

Miller, T. E., Kneitel, J. M. and J. H. Burns. 2002. Effect of community structure on invasion success and rate. Ecology 83:898–905.

Mitchell, M. 1996. An introduction to genetic algorithms. MIT Press, Cambridge, MA.

Mouquet, N. and M. Loreau. 2002. Coexistence in metacommunities: The regional similarity hypothesis. American Naturalist 159:275–288.

———. 2003. Community patterns in source-sink metacommunities. American Naturalist 162:544–557.

Pulliam, R. 2000. On the relationship between niche and distribution. Ecology Letters 3:349–361.

Reid, L. A., 2003. A multiscale study of the role of environmental variability on the diversity and abundance of rock pool communities. M.Sc. thesis. Department of Biology, McMaster University, Canada.

Ritchie, M. E. and H. Olff. 1999. Spatial scaling laws yield a synthetic theory of biodiversity. Nature 400: 557–560.

Romanuk, T. N. 1999. Spatial dependence of invertebrate assemblage structure in a rock pool metacommunity. Http://home.istar.ca/~fromanuk/spatialtoolboxframe.htm

———. 2002. The relationship between diversity and stability in tropical rock pools. Ph.D. thesis. Department of Biology, McMaster Univeristy, Canada.

Romanuk, T. N. and J. Kolasa. 2001. Simplifying the complexity of temporal diversity dynamics: A differentiation approach. Ecoscience 8:259–263.

———. 2002. Environmental variability alters the relationship between richness and variability of community abundances in aquatic rock pool microcosms. Ecoscience 9:55–62.

Therriault, T. W. 2002. Temporal patterns of diversity, abundance and evenness for invertebrate communities from coastal freshwater and brackish water rock pools. Aquatic Ecology 36:529–540.

Therriault, T. W. and J. Kolasa. 1999. Physical determinants of richness, diversity, evenness and abundance in natural aquatic microcosms. Hydrobiologia 412:123–130.

———. 2000. Patterns of community variability depend on habitat variability and habitat generalists in natural aquatic microcosms. Community Ecology 1:195–203.

———. 2001. Desiccation frequency reduces species diversity and predictability of community structure in coastal rock pools. Israel Journal of Zoology 47:477–489.

Waltho, N. and J. Kolasa. 1994. Organization of instabilities in multispecies systems: A test of hierarchy theory. Proceedings of the National Academy of Sciences, USA 91:1682–1685.

Wright, D. H. 1991. Correlations between incidence and abundance are expected by chance. Journal of Biogeography 18:463–466.

PART 3

THEORETICAL PERSPECTIVES

Richard Law and Priyanga Amarasekare

Nonspatial models of community dynamics give a simplified picture of the real world and have helped ecologists to grapple with hard problems of interactions among multiple species. However, for many species in many communities, locations of individuals in space do matter, and it is clearly important to understand the consequences this has for community dynamics. The purpose of this section on theory is to show some of the ways theorists are exploring the effects of space within an intellectual framework of metacommunities. Theoretical approaches that form the foundations of metacommunity ecology are also covered in the Core Concepts section of this book (Hoopes et al., chapter 2; Holt and Hoopes, chapter 3).

A good vantage point from which to see the relationships between the various approaches to modeling community dynamics in space is a simplified nonspatial model. Spatial features can then be added in an orderly way by systematically relaxing the simplifying assumptions of the nonspatial model. With this in mind, we begin with a spatial region containing individuals of a number of species, with individuals that move sufficiently often and over large enough distances to allow complete mixing. The density of individuals is assumed to be great enough to ignore the effects of demographic stochasticity, and the environment is uniform in time and space. In this ideal, well-mixed community, there is no internal spatial structure, and individuals encounter one another in proportion to their average density. The dynamics of such a community can be described in terms of familiar, deterministic, mean-field models, such as the generalized Lotka-Volterra equations. Clearly a metacommunity model is not called for here.

We encounter the first class of metacommunity models, *mass effect* models, by relaxing the assumption of uniformity of the environment over space. In fact, variation in the suitability of the environment is assumed to be so great that there are only certain spatially separated patches that could support species of the metacommunity; this is the assumption, familiar from metapopulations, that patches in which organisms thrive are embedded in a matrix of inhospitable environments; such an environment is illustrated by Van Nouhuys and Hanski in chapter 4 (figure 4.1). Wilson (1992) likened a metapopulation to a sea of lights winking on and off; the lights are located in the suitable patches and are switched on when the patch is occupied by the species. However, in the case of metacommunities, the environmental states are more than just good and bad. Since we are now dealing with multispecies communities, the patches can differ in their suitability for different species—one can think of Wilson's lights having a multitude of different colors corresponding to the different environmental states. Dispersal across the region remains frequent enough for losses and gains of individuals to impact directly on local population density, though not so great that abundances of species are fully homogenized in space. As Mouquet et al. describe in chapter 10, the source-sink dynamics that emerge from mass effect models enable local populations to be replenished by conspecific individuals from other patches. Under some conditions, this can promote coexistence of species in a metacommunity of competitors.

The second class of metacommunity models, the *patch dynamic* models, comes about by replacing the assumption of frequent dispersal across patches with an assumption that dispersal is rare enough to decouple within-patch dynamics from dynamics of the metacommunity. This means that within-patch dynamics take place much faster than dispersal between patches—there is, in effect, a separation of the time scales on which these processes are taking place. With this time-scale separation in place, we are no longer concerned with the transient dynamics of local communities; neither are we concerned with effects of migration on local densities; the only information carried over from the local to metacommunity are the densities of resident species in the long term—the attractor(s) of local dynamics. Patch dynamic models are the natural extension of Levins's (1969) metapopulation model to multispecies systems. For instance, the role of life history trade-offs in species coexistence has most frequently been analyzed using patch dynamics models (e.g., Levins and Culver 1971; Hastings 1980; Nee and May 1992; Tilman et al. 1994). Chapter 10 works toward a synthesis of trade-off mediated coexistence in patch dynamics models and source-sink dynamics emerging from mass effects models. In chapter 11, Law and Leibold describe the extension of Levins's model to metacommunities, and introduce a method by which arbitrary graphs of community assembly can be placed in metacommunity models. This makes it potentially possible to analyze metacommunity dynamics despite communities having a quite complicated structure.

Although not dealt with explicitly in this section of the book, there is a third class of metacommunity models, *species sorting* models, which also carry the time-scale separation of local community and metacommunity dynamics. These models differ from patch dynamic models in emphasizing differences in the environment from one patch to another. Also they assume that, in the long run, every species can get to every location, and that a species predominates once it gets into the environment most suitable to it. In this way patches gradually become sorted so that they are dominated by the species best able to live under the environmental conditions they offer; in Wilson's metaphor, patches with red lights end up with species 1, those with orange lights end up with species 2, and so on. See Leibold (1998), Holyoak et al. (chapter 1) and Chase et al. (chapter 14) for more on species sorting models.

In contrast to species sorting models are models that assume ecological equivalence of organisms in which "all individuals of every species obey exactly the same rules of ecological engagement" (Hubbell 2001, 6). This equivalence underpins *neutral* models of metacommunities, and deserves attention because it predicts correctly a variety of community and biogeographic patterns in the real world (Bell 2001; Chase et al., chapter 14). The patches occupied by organisms might still differ in their properties, including the identity of other individuals, but these differences would not be detected by the organisms themselves; abundances thus change in a manner analogous to change in gene frequencies through genetic drift. In chapter 15, McPeek and Gomulkiewicz explore the interplay between this change and the stabilizing effects of niche differentiation using the Fisher-Wright model.

A central idea behind the metacommunity concept is the spatial scale transition—the way in which rules of population dynamics change as spatial scales change from small (e.g., individual, local community) to large (e.g., metacommunity). Chesson et al. (chapter 12) investigate how the interplay between nonlinearities inherent in population dynamics models and spatial variation in the environment lead to scale-dependent changes in the outcome of species interactions. A new key idea in this regard is the fitness-density covariance, a nonlinearity that arises in all spatial models as a consequence of mechanisms such as localized dispersal that allow species to concentrate their densities in areas of the landscape that are favorable to their growth and reproduction. Chesson et al. illustrate this and other key nonlinearities that enable coexistence in the face of spatial variation with examples from competititve and host-parasitoid metacommunities. A companion chapter by Melbourne et al. (chapter 13) shows how scale transition theory can be applied to improve our understanding of empirical problems.

At the roots of metacommunity dynamics lie stochastic, multispecies, birth-death-movement processes that occur in heterogeneous environments. The modeling methods being employed to learn about metacommunity dynamics described in this section and elsewhere can be thought of as approximations to these individual-based, stochastic processes. There is a long way to go before we fully understand the formal connections between the approaches; it is our hope that, by bringing them together here, we make a step in the right direction.

Literature Cited

Bell, G. 2001. Neutral macroecology. Science 293:2413–2418.

Hastings, A. 1980. Disturbance, coexistence, history and competition for space. Theoretical Population Biology 18:363–373.

Holt, R. D. 1997. From metapopulation dynamics to community structure: Some consequences of spatial heterogeneity. Pages 149–164 *in* I. Hanski and M. Gilpin, eds. *Metapopulation biology: Ecology, genetics and evolution.* Academic Press, San Diego, CA.

Hubbell, S. P. 2001. *The unified theory of biodiversity and biogeography.* Princeton University Press, Princeton, NJ.

Leibold, M. A. 1998. Similarity and the local co-existence of species in regional biotas. Evolutionary Ecology 12:95–110.

Levins, R. 1969. Some demographic and genetic consequences of environmental heterogeneity for biological control. Bulletin of the Entomological Society of America 15:237–240.

Levins, R. and D. Culver. 1971. Regional coexistence of species and competition between rare species. Proceedings of the National Academy of Sciences, USA 68:1246–1248.

Nee, S. and R. M. May. 1992. Dynamics of metapopulations: Habitat destruction and competitive coexistence. Journal of Animal Ecology 61: 37–40.

Tilman, D., R. M. May, C. L. Lehman, and M. A. Nowak. 1994. Habitat destruction and the extinction debt. Nature 371: 65–66.

Wilson E. O. 1992. *The diversity of life.* Harvard University Press, Cambridge, MA.

CHAPTER 10

The World Is Patchy and Heterogeneous!
Trade-off and Source-Sink Dynamics in Competitive Metacommunities

Nicolas Mouquet, Martha F. Hoopes, and Priyanga Amarasekare

When we discuss communities beyond their most essential attributes as open systems, generality may elude us, except the generality of diversity.

WHITTAKER AND LEVIN, 1977

Introduction

Recognition that the world is patchy and heterogeneous has been the basis for many advances in both fundamental and applied ecology over the last thirty years (Levin 1992). This chapter reviews conceptual advances in the understanding of spatial mechanisms of competitive coexistence and places those advances in the context of metacommunity ecology. We discuss work from the past half century through extremely recent results in order to highlight unexpected links and reinterpretations. We show that apparently different mechanisms have common elements and we discuss options for integrating these elements into a broader theoretical framework. This framework could lead to a general theory of biological diversity for a natural world increasingly transformed by human activities.

For the sake of generality, ecologists frequently look for simple rules that apply to a wide range of taxa. Community ecology has provided one such simple law in the form of the competitive exclusion principle (Lack 1944; Hutchinson 1957; Hardin 1960) derived from the Lotka-Volterra models (Lotka 1925; Volterra 1926) and Gause's experiments (1934). This principle states that the number of coexisting species cannot exceed the number of limiting factors or ecological niches. It has brought community ecology into a modern synthesis led by the works of Hutchinson (1957), Hardin (1960), MacArthur (1967, 1972), and others. In reality, however, for the vast majority of communities, the competitive exclusion principle is too restrictive because it focuses only on local limiting factors in a homogeneous environment. The original theory could not explain high diversity found in natural systems with few identified limiting factors, especially for plants and aquatic communities (e.g., Hutchinson 1961). Ecology rejected niche theory reductionism in favor of complexity and entered a new era without a theory of diversity but with a diversity of theories explaining species richness (reviewed by Chesson 2000a, Barot and Gignoux 2004).

Forty years after Hutchinson discussed paradoxically high plankton diversity (Hutchinson 1961), recognition of the importance of patchiness and heterogeneity has led to more comprehensive and realistic theories of species diversity. Patchiness and heterogeneity are two interconnected forms of environmental variation. Patchy environments have spatially discrete elements either because of natural barriers (e.g., deep water between coral reefs, ponds separated by land, plant patches separated by a matrix inconducive to plant growth, etc.) or species characteristics (e.g., sessile species with a dispersal stage, territorial organisms with natal dispersal, etc.). Heterogeneity implies differences in the environmental conditions found in two patches (table 1.1), for example due to different limiting factors or historical context. Although early community ecologists recognized the discrete nature of habitats (Skellam 1951; Andrewartha and Birch 1954; Huffaker 1958), the idea became central to community ecology only after Levins defined metapopulations (Levins 1969, 1970), and Levins and Culver (1971) applied the concept to competition theory. The natural link between patchiness and regional heterogeneity has led ecologists to focus on mesoscale ecology (Roughgarden et al. 1988; Holt 1993; Ricklefs and Schluter 1993) to incorporate more realistically the dynamics of species diversity in metacommunities.

The recognition of spatial structure leads to differentiation between local (within patch) and regional (among patch) processes (see table 1.1). Since the work by Skellam (1951) there has been a separation in model formalism and assumptions between these two scales (figure 10.1). For example, models of coexistence in discrete, homogeneous environments have adopted the Levins (1969, 1970) metapopulation approach of describing dynamics in terms of extinctions and recolonization of patches. These models assume a net separation between local and regional time scales such that coexistence within a patch is impossible, and they explain regional coexistence given a trade-off between competitive and colonization abilities (Levins and Culver 1971; Horn and MacArthur 1972; Hastings 1980). In contrast, models that consider coexistence in heterogeneous environments have opened different perspectives on local species coexistence through source-sink or mass effects in metacommunities (see Holyoak et al., chapter 1 for definitions). These models have shown that local coexistence is possible through regional niche differentiation and dispersal between patches (Levin 1974; Shmida and Ellner 1984; Amarasekare and Nisbet 2001; Mouquet and Loreau 2002, 2003). This divergence has resulted for example, in competition-colonization trade-offs and source-sink dynamics being viewed as mutually exclusive mechanisms of coexistence when they should be studied in a common framework relevant to metacommunity ecology (Amarasekare et al. 2004).

Ecologists have recently begun to synthesize these two approaches into a common framework under metacommunity ecology (Wilson 1992; Holt 1993, 1997a;

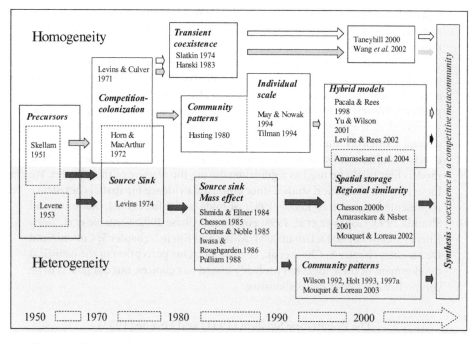

Figure 10.1 Representation of the chronology of some important papers that have studied the conditions of coexistence between competing species in patchy habitats. There was rapid separation between studies that considered homogeneous (white region) and heterogeneous (gray region) environments. Three main branches have emerged from the precursor works of Skellam (1951) and Levene (1953), respectively the source-sink hypothesis (dark gray arrows), the competition-colonization hypothesis (gray arrows) and a third branch that have studied the consequences of transient local coexistence in patch models (white arrows). We have added a chronological axis at the bottom of the figure. These papers are reviewed in the text and summarized in table 10.1. This list is not exhaustive, but rather represents what we think are the key papers.

Mouquet and Loreau 2002, 2003; Loreau et al. 2003; Leibold et al. 2004; Holyoak et al., chapter 1). Although we are not the first to make the link between these approaches (Yu and Wilson 2001; Levine and Rees 2002), a thorough clarification of the ideas underlying spatial coexistence is necessary for further progress in metacommunity ecology. Contemporary ecologists (including the authors of this chapter) have also often failed to recognize insightful contributions of early works such as Skellam (1951) and Levene (1953). This oversight is partially due to differences in focus between more recent metacommunity ecology and the early work on species coexistence in space. For example Levins and Culver (1971), only briefly considered the general topic of coexistence and species richness. Their focus instead was on the role of competition in the dynamics of rare species on island systems. The diversity of biological systems modeled also contributed to the lack of early synthesis because the questions addressed in marine versus terrestrial systems were not necessarily similar, and hence generalizations were difficult to make.

With a view toward clarification, in this chapter we review theory on coexistence in patchy environments, focusing on the competition-colonization (hereafter CC) and the source-sink (hereafter SS; also known as the mass effect) hypotheses. We retrace the chronology of these two approaches, highlighting their key results, differences, and similarities. We end with the basic elements of a more general framework that allows ecologists to examine patch dynamics and mass effects simultaneously. Table 10.1 and figure 10.1 provide an overview of these models that might be used as a guideline during the reading of this chapter. We focus only on theoretical studies since empirical evidence for spatial coexistence is presented elsewhere (Keddy 1982; Shmida and Ellner 1984; Kadmon and Shmida 1990; Gonzalez et al. 1998; Forbes and Chase 2002; Cottenie et al. 2003; Kneitel and Miller 2003; Urban 2004; Miller and Kneitel, chapter 5; Cottenie and De Meester, chapter 8; Chase et al., chapter 14). Our perception of the historical development of these ideas has probably biased our choices, but our goal is to be comprehensive rather than exhaustive.

The Precursors: Skellam (1951) and Levene (1953)

The works of Skellam and Levene are seminal to the development of metacommunity theory. Skellam used diffusion theory to study the consequences of dispersal for both local and regional species coexistence. He showed that a population could maintain itself in unfavorable regions by random dispersal from favorable regions. He stated that the stability of such a system depended "on the ability of the population in favorable habitats to make good the decline in the unfavorable ones" (Skellam 1951, 202). He also studied competition between two annual plants within the same habitat and showed that coexistence was possible given that "a disadvantage in direct competition may be offset by a superiority in reproductive capacity" (215). He further demonstrated that the advantage in reproductive capacity that an inferior competitor needs to persist depends on the density of the superior competitor (the effect of the superior competitor on the inferior competitor). Levene studied similar coexistence concepts with genes (1953; see also Levins and MacArthur 1966). He examined the conditions for the maintenance of polymorphism without any heterozygote advantage. He proposed a simple model of allele frequencies considering that "the existence of several ecological niches, with one allele favored in one niche and the other allele favored in another, might increase the possibilities for attainment of equilibrium with both alleles present in substantial proportions" (Levene 1953, 331). Under such conditions, he found that the polymorphism was maintained if the weighted arithmetic mean of the "relative fitnesses" of homozygotes was less than one. This condition, which is obviously true when the heterozygote has higher fitness, can also be obtained for particular parameter values without the heterozygote being superior to both homozygotes in any single niche.

Both authors, therefore, investigated the role of regional heterogeneity in local coexistence. Skellam's work also examines CC dynamics long before metapopulation theory (Levins 1969, 1970) brought it to mainstream attention (see next section). Skellam's work is seminal to the development of metacommunity theory because it provides the basis for both the competition-colonization trade-off and source-sink dynamics. Levene (1953) makes the fundamental link between niche theory and spatial coexistence. He is, to our knowledge, the first to propose such a regional vision of niche separation allowing local coexistence despite strong local niche overlap. Although neither author extended his results to community ecology or considered the consequences of variation in dispersal ability, these papers are the roots from which metacommunity ecology has grown, and they should be viewed as ecological masterpieces.

Competition, Patchiness and Spatial Heterogeneity

The Levins Metapopulation Model

Levins defined a metapopulation as a set of local populations with extinctions and recolonizations, giving us the most common formalism for metapopulation theory (1969, 1970). His work is based on three major assumptions: (1) The environment consists of an infinite number of patches ("localities" in the terminology of table 1.1) with identical environmental conditions, and each patch supports a local population. (2) Local dynamics operate on a faster timescale than regional dynamics, with each patch reaching carrying capacity instantaneously upon colonization. (3) Dispersal is infrequent, serving only to colonize empty patches and having no effect on within-patch population dynamics.

With these assumptions, the dynamics of patch occupation results in a balance between extinction (by stochasticity and disturbance) and recolonization from other patches in the region. Levins modeled the proportion of patches p occupied by a species at the regional scale:

$$\frac{dp}{dt} = cpV - ep$$
$$\text{with } V = 1 - p,$$

(10.1)

where c is the recolonization rate of empty patches, and e is the extinction rate of occupied patches. This model assumes that the colonization probability for any patch is proportional to the proportion of patches already colonized. In order to highlight competition for empty habitat, we have modified the original model to isolate V, the proportion of empty patches in the region. The equilibrium is given by

Table 10.1 Summary of important metacommunity models

Authors		Org	Comp	M	CCR	CCL	SSD	SST	Notes
Skellam	1951	TA, P	P	3, 5	No	Yes	Yes	No	Basis for CC and SS hypothesis.
Levene	1953	All	D	5	No	No	Yes	No	Explains allelic polymorphism without any higher adaptive value for the heterozygote.
Levins and Culver	1971	TA	D	1	Yes	No	No	No	Basis for CC models.
Horn and MacArthur	1972	All	D	1	Yes	No	No	Yes	CC model and proto-SS since a CC trade-off in habitat 1 allows species 2 to persist and maintain itself in habitat 2 via a SS effect.
Slatkin	1974	All	D	1	Yes	No	No	No	Recolonization from doubly-occupied patches enhances the potential for species coexistence.
Levin	1974	All	D	1, 2, 3	Yes	No	Yes	Yes	Perturbation theorem and first real SST.
Hastings	1980	MA	D	1	Yes	Yes	No	No	Assumes extinction is a result of disturbance and finds a hump-shaped relationship between species richness and disturbance rate.
Hanski	1981	TA	D	3	Yes	No	No	No	Spatial variance in the better competitor can be high due to inefficient dispersal.
Hanski	1983	TA	D	1	Yes	No	No	No	Rescue effect. Coexistence in patchy environment depends on the distinctness between local and regional time scales.
Shmida and Ellner	1984	P	P	4	Yes	Yes	Yes	No	Coined the term mass effect.
Chesson	1985	MA, P	P	4	No	No	Yes	No	Spatial and temporal storage effects.
Comins and Noble	1985	MA, P	P	4	No	No	Yes	No	Assuming that the environment is uniformly variable, every coexisting species must have its own transient spatial niche.
Iwasa and Roughgarden	1986	MA	P	1, 4	No	No	Yes	No	Number of locally coexisting species cannot exceed the number of communities in the region. Regional compensation between larval growth rates.
Kishimoto	1990	All	D	3	No	No	Yes	Yes	An infinite number of species can exist over two patches.
Wilson	1992	All	D	3	No	No	Yes	No	Coined the term metacommunity. Limitation to the founder effect argument proposed by Levin (1974).
Holt	1993	All	D	3	No	No	Yes	Yes	Mesoscale and metacommunity perspective.
Goldwasser et al.	1994	P	D	1	No	Yes	No	No	There is a limit to coexistence with the CC if one considers finite habitat size.
Tilman	1994	P	D	1	No	Yes	No	No	Applies the Hastings (1980) CC model at the individual scale.

		Org	Comp	M					Notes
Holmes and Wilson	1998	P	D	1	Yes	Yes	No	No	If the superior species is rare, it is possible for a long-distance disperser, which is both an inferior competitor and a bad reproducer, to coexist
Pacala and Rees	1998	P	P, D	1, 4	Yes	No	No	No	Model that combines CC and the successional niche.
Chesson	2000b	All	P, D	4, 5	No	No	Yes	No	Extends the concept of spatial storage effects.
Muko and Iwasa	2000	MA	P	4	No	No	Yes	No	Between habitat variation in mortality ratios promotes coexistence, while that of reproductive rates does not.
Taneyhill	2000	All	D	1	Yes	No	No	No	Migration from doubly occupied patches is always stabilizing.
Amarasekare and Nisbet	2001	TA	D	3	No	Yes	Yes	Yes	Perturbation theorem with spatial variance in species fitness. Limit to species competitive asymmetry.
Yu and Wilson	2001	All	P	1, 4	Yes	Yes	No	Yes	A dispersal fecundity trade-off can partition variation in patch density creating the conditions for SS based on a priority effect.
Levine and Rees	2002	P	P	5	No	Yes	Yes	No	Provide a hybrid model where coexistence occurs via SS dynamics and is facilitated by a CC trade-off.
Mouquet and Loreau	2002	All	P	1, 4	No	No	Yes	No	Definition of the regional similarity constraint. High local diversity at intermediate dispersal and regional species dissimilarity.
Wang et al.	2002	All	D	1	Yes	No	No	No	Stochastic local extinctions make it likely that species with strong competitive effects on each other can coexist.
Mouquet and Loreau	2003	All	P	1, 4	No	No	Yes	No	Community patterns in a source-sink metacommunity.
Amarasekare et al.	2004	All	P, D	1	Yes	Yes	Yes	Yes	Spatial variation in the expression of a life-history trade-off can constrain rather than promote coexistence.

Notes: *Org* refers to the type of organisms considered: (All) for all species (or undefined species), (P) for plants, (MA) for marine animals. *Comp* refers to the type of competition, either by dominance (D) or by preemption (P). *M* refers to the type of model used: (1) patch dynamic, (2) diffusion reaction, (3) Lotka-Volterra, (4) lottery model, and (5) other. The next four columns describe the coexistence mechanism addressed: competition-colonization trade-off at the local (CCL) or regional (CCR) scales as well as source-sink directional (SSD) and threshold (SST) dynamics as defined in the text. *Notes* describes the contribution within the paper that motivated our choice to include it in the table. We acknowledge that there are many other metacommunity models that we did not include due either to our oversight or because they did not add any fundamentally new result. Some of these papers are also presented in figure 10.1.

$$\hat{p} = 1 - \frac{e}{c}$$

$$\hat{V} = \frac{e}{c}. \tag{10.2}$$

The minimum proportion of available patches to maintain a species in the metapopulation is $\hat{V}$, and the basic metapopulation growth rate of a species is $r = c/e$ (Fagerström and Westoby 1997; Loreau and Mouquet 1999). This growth rate is the change in the proportion of occupied patches when a species first enters a metapopulation. The persistence of a species requires that r be greater than or equal to 1 (recolonization must compensate for extinction) and that a minimum number of patches be available at the regional scale ($V > \hat{V}$). Note also that there will always be some empty patches at equilibrium because, as $\hat{V}$ tends toward zero, c tends toward infinity or e tends toward zero, both of which are biologically unrealistic (Tilman 1994). There is a strong similarity with models of resource competition if one considers available patches as a resource consumed by a species at a rate r. In this context the quantity $\hat{V}$ can be interpreted as Tilman's R^* (1982) since it is the minimum amount of resources (available patches) required for persistence in a metapopulation (Loreau and Mouquet 1999).

One can generalize equation 10.1 to multispecies competition if species differ only in their metapopulation growth rates.

$$\frac{dp_i}{dt} = c_i p_i V - e_i p_i$$

$$V = 1 - \sum_i p_i \tag{10.3}$$

$$r_i = \frac{c_i}{e_i}$$

$$V_i^* = \frac{1}{r_i},$$

where V_i^* is the minimum proportion of available patches for a particular species to persist alone (without competition) in the metacommunity. Competition then takes the form of a lottery (Sale 1977; Chesson and Warner 1981) with an infinite number of patches. At equilibrium, only the species with the highest r_i (i.e., lowest V_i^*) will persist. This species decreases the proportion of available patches below the threshold required for the persistence of other species and thus competitively excludes other species (Loreau and Mouquet 1999). This result matches competitive exclusion by the species with the lowest R^* as proposed by Tilman (1982). When competition takes the form of a simple lottery based on extinction and recolonization, V_i^*, therefore, defines species' competitive abilities.

The Competition-Colonization Trade-off Hypothesis

The first models to consider competition in a metapopulation focused on dominance competition. These models are variations of the Levins (1969, 1970) model and all explore coexistence without classical niche partitioning mechanisms but with species differing in the way they use a single resource—space. Levins and Culver (1971) proposed the first such model. They assumed that the competitive effect between species affected their relative extinction or colonization rates. For example, for two competing species the extinction and colonization parameters of species 1 can be written as

$$e_1 = e_{12}p_2 + e_{10}(1 - p_2) \qquad (10.4)$$

and

$$c_1 = c_{12}p_2 + c_{10}(1 - p_2), \qquad (10.5)$$

where e_{12}, c_{12} are the respective extinction and colonization rates for species 1 in patches with species 2. Similarly, e_{10}, c_{10} are the respective extinction and colonization rates for species 1 in patches that do not contain species 2. The proportion of patches occupied by species 2 is p_2. When competition affects extinction (equation 10.4), they found that coexistence is possible if:

$$c_1 > e_{12} - e_{10}, \qquad (10.6)$$

which can be interpreted as a requirement that the colonization rate of species 1 must be higher than the competitive effect of species 2 on species 1. They found that when competition affects colonization (equation 10.5), coexistence is possible if

$$\frac{c_{10}}{c_{10} - c_{12}} > 1, \qquad (10.7)$$

which is always true, given that $c_{10} \neq 0$ and $c_{12} > 0$. The main result is thus that there must be a trade-off between the colonization and competitive effects for coexistence (equation 10.6) when competition affects species' extinction rates.

Following Levins and Culver (1971), Horn and MacArthur (1972) studied competition in an environment with two kinds of habitat. They used the same model and varied the fraction of habitat where each species outcompeted the other. They found no limit to the number of species that can coexist in such a heterogeneous environment as long as there is a supply of unoccupied patches and the colonization rate increases for each successive new competitor. They also found that the inferior competitor needs only to persist in one habitat to persist in both habitats. While Horn and MacArthur examined equal within- and between-habitat dispersal values, Levin (1974) then examined a similar model in which local (within-habitat) dispersal is much higher than regional (B habitats) dispersal. This case is relevant when patches tend to be clumped according to types. He showed that coexistence in this system does not require two types of patches if both species have within-habitat dispersal parameters sufficiently high that enough colonists of each species are present to replace lost populations.

Hastings (1980) developed a simpler model also adapted from the original single species metapopulation model. He assumed that competition is sufficiently rapid (or patches sufficiently small) that inferior competitors are instantaneously excluded, and no patch contains more than a single species. He acknowledged that the extreme result of this assumption is the modeling of microsites, or patches that hold only a single individual (table 1.1). Hastings (1980) also assumed a strict hierarchy of competitive dominance and added the possibility of extinction by external disturbances. For two species, the model reads

$$\frac{dp_1}{dt} = c_1 p_1 (1 - p_1) - e p_1$$

$$\frac{dp_2}{dt} = c_2 p_2 (1 - p_1 - p_2) - c_1 p_1 p_2 - e p_2. \tag{10.8}$$

This corresponds to a very strong, unidirectional competitive effect with species extinctions that are independent of competition. One can reformulate Hastings's model in terms of V^* as we did for the simple metapopulation model. The equilibrium is then

$$\hat{p}_1 = 1 - V_1^*$$
$$\hat{p}_2 = V_1^* - V_2^*$$
$$V_1^* = \frac{e}{c_1} \tag{10.9}$$
$$V_2^* = \frac{c_1 \hat{p}_1}{c_2} + \frac{e}{c_2}.$$

Species 1's performance is determined solely by its own life-history characteristics, while that of species 2 also depends on species 1's ability to exclude it ($c_1 \hat{p}_1 / c_2$). For species 2 to coexist at equilibrium with the superior competitor, the equation must satisfy $\hat{p}_2 > 0$, that is $V_2^* < V_1^*$. The competitive advantage of species 1 is then more than compensated by the ability of species 2 to get free patches, that is the worse competitor has to be a better colonizer (low V^*). This result is interesting because it helps to differentiate between two aspects of a species' competitive ability: dominance competition that is based on direct displacement of one species by a superior competitor and spatial competition that is based on the ability to colonize empty sites and linked to species V^*. Hastings (1980) also found a hump-shaped relationship between species richness and disturbance rate. At low disturbance rates the best competitor excludes all other species; at intermediate disturbance rates coexistence is most likely; but at high disturbance rates coexistence is impossible because species do not have sufficiently high colonization rates to compensate for extinction.

Hastings's (1980) model opened the way for extremely local, microsite (table 1.1) applications of CC models to plant communities and parasite strains (Nee and May 1992; May and Nowak 1994; Tilman 1994). These more recent models explicitly assume that only one individual can occupy a microsite, that is, that the ultimate level of patchiness is the individual itself (Grubb 1986). In this context the extinction rate of Hastings's (1980) model is equivalent to a mortality rate, and the colonization rate is equivalent to the species' reproductive rate. Both of these new parameters are independent of local competition. As in Hastings's (1980) model, these formulations all assume a strict competitive hierarchy and instantaneous exclusion of inferior competitors. These models predict that a virtually infinite number of species can coexist locally given an appropriate trade-off between competitive ability and fecundity or mortality (but see Adler and Mosquera 2000).

Some other models yield similar results. Models based on other types of competition yield similar results. For example, Hanski (1981) introduced spatial heterogeneity (spatial variance in abundances) into the Lotka-Volterra competition model in the context of patchy habitats. He studied regional coexistence in cases where local coexistence was not possible. His analysis suggests that an inferior competitor might survive regionally if the spatial variance in abundance of the better competitor is high, for example, due to a low rate of dispersal. Shmida and Ellner (1984) proposed a pure lottery model (Sale 1977; Chesson and Warner 1981) with plant species competing for microsites (table 1.1). They found basically the same constraint on species parameters: differences in life-history strategies in terms of competitive versus colonizing abilities explain local coexistence between competing species.

Although the CC hypothesis has received much attention from ecologists, it has also generated confusion because of its application at both local and regional scales. One source of confusion is that colonization subsumes a suite of species characteristics (e.g., fecundity, long and short distance dispersal, and colonization rate) and that competition is not always clearly separated from these characteristics, particularly fecundity. In actuality, trade-offs between any two of these features may lead to coexistence (Holmes and Wilson 1998, Yu and Wilson 2001). For example, Yu and Wilson (2001) showed that a trade-off between dispersal and fecundity could enhance the potential for species coexistence. Holmes and Wilson (1998) examined the nature of dispersal and demonstrated that, when the superior competitor is not very abundant, an inferior competitor with low fecundity but long distance dispersal can persist. This confusion about the nature of the trade-off is increased by confusion about the nature of competition. As we illustrated with Hastings's (1980) model, the notion of competitive superiority can be defined from different perspectives. A fugitive species that is supposed to be a bad competitor in models of dominance competition can be defined as a good competitor for available patches (a species with a low V^*) in models involving spatial competition. An additional source of confusion is that very different factors drive the dynamics of available patches at the different scales, and these dynamics are crucial to coexistence. At the regional scale (table 1.1), availability of localities results from stochasticity or deterministic disturbances (as in Hastings 1980), while at local scales, microsites become available through individual mortality (as in Tilman 1994). Even if these different causes of patch vacancy are necessarily related, their interpretations in terms of population dynamics are different.

Transient Local Coexistence

One of the main criticisms of patch dynamic models based on the original Levins (1969, 1970) formulation is the unrealistic assumption that colonization leads instantaneously to exclusion of inferior competitors and carrying capacity populations of dominant competitors (Hanski 1983; Levine and Rees 2002). Yu and Wilson (2001) pointed out (following Chesson and Warner 1981; Comins and Noble 1985; Chesson and Huntly 1997) that when this assumption is abandoned and seedlings only compete for available patches, as in a lottery system, local coexistence cannot occur via a CC trade-off alone. Therefore, several models have included transient patches in which inferior and superior competitors coexist. Slatkin (1974) pioneered this approach with a hybrid between Cohen's model (1970) and Levins and Culver's (1971) model. In their original paper, Levins and Culver assumed that species were distributed independently. This assumption allowed simplifications of the mathematics in their model, but Slatkin pointed out that species cannot both sort independently and affect each other's colonization or extinction rates (see also Levin 1974). Slatkin argued that Levins and Culver's assumption overestimates the number of patches where species co-occur and hence overestimates reciprocal competitive effects. Slatkin instead included four patch states: unoccupied by either species, occupied by species 1 or 2 alone, or co-occupied by both species. He found that migration from co-occupied patches increases the likelihood of coexistence by allowing rare species to colonize patches while increasing the extinction rates of common species.

Hanski (1983) found an exception to Slatkin's result if extinction is a function of patch occupancy (as colonization is for all patch dynamic models) so that the probability of extinction for a given species is negatively correlated with the proportion of sites occupied by that species in the region (but see Nee et al. 1991). He called this dependence of extinction on patch occupancy "the rescue effect" (cf. table 1.1). The rescue effect weakens intraspecific competition, thus changing the balance of inter- and intraspecific competition on the regional scale, such that the species with the wider initial distribution can outcompete the other species.

Taneyhill (2000) elaborated on the effects of co-occupied patches by integrating these two previous models into a broader framework. He showed that immigration from co-occupied patches is always stabilizing and that the rescue effect (as defined by Hanski 1983) is typically destabilizing for the less widespread competitor. Wang et al. (2000, 2002) extended these results by adding stochastic extinction and considering extinction to be a function of local abundance. They found that stochastic local extinctions make it more likely for strongly competing species to coexist.

The key result to emerge from these studies is that colonization from co-occupied patches increases the probability of coexistence between competing species. In these models, doubly-occupied patches introduce an element of spatial heterogeneity (variation in spatial abundances) because they act somewhat as a refuge for the inferior competitor. Co-occupied patches supply colonizers that are already removed from the fraction of patches occupied solely by the inferior competitor. As Chesson has pointed out (2000b), such heterogeneity in propagule production or resource-use contributes to coexistence by creating spatial niches. To simplify his model, Slatkin (1974) also examined species with similar colonization and extinction rates, an approach continued by others examining doubly-occupied patches (Taneyhill 2000; Wang, Zhang, et al. 2000; Wang, Wang, et al. 2002). These models also, therefore, reveal that ecologically similar species that would exclude each other in a uniform environment may coexist in a patchy environment (Slatkin 1974; Taneyhill 2000; Wang, Zhang, et al. 2000; Wang, Wang, et al. 2002). The following section focuses specifically on spatial heterogeneity, which is more commonly associated with SS models.

Regional Heterogeneity and Source-Sink Dynamics

The SS perspective focuses on spatial heterogeneity of the environment rather than variation in species' life-history traits. All SS models share a set of common assumptions: (1) the environment is divided into patches, and each patch has different environmental conditions (regional heterogeneity); (2) local and regional dynamics operate on the same time scale; and (3) immigration is frequent and can change the outcome of local competition (a mass effect *sensu* Shmida and Whittaker 1981). This idea has been applied to a wide range of taxa and different modes of competition, and investigated using either the Lotka-Volterra model (for both dominance and preemptive competition) or lottery models (for preemptive competition).

LOTKA-VOLTERRA MODELS (LV)

Aside from the initial work of Skellam (1951) and Levene (1953), the first person to apply SS ideas to community ecology was Levin (1974). He developed a two-patch LV model of competition in which founder effects (random differences in initial abundances) allow different species to dominate numerically in different patches (preemptive competition leading to a priority effect). Dispersal from safe areas maintains species in communities from which they would otherwise be excluded (due to lower initial abundance). Given the perturbation theorem (when different stable boundary equilibria occur in different places in space, small perturbations can create stable interior equilibria; see Amarasekare 2000 for a review) the coexistence equilibrium is stable when dispersal between communities is low. Levin suggested that dispersal has the potential to increase species richness but that high dispersal rates reduce local species richness below that in isolation. This finding suggests that intermediate dispersal maintains differences between the localities and communities.

Subsequent studies focused on the maintenance of spatial variation in fitness and the effects of this variation on diversity. Following Levin, Kishimoto (1990) showed that many species can coexist in a two-patch system if the reciprocal competitive effects between the two dominant species are sufficiently strong (high reciprocal competition coefficients) and dispersal between the two patches is intermediate. The interaction between strong competitors results in a resource surplus because these species can maintain themselves at low resource levels. The surplus becomes open to other less competitive species. As the number of species increases, the resource surplus decreases, and further invasion becomes less likely. Again, competitive superiority can be due to either differences in initial abundances for species engaged in preemptive competition, or due to differences in competitive ability for species engaged in dominance competition, for example, different limiting factors in each locality. Amarasekare and Nisbet (2001) have since clarified the conditions for species undergoing dominance competition by considering spatial variation in the strength of competition at the regional scale (spatial variance in fitness). They have shown that when competitive asymmetry between species is high, local coexistence is possible only below a critical dispersal threshold. High dispersal rates undermine the potential for coexistence by reducing spatial variance in fitness (see also Nichimura and Kishida 2001).

An important result from Levin (1974) is that spatial variance in fitness can result only from heterogeneity in initial species abundances, provided competition is preemptive. Wilson (1992) found somewhat different results by considering patches colonized at random by individuals from a disperser pool. Species then interact in a closed system; since the outcome is dependent on initial conditions, the patches end up with different combinations of species. The Wilson (1992) model then pools species at the metacommunity scale to generate a new pool of dispersers for a second metacommunity assembly process. In this context, the species composition at the regional scale converges toward the species composition at the local scale. The founder effect does not hold in the long term if one considers extinctions and recolonizations based on species abundances at the metacommunity scale. Note, however, that the difference between Levin's (1974) and Wilson's (1992) results is at least partly because Wilson does not consider a mass effect that would rescue species from local extinction, regardless of the metacommunity assembly rules.

LOTTERY SS MODELS

The SS concept has been well-integrated into lottery models (Sale 1977; Chesson and Warner 1981). Because they consider recruitment as central, immigration and emigration can be crucial in these models. For example, some authors (Shmida and Ellner 1984; Kadmon and Shmida 1990; Loreau and Mouquet 1999) have included immigration from a regional pool in the form of a mass effect, which maintains high local species richness in plant communities. In these models, plants compete for available microsites, and the species with the highest number of seeds at a site wins. Given such rules for site attribution, immigration from outside the community can maintain inferior species in the system.

Lottery models with SS dynamics have also been applied to several aquatic systems. These systems are essentially competitive metacommunities with sessile adults and mobile larvae. Planktonic larvae produced in all the local communities enter a common pool and are equally redistributed among communities. Chesson (1985) developed a stochastic model with regional heterogeneity such that each locality favors the adults of a different species (via increased survival). He showed that species can coexist locally with emigration-immigration and higher adult survival in favorable localities. As with the temporal storage effect (Chesson and Warner 1981; Chesson 1983, 1984) in which a population sustains a positive average growth rate in a temporally fluctuating environment if growth rates in good years more than compensate for bad years, here a "spatial storage effect" promotes coexistence because it buffers a species from poor recruitment in patches where it has a negative growth rate (See also Hoopes et al., chapter 2). Iwasa and Roughgarden (1986; see also Muko and Iwasa 2000) found similar results in another model of marine systems.

While the above studies suggest that spatial heterogeneity is important for coexistence, Comins and Noble (1985) showed that spatial heterogeneity does not have to be permanent to promote coexistence. In their model coexistence requires each species to be dominant in site establishment under some environmental conditions although there are no permanent differences between parts of the habitat. Rather, the environment is uniformly variable in the sense that the long-term statistical distribution of environmental conditions is the same in all patches (transient niches). They found that every coexisting species must have its own transient niche. They also found that the stabilizing effect (coexistence) of environmental variability in the basic model can be obtained either with a few patches and complete mixing or with a large number of patches and little mixing.

Chesson (2000b) recently proposed a generalization of these results where coexistence results from a spatial storage effect, as defined above (see also Hoopes et al., chapter 2). Two other mechanisms driven by spatial variation contribute to the storage effect but can also act alone to allow coexistence: spatial variation in the degree of nonlinearity in species competitive responses (spatial relative nonlinearity) and localized dispersal that concentrates an invading species in areas favorable to its growth and reproduction (growth-density covariance; see Chesson 2000b and Chesson et al., chapter 12 for more details).

Mouquet and Loreau (2002, 2003; Loreau et al., chapter 18) have recently developed a metacommunity model of lottery competition, which, because it shares some characteristics with metapopulation CC models, helps to compare the outcomes of SS and CC models. This model is based on equation 10.3 applied at the individual level and assumes that localities receive a constant number of immigrants from other localities in the metacommunity. The model is given by

$$\frac{dP_{ik}}{dt} = (I_{ik} + [1 - a]c_{ik}P_{ik})V_k - m_{ik}P_{ik}$$

$$I_{ik} = \frac{a}{N-1}\sum_{l \neq k}^{N} c_{il}P_{il} \tag{10.10}$$

$$V_k = 1 - \sum_{i=1}^{S} P_{ik}.$$

Here i refers to species and k to localities. The parameter a describes the proportion of local reproduction that emigrates into other localities, and I_{ik} is the immigration term. At the local scale, S species compete for a limited proportion of vacant microsites, V_k. The metacommunity consists of N localities that differ in their local conditions. When there is no dispersal between localities, the species with the highest local basic reproductive rate ($r_{ik} = c_{ik}/m_{ik}$) excludes all other species in the locality because it decreases the proportion of vacant sites, V_k, below the threshold required for their persistence. If however localities are linked by dispersal, and different species dominate in different localities (due to spatial heterogeneity in competitive rankings), local coexistence is possible. Individuals emigrating from source areas prevent competitive exclusion in sink areas (where they are competitively inferior). At equilibrium each individual of each species must on average during its lifetime produce one individual that survives somewhere in the metacommunity so that each species' average net reproductive rate at the scale of the metacommunity is:

$$\bar{R}_i = \frac{\sum_{k=1}^{N} \hat{V}_k r_{ik} w_{ik}}{\sum_{k=1}^{N} w_{ik}} = 1. \tag{10.11}$$

Here w_{ik} is the total quantity of propagules of species i in community k per unit of time:

$$w_{ik} = (1 - a)c_{ik}\hat{P}_{ik} + \frac{a}{N-1}\sum_{l \neq k}^{N} c_{il}\hat{P}_{il}. \tag{10.12}$$

In this lottery model, a species' competitive ability is directly related to its reproductive rate. Thus equation 10.11 can be interpreted as a constraint of regional competitive similarity (or equivalency) between coexisting species. For instance, each species' basic reproductive rate r_{ik} must be balanced over the metacommunity for equation 10.11 to hold. Local basic reproductive rates can be different locally so that different species will dominate in each locality, but they are equivalent when averaged over the region. Conditions for species coexistence are then a function of the proportion of dispersal between communities. When dispersal is too high, the metacommunity is homogenized, and species richness declines because the species that is the best competitor at the regional scale excludes all other species from the metacommunity.

The SS models discussed above assume that extrinsic spatial heterogeneity drives spatial variance in fitness. However, as shown by Levin (1974), intrinsic heterogeneity can also arise from founder effects or other random factors that generate spatial variation in initial species abundances. More recently, Yu and Wilson (2001) incorporated such spatial variation in patch density in what is essentially a CC lottery model at the microsite scale (based on appendix E in Chesson and Huntly 1997) without dominance competition. Yu and Wilson varied the fraction h of available habitat in each community. In models with complete mixing, they showed that inferior competitors can be rescued from competitive exclusion in communities with high h by emigrating from communities with low h (where the best competitor cannot persist or is at very low density) because of the CC tradeoff. Yu and Wilson also showed that including a trade-off between dispersal and fecundity can increase the potential for coexistence via spatial variation in patch density. In other words, CC itself cannot lead to coexistence in a lottery model with preemptive competition, but can allow coexistence via SS dynamics. Levine and Rees (2002) also found that a competition-colonization trade-off could not lead to coexistence when the assumption of a strict dominance hierarchy was relaxed. If coexistence could occur via spatial heterogeneity and SS dynamics, however, a competition-colonization trade-off operating simultaneously can predict the abundance patterns observed in annual plant communities (Levine and Rees 2002).

SYNTHESIS ON SOURCE-SINK DYNAMICS

Despite the differences in their mathematical formulations, LV and lottery models yield similar results. First, local coexistence in a SS metacommunity results from spatial variation in the strength of competition. Second, coexistence is most likely for intermediate dispersal rates. There is a critical dispersal threshold below which dispersal is too low to prevent local competitive exclusion and a critical dispersal threshold above which coexistence is impossible because spatial variation in fitness is homogenized by immigration. Third, species have limits to regional dissimilarity or competitive asymmetry that are a function of dispersal between communities.

Spatial heterogeneity is essential for coexistence in these models. It arises from spatial variation in abiotic or biotic factors that cause spatial variation in the strength of competition or from spatial variation in species abundances due to founder effects or other random phenomena. As pointed out by Amarasekare (2000), these two kinds of heterogeneity correspond to two different types of local dynamics—threshold and directional. Threshold dynamics occur in systems where the outcome of competition depends on initial abundances (e.g., preemptive competition) such that coexistence is possible given spatial variation in species abundances. In contrast, systems with directional local dynamics (e.g., dominance competition) are driven by spatial environmental heterogeneity and phenotypic plasticity. Threshold systems are highly sensitive to large perturbations, and it is likely that increasing dispersal will rapidly decrease species richness (following an initial increase at very low dispersal rates). Directional systems are less sensitive to perturbation, and they are more likely to have higher species richness at high dispersal rates. In both systems, however, very high dispersal values tend to homogenize spatial variance and reduce species richness. The results obtained with threshold systems (Levin 1974; Kishimoto 1990; Amarasekare 2000; Yu and Wilson 2001; Levine and Rees 2002) are striking because species can coexist only because the environment is patchy (but otherwise homogeneous), and abundances vary in space. Regional niche differentiation is not needed for species to coexist in such a metacommunity.

Because coexistence in SS models is frequently dependent on maintaining spatial variation in fitness, these models point toward generalizations about limits to similarity and regional niche differentiation. For instance, the condition of regional similarity defined by Mouquet and Loreau (2002) casts coexistence in terms of niche theory and thereby alludes to earlier models. Iwasa and Roughgarden do not interpret their results in the context of niche theory, but they state (in theorem 2) that coexistence requires compensation between larval productivity of species at the regional scale, that is, that species must have similar larval productivity at the scale of the region (1986, 198). This idea of regional compensation is also evident in Chesson's (1985) work on the spatial storage effect (see also Chesson 2000b). Models based on the LV framework produce the same regional similarity constraint. For instance, Amarasekare and Nisbet (2001) define a limit to competitive asymmetry that is equivalent to a constraint of similarity between coexisting species. These different results can be interpreted in the context of niche theory given that each locality has a different limiting factor (resulting in a different species dominating each community). Niche differentiation occurs at the regional scale and registers at the local scale through dispersal. Defining species as similar at particular spatial scales helps to reconcile niche theory with the high species diversity observed in natural systems and helps interpret niche theory in the context of metacommunity ecology. Further work to integrate classic limiting similarity concepts with the idea of regional niche differentiation will improve our understanding of competitive metacommunities (Chesson 1991).

Toward a General Framework

Although CC and the SS models have typically been regarded as two different approaches to studying spatial coexistence, elements of both classes of mechanisms are likely to operate simultaneously in natural communities. We present several key points of contact between CC- and SS-mediated coexistence and propose the basis for a general framework.

Convergence

As pointed out by Chesson (2000a), coexistence always requires intraspecific competition to be stronger than interspecific competition. This requirement allows spatial coexistence mechanisms such as CC and SS to be interpreted in the context of classical niche theory. In CC models with displacement competition, spatial niche differences arise because superior competitors can displace inferior competitors from occupied patches, but inferior competitors can only occupy patches not colonized by the superior competitors. This is equivalent to resource partitioning with inferior competitors, by virtue of their superior dispersal abilities that are better at acquiring empty patches, and superior competitors that because of their superior displacement abilities are better at acquiring occupied patches. In the case of SS models, niche differences between species arise at the regional scale due to spatial heterogeneity in the environment. For instance, if spatial variation in abiotic factors such as temperature, humidity, salinity or presence of an essential nutrient (or toxin) alters the relative strengths of intra- versus interspecific competition for a limiting resource in different locations of the landscape (for examples see Miller and Kneitel, chapter 5; Cottenie and De Meester, chapter 8; Kolasa and Romanuk, chapter 9), then space itself constitutes a second niche axis along which species exhibit differences.

A key issue in integrating CC and SS ideas in a common framework involves identifying the conditions under which both classes of mechanisms can contribute to coexistence. While the role of spatial heterogeneity and source-sink dynamics in allowing coexistence when life-history trade-offs cannot do so has been investigated (e.g., Yu and Wilson 2001; Levene and Rees 2002), how spatial variation influences trade-off mediated coexistence has not received much theoretical attention. For example, spatial heterogeneity in the biotic or abiotic environment can lead to spatial variation in the expression of a life-history trade-off. Contrary to the conventional wisdom that environmental heterogeneity promotes species coexistence, heterogeneity that influences the expression of a life-history trade-off can also constrain opportunities for coexistence (Amarasekare et al. 2004). In such a situation, source-sink dynamics can play a key role in enhancing opportu-

nities for coexistence. For example, consider an interspecific trade-off between resource exploitation and susceptibility to a predator (as in McPeek 1996). The species that allocates more energy to reproduction at the cost of reduced predator defense is competitively superior to the species that employs the opposite allocation
strategy. There is spatial variation in predator abundance such that it is present in
some localities but not others. In localities where the predator is present, the
trade-off is expressed and local coexistence is possible. In localities where the
natural enemy is absent, the inferior competitor is excluded because energy allo-
cated to predator defense (a net cost now with no benefit) gives it an overall competitive disadvantage. If there is dispersal between localities, however, coexistence
is possible everywhere (see Amarasekare et al. 2004 for a formal mathematical
analysis). This result is due to emigration from localities in which the trade-off is
expressed (sources for the inferior competitor) rescuing the inferior competitor
from exclusion in localities in which the trade-off is not expressed (sinks for the
inferior competitor). Thus, simultaneous operation of trade-offs and source-sink
dynamics can increase opportunities for coexistence.

The models we have presented in this review have considered fundamentally
different types of competitive interactions that should be carefully considered
when testing spatial theory. The key distinction between dominance and preemptive competition is that, because the former involves displacement of inferior
competitors by superior competitors, the ability of the inferior competitor to persist is an explicit function of both the superior competitor's colonization ability
and its abundance. In contrast, in preemptive competition species compete to replace individuals that have died, and no displacement occurs. The ability of any
given species to persist is a function solely of its colonization and extinction rates.
The key results we have reviewed in this chapter can be summarized as follows.
Under dominance competition coexistence can occur either via a CC trade-off in
a spatially homogeneous environment, or via SS dynamics in a spatially heterogeneous environment. When competition is preemptive, coexistence cannot occur
via a CC trade-off unless there is some spatial heterogeneity and a trade-off between species competitive ability for distinct limiting resources.

Hybrid Models

The fact that CC and SS models have been considered alternative mechanisms has
made it difficult to identify conditions under which elements of both mechanisms
operate. This underscores the need for a more general framework that accommo-
dates both approaches and allows predictions about the relative importance of
competition-colonization and source-sink in mediating spatial coexistence in nature (Amarasekare et al. 2004).

Hybrid models have opened the way for an integrated framework (figure 10.1).
The Horn and MacArthur model (1972) is a notable example because it demonstrates that a combination between CC and SS processes can explain coexistence

between competing species. More recently Pacala and Rees (1998) proposed a model that combined competition-colonization and the successional niche. They showed that coexistence between species of different successional stages can result from competition-colonization, successional niche partitioning where early successional species have a greater ability to exploit resource-rich conditions of recently disturbed habitats, or a combination of the two. The model by Yu and Wilson (2001) makes the link with source-sink for preemptive competition, and that of Levine and Rees (2002) makes the link with SS for CC models with dominance competition but no strict asymmetry. In all of these examples, a combination of competition-colonization and source-sink can elevate species richness. More recently Amarasekare et al. (2004) combined competition-colonization and source-sink in a single framework and compared their operation under different types of competition and dispersal. They showed that counterintuitive properties arise due to interactions between these two different classes of mechanisms. For instance, spatial variation in the expression of a life-history trade-off (spatial heterogeneity) can constrain rather than promote coexistence.

A general framework may be obtained most easily by incorporating hierarchical spatial structure and spatial heterogeneity into the patch occupancy framework. Patch occupancy models (following Levins and Culver 1971) are attractive because of their analytical tractability for large numbers of patches, but they lack real world applications because of the unrealistic assumption of instantaneous local dynamics. This assumption has been important in developing simple metapopulation models but "there is no reason to believe that natural systems fall into two distinct classes with respect to the difference between the local and the regional time scales" (Hanski 1983). Explicit local dynamics will be required to track local abundances and details of local competitive interactions that influence coexistence and give predictions that correspond to what ecologists can measure in the field. Such hybrid models should also be able to vary the relative contribution of CC and SS mechanisms. Because the type of competition can alter the conditions for coexistence, preemptive and dominance competition should both be considered.

Several different avenues for further research have been investigated, but many more still require investigation.

(1) Classical SS effects at the regional scale with threshold versus directional local dynamics.
(2) Classical CC trade-offs at regional or local scales as well as a combination at different scales. Consideration of whether competition-colonization can lead to coexistence when dispersal occurs on the same timescale as competition may also be important.
(3) A combination of CC trade-offs and source-sink at the regional scale as in the successional perspective discussed above. This may extend the conditions

under which competition-colonization can increase opportunities for co-existence via SS dynamics.

(4) Regional competition-colonization between groups of species that have very distinct life-history strategies and source-sink between species of the same group. This point seems promising to merge the niche and the neutral theories of species diversity (Hubbell 2001; Mouquet and Loreau 2003; Chase et al., chapter 14).

(5) Regional source-sink between groups of species that would use very different resources and competition-colonization (either local or regional) between the species with similar resource use.

Conclusions

The world is patchy and heterogeneous, the recognition of which has been central to many theories of species coexistence over the last thirty years. Only recently have ecologists integrated these ideas into the emerging field of metacommunity ecology. In this chapter we have focused on the long history of spatial models based either on competition-colonization or source-sink dynamics, and we have discussed how these two approaches provide alternative mechanisms for species coexistence in metacommunities. We have shown that these apparently different mechanisms have common elements that could be integrated into a broader theoretical framework. We have also emphasized how these mechanisms could be interpreted in the context of niche theory and how this interpretation can help broaden the niche concept to include regional phenomena. Our focus was somewhat restrictive in that we concentrated exclusively on competitive metacommunities with no consideration of trophic interactions (Nowak and May 1994; Holt 1997a; Nee et al. 1997; Hassell 2000; Hoopes et al., chapter 2; Holt and Hoopes, chapter 3) or genetic polymorphisms (Karlin and McGregor 1972a, 1972b; De Meeus et al. 1993), both of which are integral components of biological diversity in patchy environments. A complete synthesis of coexistence in spatially heterogeneous environments would require assembling these other pieces of the same puzzle. Understanding mechanisms that maintain species diversity in patchy environments will allow for the development of a general theory of biological diversity that is needed now more than ever, given the catastrophic loss of biodiversity due to increasing human encroachment.

Acknowledgments

This work was conducted as part of the Metacommunity Working Group at the National Center for Ecological Analysis and Synthesis (NCEAS), a center funded by the National Science Foundation (DEB-9421535), University of California Santa Barbara, and the state of California. The authors would like to thank Peter

Chesson, Bob Holt, Marcel Holyoak, Mathew Leibold, and an anonymous reviewer for comments on the manuscript. NM was in part supported by the School of Computational Science and Information Technology, Florida State University, USA and the MacMan European project. MFH was supported by NSF grant DEB-9806635 and NIH grant RO1ES12067-01 to C. J. Briggs.

Literature Cited

Adler, F. R., and J. Mosquera. 2000. Is space necessary? Interference competition and limits to biodiversity. Ecology 81:3226–3232.

Amarasekare, P. 2000. The geometry of coexistence. Biological Journal of the Linnean Society 71:1–31.

Amarasekare, P., M. Hoopes, N. Mouquet, and M. Holyoak. 2004. Mechanisms of coexistence in competitive metacommunities. American Naturalist 164:310–326.

Amarasekare, P., and R. M. Nisbet. 2001. Spatial heterogeneity, source-sink dynamics and the local coexistence of competing species. The American Naturalist 158:572–584.

Andrewartha, H. G., and L. C. Birch. 1954. *The distribution and abundance of animals.* University of Chicago Press, Chicago, IL.

Barot, S., and J. Gignoux. 2004. Mechanisms promoting plant coexistence: can all the proposed processes be reconciled? Oikos 106:185–192.

Chesson, P. L. 1983. Coexistence of competitors in a stochastic environment: the storage effect, Pages 188–198 in H. I. Freedman, and C. Stobeck, eds. *Population biology. Lectures notes in biomathematics.* Springer-Verlag, New York, NY.

———. 1984. The storage effect in stochastic population models. Lecture Notes in Biomathematics 54:76–89.

———. 1985. Coexistence of competitors in spatially and temporally varying environments: A look at the combined effects of different sorts of variability. Theoretical Population Biology 28:263–287.

———. 1991. A need for niches. Trends in Ecology and Evolution 6:26–28.

———. 2000a. Mechanisms of maintenance of species diversity. Annual Review of Ecology and Systematics 31:343–366.

———. 2000b. General theory of competitive coexistence in spatially-varying environments. Theoretical Population Biology 58:211–237.

Chesson, P. L., and N. Huntly. 1997. The roles of harsh and fluctuating conditions in the dynamics of ecological communities. American Naturalist 150:519–553.

Chesson, P. L., and R. W. Warner. 1981. Environmental variability promotes coexistence in lottery competitive systems. American Naturalist 117:923–943.

Cohen, D., and S. A. Levin. 1991. Dispersal in patchy environment: the effects of temporal and spatial structure. Theoretical Population Biology 39:63–99.

Cohen, J. E. 1970. A Markov contingency-table model for replicated Lotka-Volterra systems near equilibrium. American Naturalist 104:547–560.

Comins, H. N., and I. R. Noble. 1985. Dispersal, variability, and transient niches: Species coexistence in a uniformly variable environment. American Naturalist 126:706–723.

Cottenie, K., E. Michels, N. Nuytten, and L. De Meester. 2003. Zooplankton metacommunity structure: Regional versus local processes in highly interconnected ponds. Ecology 84:991–1000.

De Meeus, T., Y. Michalakis, F. Renaud, and I. Olivieri. 1993. Polymorphism in heterogeneous environments, evolution of habitat selection and sympatric speciation: Soft and hard selection models. Evolutionary Ecology 7:175–198.

Fagerström, T., and M. Westoby. 1997. Population dynamics in sessile organisms: Some general results from three seemingly different theory lineages. Oikos 80:588–594.

Forbes, A. E., and J. M. Chase. 2002. The role of habitat connectivity and landscape geometry in experimental zooplankton metacommunities. Oikos 96:433–440.

Gause, G. F. 1934. *The struggle for existence*. Williams and Wilkins, Baltimore, MD.

Goldwasser, L., J. Cook, and E. D. Silverman. 1994. The effect of variability on metapopulation dynamics and rates of invasion. Ecology 75:40–47.

Gonzalez, A., J. H. Lawton, F. S. Gilbert, T. M. Blackburn, and I. Evans-Freke. 1998. Metapopulation dynamics, abundance, and distribution in a microecosystem. Science 281:2045–2047.

Grubb, P. J. 1977. The maintenance of species-richness in plant communities: The importance of the regeneration niche. Biological Reviews 52:107–145.

———. 1986. Problems posed by sparse and patchily distributed species in species-rich plant communities. Pages 207–225 *in* J. Diamond, and T. J. Case, eds. *Community ecology*. Harper and Row, New York.

Hanski, I. 1981. Coexistence of competitors in patchy environment with and without predation. Oikos 37:306–312.

———. 1983. Coexistence of competitors in patchy environment. Ecology 64:493–500.

———. 1999. *Metapopulation ecology: Oxford series in ecology and evolution*. Oxford University Press, Oxford, U.K., New York, NY.

Hardin, G. 1960. The competitive exclusion principle. Science 131:1291–1297.

Hassell, M. P. 2000. *The spatial and temporal dynamics of host-parasitoid interaction: Oxford series in ecology and evolution*. Oxford University Press, Oxford, UK.

Hastings, A. 1980. Disturbance, coexistence, history and the competition for space. Theoretical Population Biology 18:363–373.

Holmes, E. E., and H. B. Wilson. 1998. Running from trouble: Long distance dispersal and the competitive coexistence of inferior species. American Naturalist 6:578–586.

Holt, R. D. 1993. Ecology at the mesoscale: The influence of regional processes on local communities. Pages 77–88 *in* R. E. Ricklefs, and D. Schluter, eds. *Species diversity in ecological communities: Historical and geographical perspectives*. University of Chicago Press, Chicago, IL.

———. 1997a. From metapopulation dynamics to community structure: Some consequences of spatial heterogeneity. Pages 149–164 *in* I. A. Hanski, and M. E. Gilpin, eds. *Metapopulation biology: Ecology, Genetics, and Evolution*. Academic Press, San Diego, CA.

———. 1997b. On the evolutionary stability of sink populations. Evolutionary Ecology 11:723–731.

Horn, H. S., and R. H. Mac-Arthur. 1972. Competition among fugitive species in a harlequin environment. Ecology 53:749–752.

Hubbell, S. P. 2001. *The unified neutral theory of biodiversity and biogeography*. Princeton University Press, Princeton, NJ.

Huffaker, C. B. 1958. Experimental studies on predation: Dispersal factors and predator-prey. Hilgardia 27:343–383.

Hutchinson, G. E. 1957. Concluding remarks. Cold Spring Harbor Symposium on Quantitative biology 22:415–427.

———. 1961. The paradox of the plankton. American Naturalist 95:137–145.

Iwasa, Y., and J. Roughgarden. 1986. Interspecific competition among metapopulations with space-limited subpopulation. Theoretical population biology 30:194–214.

Kadmon, R., and A. Shmida. 1990. Spatiotemporal demographic processes in plant populations: An approach and a case study. American Naturalist 135:382–397.

Karlin, S., and J. McGregor. 1972a. Application of method of small parameters to multi-niche population genetic model. Theoretical Population Biology 3:186–209.

———. 1972b. Polymorphisms for genetic and ecological Systems with weak coupling. Theoretical Population Biology 3:210–238.

Keddy, P. A. 1982. Population ecology on an environmental gradient: *Cakile edentula* on a sand dune. Oecologia 52:348–355.

Kishimoto, K. 1990. Coexistence of any number of species in the Lotka-Volterra competitive system over two-patches. Theoretical Population Biology 38:149–158.

Kneitel, J. M., and T. E. Miller. 2003. Dispersal rates affect species composition in metacommunities of Sarracenia purpurea inquilines. American Naturalist 162:165–171.

Lack, D. 1944. Ecological aspects of species formation in passerine birds. Ibis 86:260–286.

Leibold, M. A., M. Holyoak, N. Mouquet, P. Amarasekare, J. Chase, M. Hoopes, R. Holt, J. Shurin, R. Law, D. Tilman, M. Loreau, and A. Gonzalez. The metacommunity concept: A framework for multi-scale community ecology. Ecology Letters 7:601–613.

Levene, H. 1953. Genetic equilibrium when more than one ecological niche is available. American Naturalist 87:331–333.

Levine, J. M. and M. Rees. 2002. Coexistence and relative abundance in annual plant communities: the roles of competition and colonization. American Naturalist 160:452–467.

Levin, S. A. 1974. Dispersion and population interactions. American Naturalist 108:207–228.

———. 1976. Population dynamic model in heterogeneous environment. Annual Review of Ecology and Systematics 7:287–310.

———. 1992. The problem of pattern and scale in ecology. Ecology 73:1943–1967.

Levins, R. 1969. Some demographic and genetic consequences of environmental heterogeneity for biological control. Bulletin of the Entomological Society of America 15:237–240.

———. 1970. Extinction. Pages 77–107 in M. Gerstenhaber, ed. Some mathematical problems in biology. American Mathematical Society, Providence, RI.

———. 1979. Coexistence in a variable environment. American Naturalist 114:765–783.

Levins, R., and D. Culver. 1971. Regional coexistence of species and competition between rare species. Proceeding of the National Academy of Science, USA 68:1246–1248.

Levins, R., and R. MacArthur. 1966. The maintenance of genetic polymorphism in a spatially heterogeneous environment: Variations on a theme by Howard Levene. American Naturalist 100:585–589.

Loreau, M., and N. Mouquet. 1999. Immigration and the maintenance of local species diversity. American Naturalist 154:427–440.

Loreau, M., N. Mouquet, and R. Holt. 2003. Meta-ecosystem: a framework for a spatial ecosystem ecology. Ecology Letters 6:673–679.

Lotka, A. J. 1925. Elements of physical biology. Williams and Wilkins Company, Baltimore.

MacArthur, R. H. 1972. Geographical ecology. Harper and Row, New York, NY.

MacArthur, R. H., and R. Levins. 1967. The limiting similarity, convergence, and divergence of coexisting species. American Naturalist 101:377–387.

May, R. M., and M. A. Nowak. 1994. Superinfection, metapopulation dynamics and the evolution of diversity. Journal of Theoretical Biology 170:95–114.

McPeek, M. A. 1996. Trade-offs, food web structure, and the coexistence of habitat specialists and generalists. American Naturalist 148:S124–S138.

Mouquet, N., and M. Loreau. 2002. Coexistence in metacommunities: The regional similarity hypothesis. American Naturalist 159:420–426.

———. 2003. Community patterns in source-sink metacommunities. American Naturalist 162:544–557.

Mouquet, N., J. L. Moore, and M. Loreau. 2002. Plant species richness and community productivity: Why the mechanism that promotes coexistence matters. Ecology Letters 5:56–66.

Muko, S., and Y. Iwasa. 2000. Species coexistence by permanent spatial heterogeneity in a lottery model. Theoretical Population Biology 57:273–284.

Nee, S., R. D. Gregory, and R. M. May. 1991. Core and satellite species: Theory and artifact. Oikos 62:83–87.

Nee, S., and R. M. May. 1992. Dynamics of metapopulation: Habitat destruction and competitive coexistence. Journal of Animal Ecology 61:37–40.

Nee, S., R. M. May, and M. P. Hassell. 1997. Two species metapopulation models. Pages 123–147 *in* I. A. Hanski, and M. E. Gilpin, eds. *Metapopulation biology: Ecology, genetics, and evolution.* Academic Press, San Diego, CA.

Nichimura, K., and O. Kishida. 2001. Coupling of two competitive systems via density dependent migration. Ecological Research 16:359–368.

Pacala, S. W., and M. Rees. 1998. Models suggesting field experiments to test two hypotheses explaining successional diversity. American Naturalist 152:729–737.

Pulliam, H. R. 1988. Sources, sinks, and population regulation. American Naturalist 132:652–661.

Rees, M. 1993. Trade-offs among dispersal strategies in British plants. Nature 366:150–152.

Ricklefs, R. E., and D. Schluter. 1993. Species diversity: regional and historical influences. Pages 350–363 *in* R. E. Ricklefs, and D. Schluter, eds. *Species diversity in ecological communities: historical and geographical perspectives.* University of Chicago Press, Chicago, IL.

Roughgarden, J., S. Gaines, and H. Possingham. 1988. Recruitment dynamics in complex life cycles. Science 241:1460–1466.

Sale, P. F. 1977. Maintenance of high diversity in coral reef fish communities. American Naturalist 111: 337–359.

Shmida, A., and S. Ellner. 1984. Coexistence of plant species with similar niches. Vegetatio 58:29–55.

Shmida, A., and R. H. Whittaker. 1981. Pattern and biological microsite effects in two shrub communities, southern California. Ecology 62:234–251.

Skellam, J. G. 1951. Random dispersal in theoretical populations. Biometrika 38:196–218.

Slatkin, M. 1974. Competition and regional coexistence. Ecology 55:128–134.

Taneyhill, D. E. 2000. Metapopulation dynamic of multiple species: The geometry of competition in a fragmented habitat. Ecological Monographs 70:495–516.

Thompson, J. N. 1999. Specific hypothesis on the geographic mosaic of coevolution. American Naturalist 153:S1–S14.

Tilman, D. 1982. Resource competition and community structure. Princeton University Press, Princeton, N.J.

———. 1994. Competition and biodiversity in spatially structured habitats. Ecology 75:2–16.

Tilman, D., R. M. May, C. L. Lehman, and M. A. Nowak. 1994. Habitat destruction and the extinction debt. Nature 371:65–66.

Volterra, V. 1926. Variation and fluctuations of the number of individuals in animal species living together. Pages 409–448 *in* R. N. Chapman, ed. *Animal ecology.* MacGraw-Hill, New York, NY.

Urban, M. C. 2004. Disturbance heterogeneity determines freshwater metacommunity structure. Ecology 85:2971–2978.

Wang, Z. L., F. Z. Wang, S. Chen, and M. Y. Zhu. 2002. Competition and coexistence in regional habitats. American Naturalist 159:498–508.

Wang, Z. L., J. G. Zhang, and X. M. Liu. 2000. Competition and coexistence in spatially subdivided habitats. Journal of Theoretical Biology 205:631–639.

Whittaker, R. H., and S. A. Levin. 1977. The role of mosaic phenomena in natural communities. Theoretical Population Biology 12:117–139.

Wilson, D. S. 1992. Complex interactions in metacommunities, with implications for biodiversity and higher levels of selection. Ecology 73:1984–2000.

Yu, D. W., and H. B. Wilson. 2001. The competition-colonization trade-off is dead: Long live the competition-colonization trade-off. American Naturalist 158:49–63.

CHAPTER 11

Assembly Dynamics in Metacommunities

Richard Law and Mathew A. Leibold

Introduction

Much of the theory of community ecology is concerned with dynamics of inter-acting species at a local level—how interactions among the species determine transient and asymptotic properties of communities. Best known is the general-ized Lotka-Volterra model, although there are numerous elaborations on this as well. However, communities are open systems liable to be invaded and changed by species from outside; structure at a local level may depend as much on the pool of species available in the region as on local interactions (Cornell 1985; Ricklefs 1987; Leibold et al. 1997; Shurin et al. 2000). How can we extend the theory of local communities to this larger spatial scale? And how might mechanisms for species interactions at this large scale result in dynamics different from those at the local scale?

One response to the openness of communities has been to develop a theory of assembly dynamics; this invokes an external pool of species that from time to time contributes new species to the local community (Post and Pimm 1983; Rummel and Roughgarden 1985; Mithen and Lawton 1986; Drake 1990; Law and Morton 1996; Lockwood et al. 1997). New species may get established, thereby changing the structure of the local community, and may bring about further change by causing extinction of species in the community. For a finite species pool, an as-sembly graph can be constructed describing the sets of species that can persist lo-cally and the transitions between these sets caused by arrival of new species (Law and Morton 1996; Warren et al. 2003; Zimmermann et al. 2003). Assembly dy-namics then use this graph to gain insight into assembly patterns that emerge through random arrival of species (Law 1999). In the long term assembly leads to uninvasible communities (endpoints) or cyclic sequences of communities, unin-vasible by species not present within the sequence (endcycles) (Morton and Law 1997).

In many cases however, species pools are better thought of as the union of sets of species present in the local communities scattered across a region. For the pur-pose of considering assembly dynamics, communities might be better thought of as embedded in a metacommunity of many local communities linked by disper-sal (Leibold et al. 2004). This couples the regional species pool to the local com-munities so that, in the long term, the species pool contains only those species that are able to live somewhere in the region (Hastings 1980; Wilson 1992; Leibold 1999). A synthesis of assembly and metacommunity dynamics will help us to understand the relationship between local and regional diversity patterns (e.g., Shurin and Allen 2001).

Here we show how an assembly graph can be embedded in a metacommunity. The metacommunity dynamics are driven primarily by rare colonization and extinction of species within patches, assumed to take place on a timescale slower than that of births and deaths; this separation of timescales means that the main information needed for the metacommunity dynamics is the assembly graph. Additionally, this separation means that local dynamics are not influenced by dispersal or colonization. Extinctions are caused by arrival of other species (as in previous models of community assembly) and by temporary stochastic destruction of patches unrelated to such arrivals (called patch extinctions below). This extension of metapopulation dynamics to deal with interactions among species is also discussed in the context of competing species by Mouquet et al. in chapter 10, and extended to consider issues of regional diversity by Loreau et al. in chapter 18. The method we present can be used to understand community assembly in a metacommunity context in any situation where an assembly graph is available and a separation of timescales is a reasonable assumption.

We also use an example to illustrate how expectations about the outcome of species interactions can differ between the local and the metacommunity scales. The example involves nontransitive competition among three species at the local community scale; this has a heteroclinic attractor (see "Example" below for a definition), and may be illustrative of the states in which more complex communities can end up as a result of assembly (Morton and Law 1997; Steiner and Leibold 2004). In this model, patch extinctions are equivalent to extinction of local populations and we examine their role when these extinction rates are the same for all three species. In the absence of patch extinctions, this competitive system has a neutrally stable metacommunity in contrast to local dynamics that would lead to a loss of all but one species. In the presence of patch extinctions, more outcomes are possible; this includes changing the neutral cycle to a stable equilibrium. Evidently, dynamics can be changed qualitatively by switching from the local to metacommunity scale, and properties of the metacommunity can depend strongly on metacommunity-level species attributes, such as colonization rate, sometimes even more than on local attributes such as proximate niches.

Model

Assumptions

The metacommunity is thought of as a large number of local communities living in an environment made up of many identical patches. Associated with the metacommunity is a species pool, that is the set of s species present in an area encompassing the metacommunity. The metacommunity is closed: all the species have to be able to live somewhere in the area. Species are able to disperse from one patch to another; when individuals disperse, the distances they move are sufficiently long for there to be no spatial structure of local communities in neighboring patches.

For the sake of clarity, we start by dealing with extinctions that occur during assembly, when a species is driven deterministically to extinction by entry of another species into the local community. Temporary, stochastic, patch extinctions can also have major effects on the dynamics; these are introduced later on (see the example, "Long Timescale: Metacommunity Dynamics with Patch Extinctions"). Extinctions due to demographic stochasticity are less straightforward to deal with in the presence of multispecies interactions (Keeling 2002), and we therefore assume throughout that local population sizes are large enough for extinction due to random birth-death events to be negligible.

Two timescales are assumed (Law 1999): a short timescale for dynamics of population densities within local communities due to births and deaths of individuals (community dynamics), and a long timescale for turnover of local communities from one set of species to another due to arrival and extinction of species (assembly dynamics). Another way of putting this is to say that dispersal of species occurs rarely enough for species to get close to their asymptotic densities within local communities before further species arrive. Species disperse independently; in conjunction with the timescale separation, this means that steps in assembly at the local level are driven by arrival of just one species at a time. Also as a consequence of the timescale separation, dispersal events are rare enough to have a negligible effect on local community dynamics.

Short Timescale: Local Community Dynamics

The local dynamics of population density within patches (short timescale) provide qualitative information needed for metacommunity dynamics (long timescale), specifically: (1) The species that can live together (i.e., coexist) locally within patches; we term such a set of species a "persistent community." (2) The species able to invade these persistent communities. (3) The new sets of species that emerge following a successful invasion, which may include extinction of one or more species as well as the addition of the colonist species. This information is passed to the long timescale of metacommunity dynamics as an assembly graph showing all transitions allowable between communities due to invasions and extinctions. In more detail, the information is obtained as follows.

1. To establish the species able to live together locally within patches, each subset of the species pool must be checked for long-term persistence; in our example there are 2^s possible subsets, including the empty set and species pool itself. By concentrating on persistent sets of species, we are sieving the species pool, poten-

tially eliminating many communities that have only a transient existence. The persistent sets are denoted S_i, $i = 0, \ldots, r$ below, the state indexed 0 being a special empty set for patches that contain no species. In experiments, the persistent sets can sometimes be found by direct observation (Weatherby et al. 1998). In theoretical work, it is common to use the existence of an equilibrium point with various properties: (a) that all the species have positive densities, and (b) that local asymptotic stability applies (Pimm 1982). An alternative measure is permanence, requiring that orbits in the interior of the phase space are repelled by the boundary of the space (Hofbauer and Sigmund 1988; Law 1999); this has the advantage of not presupposing that densities tend to an equilibrium point in the long term. The mathematics of permanence for an arbitrary number of species have only been resolved for systems with generalized Lotka-Volterra dynamics; outside of this framework, numerical integration of the system would be needed to investigate persistence.

2. To establish which species are able to invade a persistent local community, each species that is absent from the community has to be tested for its capacity to increase at the asymptotic state of the local community. In experiments, this can be achieved by inoculating the persistent communities with each of the other species from the pool (Law et al. 2000). In theoretical work, tests are straightforward if the asymptotic state is an equilibrium point: a species invades if its per capita rate of increase is positive when evaluated at the equilibrium point of the resident species. Matters are not so straightforward if the asymptotic state is a nonequilibrium attractor, because invasion then depends on a time average of the initial rate of increase of the new species along the attractor of the resident species. However, progress is possible in the case of generalized Lotka-Volterra dynamics because the time average of the new species' per capita rate of increase tends to its value at the equilibrium point of the resident species (Hofbauer and Sigmund 1988, 62). Numerical integration could be used if the asymptotic state is not an equilibrium point and if the system does not have Lotka-Volterra dynamics.

3. Following invasion, a new species may simply be absorbed into the resident community, or one or more species may be driven to extinction. The new community can be found experimentally by following the state of the postinvasion community until the species composition stops changing (Warren et al. 2003). In theoretical work, one can follow the transient dynamics by numerical integration until a new attractor is reached (e.g., Case 1990). Alternatively, an algorithm based on permanence is available to jump in a single step to the new community (Law and Morton 1996).

Long Timescale: Metacommunity Dynamics

As long as the timescales are separate, metacommunity dynamics depend principally on information in the assembly graph. This is with the caveat that the dispersal rate of a species from a patch may depend on its density within the patch, and this density may in turn depend on what other species are also in the patch.

Such effects of local density can be dealt with by indexing dispersal rate according to the patch type from which dispersal occurs, as explained below.

The state of the metacommunity at time t on the long timescale is given by the vector $p(t)$ of the proportion of patches containing each of the persistent communities: $p = (p_0, p_1, \ldots, p_r)'$, with $\Sigma_i p_i = 1$. The dynamics are described by a system of differential equations with orbits that remain on the simplex $\Sigma_i p_i = 1$ in a nonnegative phase space of dimension $r + 1$. The dynamics depend first on dispersal, and second on the changes that follow from successful invasion of species that have dispersed into persistent communities. For the ith persistent community, the flux is given by the following equation:

$$\dot{p}_i = (TCp)_i. \tag{11.1}$$

Here C is a matrix of dimension s by $r + 1$ describing the intrinsic rates at which species disperse from patches of different kinds, and T is a matrix of dimension $r + 1$ by s describing transitions between persistent communities associated with arrival of species.

The species colonization matrix C has the form

$$
\begin{array}{c}
\text{species 1} \\
\vdots \\
\text{species } s
\end{array}
\left(
\begin{array}{cccc}
0 & c_{11} & \cdots & c_{1r} \\
\vdots & \vdots & & \\
0 & c_{s1} & & c_{sr}
\end{array}
\right).
$$
$$\text{com 1} \quad \cdots \quad \text{com } r$$

(Strictly speaking, it would be more accurate to call this a dispersal matrix, but we use the notation from metapopulations because this is more familiar.) The parameter c_{kl} is a constant nonnegative rate (dimensions T^{-1}) at which propagules of species k disperse from patches of type l. The first column of C must be zero because there are no species in empty patches; elements in other columns are nonzero only if community l contains species k. A simple assumption would be that species k disperses at the same rate from every type of patch in which it occurs. However, the abundance of species k is likely to be different in local communities with different species composition and this would affect the rate at which propagules of species k disperse from different kinds of patches in which it lives; this is allowed for by indexing c according to patch type l from which species k disperses. The total rates at which species are arriving in each patch type are the elements of Cp.

The matrix T deals with transitions from one persistent community to another caused by the arrival of species, and has the general form

$$
\begin{array}{c}
\text{community 0} \\
\text{community 1} \\
\vdots \\
\text{community } r
\end{array}
\left(
\begin{array}{cccc}
\bullet & \cdots & & \bullet \\
\bullet & & & \bullet \\
\vdots & \vdots & & \\
\bullet & & & \bullet
\end{array}
\right).
$$
$$\text{sp 1} \quad \cdots \quad \text{sp } s$$

A transition from local community j to i due to arrival of species k gives a negative term $-p_j$ in row j and a positive term $+p_j$ in row i, both of these in column k of the matrix. The columns of T thus sum to zero and $\Sigma_i \dot{p}_i = 0$. Consider, for instance, the flux involved in producing patches of type i from j, when species k disperses from patches of type l into j; this flux is $+ p_j c_{kl} p_l$ in row i and $-p_j c_{kl} p_l$ in row j of the dynamical system. Notice that each flux term is the product of the relative abundance of the patch type from which the species disperses and the patch type into which it falls. This second-order dependence on p is essentially a law of mass action, that propagules from patches of type l encounter patches of type j in proportion to the abundance of the j and l patches.

Example

To illustrate the model above, we consider a species pool of three competing species with a property of nontransitive competition, first given by May and Leonard (1975); properties of the dynamics are discussed by Hofbauer and Sigmund (1988, 66 et seq.). Nontransitivity means that species 2 can outcompete species 1; species 3 can outcompete species 2; and species 1 can outcompete species 3. This leads to a cyclic sequence of species present at a local level. This example has its own intrinsic interest and empirical examples have been studied by Buss and Jackson (1979), Paquin and Adams (1983), and Kerr et al. (2002). We have picked this example also because it illustrates how the outcomes of dynamics at the large scale of the metacommunity can differ from those at the level of the local community.

Short Timescale: Local Community Dynamics

The local community dynamics are given by the following system of equations (Hofbauer and Sigmund 1988, 66)

$$\dot{x}_1 = x_1(1 - x_1 - \alpha x_2 - \beta x_3)$$
$$\dot{x}_2 = x_2(1 - \beta x_1 - x_2 - \alpha x_3)$$
$$\dot{x}_3 = x_3(1 - \alpha x_1 - \beta x_2 - x_3),$$

where x_i is the local population density of species i, and the parameters α and β satisfy the inequalities $0 < \beta < 1 < \alpha$ and $\alpha + \beta > 2$. Out of the 2^3 possible local communities, only $\{\}, \{1\}, \{2\}, \{3\}$ are permanent; the local, two-species communities have no equilibrium points, and the local community $\{1, 2, 3\}$ is not permanent with the inequalities that apply to α and β above. The attractor here is in fact a heteroclinic cycle (Hofbauer and Sigmund 1988, 211); starting with the densities of all species positive, orbits spiral out toward the boundary of the phase space of the local community, spending longer and longer periods of time at each of the one-species equilibria before being taken over by the next species in the sequence. In practice all but one species would eventually become extinct; the species would not coexist locally.

To construct the assembly graph, we note that the persistent sets are the com-

Table 11.1 Lotka-Volterra model of nontransitive competition among three species in a local community

	$f_1(\hat{x})$	$f_2(\hat{x})$	$f_3(\hat{x})$
{}	+	+	+
{1}	0	+	−
{2}	−	0	+
{3}	+	−	0

Note: Table elements are signs of per capita rate of increase $f_i(\hat{x})$ of each species at each boundary equilibrium point $\hat{x}$.

munities $S_0 = \{\}$, $S_1 = \{1\}$, $S_2 = \{2\}$ and $S_3 = \{3\}$. Initial rates of increase of the species at the equilibrium points of these communities have signs as shown in table 11.1. Clearly each species invades the empty community; apart from this we see the cyclic competition in which species 2 replaces species 1, 3 replaces 2, and 1 replaces 3, giving the assembly graph in figure 11.1. This assembly graph carries the primary information on transitions between states needed for the metacommunity dynamics.

Long Timescale: Metacommunity Dynamics without Patch Extinctions

At the level of the metacommunity, the assembly graph leads to the following colonization and transition matrices:

$$C = \begin{pmatrix} 0 & c_1 & 0 & 0 \\ 0 & 0 & c_2 & 0 \\ 0 & 0 & 0 & c_3 \end{pmatrix}.$$

$$T = \begin{vmatrix} -p_0 & -p_0 & -p_0 \\ p_0 + p_3 & -p_1 & 0 \\ 0 & p_0 + p_1 & -p_2 \\ -p_3 & 0 & p_0 + p_2 \end{vmatrix}.$$

The second indices of the c's are not needed here because each species occurs in only one kind of patch. From equations 11.1, the dynamical system then takes the form

$$\dot{p}_0 = -p_0(c_1 p_1 + c_2 p_2 + c_3 p_3)$$
$$\dot{p}_1 = p_1(c_1[p_0 + p_3] - c_2 p_2)$$
$$\dot{p}_2 = p_2(c_2[p_0 + p_1] - c_3 p_3)$$
$$\dot{p}_3 = p_3(c_3[p_0 + p_2] - c_1 p_1)$$

One of the equations can be eliminated using the relation $\Sigma_i p_i = 1$, to get

$$\dot{p}_1 = p_1(c_1[1 - p_1 - p_2] - c_2 p_2)$$
$$\dot{p}_2 = p_2(c_2[1 - p_2 - p_3] - c_3 p_3) \qquad (11.2)$$
$$\dot{p}_3 = p_3(c_3[1 - p_1 - p_3] - c_1 p_1)$$

(a)

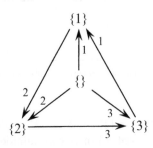

(b)

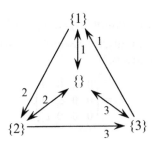

Figure 11.1 Assembly graphs for a Lotka-Volterra model of cyclic competition among three species in a local community. Arrows indicate allowable transitions between communities, the number beside each arrow being the species responsible for the change: (a) assembly graph without patch extinctions; (b) assembly graph with patch extinctions.

This system has five feasible equilibrium points:

$$\hat{p}^{\mathrm{I}} = (1\ \ 0\ \ 0\ \ 0)$$
$$\hat{p}^{\mathrm{II}} = (0\ \ 1\ \ 0\ \ 0)$$
$$\hat{p}^{\mathrm{III}} = (0\ \ 0\ \ 1\ \ 0)$$
$$\hat{p}^{\mathrm{IV}} = (0\ \ 0\ \ 0\ \ 1)$$
$$\hat{p}^{\mathrm{V}} = (0\ \ c_3\ c_1\ c_2)\frac{1}{A},$$
with $A = c_1 + c_2 + c_3$.

Because there is no path back to the empty community (figure 11.1), eventually all patches are filled. Equilibrium points II, III and IV are states in which the meta-community is completely dominated by one species. But V is an equilibrium

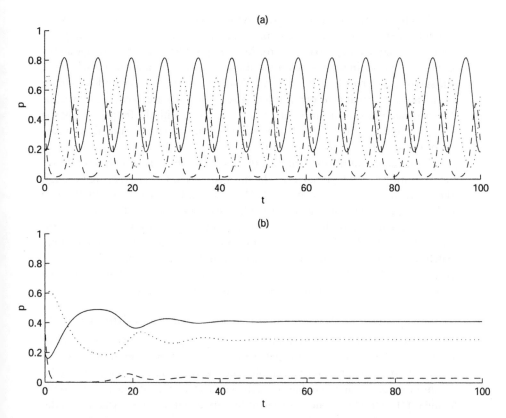

Figure 11.2 Metacommunity dynamics with cyclic competition within local communities, based on colonization and extinction graphs in figure 11.1. Parameter values are $c_1 = 1$, $c_2 = 2$, $c_3 = 3$; (a) extinction absent; (b) extinction present ($e = 0.5$). Continuous, dashed, and dotted lines are respectively the proportion of patches occupied by species 1, 2, and 3.

point at which all species are present, scattered across the metacommunity in different patches. An eigenvalue analysis shows that equilibrium points I, II, III, and IV cannot be attractors as long as dispersal rates are all greater than zero. The eigenvalues of the Jacobian at V comprise one real negative root, and a pair of purely imaginary roots

$$A\lambda_1 = -(c_1c_2 + c_1c_3 + c_2c_3)$$
$$A\lambda_2 = + i[c_1c_2c_3(c_1 + c_2 + c_3)]^{1/2}$$
$$A\lambda_3 = - i[c_1c_2c_3(c_1 + c_2 + c_3)]^{1/2}.$$

Thus the system has neutral stability in a neighborhood of V. In the neighborhood of V, the proportion of patches in each state oscillates over time with an amplitude set by the initial value of p (figure 11.2a), and numerical integrations from initial conditions that are further from V show the same behavior.

The simple form of equations 11.2 has the effect of making the time average of patch frequencies (p_is) the same as their equilibrium values in the long term, as long as no frequency approaches zero. Surprisingly, this means the average abundance of a species in the metacommunity is proportional to the dispersal rate of the species that precedes it in the competition cycle. This is less paradoxical than it might seem once it is recognized that, after all patches are filled, the only positive per-patch flux terms in equations 11.2 are those with the frequencies of the preceding species in the competition cycle; this can be seen by noting the replacement $1 = p_1 + p_2 + p_3$ in equations 11.2.

The metacommunity dynamics are interesting to compare with those of the local community. At a regional level, the three species in fact coexist, even though the heteroclinic cycle within an isolated local community would eventually leave only one species. Standing within a single patch, one would still just see a single species present most of the time, although the identity of this species would shift from 1 to 2 to 3 to 1 and so on, as time goes on. However, at a single point in time, patches dominated by each of the species would be found scattered across the region as a whole, the frequencies of patches in each state oscillating over the course of time to an extent that depends on the initial conditions.

Long Timescale: Metacommunity Dynamics with Patch Extinctions

The neutrality of the interior equilibrium point disappears if there are complete extinctions of populations within patches from time to time in addition to extinctions caused by the process of assembly. These extinctions add new paths to the assembly graph, as shown in figure 11.1b. Suppose the rate of patch extinction is e, still on the long timescale of metacommunity dynamics. The dynamics of the metacommunity then become

$$\dot{p}_1 = p_1(c_1[1 - p_1 - p_2] - c_2 p_2 - e)$$
$$\dot{p}_2 = p_2(c_2[1 - p_2 - p_3] - c_3 p_3 - e) \tag{11.3}$$
$$\dot{p}_3 = p_3(c_3[1 - p_1 - p_3] - c_1 p_1 - e)$$

These equations thus take on the form of more conventionally studied metapopulation models (e.g., Levins and Culver 1971; Hastings 1980), although the cause of extinction is not demographic stochasticity. Because of extinctions, the metacommunity in general contains some empty patches.

Any of the 2^3 equilibria, corresponding to any combination of species in the metacommunity (the power set of the species pool), can now be feasible:

$$\hat{p}^{\mathrm{I}} = (1\ \ 0\ \ 0\ \ 0)$$

$$\hat{p}^{\mathrm{II}} = \left(\frac{e}{c_1}\ \ \frac{c_1 - e}{c_1}\ \ 0\ \ 0\right)$$

$$\hat{p}^{\mathrm{III}} = \left(\frac{e}{c_2}\ \ 0\ \ \frac{c_2 - e}{c_2}\ \ 0\right)$$

$$\hat{p}^{IV} = \left(\frac{e}{c_3} \ \ 0 \ \ 0 \ \ \frac{c_3 - e}{c_3} \right)$$

$$\hat{p}^{V} = \left(\frac{c_2}{c_1} \ \ \frac{c_1 e - c_2^2}{c_1 c_2} \ \ \frac{c_2 - e}{c_2} \ \ 0 \right)$$

$$\hat{p}^{VI} = \left(\frac{c_3}{c_2} \ \ 0 \ \ \frac{c_2 e - c_3^2}{c_2 c_3} \ \ \frac{c_3 - e}{c_3} \right)$$

$$\hat{p}^{VII} = \left(\frac{c_1}{c_3} \ \ \frac{c_1 - e}{c_1} \ \ 0 \ \ \frac{c_3 e - c_1^2}{c_3 c_1} \right)$$

$$\hat{p}^{VIII} = (A^2 e \ \ Bc_3 - Ac_2 e \ \ Bc_1 - Ac_3 e \ \ Bc_2 - Ac_1 e) \frac{1}{AB}$$

$$\text{with } A = c_1 + c_2 + c_3$$
$$B = c_1 c_2 + c_2 c_3 + c_3 c_1.$$

The cyclic structure of equations 11.3 precludes the simultaneous existence of more than two equilibria with two species present. With appropriate choice of c's and e, any of these equilibria can be made asymptotically stable.

To demonstrate how metacommunity dynamics are enriched by the small change from equations 11.2 to 11.3 (the patch extinction rate e), we give some results from a numerical analysis. In this analysis, parameters of species 1 and 3 were held constant at values that permitted them to invade the empty metacommunity and to coexist in the absence of species 2 ($c_1 = 1$, $c_3 = 3$, $e = 0.5$) at equilibrium point VII. This coexistence is caused by a competition-colonization trade-off (in the presence of patch extinctions), the dispersal rate of species 3 on the long timescale being large enough to compensate for its competitive disadvantage on the short timescale. The dispersal rate c_2 of species 2 was then gradually increased from a small value.

The numerical analysis shows three main types of dynamics associated with increasing c_2. (1) If c_2 is small enough, species 2 cannot remain in the metacommunity. Obviously loss of species 2 would apply if $c_2 < e$; but it also applies if $c_2 > e$ and still not great enough to invade the metacommunity at equilibrium point VII. (2) Larger values of c_2 make equilibrium point VIII feasible and asymptotically stable. Thus, in this case, the metacommunity moves to a state that maintains all three species in stable proportions, as illustrated in figure 11.2b. (3) Increasing c_2 still further makes equilibrium point VI feasible; at first VI is unstable, but eventually VIII disappears and VI is asymptotically stable. The reason for this new attractor lacking species 1 is that the dispersal rate of species 2 has become large enough to take over patches containing species 1 faster than such patches can be created. In this attractor species 2 and species 3 now coexist via a competition-colonization trade-off and species 3 is now the competitive dominant species instead of the colonizing specialist.

Evidently, patch extinctions have a major effect on the behavior of the metacommunity. This is because there are always some empty patches available for colonization, in contrast to the case where extinctions occur only as a result of arrival of species in local communities, which leads eventually to every patch being filled. Resetting patches to the empty state, creates assembly paths that would not otherwise occur in the long term, fundamentally changing the dynamics of the metacommunity.

Discussion

Synthesis of Assembly and Metacommunity Dynamics

Our first main goal in this chapter was to couple species pools to local communities, on the basis that, unless a species can live in a local community somewhere, the species is unlikely to be in the species pool. Metacommunities, comprising many patches of local communities, provide a convenient framework for doing this. We have therefore put in place a general dynamic for metacommunities, appropriate in cases where the fast timescale of births and deaths within local communities is sufficiently different from the slow timescale over which species disperse among local communities and become locally extinct.

The metacommunity dynamics that emerge from this study can potentially be quite different from local dynamics of a single, closed community for at least two reasons. First, the metacommunity dynamics do not deal directly with birth and death events of individuals. It is just asymptotic information about the set of species eventually in the local community that gets passed on to the longer timescale of the metacommunity dynamics. Second, dynamics of the metacommunity make use of information about dispersal and extinction, which is quite different and distinct from information on birth and death rates at the faster timescale and only indirectly affected by them.

The uncoupling of local and metacommunity dynamics contrasts with situations where local dynamics have strong stabilizing effects on metacommunity dynamics. In similar situations when local dynamics have strong stabilizing effects (e.g., via patch-type specialization in metacommunities with heterogeneous patch types), metacommunity dynamics are primarily driven by this specialization. Dispersal and patch extinctions then play a minor role, unless dispersal rates become so large that they operate on the same timescale as species interactions, leading to homogenization and reduced diversity of the metacommunity (see Amarasekare and Nisbet 2002; Mouquet and Loreau 2002).

Metacommunity dynamics can also potentially be quite different from assembly dynamics of a single open community with an external species pool. First, the species pool is obviously reduced by elimination of species unable to live in any kind of local community. Second, the species pool may be gradually eroded as time goes on and as local communities reach states uninvadable by other species from the pool (endstates of assembly; see Law and Morton 1996). If the number of endstates is small, the number of species in the pool could fall to a small proportion of those present earlier on. To rescue persistent communities other than endstates, an extinction process is needed—beyond that caused by arrival of new species—resetting some local communities to earlier stages in the assembly sequence.

Nontransitive Competition

The uncoupling of local and metacommunity dynamics is clearly seen in the case of nontransitive competition. On the one hand, an isolated local community eventually collapses to only one species. On the other hand, at the level of the metacommunity, all three species may persist. The outcome for the metacommunity is strongly regulated by the dispersal rates and rate at which patches are temporarily destroyed (patch extinctions). Dispersal without patch extinction is relatively straightforward, leading to a metacommunity with all three species and an equilibrium point with neutral stability. But it is remarkable how much richer the metacommunity dynamics become, simply as a result of introducing one further parameter, a patch extinction rate. In principle, this parameter makes it possible for any of the eight equilibria, corresponding to zero, one, two, or three species, to be an attractor; the three-species equilibrium point can now be asymptotically stable, rather than neutral. Patch extinctions cause this richness by keeping open assembly paths that would otherwise disappear in the long term.

It is also notable that coexistence in the two-species subsystems of the heteroclinic cycle is achieved by a competition-colonization trade-off (Hastings 1980; Tilman 1994). We would not expect the competition-colonization trade-off to lead to coexistence of arbitrary subsets of species in longer nontransitive loops because the Jacobian matrices would not in general have the property of triangularity (cf. Tilman 1994). However, we conjecture that the complete loops competing species would retain the potential for coexistence at the metacommunity level, although the likelihood of long loops in reality is debatable. However, Holt has pointed out a sequence of resource-consumer replacements which could generate a heteroclinic cycle of arbitrary length (pers. comm.).

The neutral stability of nontransitive competition in the absence of patch extinctions illustrates a wider phenomenon that might be applicable to assembly dynamics associated with cyclic endstates. If there are no extinctions apart from those caused by arrival of species, and if there is more than one community comprising an endstate of assembly, the metacommunity model should be an equilibrium with at least a pair of imaginary eigenvalues, giving neutral stability. These dynamics are structurally unstable and an intrinsic tendency for local extinction would lead to quite different metacommunity dynamics. Depending on the rates of colonization and patch extinction of the component species this could either stabilize the dynamics or destabilize them (leading to a collapse of the metacommunity).

Extensions and Caveats

We have made much use of an assumption that births and deaths occur on a shorter timescale than dispersal and extinction events. Removing the separation of the timescales would make the theory more applicable, more interesting, and more difficult to analyze. At the limit, with dispersal so frequent that individuals are evenly spread across all sites, we would expect the metacommunity to become equivalent a single local community, and to have dynamics the same as those of a local community. It is the intermediate ground that is particularly difficult, where dispersal is reduced enough to allow some separation of local populations, but not enough to ignore mass effects (Mouquet et al., chapter 4). However, it is rather striking that, even when dispersal is low enough to justify the timescale separation, dispersal still has fundamental effects on the dynamics.

Extinctions evidently have a major impact on the metacommunity dynamics, because they open up space in which assembly can start afresh. We have dealt with two classes of extinction, caused (deterministically) by arrival of species and by temporary stochastic destruction of patches. But it is important to bear in mind a third class of extinction: that due to demographic stochasticity in small populations, which is often assumed to be the mechanism underpinning Levins's metapopulation model (1969). The effects of demographic stochasticity are more intricate. Initially, there might be extinction of just one (particularly rare) species in a local community, but the loss of this species might then lead to a cascade of further extinctions, as other species that interact directly and indirectly with the first species are disrupted (Ebenman et al. 2004). Effects of demographic stochasticity on metacommunity dynamics need further work because there is evidence that the standard linear and quadratic dependencies of fluxes in Levins's model do not necessarily apply in multispecies metacommunities with small (finite) local populations (Keeling 2002).

Conclusions

This chapter has taken the subjects of metapopulation dynamics and community assembly that have previously been treated independently in community ecology and put them together in a common formal framework of metacommunity dynamics. The key to doing this is to separate the timescale of births and deaths within local communities from a slower timescale of colonizations and extinctions of species across the metacommunity. With this synthesis in place, the behavior of a broad range of food webs scattered across spatial regions can be examined. Not surprisingly, the outcome of metacommunity dynamics can be quite different from the outcome of local community dynamics; in the example of nontransitive competition, coexistence can be achieved at the regional level where this would not be possible at the local level. Also, the outcome of community assembly is changed by internalizing the species pool from which local communities are assembled. Evidently the spatial extension of community assembly has important implications for community ecology.

We see this chapter very much as a first glimpse of assembly dynamics in a spatial setting. The chapter points toward future work that deals with the effects of space on assembly of food webs; in principle any assembly graph could be used, although the dynamical system would typically become complicated as the number of persistent states increases. The ideas given here could also readily extend to heterogeneous local environments, giving rise to some species sorting across the metacommunity. In addition, the chapter raises some technical issues needing more research, such as the formal derivation of a macroscopic dynamic for metacommunities when demographic stochasticity is the driver of local extinctions. As in other areas of spatial ecology, there is a great deal to be learnt; the relatively simple metacommunity framework developed here holds some promise as a tool for doing this.

Acknowledgments

We thank the participants of the NCEAS working group on metacommunities for the stimulating discussions that led to this work. We are grateful to NCEAS and the University of Chicago for supporting this research.

Literature Cited

Amarasekare, P., and R. M. Nisbet. 2001. Spatial heterogeneity, source-sink dynamics, and the local coexistence of competing species. American Naturalist 158:572–584.

Buss, L. W. and J. B. C. Jackson. 1979. Competitive networks: Nontransitive competitive relationships in cryptic coral reef communities. American Naturalist 113:223–234.

Case, T. J. 1990. Invasion resistance arises in strongly interacting species-rich model competition communities. Proceedings of the National Academy of Sciences, USA 87:9610–9614.

Cornell, H. V. 1985. Species assemblages of cynipid gall wasps are not saturated. American Naturalist 126:565–569.

Drake, J. A. 1990. The mechanics of community assembly and succession. Journal of theoretical Biology 147:213–233.

Ebenman, B., R. Law, and C. Borvall. 2004. Community viability analysis: The response of ecological communities to species loss. Ecology 85:2591–2600.

Hastings, A. 1980. Disturbance, coexistence, history and competition for space. Theoretical Population Biology 163:491–504.

Hofbauer, J. and K. Sigmund. 1988. *The theory of evolution and dynamical systems.* Cambridge University Press, Cambridge, UK.

Keeling, M. J. 2002. Using individual-based simulations to test the Levins metapopulation paradigm. Journal of Animal Ecology 71:270–279.

Kerr, B., M. A. Riley, M. W. Feldman and B. J. M. Bohannan. 2002. Local dispersal promotes biodiversity in a real-life game of rock-paper-scissors. Nature 418:171–174.

Law, R. 1999. Theoretical aspects of community assembly. Pages 143–171 *in* J. M. McGlade, ed. *Advanced ecological theory.* Blackwell Science, Oxford, UK.

Law, R. and R. D. Morton. 1996. Permanence and the assembly of ecological communities. Ecology 74: 1347–1361.

Law, R., A. J. Weatherby and P. H. Warren. 2000. On the invasibility of persistent protist communities. Oikos 88:319–326.

Leibold, M. A. 1997. Biodiversity and nutrient enrichment in pond plankton communities. Evolutionary Ecology Research 1:73–95.

Leibold, M. A., M. Holyoak, N. Mouquet, P. Amarasekare, J. M. Chase, M. F. Hoopes, R. D. Holt, J. B. Shurin, R. Law, D. Tilman, M. Loreau, and A. Gonzalez, A. 2004. The metacommunity concept: A framework for multi-scale community ecology. Ecology Letters 7:601–613.

Levins, R. 1969. Some demographic and genetic consequences of environmental heterogeneity for biological control. Bulletin of the Entomological Society of America 15:237–240.

Levins, R., and D. Culver. 1971. Regional coexistence of species and competition between rare species. Proceedings of the National Academy of Science, USA 68:1246–1248.

Lockwood, J. L., R. D. Powell, M. P. Nott and S. L. Pimm. 1997. Assembling ecological communities in time and space. Oikos 80:549–553.

May, R. M. and W. Leonard. 1975. Nonlinear aspects of competition between three species. SIAM Journal of Applied Mathematics 29:243–252.

Mithen, S. J. and J. H. Lawton. 1986. Food-web models that generate constant predator-prey ratios. Oecologia 69:542–550.

Morton, R. D. and R. Law. 1997. Regional species pools and the assembly of local ecological communities. Journal of theoretical Biology 187:321–331.

Mouquet, N., and M. Loreau. 2002. Coexistence in metacommunities: The regional similarity hypothesis. American Naturalist 159:420–426.

Paquin, C. E., and J. Adams. 1983. Relative fitness can decrease in evolving asexual populations of S. cerevisiae. Nature 306:368–371.

Pimm, S. L. 1982. Food webs. Chapman and Hall, London, UK.

Post, W. M. and S. L. Pimm. 1983. Community assembly and food web stability. Mathematical Biosciences 64:169–192.

Ricklefs, R. E. 1987. Community diversity: Relative roles of local and regional processes. Science 235: 167–171.

Rummel, J. D. and J. Roughgarden. 1985. A theory of faunal build up for competition communities. Evolution 39:1009–1033.

Shurin, J. B. and E. G. Allen. 2001. Effects of competition, predation and dispersal on local and regional species richness. American Naturalist 158:624–637.

Shurin, J. B., J. E. Havel, M. A. Leibold and B. Pinel-Alloul. 2000. Local and regional zooplankton species richness: A scale-independent test for saturation. Ecology 81:3062–3073.

Steiner, C. F. and M. A. Leibold. 2004. Cyclic assembly trajectories and the generation of scale-dependent productivity-biodiversity relationships. Ecology 85:107–113.

Tilman, D. 1994. Competition and diversity in spatially structured habitats. Ecology 75:2–16.

Warren, P. H., Law, R. and A. Weatherby. 2003. Mapping the assembly of protist communities in microcosms. Ecology 84:1001–1011.

Weatherby, A. J., P. H. Warren and R. Law. 1998. Coexistence and collapse: An experimental investigation of the persistent communities of a protist species pool. Journal of Animal Ecology 67:554–566.

Wilson, D. S. 1992. Complex interactions in metacommunities, with implications for biodiversity and higher levels of selection. Ecology 73:1984–2000.

Zimmermann, C. R., T. Fukami and J. A. Drake. 2003. An experimentally-derived map of community assembly space. In A. Minai and Y. Bar-Yam, eds. Unifying themes in complex systems II. Proceedings of the second international conference on complex systems. New England Complex Systems Institute, Perseus Press.

CHAPTER 12

Scale Transition Theory for Understanding Mechanisms in Metacommunities

Peter Chesson, Megan J. Donahue, Brett A. Melbourne,
and Anna L. W. Sears

Introduction

Ecological findings on small spatial or temporal scales often do not extrapolate to larger scales (Wiens 1989; Kareiva 1990; Horne and Schneider 1995; Englund and Cooper 2003). This empirical pattern has a parallel in ecological theory. In many model systems, the rules devised for population dynamics on small spatial or temporal scales can be quite different from those that emerge on large scales (Levin 1992). For example, in models of host-parasitoid systems, the highly unstable Nicholson-Bailey model might describe local population dynamics, but combined with spatial environmental variation, these local-scale instabilities may give rise to dynamics on the larger spatial scale described by the potentially highly stable host-parasitoid models of Bailey et al. (1962) and May (1978). Indeed, theoretical examples abound where the rules assumed on a small scale lead to contrasting outcomes on a larger scale, including changes in stability properties of single species models, conversion of competitive exclusion into competitive coexistence, and more mundane cases simply involving changes in quantitative features of models, such as mean densities (Chesson 2001).

These results from theory are interesting for two reasons. First, they may explain at least some failures of small-scale experimental findings to translate to larger scales. Second, they may provide a means of extrapolating, or scaling-up, findings for small scales to larger scales. The metacommunity concept owes its importance primarily to these issues of changes in rules with changes in scale, as exemplified by early metacommunity models (Slatkin 1974; Hastings 1977, 1980; see review by Mouquet et al., chapter 10; Caswell 1978).

A complete ecological theory should explain community dynamics on all scales (Levin 1992). Critical interactions between individual organisms occur over a range of scales (Chesson 1998a; Murrell and Law 2003), but these scales are mostly small. Scaling up theoretical findings from small scales is therefore essential. In classical (Hanski 1999), or equivalently, strict metacommunity models (Chesson 2001), it is assumed that interactions between individuals take place within local communities, but also that local communities do not persist. It follows that the system as a whole cannot be understood without reference to the larger scale of many local communities, which we refer to here as the regional or metacommunity scale. This strict metacommunity scenario represents one extreme of spatial structure that pervades ecological systems. In spatially-structured systems generally, the implications of local-scale interactions are necessarily modified by dispersal between localities in different states.

Scale transition theory (Chesson 1998a) is an attempt to systematize the study of how local-scale dynamical rules become modified as larger scales are taken into account. It unites analytical, simulation, and empirical approaches to an ecological question through key explanatory quantities accessible in all three approaches. The broad outline of scale transition theory was formulated many years ago (Chesson 1978, 1981), but rising theoretical and empirical interest in scale-dependent phenomena implies broad potential for its use. In essence, scale transition theory explains changes in dynamical rules with changes in scale in terms of interactions between spatial or temporal variation and nonlinear dynamics on local spatial scales. Many ecological processes lead to nonlinear dynamics including intra- and interspecific competition, predation, and mutualism. At the same time, spatial heterogeneity is a pervasive feature of ecological systems. It should therefore not be surprising that their combined effects are important.

The scale transition approach describes local ecological interactions mathematically in order to characterize their nonlinear nature. It then scales-up by averaging the formulae describing local interactions over space. Regional dynamics are then described by the *mean field*, that is, regional dynamics in the absence of spatial variation, plus the *scale transition*, which quantifies how spatial variances and covariances of population densities and environmental factors interact with local nonlinearities to modify regional dynamics.

Traditional approaches to metapopulations emphasize colonization and extinction dynamics of local populations connected by dispersal, with limited attention to the details of local community dynamics (Nee et al. 1997, Mouquet et al., chapter 10). Scale transition theory provides an alternative framework to understanding metapopulations and metacommunities in which the details of local dynamics are central. Scale transition theory can be applied to systems in which colonization and extinction are important features of local communities, but it also applies broadly to spatially structured systems regardless of the importance of local extinction and recolonization. Thus, scale transition theory avoids the criticism of traditional metapopulation models that the basic model structure is too limited to apply to many systems in nature (Harrison and Taylor 1997).

To introduce the basic concepts of scale transition theory, we open with a discussion of how spatial variation in parasitism can stabilize host-parasitoid interactions. This example illustrates a fundamental formula for discrete-time models whereby average individual fitness over a spatial region is expressed in terms of a mean-field component, interactions between nonlinear fitness and spatial variance, and covariation between fitness and density. Next, we show how this formula applies to patch models, which we illustrate using a model with local logistic dynamics, and discuss the role of dispersal in generating the spatial variances and covariances critical to the scale transition. We then extend the discrete-time formulation to continuous time and use it to understand coexistence of competitors in spatial Lotka-Volterra and related models. Our final model development illustrates the full power of scale transition concepts in the study of coexistence of annual plant species in a spatially varying environment. In conclusion, we relate the scale-transition approach to other developments of spatial dynamics as applied to metacommunities. In chapter 13, Melbourne et al. apply the results presented here to three empirical systems.

General Discrete-Time Formulation of the Scale Transition
for Variation in Space

An individual at some location x in space has a fitness λ_x, which measures the average contribution of an individual at x to the population after some defined interval of time. Space can be continuous or discrete. Thus, x might represent a patch, a point in continuous space, or a point on a discrete lattice. Time is assumed discrete. Continuous-time formulations are discussed later and replace λ_x with a per capita growth rate.

We use the example of host-parasitoid models, as presented in box 12.1, to introduce the fundamentals of scale transition theory. Host-parasitoid models have had a key role in the many decades of theoretical research that attempted to explain the persistence of exploiter-victim interactions (Hassell 2000; Murdoch et al. 2003). In host-parasitoid models, the primary factor affecting host fitness is the spatially-varying parasitoid density, P_x, and so $\lambda_x = f(P_x)$. Under Nicholson-Bailey assumptions for local dynamics, f is an exponentially declining nonlinear function, as illustrated by the solid line in figure 12.1. If the parasitoid were homogeneously distributed in space, this exponential decline in host fitness with parasitoid density would lead to violent fluctuations in regional-scale host and parasitoid densities, matching the local scale dynamics. In contrast, with enough spatial variation in parasitoid density, these violent fluctuations are replaced by convergence of the regional population densities on a stable equilibrium point, as discussed in box 12.1.

The Effects of Spatial Variation and Nonlinearity on Spatial Average Fitness

To understand how stability results from spatial variation in the host-parasitoid model, note that the spatial averages, $\bar{N}$ and $\bar{P}$, of host and parasitoid densities are the regional-scale densities. We then ask how the spatial average, $\bar{\lambda}$, of λ_x depends on these regional-scale densities. Particular formulae are given in box 12.1, but figure 12.1 gives a graphical illustration for the case where P_x varies between just two values equal to a fraction of $\bar{P}$ and a multiple of $\bar{P}$, which might occur if the

Box 12.1. Spatial variation in host-parasitoid systems

The traditional discrete-time Nicholson-Bailey host-parasitoid model takes the form

$$N_{t+1} = Re^{-aP_t}N_t$$
$$P_{t+1} = N_t - R^{-1}N_{t+1}$$

(12.B1.1)

where N and P are respectively host and parasitoid density, R is fitness of unparasitized hosts, and a is the average rate at which individual parasitoids discover hosts (Hassell 2000). The negative exponential above, $\exp(-aP_t)$, is the fraction of hosts escaping parasitism. The second line of the equation simply says that the number of parasitoids in the next generation is equal to the number of hosts killed by them in the current generation. It is more commonly written $P_{t+1} = [1 - \exp(-aP_t)]N_t$, but our form is more general, and being linear in densities, does not change with spatial scale.

This model is notoriously unstable and much theoretical research has been directed toward modifications that predict stable coexistence of host and parasitoid. Of the many such modifications (Hassell 2000), we are concerned here only with those relating to spatial variation in the density of parasitoids. These modifications lead to changes in the first line of equation 12.B1.1, but not the more general second line.

The first line of equation 12.B1.1 implies that host fitness is $\lambda_t = Rexp(-aP_t)$. We can modify this equation by having the parasitoid density vary in space, denoting by P_x the density of parasitoids at location x. This density will change with time, but for simplicity of notation, a subscript indicating time is omitted. Host fitness as a function of local parasitoid density is then

$$\lambda_x = Re^{-aP_x}.$$

(12.B1.2)

Here R is the host fitness in the absence of parasitism, P_x is the local parasitoid density, and a is the average rate at which individual parasitoids discover hosts.

Scale Transition for Independent Host and Parasitoid Distributions

May (1978) pointed out that when P_x is distributed in space according a gamma distribution, the average of λ_x over space takes the form

$$\bar{\lambda}_t = R(1 + a\bar{P}_t/k)^{-k}$$

(12.B1.3)

where $\bar{P}_t$ is the average of P_x over space in year t, and $1/k$ is equal to the square of the coefficient of variation (CV) of the parasitoid distribution in space. When the hosts and parasitoids are distributed independently of one another, then, as explained in the text, the first line of equation 12.B1.1 can be validly replaced by

$$\overline{N}_{t+1} = \overline{\lambda}_t \overline{N}_t. \tag{12.B1.4}$$

The model then predicts stable host and parasitoid coexistence provided the parasitoid CV is greater than 1.

For arbitrary spatial distributions of parasitoids, equation 12.B1.3 is replaced by

$$\overline{\lambda}_t = F\varphi(a\overline{P}_t), \tag{12.B1.5}$$

(Chesson and Murdoch 1986) where φ is a function dependent on the relative distribution of parasitoids in space (the distribution of $P_x/\overline{P}$) and is called the Laplace transform in statistical theory. (Figure 12.1 provides a geometrical construction of φ for the simple case where parasitoid densities take on only two values.) The chief finding for the general case given by equation 12.B1.5 is that the CV criterion for stability applicable to gamma distributed hosts remains approximately applicable in this more general case (Hassell et al. 1991).

Scale Transition for Correlated Host and Parasitoid Distributions

Several implicit assumptions are involved in the above conclusions. First, as explained in the text, equation 12.B1.4 is only valid when the fraction of hosts parasitized is uncorrelated with host density. In other cases, the applicable equation for population dynamics is

$$\overline{N}_{t+1} = \tilde{\lambda}_t \overline{N}_t \tag{12.B1.6}$$

where $\tilde{\lambda}_L$ is host individual average fitness, rather than host spatial average fitness ($\overline{\lambda}$). Second the distribution of $P_x/\overline{P}$ is assumed constant over time, which might not be true depending on the dispersal scenarios. The conclusions above about stability, however, continue to apply to the CV of the distribution of parasitoids per host, provided this CV does not vary importantly over time (Hassell et al. 1991). Equations 12.B1.3 and 12.B1.5 are valid for $\tilde{\lambda}$ replacing $\overline{\lambda}$ when the probability distribution of P_x is the per host distribution (Chesson and Murdoch 1986), but if the distribution of $P_x/\overline{P}$ over hosts varies with time, then k and φ must be given subscripts t to indicate that variation. How the distribution of $P_x/\overline{P}$ might vary over time depends on the nature of dispersal, which we have not addressed explicitly here.

Scale Transition Approximations

We can use formulae 12.6–12.8 in the text to obtain approximations to $\tilde{\lambda}$. As $f(P_x) = Re^{-aP_x}$, $f'(\bar{P}) = -Rae^{-a\bar{P}}$ and $f''(\bar{P}) = Ra^2e^{-a\bar{P}}$. Hence, $\tilde{\lambda}$ is approximated as

$$\tilde{\lambda} \approx Re^{-a\bar{P}} + \frac{1}{2}Ra^2e^{-a\bar{P}}\,\mathrm{Var}(P), \qquad (12.B1.7)$$

and the fitness-density covariance is approximately,

$$\mathrm{Cov}(\lambda, v) \approx -Rae^{-a\bar{P}}\,\mathrm{Cov}(P, N)/\bar{N}. \qquad (12.B1.8)$$

These combine to give an overall value of $\tilde{\lambda}$ equal to

$$\tilde{\lambda} \approx Re^{-a\bar{P}}\left\{1 + \frac{1}{2}a^2\,\mathrm{Var}(P) - a\,\mathrm{Cov}(P, N)/\bar{N}\right\}. \qquad (12.B1.9)$$

These approximations serve to show the moderating effect of variation in parasitism on the relationship between $\tilde{\lambda}$ and $\bar{P}$, and how this relationship is affected by covariance between parasitoids and host density. Approximations like these, coupled with the corresponding approximations for $d\tilde{\lambda}/d\bar{P}$, allow the stability rule discussed here to be derived for cases where R is near 1, that is, when host fitness in the absence of parasitism is not too large (Hassell et al. 1991).

system consisted of just two patches. The dashed curve gives the relationship between $\tilde{\lambda}$ and $\bar{P}$ that is traced out as $\bar{P}$ varies. The diamonds illustrate how this curve is calculated. The diamonds lie at the midpoints of line segments connecting (P_x, λ_x) points for pairs of low and high parasitoid localities. The midpoint of a line segment is necessarily the simple average of the endpoints of the same line segment. Because the function $f(P_x)$ curves up (is concave up; see box 12.2), the midpoints of the line segments are necessarily above the original curve. Thus, $\tilde{\lambda} > f(\bar{P})$, that is, spatial average fitness ($\tilde{\lambda}$, which equals $\bar{f(P)}$) exceeds that predicted by the local relationship between fitness and parasitoid density ($f(\bar{P})$). Of most importance for regional stability, the decline in $\tilde{\lambda}$ as a function of $\bar{P}$ is much more moderate than the decline in λ_x as a function of P_x.

This effect of variation in parasitoid density on mean fitness is a particular illustration of the phenomenon of nonlinear averaging (box 12.2) where a nonlinear function of some variable quantity is averaged, and the result differs from the nonlinear function of the average of that variable quantity. Here $\lambda_x = f(P_x)$, but $\tilde{\lambda} \neq f(\bar{P})$. When the nonlinear function f curves in a consistent direction (i.e., is concave up or concave down), the direction of the deviation between the average

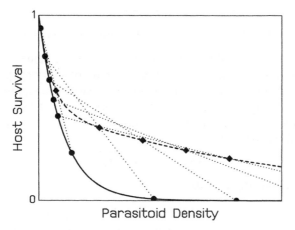

Figure 12.1 Decline in host density as a function of parasitoid density, locally in space (solid line: λ_x = Fe^{-aP_x}), and regionally (dashed line: $\bar{\lambda} = \overline{Fe^{-aP}} = F\varphi(a\bar{P})$) for the special case of variation in parasitoid density in space between just two values of P_x, one 5% of the mean and other 195% of the mean (i.e., $P_x = 0.05\,\bar{P}$ and $P_x = 1.95\,\bar{P}$). Note that the decline in host survival as parasitoid average density increases is much more moderate in the presence of spatial heterogeneity (dashed line) than in its absence (solid line). The ◆'s are at particular $(\bar{P}, \bar{\lambda})$ pairs illustrating the effect of Jensen's inequality on the relationship between $\bar{\lambda}$ and $\bar{P}$. The average of the two (P_x, λ_x) at the ●'s on each dotted line equals the particular $(\bar{P}, \bar{\lambda})$ pair at the ◆ in the center of the line. Because of the curvature of λ_x as a function of P_x, these averages necessarily lie above the values that would be given by the local λ-P relationship.

of the function $\overline{f(P)}$ and the function of the average $f(\bar{P})$ is predictable and is known as Jensen's inequality (box 12.2).

The Importance of the Fitness-Density Covariance and Individual Average Fitness

Given the more moderate behavior of the spatial average fitness as a function of average parasitoid density in the host-parasitoid model, it is not surprising that the regional-scale dynamics are more stable in the presence of spatial variation. To draw firm conclusions, however, we must determine the extent to which $\bar{\lambda}$ determines regional-scale population dynamics. If λ_x is fitness for the period t to $t + 1$, the $N_{x,t}$ individuals at location x and time t, give rise by reproduction and survival to $\lambda_x N_{x,t}$ individuals at time $t + 1$. If some of these disperse, then $N_{x,t+1} \neq \lambda_x N_{x,t}$, but assuming dispersers do not leave the region, the total density in the system at time $t + 1$ is nevertheless

$$\sum_x N_{x,t+1} = \sum_x \lambda_x N_{x,t}.$$

(12.1)

Dividing equation 12.1 by the number of local sites x in the system now shows

Box 12.2   Properties of Nonlinear Averages

Basic to the scale transition is the idea of nonlinear averaging, that is, taking an average of some nonlinear function $f(W)$ of a varying quantity W (Chesson 1998a). Most introductions to statistics give some appreciation of nonlinear averages from the two different formulae for the variance (see, e.g. Ross 1997). We know that the variance $\text{Var}(W)$ is defined as the average squared deviation from the mean:

$$\text{Var}(W) = \overline{(W - \overline{W})^2}, \tag{12.B2.1}$$

but it also obeys the formula

$$\text{Var}(W) = \overline{W^2} - (\overline{W})^2, \tag{12.B2.2}$$

that is, the variance of W is the difference between the mean of the square of W and the square of the mean of W. This formula rearranges to show that the mean of the square of W exceeds the square of the mean of W by an amount exactly equal to the variance:

$$\overline{W^2} = (\overline{W})^2 + \text{Var}(W). \tag{12.B2.3}$$

This finding is a particular case of the result that the mean, $\overline{f(W)}$, of a nonlinear function, generally differs from the function of the mean, $f(\overline{W})$; here $f(W) = W^2$.

Jensen's Inequality

Unfortunately, only in special cases such as $f(W) = W^2$ do we know exactly how much the mean of the function and the function of the mean differ from each other. However, we can tell the direction of the difference if the function curves in a consistent direction, that is, it is concave, either up or down. If $f(W)$ is concave up, as indicated by $f''(W) > 0$, Jensen's inequality (Ross 1997) says that

$$\overline{f(W)} > f(\overline{W}), \tag{12.B2.4}$$

that is, the mean of the function is greater than the function of the mean. Naturally, for concave down functions ($f''(W) < 0$), the reverse inequality to inequality 12.B2.4 applies. For the particular case where $f(W) = W^2$, equation 12.B2.3 tells us that the amount by which $\overline{f(W)}$ exceeds $f(\overline{W})$ is simply the variance. Another example of an exact formula for $\overline{f(W)}$ is given in box 12.1 for the host-parasitoid. Figures 12.1 and 12.2 give graphical illustrations of Jensen's inequality, which are available for the simple case where W (respectively P and N in these figures) vary over just two values.

Quadratic Approximations

Although the exact difference between $\overline{f(W)}$ and $f(\overline{W})$ is not often known, the following approximation

$$\overline{f(W)} \approx f(\overline{W}) + \frac{1}{2} f''(\overline{W})\text{Var}(W), \qquad (12.\text{B}2.5)$$

holds generally whenever $f(W)$ can be satisfactorily approximated by its second order Taylor expansion, and can be derived simply from the special case, equation 12.B2.3 (Chesson 1998a).

Multidimensional Nonlinearities

Ecological models often involve multiple quantities varying simultaneously. With two varying quantities, U and W, we need to understand averages, $\overline{f(U, W)}$, of functions nonlinear jointly in these varying quantities. The simplest two-variable nonlinear function with a joint nonlinearity is the product $f(U, W) = UW$. A simple extension of the variance formula (12.B2.3) leads to the relationship

$$\overline{UW} = \overline{U} \cdot \overline{W} + \text{Cov}(U, W), \qquad (12.\text{B}2.6)$$

that is, the mean of the product differs from the product of the means by an amount equal to the covariance of the two quantities (Ross 1997).

Special Nonlinear Forms in Spatial Models

In spatial models, we encounter nonlinear functions of two variables of the form $f(W)U$, where $f(W)$ is a nonlinear function of W separately. Application of formula 12.B2.6 shows that

$$\overline{f(W)U} = \overline{f(W)} \cdot \overline{U} + \text{Cov}(f(W), U). \qquad (12.\text{B}2.7)$$

This covariance in this formula can be approximated by the formula

$$\text{Cov}(f(W), U) \approx f'(\overline{W})\text{Cov}(W, U) \qquad (12.\text{B}2.8)$$

to the same order of accuracy as the approximation 12.B2.5 by linearly approximating $f(W)$ as $f(\overline{W}) + f'(\overline{W})(W - \overline{W})$, using standard statistical arguments (Rao 1973). Substituting the approximation 12.B2.5 for $\overline{f(W)}$, we get the overall approximation

$$\overline{f(W)U} \approx \left\{ f(\overline{W}) + \frac{1}{2} f''(\overline{W})\text{Var}(W) \right\} \overline{U} + f'(\overline{W})\text{Cov}(W, U). \quad (12.\text{B}2.9)$$

Nonlinearities in Patch Models

In the text, under the heading of patch models, we use expression 12.B2.9 in the special case where $N = W = U$, and so $\text{Cov}(W, U) = \text{Var}(N)$. Thus, expression 12.B2.9 implies

$$\overline{f(N)N} \approx \left\{ f(\overline{N}) + \frac{1}{2} f''(\overline{N})\text{Var}(N) \right\} \overline{N} + f'(\overline{N})\text{Var}(N)$$

$$= \left\{ f(\overline{N}) + \frac{1}{2} f''(\overline{N})\text{Var}(N) + f'(\overline{N})\text{Var}(N)/\overline{N} \right\} \overline{N}. \quad (12.\text{B2.10})$$

But applying approximation 12.B2.5 to the function $F(N) = f(N)N$, we see also that

$$\overline{F(N)} \approx F(\overline{N}) + \frac{1}{2} F''(\overline{N})\text{Var}(N). \quad (12.\text{B2.11})$$

However, as $F''(N) = f''(N)N = 2f'(N)$, we see that expressions 12.B2.10 and 12.B2.11 are in fact identical.

Checking Approximations

The approximations given here and used in the text work best when spatial variation is small in magnitude. Thus, they give the initial trends in the scale transition as spatial variation is introduced to a system. For any particular application with large spatial variation, it is necessary to check the accuracy of these approximations by some means. It is not difficult to calculate nonlinear averages numerically when the probability distributions for spatial variation are known, and thus check these formulae. When these distributions are not known, simulation or other numerical methods are needed to determine them.

---

that the dynamics of the regional-scale population density $\overline{N}$ are given by the equation

$$\overline{N}_{t+1} = \overline{\lambda N_t} = \tilde{\lambda}\overline{N}_t, \quad (12.2)$$

where $\tilde{\lambda}$ is the average of λ_x over all individuals in the population, and is given by the formula $\tilde{\lambda} = \overline{\lambda N_t}/\overline{N}_t = \sum_x \lambda_x N_{t,x}/\sum_x N_{t,x}$. Of most importance, average individual fitness determines regional-scale population dynamics, because equation 12.2 says $\overline{N}_{t+1} = \tilde{\lambda}\overline{N}_t$.

To relate average individual fitness, $\tilde{\lambda}$, to the spatial average fitness, $\overline{\lambda}$, we define the relative density at location x as $v_x = N_x/\overline{N}_t$, that is, the density at location x compared with the average density in the system. Then we see that $\tilde{\lambda}$ can be expressed as

$$\tilde{\lambda} = \overline{\lambda v}, \tag{12.3}$$

the spatial average of $\lambda_x v_x$. Expression 12.3 is the average of a product, which is therefore equal to the product of the average, $\bar{\lambda} \cdot \bar{v}$, plus the covariance, $\mathrm{Cov}(\lambda, v)$, between them (box 12.2). As $\bar{v}$ is necessarily 1, this means that

$$\tilde{\lambda} = \overline{\lambda v} = \bar{\lambda} + \mathrm{Cov}(\lambda, v), \tag{12.4}$$

(Chesson 2000a). Thus, we see that $\tilde{\lambda}$ differs from $\bar{\lambda}$ by the spatial covariance, $\mathrm{Cov}(\lambda, v)$, between fitness of an individual at location x and the relative density there, which is called the fitness-density covariance (or growth-density covariance in Snyder and Chesson 2003).

For any system, equation 12.4 says that the spatial average fitness $\tilde{\lambda}$ is equal to the average individual fitness $\bar{\lambda}$ when the fitness-density covariance is zero. For the host-parasitoid model, the fitness-density covariance is zero whenever host density is uncorrelated with the incidence of parasitism, which is in fact not uncommon in nature (Pacala and Hassell 1991). In such cases, the conclusions above about $\bar{\lambda}$ are also conclusions about $\tilde{\lambda}$; that is, sufficient spatial variation in P_x stabilizes the host-parasitoid interaction (box 12.1). In cases where the fitness-density covariance is not zero, it is still possible for spatial variation to be stabilizing (box 12.1), but the relevant calculations have to include this covariance.

To generalize these ideas, fitness can be simply a nonlinear function of some factor W_x determined by the point x in space:

$$\lambda_x = f(W_x). \tag{12.5}$$

Ideally, W_x is some quantity whose spatial average is meaningful. In the host-parasitoid example, where $W_x = P_x$, the spatial average is simply parasitoid density on the larger spatial scale, which is both perfectly meaningful and useful. Such utility need not apply to other fitness factors; for example, choosing $W_x = \ln P_x$ would not help, because the spatial average of $\ln P_x$ is not meaningful.

General Scale Transition Formulae

For general fitness factors, a simple and standard approximation to Jensen's inequality allows us to see just how the scale transition works (Chesson 1998b; box 12.2). Here, equation 12.5 means that

$$\bar{\lambda} = \overline{f(W)} \approx f(\bar{W}) + \frac{1}{2} f''(\bar{W}) \mathrm{Var}(W) \tag{12.6}$$

Thus, the relationship of average fitness to average W differs from the relationship that applies locally in space by an amount that is approximately proportional to the variance $\mathrm{Var}(W)$ of the fitness factor W. The proportionality constant $1/2\, f''(\bar{W})$ is a measure of the nonlinearity of the local-scale relationship between fitness and W_x at the mean of W_x.

To obtain the relationship between average individual fitness $\tilde{\lambda}$ and $\overline{W}$, we must add the fitness-density covariance to equation 12.6. As explained in box 12.2, this covariance can be approximated as

$$\text{Cov}(f(W), v) \approx f'(\overline{W})\text{Cov}(W, v). \tag{12.7}$$

Combining equations 12.6 and 12.7, we obtain a general purpose approximation to $\tilde{\lambda}$:

$$\tilde{\lambda} \approx f(\overline{W}) + \frac{1}{2}f''(\overline{W})\text{Var}(W) + f'(\overline{W})\text{Cov}(W, v). \tag{12.8}$$

The first term, $f(\overline{W})$, is simply the nonspatial formula for λ, and is commonly referred to in the literature as the "mean field." We define the "scale transition" to be the difference $\tilde{\lambda} - f(\overline{W})$, which is expressed here as the effects of variance on local nonlinear dynamics, $1/2\ f''(\overline{W})\text{Var}(W)$, and fitness-density covariance, $f'(\overline{W})\text{Cov}(W, v)$.

In box 12.1, this formula is applied to the host-parasitoid model to give a simple understanding of how $\tilde{\lambda}$ is changed by various spatial relationships. It applies generally, however, to spatial models represented in discrete time, whether they are spatially explicit, spatially implicit, patch models, lattice models, or continuous space models. Moreover, it does not matter whether population sizes, N_x, are represented as continuous variables, or are discrete. In continuous-space discrete-N_x models, for example, N_x is zero at most points in space, and takes the value one at places where an individual is present. In continuous space models, and in many lattice models, W_x is not simply a function of population densities and environmental factors at the point x, but instead depends on conditions around x expressed as a summary of the densities and environmental factors in a neighborhood centered at x (Neuhauser and Pacala 1999), or as averages of these factors weighted inversely with distance from x using functions called competition kernels (Pacala and Silander 1985). It should be noted, however, that approximations like those above work best when spatial variation in W_x is small, and so should be regarded primarily as giving the initial trends for the scale transition as spatial variation is increased from small values (box 12.2).

Patch Models

In patch-model approaches to spatial dynamics, the location of an individual has the resolution of a patch, and the dynamical equations are formulated in terms of the output of a patch, $\lambda_x N_x$. In the simplest models this output is just a function of N_x:

$$\lambda_x N_x = F(N_x). \tag{12.9}$$

Previous discussions of the scale transition have emphasized patch models pointing out that

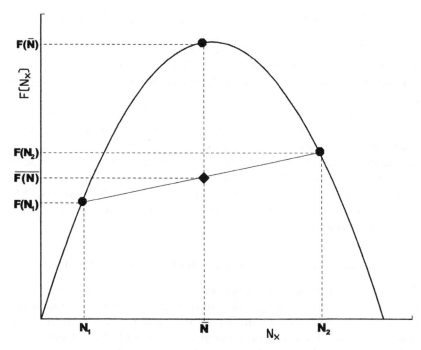

Figure 12.2 Scale transition for a patch model with $\lambda_x N_x = F(N_x)$, and density varying between two values with equal frequency, showing the construction of one point (◆) on the relationship between $\overline{F(N)}$ and $\overline{N}$. Note that the regional-scale output, $\overline{N}_{t+1} = \overline{F(N_t)} = \hat{\lambda} \overline{N}_t$, lies below the output given by the local relationship, $F(\overline{N}_t)$. The particular curve represented here is the logistic.

$$\overline{N}_{t+1} = \overline{\lambda N} \approx F(\overline{N}_t) + \frac{1}{2} F''(\overline{N}_t) \mathrm{Var}(N). \qquad (12.10)$$

Figure 12.2 shows how this works when N_x varies between just two values, for example if the regional scale consists of just two local-scale patches.

The curve in figure 12.2 is the logistic equation, that is, $F(N) = N[1 + r(1 - N/K)]$, for which $F''(\overline{N}) = -2r/K$. From equation 12.10, we obtain an equation for population dynamics in the form

$$\overline{N}_{t+1} = \overline{N}_t \left\{ 1 + r \left(1 - \frac{\overline{N}_t}{K} \right) \right\} - \frac{r}{K} \mathrm{Var}(N), \qquad (12.11)$$

which is exact because the logistic equation is quadratic and so approximation (12.10) is exact. In box 12.2 it is shown that these patch-model approximations for the scale transition are special cases of the previous approximations focusing on $\tilde{\lambda}$. Note that here $F(N) = f(N)N$, $W = N$, $\mathrm{Var}(W) = \mathrm{Var}(N) = \mathrm{Cov}(W, N) = \overline{N}\mathrm{Cov}(W, v)$, and so equation 12.8 for $\tilde{\lambda}$ implies

$$\overline{N}_{t+1} = \hat{\lambda}\overline{N}_t \approx f(\overline{N}_t)\overline{N}_t + \frac{1}{2}f''(\overline{N}_t)\text{Var}(N)\overline{N}_t + f'(\overline{N}_t)\text{Var}(N), \quad (12.12)$$

where the first term of equation 12.12 is the mean field, and the second two terms are scale transition terms summing to $1/2$ $F''(\overline{N})\text{Var}(N)$ but dividing this quantity into the effect of variance in nonlinear fitness and the effect of fitness-density covariance.

For the logistic, the fitness function f is linear and so $f''(\overline{N}) = 0$. Thus, in the logistic all the effects of the scale transition come from the fitness-density covariance. Substituting the logistic formula $f'(\overline{N}) = -r/K$ into equation 12.12 reproduces the previous patch-model formula (equation 12.11) exactly. Alternatives to the logistic, such as the Ricker model (Chesson 1998a), and the Ayala-Gilpin θ-logistic model (Chesson 1991b) have nonlinear fitness functions, that is, $f''(\overline{N}) \neq 0$, and all terms of the regional equation 12.12 are nonzero with these models for local population dynamics.

Spatial Variance, Population Dynamics, and Dispersal

The development above omits explicit consideration of dispersal. By focusing on λ_x, or the output of a patch, $\lambda_x N_x$, the equations are correct for any mode of dispersal provided any mortality that occurs during dispersal is factored into λ_x (Chesson 1998a). The effects of dispersal are expressed by the relevant spatial variances and covariances, which we have left unspecified. To allow population dynamics to be fully understood, the way these variances and covariances change over time, and especially how they depend on mean population densities, needs to be known. Determining such relationships is the subject of moment closure approximation techniques (Bolker and Pacala 1997), and can be difficult and highly complex because these relationships depend not only on dispersal but also on local dynamics and environmental variation. However, in certain cases, these relationships are simply determined (Chesson 1998a). For example, if most individuals disperse, and dispersal is over long distances with the physical environment determining where individuals concentrate, then $N_{x,t}$ can be modeled as

$$N_{x,t} = U_x \overline{N}_t, \quad (12.13)$$

where U_x varies in space (and possibly in time also) encoding features of the physical environment that influence dispersal to location x. Here the relative density, v_x, is simply U_x. With this model $\text{Var}(N) = \text{Var}(U)\overline{N}_t^2$, that is, the variance is proportional to the square of mean density, and so the coefficient of variation of density in space is constant over time, and determined by spatial variation in the physical environment. In contrast, when patchiness in space is determined by demographic stochasticity rather than environmental variation, long-distance dispersal implies that $\text{Var}(N)$ is proportional to $\overline{N}_t$ (Chesson 1998a).

Substituting the model (equation 12.13) for the variance into the logistic formula (equation 12.11), leads to regional-scale dynamics that are logistic like the local-scale dynamics. Spatial variance has the effect of strengthening the density dependence at the regional scale, which is revealed as a lowered carrying capacity: The new value of K equals the old value divided by $(1 + \text{Var}(U))$. The regional dynamics of the system are altered quantitatively by spatial variation, but the rules for regional dynamics do not differ qualitatively from the rules applicable to inputs and outputs of a patch at the local scale. However, when the linear fitness function, $f(N)$, of the logistic model is replaced by the more realistic nonlinear $f(N)$ of the Ricker model, regional-scale dynamics can be changed qualitatively by environmentally-dependent dispersal (Chesson 1998a). Although local scale dynamics may be chaotic in the Ricker model, environmentally-dependent dispersal can lead to highly stable regional dynamics. Similar, though much weaker effects, are possible from demographic stochasticity ($\text{Var}(N)$ proportional to $\overline{N}_t$), in both the logistic model (Chesson 1998b) and the Ricker model (Chesson 1998a).

Continuous Time

From the perspective of the scale transition, the distinction between discrete time and continuous time is not large. In continuous-time models, λ_x is replaced by the local per capita growth rate $r_x = (dN_x/dt)/N_x$ as the fitness measure, with average individual fitness now denoted $\tilde{r}$. The growth of a population at the regional scale then takes the form

$$\frac{d\overline{N}}{dt} = \overline{rN} = \tilde{r} \cdot \overline{N} = \{\tilde{r} + \text{Cov}(r, v)\}\overline{N}. \qquad (12.14)$$

In cases where r_x has the representation $r_x = f(W_x)$, we have the general quadratic approximation

$$\frac{d\overline{N}}{dt} \approx f(\overline{W})\overline{N} + \left\{\frac{1}{2}f''(\overline{W})\text{Var}(W) + f'(\overline{W})\text{Cov}(W, v)\right\}\overline{N}, \qquad (12.15)$$

which specializes to forms analogous to equations 12.10 and 12.12 in the appropriate patch-model circumstances. Here $f(\overline{W})\overline{N}$ is the mean field, and the remaining terms give the scale transition in terms of local nonlinearity and fitness-density covariance.

Lotka-Volterra Competition Models

A particular illustration of the formula (12.15) has been given by Bolker and Pacala (1999) for a continuous space Lotka-Volterra competition model. In their model, W_x represents the total of intraspecific and interspecific competition experienced by an individual at x, which is assumed to be linearly related to the densities of competitors in neighborhoods centered on the point x. Their model can be simplified to a patch model without loss of information for our purposes. Then W_{ix} can be defined as $\alpha_{ii}N_{ix} + \alpha_{ij}N_{jx}$, where α denotes competition coefficients in the form advocated in Chesson (2000b), and $r_{ix} = f_i(W_{ix}) = r_i(1 - W_{ix})$, where r_i is the intrinsic rate of increase for species i. Thus, the local dynamical equation in the absence of dispersal is

$$\frac{dN_i}{dt} = r_i(1 - W_i)N_i = r_i\{1 - \alpha_{ii}N_i - \alpha_{ij}N_j\} N_i. \qquad (12.16)$$

When expressed in this form, the criteria for species coexistence on the local scale are simply

$$\alpha_{jj}/\alpha_{ij} > 1, i \neq j, \qquad (12.17)$$

that is, local-scale intraspecific competition must exceed local-scale interspecific competition Chesson (2000b).

As f_i is linear here, equation 12.15 is exact and reduces to

$$\frac{d\overline{N}_i}{dt} = r_i\{1 - \overline{W}_i\}\overline{N}_i - r_i\{\text{Cov}(W_i, v_i)\}\overline{N}_i$$

$$= r_i\{1 - \alpha_{ii}\overline{N}_i - \alpha_{ij}\overline{N}_j\} \overline{N}_i - r_i\{\alpha_{ii}\text{Var}(N_i) + \alpha_{ij}\text{Cov}(N_i, N_j)\}. \qquad (12.18)$$

(Lloyd and White 1980 produced a similar expression using the concept of "mean crowding," which is an important precursor to scale transition concepts, but applies only to linear fitness functions.) In equation 12.18, the mean field is $r_i\{1 - \alpha_{ii}\overline{N}_i - \alpha_{ij}\overline{N}_j\} \overline{N}_i$, that is, the Lotka-Volterra equations (12.16) with regional variables substituted for local variables. The linearity of f_i means that the scale transition is simply the fitness-density covariance. Bolker and Pacala (1999) focus on spatial variation that arises from demographic stochasticity and spatially local dispersal, and uncover a variety of situations where the fitness-density covariance is responsible for species coexistence. For example, they note that a locally dispersing species with a high r_i can coexist with a superior competitor that lacks these traits even though the superior species excludes the other in the nonspatial case.

The most difficult feature of the Bolker and Pacala analysis is derivation of dynamical equations for the relevant spatial variances and covariances using moment closure approximation (see "Spatial Variance, Population Dynamics, and Dispersal" above). For a much simpler illustration of how coexistence may result from the fitness-density covariance terms in equation 12.18, assume that dispersal is rapid, and influenced by the physical environment so that N_{ix} is well approximated by $U_{ix}\overline{N}_i$, as in the previous section. Here the physical environmental effects encoded in U_{ix} may differ between species.

Defining $\sigma_{ii} = \text{Var}(U_{ix})$ and $\sigma_{ij} = \text{Cov}(U_{ix}, U_{jx})$ we have $\text{Var}(N_i) = \sigma_{ii}\overline{N}_i^2$, and $\text{Cov}(N_i, N_j) = \sigma_{ij}\overline{N}_i\overline{N}_j$. Substituting into equation 12.18, we obtain

$$\frac{d\overline{N}_i}{dt} = r_i\{1 - \alpha_{ii}(1 + \sigma_{ii})\overline{N}_i - \alpha_{ij}(1 + \sigma_{ij})\overline{N}_j\}\overline{N}_i. \tag{12.19}$$

Here the dynamics at the regional scale are again given by Lotka-Volterra equations, but the competition coefficients differ from the local scale coefficients: α_{ii} is replaced by $\alpha_{ii}(1 + \sigma_{ii})$ and α_{ij} is replaced by $\alpha_{ij}(1 + \sigma_{ij})$. Applying standard Lotka-Volterra coexistence criteria from Chesson (2000b), it follows that the species coexist if

$$\frac{\alpha_{jj}(1 + \sigma_{jj})}{\alpha_{ij}(1 + \sigma_{ij})} > 1, \text{ or equivalently } \frac{\alpha_{jj}}{\alpha_{ij}} > \frac{1 + \sigma_{ij}}{1 + \sigma_{jj}}, i \neq j. \tag{12.20}$$

The left hand inequality simply says that regional-scale intraspecific competition must exceed regional-scale interspecific competition. Variation in space modifies the regional-scale coefficients over the local coefficients, necessarily increasing regional-scale intraspecific competition, $\alpha_{jj}(1 + \sigma_{jj})$, because σ_{jj} is a variance and must be positive. However, regional-scale interspecific competition, $\alpha_{ij}(1 + \sigma_{ij})$, need not be increased because σ_{ij} can be negative or zero. Moreover, the covariance σ_{ij} equals $\rho\sqrt{\sigma_{ii}\sigma_{jj}}$, where ρ is the correlation in space between N_{ix} and N_{jx}.

A zero value of ρ means that the species use space independently, and the covariances are also zero. A ρ of 1, on the other hand, means identical use of space by the two species; and if the variances are equal, the covariances are the same as the variances. Then spatial variation terms cancel out of criterion 12.20 indicating for that case an absence of any effect of spatial variation on coexistence. However, identical use of space is unlikely, meaning that correlations in space are usually less than 1. Thus, whenever the variances σ_{jj} are similar for the two species (i.e., the species have similar coefficients of variation of density) the ratio $(1 + \sigma_{ij})/(1 + \sigma_{jj})$ is less than 1, and the criterion 12.20 is more easily satisfied. In particular, coexistence can occur at the region-scale even when local-scale interspecific competition exceeds local-scale intraspecific competition.

This example is just one illustration of how differences between species in the use of space can contribute to species coexistence. This possibility was first elucidated in spatial Lotka-Volterra models with rapid dispersal by Shigesada and Roughgarden (1982). Analogous to these spatial Lotka-Volterra competition models are discrete-time spatial models of insects competing for patchily distributed resources (Atkinson and Shorrocks 1981; Ives 1988). Coexistence is known to result in these models when intraspecific spatial aggregation is stronger than interspecific aggregation. Such spatial aggregation can be measured by variances and covariances analogous to σ_{ii} and σ_{ij} given here (Ives 1991). With few exceptions (Heard and Remer 1997; Remer and Heard 1998), the spatial coexistence mechanism in these models has been found to be equivalent to spatial niche differences (Green 1986; Chesson 1991; Hartley and Shorrocks 2002) or the spatial storage effect (Chesson 2000a).

However, spatial variation need not always promote coexistence in these sorts of models. For example, the criteria (12.20) imply that coexistence might be converted into competitive exclusion when species covary positively in space and have very different spatial variances. Moreover, in other models for dispersal, such as slow short-distance dispersal, heavily influenced by demographic stochasticity (Bolker and Pacala 1999), the Lotka-Volterra equations are not recovered at the regional level, which means that regional dynamics are qualitatively different from local dynamics. In particular, the regional-level competition coefficients are density-dependent, and the regional-level coexistence criteria are far more complex than those above.

Coexistence Mechanisms in a Spatially Variable Environment

Scale transition ideas have been developed most extensively in the formulation and analysis of mechanisms of competitive coexistence in a variable environment (Chesson 2000a; Snyder and Chesson 2003). In variable-environment competition theory, particular biological processes generate particular types of nonlinearity and define distinct classes of coexistence mechanism. Aspects of this theory are illustrated here using the annual plant model defined in table 12.1.

As defined by equation 12.T1.1, the fitness of an individual seed involves two pathways in the life cycle. First, in any given year a seed may remain dormant, and either survive to the next year or perish. Second, it may germinate, then potentially survive as a seedling, grow, experience competition, and produce seed, some of which may perish before the next year. Each of these events occurs with a probability or magnitude determined by the life-history parameters defined in the table.

Each of the life-history parameters in the table could vary spatially with the environment in ways that differ between species, defining species-specific responses to environmental conditions. Competition is also spatially variable, and its magnitude at location x is expressed in the model by the quantity C_x (the competitive response), which combines the effects of intraspecific and interspecific competition for any species. However, we assume that C_x is the same for each species, precluding the possibility of coexistence in the absence of variation in fitness in space or time. Although not varying with species, C_x varies in space because population densities vary in space as a consequence of spatial variation in life-history parameters. In addition, and more importantly for species coexistence, C_x may be spatially variable because it is directly a function of two life-history parameters that may vary in space. These life-history parameters are the germination fraction G, which affects the initial number of competing seedlings, and V, which expresses seedling survival and growth, and hence demand for resources by an individual seedling as a function of the local physical environmental conditions. Thus, variation in G and V, due to spatial variation in the environment, of necessity leads to

Table 12.1 Model of an annual plant community

$N_{jx}(t)$	The number of seeds in the seed bank of species j in patch x at the beginning of year t.
$\lambda_{jx}(t)N_{jx}(t)$	The amount of new seed produced by seeds that germinate, plus the survivors of the seeds that do not germinate by the end of year t for species j and patch x.

$$\lambda_{jx}(t) = s_j(1 - G_j) + \frac{U_jY_jV_jG_j}{C_x}, \qquad (12.\text{T}1.1)$$

with the following notation:

$\lambda_{jx}(t)$	The total output of a seed of species j at site x during year t (survival plus new production).
s_j	Survival of ungerminated seed.
$1 - G_j$	Fraction of seed not germinating.
U_j	Fraction of seed production of species j successfully incorporated into the seed bank by the beginning of the next year (seed not lost to predation, pathogens, or other means before being mixed into the soil surface).
Y_j	Seed production per surviving seedling of unit vigor in the absence of competition.
V_j	Seedling survival and vigor (size).
G_j	Fraction of seed of species j germinating.
a_j	Competitive effect of one seedling of species j.
$C_x = 1 + \sum_j a_j V_j G_j N_{jx}(t)$	Reduction in per capita seed output due to competition.
$f(E_{jx}, C_x) = s(1 - G) + UYE_{jx}G/C_x$	Example of f for the case where E_{jx} is V_j.
E^*, C^*	Reference values of E_{jx} and C_x chosen near their means (Chesson 2000a).
$\mathcal{E}_{jx} = f(E_{jx}, C^*) - 1$	Standard environmental response.
$\mathcal{C}_x = 1 - f(E^*, C_x)$	Standard competitive response.

the phenomenon of covariance between environment and competition (Chesson 2000a), which has a critical role in species coexistence, as we shall see.

To illustrate the behavior of the model under a variety of contrasting circumstances, we consider some simple extreme environmental scenarios (Chesson 1985). The first and simplest scenario is the case of pure spatiotemporal environmental variation where the environment fluctuates independently in space and time and, as a consequence, averages out to a constant in space and in time. Second is the case of pure spatial variation where the physical environment varies only spatially.

Several simplifying assumptions are introduced to facilitate analysis. First we consider spatial variation in just one parameter at a time, and we assume that the species have identical parameters except for those varying with the environment, which we term "environmental responses." Environmental responses are assumed here to differ between species in their means but not in their variances. Second, we assume that environmental responses vary independently between species over space or space and time. Third, we assume a patch model where plants interact with each other homogeneously within patches, but disperse between patches. Fourth, just two kinds of dispersal are considered: (a) widespread dispersal where all new seed is divided evenly between patches, and (b) widespread dispersal with local retention where a fixed fraction p of the new seed is retained in the natal patch. The number of patches will be assumed to be effectively infinite. Finally, we restrict attention to just two species. Much more general treatments of species coexistence in spatially variable environments are possible (Chesson 2000a), but require more detailed analysis than can be given here.

To analyze the model, we must first express the fitness of an individual in terms of spatially varying factors. Thus, using the equation 12.T1.1 in table 12.1, we write the fitness of an individual of any species j as a function, $f(E_{jx}, C_x)$, of the environmental response, which we denote generically as E_{jx}, and the competitive response, C_x. Then, to provide a unified treatment of models with different spatially varying life-history parameters, and indeed different basic models, these variables are transformed into the new variables, $\mathscr{E}_{jx}$ and $\mathscr{C}_x$, defined in table 12.1. These new variables are increasing functions respectively of E_{jx} and C_x, reflect the same underlying biological processes, and contain the same information. To define them (Chesson 2000a), one first chooses values E^* and C^* of E_{jx} and C_x near their means with the property that $\lambda_{jx} = 1$ when $(E_{jx}, C_x) = (E^*, C^*)$. Then $\mathscr{E}_{jx}$ is the change in λ_{jx} as E_{jx} moves away from E^* and $\mathscr{C}_x$ is the change in $-\lambda_{jx}$ as C_x moves away from C^*. These definitions mean that the effects of E_{jx} and C_x are translated into the separate effects of environment and competition on λ_{jx}. To obtain the joint effects of environment and competition on λ_{jx} all that is needed in addition to $\mathscr{E}_{jx}$ and $\mathscr{C}_x$ is the interaction between environment and competition. It follows that λ_{jx} is given in terms of $\mathscr{E}_{jx}$ and $\mathscr{C}_x$ by the generic equation,

$$\lambda_{jx} = 1 + \mathscr{E}_{jx} - \mathscr{C}_x + \gamma \mathscr{E}_{jx} \mathscr{C}_x, \qquad (12.21)$$

where the single parameter γ is negative and defines how strongly E_{jx} and C_x (equivalently, $\mathscr{E}_{jx}$ and $\mathscr{C}_x$) interact in their determination of λ_{jx} (Chesson 2000a). This equation is true regardless of which life history parameter is represented by E_{jx} and regardless of the form of competition.

A product such as $\mathscr{E}_{jx} \mathscr{C}_x$ in equation 12.21 is the mathematically simplest form of an interaction between two variables in their determination of a third variable. An interaction between two variables is also a special form of multidimensional nonlinearity. The particular interaction in equation (12.21) stems directly from the product of E_{jx} and $1/C_x$ that occurs in λ_{jx} when E_{jx} is any of the life-history parameters U, Y, V, or G defined in table 12.1. This interaction occurs because the product $UYVG$ determines how much new seed an individual seed would return to the seed bank the following year in the absence of competition, and this amount is assumed to be reduced by a given proportion for a given magnitude of competition. Thus, if two locations differ in the value of $UYVG$, the same magnitude of competition would have a larger effect at the location with the higher value of $UYVG$.

To understand regional-scale dynamics of this model we need $\tilde{\lambda}$, which of course splits into the spatial average fitness and the fitness-density covariance: $\tilde{\lambda}$ = $\bar{\lambda}$ + Cov(λ, v) (equation 12.4). For $\bar{\lambda}$, equation 12.21 must be averaged over space. As discussed in box 12.2, the product nonlinearity $\mathscr{E}_{jx}\mathscr{C}_x$ introduces a co-variance term into $\bar{\lambda}$, so that

$$\bar{\lambda} = 1 + \overline{\mathscr{E}}_j - \overline{\mathscr{C}} + \gamma\overline{\mathscr{E}}_j \cdot \overline{\mathscr{C}} + \gamma\mathrm{Cov}(\mathscr{E}_j, \mathscr{C}). \qquad (12.22)$$

Equation 12.22 simplifies if we assume that environmental fluctuations are small in magnitude with small average differences between species, for then the term $\overline{\mathscr{E}}_j \cdot \overline{\mathscr{C}}$ is small in comparison with the other terms (Chesson 2000a), and can be neglected. Adding in the fitness-density covariance, we obtain

$$\tilde{\lambda}_j \approx 1 + \overline{\mathscr{E}}_j - \overline{\mathscr{C}} + \gamma\mathrm{Cov}(\mathscr{E}_j, \mathscr{C}) + \mathrm{Cov}(\lambda_j, v_j). \qquad (12.23)$$

Expression 12.23 can now be used to study species coexistence by invasibility analysis (Chesson 2000a). In an invasibility analysis, one of the species, "the invader" labeled i, is set to zero density everywhere in space, and the other species, "the resident" labeled r, has dynamics that are independent of the invader. We wish to see if the invader can increase and enter the system, which means that it must have a value of $\tilde{\lambda}$ greater than 1. The species coexist according to the invasibility criterion if they can each increase as invaders.

Differences between resident and invader in the two covariance terms of expression 12.23 are critical to species coexistence. Table 12.2 gives the signs of these covariance terms for various scenarios where the sign can be determined, without further information, by the methods of Chesson (2000a). Full details will be published elsewhere, but in many cases, as shown here, the entries in the table can be determined quite simply and understood intuitively. Note that under the small variation assumption invoked here, the signs of covariances involving the competitive and environmental responses, $\mathscr{E}_{jx}$ and $\mathscr{C}_x$, are the same as those involving the original responses, E_{jx} and C_x. Thus, arguments relating to E_{jx} and C_x carry over to $\mathscr{E}_{jx}$ and $\mathscr{C}_x$.

Resident covariance between environment and competition, Cov($\mathscr{E}_r$, $\mathscr{C}$), is always positive when the competitive response C_x is directly a function of the environmental response E_{rx}. However, it can also be positive when Cov($\mathscr{E}_r$, v_r) is positive, that is, when population density builds up in locations that are favorable based on the response of the organisms to the physical environment. The invader always has zero covariance between environment and competition because the invader's environmental response is never related to the cause of competition—when it has zero density, C_x is not a function of the invader's environmental response, and we have assumed that its environmental response is not correlated with that of the resident. However, covariance between fitness and density can be

Table 12.2 Results of the annual plant competition model

Variable parameter	Type of environmental variation	Type of dispersal	Covariance between environment and competition		Covariance between fitness and density	
			resident	invader	resident	invader
V	spatiotemporal (st)	widespread (w)	+	0	0	0
	st	local retention (l)	+	0	−	+
	Pure spatial (ps)	w	+	0	0	0
	ps	l	+	0	+	+
G	st	w	+	0	−	0
		l	+	0	−	?
	ps	w	+	0	−	−
		l	+	0	?	?
YU	st	w	0	0	0	0
		l	0	0	−	+
	ps	w	0	0	0	0
		l	+	0	+	+

positive for the invader in cases with local retention and either sort of spatial variation. Local retention allows invader density to build up in environmentally or competitively favorable locations. (Although absolute invader density is zero for the invader, relative invader density is not: it is the limit of the ratio of two quantities each approaching zero.) For the resident, local retention mostly gives negative fitness-density covariance when environmental variation is spatio-temporal. This occurs because density builds up locally by chance runs of favorable local environments, increasing local competition. Since the current density is determined by previous environments, which are uncorrelated with the current environment, the component of the growth rate that is correlated with density is simply competition; therefore fitness-density covariance is necessarily negative.

To see if $\tilde{\lambda}_i$ is greater than 1 so that an invader can be successful, one notes that $\tilde{\lambda}$ for a resident ($\tilde{\lambda}_r$) must be equal to 1, assuming that the species comes to equilibrium in the region as a whole (but not necessarily locally in space). Note also that $\mathscr{C}$ and γ are the same for the invader and resident under the assumptions above, and so subtracting the resident $\tilde{\lambda}$ from the invader $\tilde{\lambda}$ leads to the equation

$$\tilde{\lambda}_i - 1 = \Delta E + \Delta I + \Delta\kappa \qquad (12.24)$$

Where

$$\Delta E = \overline{\mathscr{C}}_i - \overline{\mathscr{C}}_r, \qquad (12.25)$$

is a comparison of mean responses to the environment,

$$\Delta I = (-\gamma)\{\text{Cov}(\mathcal{E}_r, \mathcal{C}) - \text{Cov}(\mathcal{E}_i, \mathcal{C})\} \qquad (12.26)$$

measures the spatial storage effect, and

$$\Delta\kappa = \text{Cov}(\lambda_i, v_i) - \text{Cov}(\lambda_r, v_r), \qquad (12.27)$$

compares the fitness-density covariance for the invader and resident.

The three different terms in equation 12.24 have different effects on species co-existence, representing different classes of mechanism. The first term, ΔE, would yield competitive exclusion in the absence of spatial variation. It is an average fitness comparison between species (Chesson 2000b) and is the mean-field component. The second two terms, which are scale transition terms, have the potential to be positive. Indeed, from table 12.2 there are many instances where ΔI is clearly positive and several where $\Delta\kappa$ is clearly positive. These terms therefore have the capability of negating the average fitness differences, thereby counteracting the mean-field component and leading to situations where $\tilde{\lambda}_i - 1$ is positive for both species, permitting them to coexist as determined by the invasibility criterion.

The clearest situation applies to spatial variation in the vigor parameter V. Spatial environmental variation affecting the growing plant feeds into competition because bigger plants use more resources. Thus, for the resident species, there is always positive covariance between its environmental response and competitive response. The assumption that the two species have statistically independent responses to the environment means that the invader's environmental response is independent of its competitive response, which is determined by the density and environmental response of the resident. Thus, $\Delta I = (-\gamma)\text{Cov}(\mathcal{E}_r, \mathcal{C})$, which is positive. With widespread dispersal of seeds, the local environment leaves no signature on population density, and so in that case $\text{Cov}(\lambda_i, v_j)$ is simply zero. Thus, a positive spatial storage effect is found; if this is large enough, it would overcome average differences in the seedling survival rate, measured by ΔE, and permit the two species to coexist.

With local retention of seed, some effects on $\text{Cov}(\lambda_i, v_j)$ are possible. For example, with spatiotemporal variation, the local resident density will reflect some past values of V, which will affect competition. However, density will not be correlated with the environmental component of λ_{rx} because of the independence of the environment over time. The net result is negative fitness-density covariance for the resident. The invader's relative density will be negatively related to that of the resident, reflecting previous competition. As a consequence, a weak positive $\text{Cov}(\lambda_i, v_i)$ is expected. Thus, $\Delta\kappa$ will be positive, promoting coexistence.

Germination fraction variation is similar to variation in vigor in that it always leads to a positive storage effect through its effects on covariance between environment. However, its effects on fitness-density covariance are more complicated because higher local values of G cause local depletion of the seed bank. These cause changes in relative density regardless of local retention. Thus, only in the case of spatiotemporal variation with widespread dispersal is a clear conclusion possible. In that case the positive storage effect (ΔI) combines with positive $\Delta\kappa$ to give an overall fitness promoting effect of variation in the germination fraction.

The final particular case of note is variation in the final yield parameter Y (given vigor) or survival of seed predation, U. Variation in these parameters is assumed not to directly affect competition. This means that covariance between environment and competition does not occur in the case of spatiotemporal variation regardless of dispersal. However, it does occur in the case of pure spatial variation with local retention because local resident density increases with the fixed favorability of the environment, increasing local competition, and causing covariance between environment and competition. However, covariance between fitness and density also results. Although this is positive for both resident and invader, it is shown in Chesson (2000a) that in the absence of a persistent seed bank ($G = 1$, or $s = 0$), $\Delta\kappa$ is approximately $2[p/(1 - p)]\{\mathrm{Cov}(\mathcal{E}_r, \mathcal{C}) - \mathrm{Cov}(\mathcal{E}_i, \mathcal{C})\}$, where p is the fraction of seed retained at the site. (Note that in Chesson (2000a) the factor 2 was inadvertently omitted.) Thus, $\Delta\kappa$ is positive, is proportional to the storage effect, and reinforces it, promoting coexistence.

Conclusions

Scale transition theory focuses on the mechanisms by which the rules for population dynamics on local scales become modified to produce different rules for dynamics on larger spatial scales. Many of the specific issues discussed here as part of scale-transition theory pervade analyses of spatial ecological models, including models formally identified as metacommunity models. The difference here is the focus on the interaction between nonlinearity in local population dynamics and spatial variation as the explanation of the important outcomes on larger spatial scales. In essence, the material presented here implies a research program in which the interaction between nonlinearities and spatial variation is explored for its mechanistic and biological content. Nonlinearities often arise from specific biological postulates such as the nature of interactions within and between species (Chesson 2001). The properties of these nonlinearities identify the kinds of patterns of spatial variation that are important to outcomes at the scale of the whole system.

In the metacommunity context, interactions between species, and between species and their environment, lead to particular kinds of nonlinearity. For example, in host-parasitoid systems, we have seen how the relationship between percent parasitism and parasitoid density is a critical nonlinearity arising from assumptions about parasitoid foraging (Hassell 2000) that implicates the coefficient of variation of parasitoid density as a critical aspect of spatial variation for predicting the dynamics of the metacommunity (box 12.1). In studies of competition in a spatially variable environment, we have seen that a two-dimensional nonlinearity expressing the interaction between responses to the physical environment and responses to competition arises from life-history postulates. This nonlinearity implicates the covariance in space between the response to the environment and the response to competition as a critical aspect of spatial variation for coexistence at the regional scale. This covariance, together with the nonlinearity that makes it important, defines the species coexistence mechanism called the spatial storage effect.

We have seen here also an important distinction between average individual fitness, denoted by $\tilde{\lambda}$, and the spatial average of fitness $\bar{\lambda}$. This distinction was first discussed in the scientific literature by Lloyd (1967), who introduced the concept of mean crowding, applicable to the logistic model. Fitness-density covariance discussed here generalizes Lloyd's concept, and itself reflects a two-dimensional nonlinearity, the product of fitness and local density. This nonlinearity is present in all spatially structured systems, and we have seen how the fitness-density covariance arising from it modifies the stability conditions in host-parasitoid models, supplements the spatial storage effect in spatial competition models, and introduces a coexistence mechanism with properties very similar to the spatial storage effect in spatial Lotka-Volterra and similar models.

The role of the interaction between nonlinearities and spatial variation is less apparent in the approach to metacommunities that describes local densities simply in terms of presence and absence of a species (Nee et al. 1997). Nonlinearities in local population dynamics can only be represented in these models in limited and somewhat extreme ways because of the limitations on the state variables for local population densities. Nevertheless, nonlinearities in these presence-absence models have critical roles in the outcomes at the metacommunity scale (Chesson 2001).

The full scale transition program involves linkages between analytical methods, simulation and numerical methods, and experimental and observational approaches to understanding metacommunities. Analytical theory, as presented here, identifies the key nonlinearities and spatial variation associated with it. This stage is important especially in exploring the biological origin of nonlinearities and their mechanistic role. The nature of spatial variation has a critical mechanistic role, as most strikingly illustrated by the discussion of competition in a variable environment where the pattern of covariance between environment and competition arises from the way species relate to each other and to the physical environment. Techniques of moment closure (Bolker and Pacala 1999), pair approximation (Ellner et al. 1998), and Fourier analysis (Snyder and Chesson 2003) have the potential to expand understanding in this area by approximate analytical and numerical techniques.

In areas where approximations fail or need to be supplemented, simulation approaches can provide knowledge of the relevant variances and covariances or other measures of spatial variation applicable to the relevant nonlinearities. When simulation must be used for this purpose, it will generally be in association with analytical understanding of the nonlinearities and the mechanisms, as presented here. Thus, rather than simply use simulation to solve a problem, such as whether the species coexist, one can ask instead which mechanisms are involved. These mechanisms are quantified by calculating the measures of these mechanisms, such as the quantities ΔI and $\Delta \kappa$ in the assessment of species coexistence in a variable environment. Thus, the relative importance of different mechanisms is assessed quantitatively, and the comparison between different situations and different models is greatly facilitated.

Of most importance, this program extends to the empirical level, as illustrated by Melbourne et al. in chapter 13. Elements of models can be fitted to data to quantify the relevant nonlinearities. Spatial variances and covariances are found using variations on standard experimental and sampling designs. The same concepts and same quantities are used at all levels in the scale transition program, and so the ability to measure and test mechanisms quantitatively is greatly enhanced.

Acknowledgments

This manuscript has benefited greatly from the comments of several anonymous reviewers. We are especially appreciative also of the editorial guidance provided by Marcel Holyoak. This work was supported by NSF grant DEB-9981926.

Literature Cited

Atkinson, W. D., and B. Shorrocks. 1981. Competition on a divided and ephemeral resource. Journal of Animal Ecology 50:461–471.

Bailey, V. A., A. J. Nicholson, and E. J. Williams. 1962. Interaction between hosts and parasites when some host individuals are more difficult to find than others. Journal of Theoretical Biology 3:1–18.

Bolker, B., and S. Pacala. 1999. Spatial moment equations for plant competition: Understanding spatial strategies and the advantages of short dispersal. American Naturalist 153:575–602.

———. 1997. Using moment equations to understand stochastically driven spatial pattern formation in ecological systems. Theoretical Population Biology 52:179–197.

Caswell, H. 1978. Predator-mediated coexistence: A nonequilibrium model. American Naturalist 112: 127–154.

Chesson, P. 1978. Predator-prey theory and variability. Annual Review of Ecology and Systematics 9: 323–347.

———. 1991. A need for niches? Trends in Ecology and Evolution 6:26–28.

———. 1998a. Making sense of spatial models in ecology. Pages 151–166 in J. Bascompte, and R. V. Sole, eds. Modeling spatiotemporal dynamics in ecology. Springer, New York.

———. 1998b. Spatial scales in the study of reef fishes: A theoretical perspective. Australian Journal of Ecology 23:209–215.

———. 2000a. General theory of competitive coexistence in spatially-varying environments. Theoretical Population Biology 58:211–237.

———. 2000b. Mechanisms of maintenance of species diversity. Annual Review of Ecology and Systematics 31:343–366.

———. 2001. Metapopulations. Pages 161–176 in S. A. Levin, ed. Encyclopedia of biodiversity. Academic Press, San Diego, CA.

Chesson, P. L. 1981. Models for spatially distributed populations: The effect of within-patch variability. Theoretical Population Biology 19:288–325.

———. 1985. Coexistence of competitors in spatially and temporally varying environments: A look at the combined effects of different sorts of variability. Theoretical Population Biology 28:263–287.

———. 1991b. Stochastic population models. Pages 123–143 in J. Kolasa, and S. T. A. Pickett, eds. Ecological heterogeneity. Springer-Verlag, New York.

Chesson, P. L., and W. W. Murdoch. 1986. Aggregation of risk: Relationships among host-parasitoid models. American Naturalist 127:696–715.

Ellner, S. P., A. Sasaki, Y. Haraguchi, and H. Matsuda. 1998. Speed of invasion in lattice population models pair-edge approximation. Journal of Mathematical Biology 36:469–484.

Englund, G., and S. D. Cooper. 2003. Scale effects and extrapolation in ecological experiments. Pages 161–213 in H. Caswell, ed. Advances in Ecological Research. Vol. 33. Academic Press, Amsterdam, The Netherlands.

Green, P. F. 1986. Does aggregation prevent competitive exclusion? A response to Atkinson and Shorrocks. American Naturalist 128:301–304.

Hanski, I. 1999. Metapopulation ecology. Oxford University Press, Oxford, U.K.

Harrison, S., and A. D. Taylor. 1997. Empirical evidence for metapopulation dynamics. Pages 27–42 in I. Hanski, and M. E. Gilpin, eds. Metapopulation biology: Ecology, genetics, evolution. Academic Press, San Diego, CA.

Hartley, S., and B. Shorrocks. 2002. A general framework for the aggregation model of coexistence. Journal of Animal Ecology 71:651–662.

Hassell, M. P. 2000. The spatial and temporal dynamics of host-parasitoid interactions: Oxford series in ecology and evolution. Oxford, Oxford University Press.

Hassell, M. P., R. M. May, S. W. Pacala, and P. L. Chesson. 1991. The persistence of host-parasitoid associations in patchy environments. I. A general criterion. American Naturalist 138:568–583.

Hastings, A. 1977. Spatial heterogeneity and the stability of predator-prey systems. Theoretical Population Biology 12:37–48.

———. 1980. Disturbance, coexistence, history, and competition for space. Theoretical Population Biology 18:363–373.

Heard, S. B., and L. C. Remer. 1997. Clutch-size behavior and coexistence in ephemeral-patch competition models. American Naturalist 150:744–770.

Horne, J. K., and D. C. Schneider. 1995. Spatial variance in ecology. Oikos 74:18–26.

Ives, A. R. 1988. Covariance, coexistence and the population dynamics of two competitors using a patchy resource. Journal of Theoretical Biology 133:345–361.

———. 1991. Aggregation and coexistence in a carrion fly community. Ecological Monographs 61: 75–94.

Kareiva, P. 1990. Population dynamics in spatially complex environments: Theory and data. Philosophical Transactions of the Royal Society of London, series B, Biological Sciences 330:175–190.

Levin, S. A. 1992. The problem of pattern and scale in ecology. Ecology 73:1943–1967.

Lloyd, M. 1967. Mean crowding. Journal of Animal Ecology 36:1–30.

Lloyd, M., and J. White. 1980. On reconciling patchy microspatial distributions with competition models. American Naturalist 115:29–44.

May, R. M. 1978. Host-parasitoid systems in patchy environments: Phenomenological model. Journal of Animal Ecology 47:833–843.

Murdoch, W. W., C. J. Briggs, and R. M. Nisbet. 2003. Consumer-resource dynamics. Princeton University Press, Princeton, NJ.

Murrell, D. J., and R. Law. 2003. Heteromyopia and the spatial coexistence of similar competitors. Ecology Letters 6:48–59.

Nee, S., R. M. May, and M. P. Hassell. 1997. Two-species metapopulation models. Pages 123–147 *in* I. A. Hanski, and M. E. Gilpin, eds. *Metapopulation biology: Ecology, genetics, and evolution.* Academic Press, San Diego, CA.

Neuhauser, C., and S. W. Pacala. 1999. An explicitly spatial version of the Lotka-Volterra model with interspecific competition. Annals of Applied Probability 9:1226–1259.

Pacala, S. W., and M. P. Hassell. 1991. The persistence of host-parasitoid associations in patchy environments. II Evaluation of field data. American Naturalist 138:584–605.

Pacala, S. W., and J. A. Silander, Jr. 1985. Neighborhood models of plant population dynamics. I. Single-species models of annuals. American Naturalist 125:385–411.

Rao, C. R. 1973. *Linear statistical inference and its applications.* John Wiley, New York.

Remer, L. C., and S. B. Heard. 1998. Local movement and edge effects on competition and coexistence in ephemeral patch models. American Naturalist 152:896–904.

Ross, S. 1997. *A first course in probability.* Prentice Hall, Upper Saddle River, NJ.

Shigesada, N., and J. Roughgarden. 1982. The role of rapid dispersal in the population dynamics of competition. Theoretical Population Biology 21:253–373.

Slatkin, M. 1974. Competition and regional coexistence. Ecology 55:128–134.

Snyder, R. E., and P. Chesson. 2003. Local dispersal can facilitate coexistence in the presence of permanent spatial heterogeneity. Ecology Letters 6:301–309.

Wiens, J. A. 1989. Spatial scaling in ecology. Functional Ecology 3:385–397.

CHAPTER 13

Applying Scale Transition Theory to Metacommunities in the Field

Brett A. Melbourne, Anna L. W. Sears,
Megan J. Donahue, and Peter Chesson

Introduction

A metacommunity consists of local communities linked by dispersal. This concept distinguishes between community processes that operate at local spatial scales and dispersal processes that link local communities together. The dynamics of the metacommunity as a whole result from an interplay between these local and regional processes, affecting dynamical outcomes such as stability, competitive exclusion, and persistence (De Jong 1979; Ives 1988; Hassell et al. 1991; Bolker and Pacala 1997; Pacala and Levin 1997; Chesson 2000). The major issue in metacommunity studies is to understand how regional dynamics arise from local dynamics. In other words, to understand how dynamics scale up from local communities to the larger spatial scale of the metacommunity. In the chapter on scale transition theory (Chesson et al., chapter 12), we saw that the rules that determine community dynamics change as our view shifts from the small spatial scale of the local community to the larger spatial scale of the metacommunity. The key determinants of these changes were shown to be spatial variation between local communities and nonlinearity in local processes, neither being important without the other. Indeed, it is the interaction between spatial variation and nonlinearity that underpins the changes. Spatial heterogeneity is ubiquitous to ecological systems: we know from field experience that the density of each species, their resources, competitors, and predators vary over space. By nonlinearities, we mean that the relationship between population growth rate and the density of conspecifics, resources, competitors, or predators is nonlinear. Nonlinearities are also ubiquitous to ecological systems. For example, local per capita growth rates that depend on environmental conditions, or the simple act of one species feeding on another, involve nonlinearities. Any source of density dependence, such as intra- or interspecific competition, also involves a nonlinearity.

The critical insight of scale transition theory is that spatial variation between local communities as well as nonlinearities within local communities interact to determine the dynamics of the metacommunity. Through this insight, scale transition theory systemizes the study of metacommunities by identifying key quantities, which represent spatial mechanisms that contribute to change with spatial scale. The spatial storage effect and fitness-density covariance are two examples of such mechanisms (Chesson 2000; Chesson et al., chapter 12). The presence and magnitude of these mechanisms also provides evidence that metacommunity processes are important to the dynamics of the metacommunity. If these mechanisms are absent we would expect little difference in dynamics between local-scale communities and metacommunities, that is, linking local communities together by dispersal would have little consequence.

In this chapter we show how to apply the scale transition ideas presented in chapter 12 using examples from three quite different field systems. There are three steps to implement this research program: (1) derive a model to translate the effects of local dynamics to the metacommunity scale and to identify key interactions between nonlinearity and spatial variation; (2) measure nonlinearities at local scales; and (3) measure spatial variances and covariances. We show how to work through each of these steps and how to combine the models and data to calculate effects on dynamics at the metacommunity scale. Our field examples demonstrate a range of dynamical outcomes that result from a metacommunity structure. The first two field examples focus on a single species or functional group within a metacommunity but include interactions with other trophic levels. These first two examples are perhaps best considered metapopulation studies but they serve as a bridge to demonstrate how scale transition tools are applied to metacommunities. The final example is for competition between two species of annual plants. In all three examples, considering the spatial context of the metacommunity allows a deeper understanding of the mechanics of the system.

Example 1: Benthic Algae and Grazers on Stream Cobbles

The first field example is for the dynamics of benthic algae growing on stream cobbles and grazed by insect larvae. Recently, stream systems, and the algae-grazer system in particular, have inspired much interest in the issue of heterogeneity and scaling because of high spatial variation in the density of organisms and physical conditions (Downes et al. 1993; Cooper et al. 1997, 1998; Nisbet et al. 1997). Melbourne and Chesson (submitted) described a study of the scale transition for algae (more accurately periphyton dominated by algae) and grazers in a stream system in southeastern Australia. In these streams, algal assemblages were dominated by diatoms and green algae and were consumed mainly by the aquatic larval stages of caddisfly, mayfly, and waterpenny species. Interactions between algae and grazers occur in local patches, on centimeter scales, yet we wish to scale up to a metacommunity scale that encompasses the entire study area, some 42 km of stream. What role does spatial structure play in this system? How big might the effects of spatial structure and spatial variation be on algal-grazer dynamics at the larger scale of the stream?

Model to Scale Up Algae and Grazers

It is worth going into some detail to show how models were derived to scale up from local processes at the scale of centimeter-sized patches to the regional scale of the stream. We begin by formulating a model for the dynamics of algae and grazers at the local scale. We treat algae and grazers as functional groups, not differentiating the separate dynamics of each species but instead examining the overall dynamics of algae and grazers. At the local scale, algal dynamics are determined by three processes: algal growth, consumption by grazers, and dispersal of algae into and out of a location. Here we use the patch model formulation of the scale transition (Chesson et al., chapter 12, equation 12.10). A simple model for algal dynamics at location x (a small patch, say 2 cm in diameter) is thus

$$\frac{dA_x}{dt} = g(A_x) - f(A_x)G_x + I_{A,x} - \varepsilon_{A,x,} \tag{13.1}$$

where A is the density of algae (biomass per unit area), t is time, and G is the density of grazers (biomass per unit area). The model is a standard balance equation plus dispersal terms, where $g(A)$ is a function that describes algal growth rate (biomass per unit area per unit time) as a function of algal density, and $f(A)G$ describes the rate of algal consumption by grazers, where $f(A)$ is the functional response of a grazer. In writing consumption as $f(A)G$, we are assuming that grazer consumption is linear in G, which means that grazers do not interact with each other directly through interference but only indirectly by consuming algae, a standard assumption for this system (Nisbet et al. 1997). Algal dispersal to and from location x is modeled by the generic functions $I_{A,x}$ and $\varepsilon_{A,x}$, to which we can assign any set of dispersal rules (provided that dispersal mortality is factored into the growth rate terms [Chesson et al., chapter 12]). For example, algal immigration rate, I_A, at location x might depend on the output from nearby patches or the vagaries of stream currents, and algal emigration rate ε_A at location x might depend on algal density at that location. Equation 13.1 is in continuous time, which matches the nature of algal dynamics. A simple model for grazer dynamics at location x is

$$\frac{dG_x}{dt} = [cf(A_x) - m]G_x + I_{G,x} - \varepsilon_{G,x,} \tag{13.2}$$

where c is the conversion efficiency of grazers, m is grazer mortality, and grazer movement is modeled by the generic functions $I_{G,x}$ and $\varepsilon_{G,x}$. Thus, a model for the local dynamics of algae and grazers consists of the coupled equations 13.1 and 13.2. To match the timescale that can be achieved in field experiments (i.e., two to three weeks), we reformulate the model for grazer dynamics to a short timescale, over which the net gain in grazer density due to the consumption of algae is trivial (Nisbet et al. 1997). Thus, we set $[cf(A_x) - m]G_x = 0$, and the local dynamics of grazers is determined only by movement into and out of the location. For example, such movement might be a result of foraging decisions made by individual grazers (recall that a location is a 2 cm diameter patch), or a response to environmental conditions or predators at the location, or upstream or downstream movement in response to stream flow. To fully specify the model, we need to put forward specific functions for $g(A)$ and $f(A)$ but for now we leave the model in its general form and proceed to the scaling-up step. As we will see, we do not need to specify the algal dispersal and grazer movement functions.

We wish to scale up to find the dynamics for the entire stream. At this regional scale, we consider the system to be effectively closed, meaning that the rate of dispersal across the boundaries is so small that it can have no effect on internal dynamics. Over short timescales, this is an especially realistic assumption for streams. The entire stream encompasses the headwaters, so there is no immigration from upstream and there is no immigration from the surrounding terrestrial environment. The only dispersal that occurs across the stream boundaries is at the downstream end, where the amount of dispersal across the stream boundary is tiny compared to that within the system upstream. We consider only short timescales here but over longer timescales we might need to deal with flood events, when there can be a large export of algae and grazers from the system at the downstream end.

To scale up from local, centimeter sized patches, to the regional scale of a stream where patches (indexed by x) are now linked by dispersal of algae and movement of grazers, we average the local dynamics over all patches. This regional average is a foundation of the scale transition: the average of the local densities is in fact the regional density (Chesson et al., chapter 12). Thus, averaging over both sides of equations 13.1 and 13.2, the regional dynamics are

$$\frac{\overline{dA}}{dt} = \overline{g(A) - f(A)G + I_A - \varepsilon_A} = \overline{g(A)} - \overline{f(A)G} + \overline{I_A} - \overline{\varepsilon_A}$$

$$= \overline{g(A)} - \overline{f(A)G} \tag{13.3}$$

$$\frac{\overline{dG}}{dt} = \overline{I_G - \varepsilon_G} = \overline{I_G} - \overline{\varepsilon_G} = 0,$$

where the overbars indicate an average over space, and the subscript x disappears because that is what we are averaging over. The overbar extends all the way across the right hand side, indicating that the rate of change in algal density at the regional scale is an average of the local functions for growth, consumption, and dispersal. As averages are additive, the regional average simplifies to the separate averages for growth, consumption, and dispersal (equation 13.3). Since the system is effectively closed at the regional scale, individuals leaving one location within the region enter another location, so that the total number of individuals leaving locations is equal to the total number of individuals entering locations (i.e., $\bar{I} - \bar{\varepsilon} = 0$). Thus the dispersal functions disappear and dispersal within the system becomes an implicit part of the model, rather than appearing explicitly at the regional scale.

Equation 13.3 is our fundamental equation for the regional dynamics and we can now work through it to find a useful expression for the scale transition by using the properties of nonlinear averages given in chapter 12, box 2. The average of the product $\overline{f(A)G}$ in equation 13.3 splits into the product of the averages and the covariance in space between them (chapter 12, equation 12.B2.6), so

$$\frac{d\bar{A}}{dt} = \overline{g(A)} - \overline{f(A)}\,\overline{G} - \mathrm{Cov}[f(A), G]. \tag{13.4}$$

At this stage, the equation still gives the exact dynamics for the region. Thus, in equation 13.4 it can be seen that the dynamics for the region involves the covariance over space between grazer functional response and grazer density. The grazer functional response will vary in space according to variation in algal density between locations. By approximating $g(A)$ and $f(A)$ as second order Taylor polynomials (chapter 12, equation 12.B2.5), we get

$$\frac{d\bar{A}}{dt} = [g(\bar{A}) + \frac{1}{2}g''(\bar{A})\mathrm{Var}(A)]$$
$$- [f(\bar{A}) + \frac{1}{2}f''(\bar{A})\mathrm{Var}(A)]\bar{G} - \mathrm{Cov}[f(A), G], \tag{13.5}$$

where $\mathrm{Var}(A)$ is the variance of A over space and $g''(\bar{A})$ and $f''(\bar{A})$ are the second derivatives of $g(\bar{A})$ and $f(\bar{A})$. Finally, rearranging and taking a linear approximation for $f(A)$ in the covariance term (chapter 12, equation 12.B2.8) we get

$$\frac{d\bar{A}}{dt} = \underbrace{g(\bar{A}) - f(\bar{A})\bar{G}}_{\text{mean-field model}} + \underbrace{\frac{1}{2}g''(\bar{A})\mathrm{Var}(A)}_{a}$$
$$- \underbrace{\frac{1}{2}f''(\bar{A})\bar{G}\mathrm{Var}(A)}_{b} - \underbrace{f'(\bar{A})\mathrm{Cov}(A, G)}_{c}, \tag{13.6}$$

where $f'(\bar{A})$ is the first derivative of $f(\bar{A})$. The structure of this model for the regional dynamics is very informative and empirically useful. The first two terms are the mean field model, which is simply the equation for local dynamics (equation 13.1 without dispersal) with the mean density across patches substituted for local density. We can think of the mean-field model as representing local dynamics. The regional dynamics are thus given by the mean field model plus three scale transition terms (a–c) that correct the mean-field model to account for spatial structure. Most importantly, equation 13.6 tells us what we need to measure in field studies: the nonlinearities in algal growth and grazer consumption expressed by $g''(A)$ and $f''(A)$, the spatial variance of algal density, Var(A), and the spatial covariance of algal and grazer density, Cov(A, G). Here nonlinear means that the rate of algal growth or consumption varies as a nonlinear function of algal density (discussed below).

To analyze the consequences of the metapopulation structure, we consider the magnitude and sign of the terms (a–c) relative to the mean-field model and to each other. Terms (a) and (b) are a consequence of spatial variation in algal density between patches and clearly demonstrate how changing scale involves an interaction between nonlinearity and spatial variation, since they involve the product of 1) the second derivative $g''(A)$ or $f''(A)$, which measure nonlinearities in the local processes of algal growth or grazer functional response respectively, and 2) the variance in algal density between locations, Var(A). For algal growth $g(A)$, local nonlinearities might result from competition for resources within the algal film. For example, as algal density increases, shading by overlying algal cells reduces the photosynthetic rate of cells in layers below, resulting in density dependent growth. For the functional response of grazers $f(A)$, local nonlinearities might result from a handling time requirement or an attack rate that varies with algal density, for example, as in a classic Type II or Type III functional response. Depending on the sign of the second derivatives $g''(\bar{A})$ and $f''(\bar{A})$, which are determined by whether each function is concave down (negative second derivative) or concave up (positive second derivative), the nonlinearities may work either in opposing directions, thus moderating the effect of variation, or in the same direction, thus compounding the effect of variation on regional dynamics. The third term (c) also results from an interaction between nonlinearity and spatial variation but the interaction is now due to the two-dimensional nonlinearity of algae and grazer density. This nonlinearity is simply because grazers eat algae. Spatial variation in G alone does not appear in the regional model and the effect of spatial variation in G enters only through spatial covariance with A. This is because we assume that grazer consumption is linear in G (equation 13.1). The influence of spatial variation in grazer density on the scale transition depends on the covariance between algae and grazer density (term c), in other words, on how algae and grazers are distributed in space in relation to one another.

While dispersal of algae and movement of grazers within the system are explicit in the construction of the model, they do not appear in equation 13.6. Instead, the effects of dispersal and movement on the spatial pattern are represented by the variance and covariance terms. For example, the movement of grazers affects these terms because grazers move around in the stream eating algae, thus creating (or destroying) variation in algal density. Similarly, movement and consumption could set up covariance between algae and grazer density. Thus, while dispersal of algae and movement of grazers does not appear in the regional model (indeed an equation for the dynamics of grazers does not appear at all), the model incorporates the effects of algal dispersal and grazer movement within the system.

We know that this system consists of local patches in the stream that are linked by dispersal of algae and movement of grazers. If this metapopulation structure is important in explaining dynamics at the larger scale of the stream, then we should be able to detect and quantify the interactions (a–c) in the field. That requires measuring the nonlinearities associated with algal growth and grazer consumption as well as the variances and covariance in equation 13.6.

Measuring Nonlinearities in Algae and Grazers

In general, local nonlinearities can be estimated by varying levels of the nonlinear factors (usually population density) in an experiment and fitting models to the experimental data to describe the shape of the nonlinearities. This same general approach is applied to all three field examples in this chapter, although the methodological details are peculiar to the processes and experimental constraints of the different systems. To measure nonlinearities in the algae-grazer system, we need to put forward specific functional forms for algal growth and grazer foraging. Recall that nonlinear means that the rate of algal growth or consumption varies as a nonlinear function of algal density. It is useful in this step to put forward alternative models as hypotheses for the functional form and to test between models by confronting them with experimental data, as described in Melbourne and Chesson (submitted). To measure nonlinearities in algal growth, the experiment involved excluding grazers from rocks in the stream and examining the growth of algae in centimeter-sized patches in the absence of grazers, over a range of initial density for the algae. The functional response of grazers was determined similarly by examining consumption over a range of algal densities in the stream. The best-fitting models were the logistic model for algal growth and the Type III functional response for grazers. These models, parameterized from the local scale data, characterize the nonlinearities in the process of algal growth $g(A)$ and grazer consumption $f(A)$ at local scales.

Measuring Spatial Variances and Covariances

Measuring spatial variation at the scale of local nonlinearities is critical to estimating the scale transition. We need to first identify the scale of the local nonlinearity, which is defined as the range of space over which the influence of the nonlinear process is homogeneous (the ecological neighborhood sensu Addicott et al. 1987). For a density dependent process, like algal growth, this scale is the scale of density dependence (Chesson 1996, 1998a). These scales can be tricky to determine precisely but their order of magnitude is often obvious. In the algae-grazer system, the different scales are the scale of competition between algal cells (on the order of micrometers to millimeters) and the scale of the grazer consumption process (on the order of centimeters to meters). We set the scale of algal growth to the smallest scale we could measure, which corresponded to our algal sampling device (2.4 cm diameter). For the scale of grazer consumption, we also assume the smallest scale we could measure (2.4 cm diameter), meaning that grazers can perceive and respond to variation in algal density between centimeter-sized patches. To measure the spatial variance of algal density and the spatial covariance for algal and grazer density, samples of algae and grazer density were taken in a hierarchical sampling design and the total variance and covariance at the scale of the local nonlinearity was estimated by analysis of the variance components (Searle et al. 1992; Underwood and Chapman 1996; Melbourne and Chesson, submitted).

Combining the Model with Field Data to Scale Up

Having identified models and estimated parameters for the nonlinear functions for algal growth and grazer foraging, and estimated the spatial variances and covariance in equation 13.6, we can now scale up the dynamics from the local scale of a centimeter-sized patch, where interactions between individuals take place, to the scale of the stream, which is made up of patches linked by dispersal of algae and movement of grazers. A note of caution is required here. The approximations in equation 13.6 give the trend as the variance increases from zero, but if the equations are not quadratic, the approximation may become inaccurate when the spatial variance is large. The logistic model for algal growth is exactly quadratic but

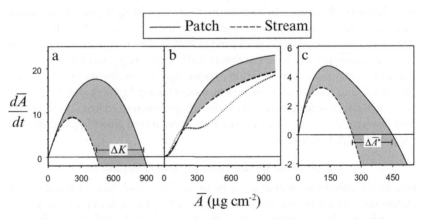

Figure 13.1 Comparison of the patch-scale model (mean-field model) and the stream-scale model for the algae-grazer system (equation 13.6). The calculations are based on fitted functions, parameter estimates, and spatial variances and covariances from field data. The shaded areas represent the reduction in the instantaneous rate of change that occurs in the stream-scale model: (a) Logistic algal growth, $g(A)$; ΔK change in carrying capacity in the logistic model. Since the logistic is quadratic, the stream-scale model is exact; (b) Type III grazer removal rate, $f(A)G$. Dashed line is the exact stream-scale model. The dotted line is the quadratic approximation for the stream-scale model and deviates markedly from the exact model; (c) Full periphyton-grazer model, $g(A) - f(A)G$. The stream-scale model is exact. $\Delta \bar{A}^*$ change in equilibrium biomass.

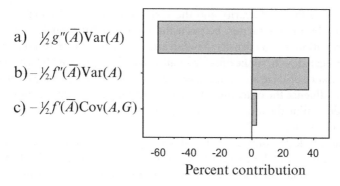

a) $\frac{1}{2}g''(\overline{A})\mathrm{Var}(A)$

b) $-\frac{1}{2}f''(\overline{A})\mathrm{Var}(A)$

c) $-\frac{1}{2}f'(\overline{A})\mathrm{Cov}(A,G)$

Percent contribution

Figure 13.2 Contribution of each scale transition term to the total change in the equilibrium density of algae at the stream scale (equation 13.6, a–c). The calculations are based on fitted functions, parameter estimates, and spatial variances and covariances from field data. Term (a) reduced the equilibrium biomass at the stream scale and was responsible for about 60% of the total change, whereas term (b) counteracted the effect of term (a), and term (c) was of little importance. Although the terms are labeled by their quadratic approximations, the exact scale transition is shown for term (b) and (c). Term (a) is also exact, since $g(A)$ is the logistic model and is quadratic.

the Type III functional response for grazers departs markedly from a quadratic and the approximation breaks down because we observe high variance (figure 13.1). For the Type III response, we calculated the exact scale transition for terms (b) and (c) in equation 13.6, as described in Melbourne and Chesson (submitted).

The effect of scaling up is to reduce the instantaneous rate of growth at the stream scale compared to local patches (figure 13.1). That is, for a given algal density at the stream scale, growth is reduced compared to growth for the same algal density at the patch scale. The area shaded in gray in figure 13.1 represents the scale transition: the difference in output from the patch-scale population dynamics (given by the mean-field curve) and the spatial model for the stream that takes into account the metapopulation structure. The metapopulation structure doesn't qualitatively alter aspects of the short-term dynamics, such as the stability because the general form of the spatial model is still a hump-shaped curve (figure 13.1). However, an important basic property that is quantitatively altered is the equilibrium density of algae A^*, the balance of growth and consumption. This short term equilibrium density is reduced at the stream scale by about 41%, a substantial reduction (figure 13.1). Thus, spatial variation and nonlinearity interact to reduce algal density at the metapopulation scale.

Recall that the important components that arise because of the metapopulation structure are the terms (a–c) in equation 13.6 that represent interactions between nonlinearity and spatial variance. When we look at the three components of the scale transition, the main effect is seen to be from the interaction of nonlinearity in algal growth with spatial variation in algal density (figure 13.2), because both the nonlinearity and the variance were high. This effect is offset somewhat by the interaction between the nonlinear functional response with spatial variation in algal density. Nonlinearity in the functional response was less than for algal growth, so the offset is smaller than the algal growth term. The spatial covariance of algae and grazers was negative but the covariance term was too small to influence the outcome. Thus, by examining these relative contributions we can determine the most important spatial mechanisms that alter dynamics at the regional scale.

This example demonstrates the basic concepts and properties of the scale transition in translating the effects of local scale processes to the scale of the metapopulation, where locations are linked by dispersal. Scale transition theory identified the important component mechanisms that change dynamics at the metapopulation scale. These components were then possible to measure relatively easily by applying standard field experiments and sampling designs. We confirmed that metapopulation processes were important to the dynamics of the system because we detected large scale transition components. Had we not detected such large components, we would not expect the metapopulation structure to have any important effects on dynamics at the larger scale of the stream; the stream would act much like a patch. The next field example involves a greater variety of local processes than the simple example above, demonstrating that more complicated cases are approached in the same way.

Example 2: Dynamics of an Intertidal Crab

The second field example is for the dynamics of an intertidal crab that, like many other marine organisms, has a planktonic larval stage. Current research in marine ecology recognizes that marine systems are metacommunities in which local populations and communities are interconnected through larval dispersal (Chesson 1998b; Jones et al. 1999; Swearer, Caselle, et al. 1999; Swearer, Shima, et al. 2002; Armsworth 2002; Thorrold et al. 2002). This metacommunity view contrasts with a "supply-side" perspective of marine systems in which larval supply is treated as an external force that sets the stage for subsequent local dynamics but does not feed back to the larval pool (Gaines and Roughgarden 1985; Grosberg and Levitan 1992). When marine systems are viewed as a metacommunity, larval supply is not an external force on local communities but interacts with local processes to influence local and metacommunity dynamics (Chesson 1998b; Armsworth 2002).

In our example system, the porcelain crab, *Petrolisthes cinctipes*, lives under rocks in intertidal cobblefields in the northeast Pacific. It releases larvae that feed and develop in the plankton for one to two months. Upon returning to adult habitat, the larvae settle and, if they survive high postsettlement mortality, recruit to the adult population (Jensen 1989, 1991; Donahue 2003). Adults are filter-feeders and live in aggregations under rocks. Rocks are relatively discrete habitat patches because *P. cinctipes* move little unless their rock is disturbed. There are two natural scales of variation to consider: rocks (rock area $\approx 0.1 \text{m}^2$) within sites, and sites (site length ≈ 1 km) within the metacommunity that encompasses the coastline. Competitive interactions between crabs occur at the rock scale. Predators forage across rocks within sites. Sites are connected by larval dispersal. We model larval distribution in the following way: larvae generated at all sites enter a larval pool and the proportion of this larval pool distributed to a particular site is a site-scale characteristic. We wish to scale up from rock- and site-scale interactions to the metapopulation that encompasses the coast.

Model to Scale Up Crab Dynamics

We begin by formulating a model for the dynamics of *P. cinctipes* at the cobble scale, where interactions between individuals take place. The important population processes can be divided into recruitment and adult growth. Juveniles settle gregariously in response to adult density (Jensen 1989; Donahue 2003). To recruit, settlers must survive predation from resident predators (several species of fish and crabs) (Jensen 1991; Donahue 2003) and competition from resident adults (Donahue 2004). For adult crabs, growth rate and fecundity depend on conspecific density (Donahue 2004). These processes lead to a general model for the dynamics of crab biomass, N_x under cobble x from the release of larvae in one season, t, to the release of larvae in the next $t + 1$:

$$N_x(t+1) = N_x(t) + g[N_x(t)] + h[N_x(t), P_x(t), L_x(t)]. \qquad (13.7)$$

In this model, $g[N_x(t)]$ is a function that describes the density dependent growth and mortality of adult crabs and h is a function that describes recruitment. Recruitment is determined by the biomass of adults under the rock $N_x(t)$, the biomass of predators $P_x(t)$, and larval supply $L_x(t)$, which is the total biomass of larvae delivered to the site. The larvae $L_x(t)$ are the dispersed offspring of the adults present in the system at time t. Adults from cobble x produce $bN_x(t)$ new larvae, which disperse and become the inputs $L_x(t)$ to equation 13.7. The equation implicitly assumes that dispersal is instantaneous. In reality it is spread out over time during the interval t to $t + 1$, with equation 13.7 being most accurate if dispersal is concentrated early in the time interval. Dispersal of larvae between sites could follow any mode of dispersal. For example, larvae could enter a global pool and disperse to sites according to $L_x(t) = U_x b N(t)$, where U_x is the proportion of the regional larval pool delivered to the site (see also equation 12.13 in chapter 12). As the transition from adults to larvae is linear, we focus on the stage transition from larvae to adults in equation 13.7. Adult growth and competition take place at the rock scale, while recruitment has both site-scale (larval supply, predator abundance) and rock-scale (competition, gregarious settlement) factors. To fully specify the model, we need specific functions for g and h but as in the previous example, we leave the model in its general form and proceed to the scaling-up step.

To understand regional dynamics in this system, we take spatial averages of both sides of this equation:

$$\overline{N(t+1)} = \overline{N(t) + g[N(t)] + h[N(t), P(t), L(t)]}. \tag{13.8}$$

The spatial average of the right-hand side can be approximated by a Taylor expansion. Since our general model is a function of three variables, we perform our Taylor expansion around mean adult density, mean predator density, and mean larval supply, each of which generate variance and covariance terms. Note that we are not tracking the dynamics of predators, but predator density influences attack rates.

$$
\overline{N(t+1)} \approx \underbrace{\overline{N(t)} + g[\overline{N(t)}] + h[\overline{N(t)}, \overline{P(t)}, \overline{L(t)}]}_{\text{mean-field model}}
$$

$$
+ \underbrace{\frac{1}{2}\left.\frac{\partial^2 g}{\partial N^2}\right|_{\overline{N}} \mathrm{Var}(N)}_{a} + \underbrace{\frac{1}{2}\left.\frac{\partial^2 h}{\partial N^2}\right|_{\overline{N}} \mathrm{Var}(N)}_{b} + \underbrace{\frac{1}{2}\left.\frac{\partial^2 h}{\partial P^2}\right|_{\overline{P}} \mathrm{Var}(P)}_{c}
$$

$$
+ \underbrace{\frac{1}{2}\left.\frac{\partial^2 h}{\partial L^2}\right|_{\overline{L}} \mathrm{Var}(L)}_{d} + \underbrace{\left.\frac{\partial^2 h}{\partial N \partial P}\right|_{\overline{N},\overline{P}} \mathrm{Cov}(N, P)}_{e} \tag{13.9}
$$

$$
+ \underbrace{\left.\frac{\partial^2 h}{\partial N \partial L}\right|_{\overline{N},\overline{L}} \mathrm{Cov}(N, L)}_{f} + \underbrace{\left.\frac{\partial^2 h}{\partial P \partial L}\right|_{\overline{P},\overline{L}} \mathrm{Cov}(P, L)}_{g}.
$$

On first impression, this equation looks rather complicated because of the need for partial derivatives to account for nonlinearities. However, the equation has the familiar form that we saw in the first example and in chapter 12. The first line of equation 13.9 is the mean-field model. If there were no spatial variation in adult density, predator abundance, or larval supply, the mean-field model would be the correct model at the metapopulation scale. Of course, there is spatial variation in these variables, and the next three lines of the equation take this into account. Thus, the regional dynamics are given by the mean-field model plus the seven terms (a–g) that correct the mean-field model to account for spatial structure. To scale up, we need to measure the scale transition terms from field studies.

The first scale transition term (a) of equation 13.9 is the interaction of nonlinearity in adult growth with spatial variation in adult density. The nonlinearity measures the strength of competition between adults. Since *P. cinctipes* have a clumped distribution, term (a) accounts for higher competition at higher densities. The second term (b) includes the nonlinearity of recruitment with respect to adult density measured by the partial derivative. This nonlinearity is influenced by the competitive effect of adults on settlers, the protection from predation that settlers gain from adults, and the tendency for settlers to choose rocks with higher adult densities. Each of these processes interacts with variation in adult density. The third term (c) is nonzero only when predation rate has a nonlinear dependence on predator density, for example, if there were interference between predators. In our study of *P. cinctipes*, we have assumed that predator effects are additive and this term is zero. The fourth term (d) is the nonlinearity of recruitment with respect to larval supply times the variance of larval supply. Spatial variation in larval supply is notoriously high among marine organisms, ranging several orders of magnitude. This term could be important even if the nonlinearity with respect to larval supply is comparatively small.

Terms (e) to (g) of equation 13.9 result from the two-dimensional nonlinearities in recruitment. There are no two-dimensional nonlinearities in adult growth because it is a function of a single variable—adult density. While the predator term (c) equals zero, there can be two-dimensional nonlinearities between predators and adult density and between predators and larval supply. Term (e) arises because adult density influences the abundance of settlers (through gregarious settlement, competition, and protection from predation) and predators respond to settler abundance. Therefore, there is a two-dimensional nonlinearity between predators and adult density. Term (f) arises because of gregarious settlement, in which larval supply interacts with adult density. Finally, term (g) arises because predators respond to prey density, and larval supply influences prey density. Each of the two-dimensional nonlinearities is multiplied by its respective covariance term.

Measuring Local Nonlinearities

Our next goal was to empirically evaluate each component of this model. Since we wanted to estimate the functions for adult growth, $g[N(t)]$, and larval recruitment, $h[N(t), P(t), L(t)]$, we needed to perform growth and recruitment experiments over a range of densities. Starting with the growth component, we tested for intraspecific competition by tracking individual growth rates over a range of crab densities in field and laboratory experiments. We found that growth rate declined with increasing density and that smaller individuals were more strongly impacted by competition than large individuals (Donahue 2004). Since our general model has only two size classes, we incorporated size-specific competition by fitting a different competition coefficient for adults on settlers than adults on other adults. Competition is a nonlinearity with respect to adult density in both the growth and recruitment components of equation 13.9.

Previous studies indicated that *P. cinctipes* settle preferentially with conspecifics (Jensen 1989). To quantify gregarious settlement, we manipulated adult density in field enclosures and counted the number of settlers to each enclosure (Donahue 2003). We performed this experiment during two settlement pulses to see how larval supply influenced the settlement pattern. We found that the per capita rate of settlement increased then saturated with adult density. This generates nonlinearity with respect to adult density.

Settlers remain under adults and are protected from predation by this behavior (Jensen 1991). One model for this interaction is a decline in attack rate as the ratio of adults to settlers increases (Donahue 2003). We assume there is no interference between predators. This generates nonlinearity with respect to adult density and larval supply, but only two-dimensional nonlinearities with respect to predators.

Measuring Spatial Variances and Covariances

With models of density dependence established and parameterized, we measured variation in adult density, predator density, and supply at the appropriate scales. Variation in adult density is at the rock scale, while variation in predator density and larval supply is at the site scale. At fourteen cobblefield sites, we sampled forty rocks for adult crabs and predators. Since predators were less abundant, we sampled an additional eighty rocks for predators. From this, we could estimate the rock-scale and site-scale variance in adult density $Var(N)$, site-scale variance in predator abundance $Var(P)$, and site-scale covariance between adults and predators $Cov(N, P)$. At eight sites, we sampled larval supply using nearshore collectors and estimated the site-scale variance in larval supply $Var(L)$, and the site-scale covariances between density and supply $Cov(N, L)$, and between predator abundance and supply $Cov(P, L)$. Within a site, we consider supply to rocks and predator density at each rock within a single site to be uniform and, therefore, there is no contribution of variance (or covariance) from the rock-scale that contributes to site-scale variance.

Combining Nonlinearities and Spatial Variances to Scale Up

From the empirically derived functions and parameter estimates for growth, competition, predation, and gregarious settlement, we calculated values for each of the seven terms (equation 13.9, a–g) in the regional model (figure 13.3) Spatial variation in adult abundance has a negative effect on adult growth (figure 13.3a). Since *P. cinctipes* are highly aggregated, a large proportion of adults live at higher than average density and the overall effect of variation is to almost cancel out the growth of adults. The overall effect of spatial variation in adult density, predator abundance, and larval supply is to decrease recruitment per rock (figure 13.3, the sum of scale transition terms b–g). Of the five components of equation 13.9 that contribute to the scale transition, only three are large enough to see in figure 13.3. First, adults compete with settlers, which results in an overall negative effect of adult aggregation on recruitment. Therefore, $1/2\ \partial^2 h/\partial N^2\ Var(N)$ (equation 13.9b) contributes to the scale transition. Second, larvae settle gregariously with adults, and adults and larvae positively covary, resulting in a positive contribution to the scale transition through the joint nonlinearity of adults and larvae,

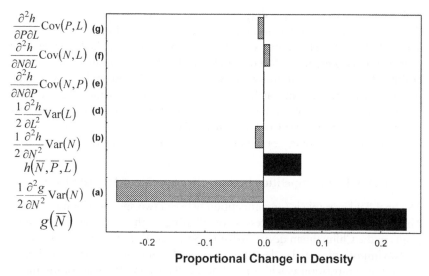

$$\frac{\partial^2 h}{\partial P \partial L} \text{Cov}(P,L) \text{ (g)}$$

$$\frac{\partial^2 h}{\partial N \partial L} \text{Cov}(N,L) \text{ (f)}$$

$$\frac{\partial^2 h}{\partial N \partial P} \text{Cov}(N,P) \text{ (e)}$$

$$\frac{1}{2}\frac{\partial^2 h}{\partial L^2} \text{Var}(L) \text{ (d)}$$

$$\frac{1}{2}\frac{\partial^2 h}{\partial N^2} \text{Var}(N) \text{ (b)}$$

$$h(\overline{N},\overline{P},\overline{L})$$

$$\frac{1}{2}\frac{\partial^2 g}{\partial N^2} \text{Var}(N) \text{ (a)}$$

$$g(\overline{N})$$

-0.2   -0.1   0.0   0.1   0.2

Proportional Change in Density

Figure 13.3 Scale transition for *P. cinctipes*. The scale transition terms (equation 13.9 a–g) are shown in gray and the mean-field model components in black. The calculations are based on fitted functions, parameter estimates, and spatial variances and covariances from field and laboratory data. All components are densities (biomass per rock) relative to the regional mean. Therefore, adult growth in the mean-field model is an increase of 24%, which is cancelled out by the scale transition for growth (a). Overall, the scale transition terms for recruitment (b–g) reduce recruitment at the regional scale compared to the mean-field model for recruitment.

$\partial^2 h/\partial N \partial L$ Cov(*N*, *L*) (equation 13.9f). Third, increasing larval supply increases predation as the number of settlers outstrips the protective effect of adults, resulting in a negative effect of $\partial^2 h/\partial P \partial L$ Cov(*P*, *L*) (equation 13.9g). Nonlinearity with respect to larval supply is small so that even with the large variance in larval supply $1/2\ \partial^2 h/\partial L^2$ Var(*L*) (equation 13.9d) does not contribute substantially to the scale transition for recruitment. Adults decrease predation on settlers and there is negative covariance between adults and predators; this results in a negative but negligible contribution to recruitment through $\partial^2 h/\partial N \partial P$ Cov(*N*, *P*) (equation 13.9e). Surprisingly, the pronounced spatial variation in recruitment does not have a strong effect on regional dynamics. Though spatial variation in larval supply can be extremely large, an average adult biomass is ~200× larger than settler biomass. The net effect is that competition between adults is more important than recruitment in determining regional biomass.

From this empirical exploration of scale transitions in *P. cinctipes*, we can draw several conclusions. First, spatial variation in recruitment decreases regional recruitment compared to the mean-field model. Supply-side theory and experiments demonstrate large differences in local dynamics due to spatial variation in larval supply; our results demonstrate the importance of these local changes to regional recruitment. Second, despite enormous spatial variation in larval supply demonstrated for this and many other marine species, recruitment has a relatively small effect on overall standing stock for iteroparous species. Finally, while adult aggregation has a negative effect on adult growth rate, it has positive effects on recruitment through joint nonlinearities with larval supply and predators as a result of gregarious settlement and facilitation. Since gregarious settlement (Burke 1986; Pawlik 1992) and facilitation (Bertness and Grosholz 1985; Bertness and Callaway 1994; Bertness and Leonard 1997) are relatively common among marine organisms, the importance of these joint nonlinearities may be a general phenomenon.

Example 3: Competition between Two Species of Annual Plants

The final field example is for competition between two species of annual plants. The aim was to quantify processes contributing to the coexistence of annual plants in the Chihuahuan desert, Arizona (Sears 2004). In this system, we envisage two important spatial scales: the scale of local communities, corresponding in size to the competition neighborhood of the plants, and the scale of the metacommunity where local communities are linked by dispersal of seed. The spatial scale of the metacommunity is perhaps the most natural one for examining coexistence processes, because net population dynamics at the metacommunity scale will reflect spatial variation in density and in the abiotic environment. However, the importance of metacommunity processes in providing opportunities for species coexistence, such as competition / colonization trade-offs and environmental heterogeneity, have been difficult to test, given the complexity of natural systems. Here we show how community models can be designed to be parameterized with readily available field data to quantify an important metacommunity process contributing to coexistence, the spatial storage effect (Chesson et al., chapter 12).

We compared the contribution of local-scale and metacommunity-scale processes to species coexistence by comparing the effects of local competition with the influence of the spatial storage effect. The focal species were *Erodium cicutarium* and *Phacelia popeii* (hereafter *Erodium* and *Phacelia*), rosette-forming annual plants, that germinate with the winter rains, and complete their lifecycle in the spring. In the years immediately before this study, the population of *Erodium*, a nonnative, went from low abundance at the study site to being the dominant winter annual species. *Phacelia* is native to the region, and has highly variable abundance depending on rainfall. Though sparse, *Phacelia* was the second most common annual plant during the course of the study.

The spatial storage effect can arise when there is spatial environmental variation, each species has a particular patch type that is most favorable to it (i.e., each species has an environmental niche), and species disperse between locations. If a species has a positive covariance between the favorability of the environment and the competition that it experiences at different locations, variation in the environment acts to reduce the population growth rate of the species at the meta-community scale (Chesson 2000). When the growth rates of high density species are more limited by this mechanism than are the growth rates of low-density species, then variation in the environment acts to promote coexistence. Many studies have shown that competition intensity changes with variation in the environment (see review in Goldberg et al. 1999). But when does environmental variation make a difference for community dynamics? Using carefully designed models, we can make progress in understanding how metacommunity processes contribute to diversity maintenance in complex environments.

Model to Scale Up the Dynamics of Annual Plants

In chapter 12, we formulated a model for the local dynamics of an annual plant community and scaled that model up to give the dynamics of the metacommunity. In that model, we saw that the dynamics of the metacommunity depend on a number of spatial covariances (Chesson et al., chapter 12, equation 12.23). Sears (2004) focuses in on one component of that model—the effect of spatial variation on plant yield and how this contributes to species coexistence through the spatial storage effect. The spatial storage effect involves $\text{Cov}(E, C)$, the covariance of a species' response to the environment E, and response to competition C (Chesson et al., chapter 12, equation 12.26). We begin by formulating a local model for plant yield (quantified in the field as the number of inflorescences produced per adult, a surrogate for seed production), which is the result of two local processes, the response of the plant to the local environment and the response of the plant to the presence of conspecific or interspecific neighbors. In this case, the driver of the plant environmental response is the vigor (V_j) parameter described in chapter 12, table 12.1. We use the symbol Y_{jx} as the final yield combining vigor (V) with the yield per unit biomass (Y) and competition (C) of table 12.1. To facilitate data analysis and improve the numerical accuracy of the approximations we use a log scale. The model is defined in chapter 12 in the way most conducive to understanding, but the log scale works better for data analysis and numerical calculations. Thus, we write the yield for species j at each location x as

$$\ln Y_{jx} = E_{jx} - C_{jx}, \tag{13.10}$$

where E_{jx} is the log yield (as growth or fecundity) of an individual in response to environmental conditions alone, and C_{jx} is the competitive effect of neighbors on focal plants, defined as the reduction in log yield. Thus, E_{jx} is the log of the product of the vigor and yield per unit biomass parameters of chapter 12, and C_{jx} is the log of the C of chapter 12. While we generally define C_{jx} as a competition term, there is no sign restriction, and thus the theory permits facilitation when C_{jx} is negative. Yield is the antilog of equation 13.10,

$$Y_{jx} = e^{(E_{jx} - C_{jx})} = e^{E_{jx}} e^{-C_{jx}}. \tag{13.11}$$

Thus, equation 13.11 is our local model for plant yield.

To scale up from local community dynamics to the regional scale of the meta-community where locations are now linked by dispersal of seed, we take the spatial average of the local dynamics over all locations. We expect both the environmental response E and competition C to vary in space, so averaging over both sides of equation 13.11, and expanding the average of the product (chapter 12, equation 12.B2.6), the regional yield is

$$\overline{Y}_j = \overline{e^{E_j}e^{-C_j}} = \overline{e^{E_j}}\,\overline{e^{-C_j}} + \text{Cov}(e^{E_j}, e^{-C_j}). \qquad (13.12)$$

This equation gives the exact yield for the metacommunity scale. Approximating e^{E_j} and e^{-C_j} of the mean terms as second order Taylor polynomials (chapter 12, equation 12.B2.5), and taking a linear approximation for e^{E_j} and e^{-C_j} in the co-variance term (chapter 12, equation 12.B2.8) we get

$$\overline{Y}_j = \left(e^{\overline{E}_j} + \frac{1}{2}e^{\overline{E}_j}\,\text{Var}(E_j)\right)\left(e^{-\overline{C}_j} + \frac{1}{2}e^{-\overline{C}_j}\text{Var}(C_j)\right) - e^{(\overline{E}_j - \overline{C}_j)}\text{Cov}(E_j, C_j), \quad (13.13)$$

where $\text{Var}(E_j)$ is the spatial variance of the environmental response and $\text{Var}(C_j)$ is the spatial variance of the competitive response. We use equation 13.13 for analyses but a more intuitive form is obtained by expanding the products and dropping a higher order term:

$$\overline{Y}_j = \underbrace{e^{(\overline{E}_j - \overline{C}_j)}}_{\substack{\text{mean-field} \\ \text{model}}} + \underbrace{\frac{1}{2}e^{(\overline{E}_j - \overline{C}_j)}\,\text{Var}(E_j)}_{a} + \underbrace{\frac{1}{2}e^{\overline{E}_j - \overline{C}_j}\text{Var}(C_j)}_{b} - \underbrace{e^{(\overline{E}_j - \overline{C}_j)}\text{Cov}(E_j, C_j)}_{c}. \quad (13.14)$$

Like the regional models of the previous examples, the structure of this model tells us how plant response to competition is modified at the metacommunity scale and it tells us what we need to measure in field studies to quantify the importance of metacommunity-scale processes. The dynamics of the metacommunity are given by the mean-field model, which represents the average local response of each species to environment and competition, plus terms (a–c) that correct the mean-field model to account for the metacommunity structure.

Thus, spatial variation potentially plays an important role in the metacommunity. In addition to local responses (represented by the mean-field model), average per capita yield for each species depends on spatial variation in its response to the environment (equation 13.14a) and to competition, (equation 13.14b) and on the spatial covariance of its response to environment and competition (equation 13.14c). Equation 13.14 tells us that spatial variance in response to the environment and competition increases the average local response to the environment, but that a positive covariance between these responses reduces a species' yield at the metacommunity scale. However, of most interest here is how these quantities change with relative abundance of the species for then we can determine how they contribute to species coexistence (Chesson et al., chapter 12). The most important effects come from the behavior of Cov(E, C).

At a metacommunity-scale, Cov(E, C) can provide a brake to population growth rates that is, in a sense, independent of local processes that promote coexistence (such as resource partitioning). Positive Cov(E, C) means that the response of a species to competition is greatest where the environment is most favorable. To promote coexistence between species, the numerically dominant competitor must have a greater positive Cov(E, C) than that of the low-density species. We expect that high density species may often have a greater Cov(E, C) than low-density species because they are likely to experience intense intraspecific competition in their most favorable locations in the environment. When there are species-specific differences in environmental preference, low-density species are more likely than high density species to experience competitive release in their best growing locations, and for low-density species there is never likely to be a strong relationship between competition intensity and the favorability of the environment (see chapter 12 for a comprehensive discussion of how this works for different types of spatial environmental variation).

There are a number of reasons why it is essential to think of Cov(E, C) as an independent, emergent process. In the first place, Cov(E, C) requires environmental heterogeneity, and thus cannot be considered at small spatial scales. The effects of Cov(E, C) are also somewhat independent of neighborhood-scale resource partitioning, such as light-nutrient limitation differences or differences in rooting depth. While resource partitioning at the neighborhood scale can promote coexistence in the absence of environmental variation, likewise Cov(E, C) can promote coexistence when the relative intensity of intraspecific and interspecific neighborhood competition predicts competitive exclusion, as we show here. Finally, it is convenient to think of it as an independent mechanism because it occupies a separate position in population models, allowing us to analyze how this interaction contributes independently to coexistence. In this form, local competition intensity and Cov(E, C) are community-level parameters, allowing comparison between species, and between systems.

If a metacommunity structure is important in explaining the regional dynamics in this system, such as in promoting the coexistence of *Erodium* and *Phacelia*, then we should be able to detect and quantify the variation-dependent terms of equation 13.14 (a–c) in the field, in addition to the local response (the mean-field model) of each species in the community.

Measuring Nonlinearities in Annual Plants

In contrast to the previous two field examples, the nonlinearities here have been defined explicitly instead of being expressed as general functions. That is, yield is a nonlinear function of environment and competition (equation 13.11). As a result of these nonlinearities, the quantities to measure (equation 13.14) are E_j and

C_j. Sears (2004) used a field experiment to quantify these terms. Metacommunity dynamics were estimated within a 400 m² field site in the Chihuahuan desert of Arizona. Here, we envisage local-scale communities in 1 m² quadrats, linked by dispersal to make up the metacommunity at the 400 m² site scale. This habitat was dominated by *Erodium*, but included a sparse distribution of *Phacelia* and negligible densities of other annual species. A standard neighborhood competition experiment was used to compare the yield of *Erodium* and *Phacelia* with and without *Erodium* competitors. The experiment did not look at the intraspecific competition of *Phacelia*, or at its effects on *Erodium*, because *Phacelia* was at very low densities and unlikely to have strong neighborhood effects.

The experiment was done in a randomized block design, where in each of ten, 1 m² blocks distributed across the study site, two paired *Erodium* and two paired *Phacelia* plants were chosen to represent responses to the local (block) conditions. Plants in each pair were randomly assigned to removal or control treatments, where immediate neighbors were either removed or retained. The plant response to the environment (E_j) is estimated as the ln-transformed yield of plants in the removal treatment, $\ln(Y_{j,removal})$, thus favorable environments for either species were those blocks in which they had the most positive response in the absence of neighbors. The response to competition (C_j) is estimated as the log response-ratio of plants in the removal and control treatments, $\ln(Y_{j,removal}/Y_{j,control})$. The relevant nonlinearities (equation 13.14) are derived from corrected values of E_j and C_j, described below. Local-scale competition was estimated as the mean-field response to neighbors (equation 13.14), which excludes the effects of variation in E_j or C_j.

Measuring Spatial Variances and Covariances

In previous examples, variances and covariances involved densities that are straightforward to measure directly. Here, these quantities are measured in a field experiment by assaying the responses of individual plants to the presence and absence of neighbors. With this approach, there is no independent estimate of response to the environment and response to competition, because competitive response is estimated as the log response ratio of plants in the removal and control treatments. Thus, to accurately estimate the variances and covariance, we must correct our estimates of E_j and C_j to remove their common error. This requires a customized approach (Sears 2004; Chesson and Sears, in prep.). In this approach, the statistical model assumes that the environmental response varies between blocks within the metacommunity, but that plants of the same species within a block (the local community) share a common environmental response. The statistical model simultaneously estimates E_j, C_j, Var(E_j), Var(C_j), and Cov(E_j, C_j).

Combining the Model with Field Data to Scale Up

With measurements for the nonlinear function for yield and estimates of spatial variances and covariances in hand, we can now scale up the dynamics for yield from the local scale of the competition neighborhood to the scale of the

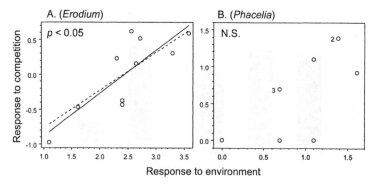

Figure 13.4 Relationship between plant responses to the environment and competition for *Erodium* and *Phacelia*. Each point gives the raw data from individual blocks, in units of ln(inflorescence number). One was added to the *Phacelia* inflorescence number before ln-transformation. Response to the environment is estimated as the response of target plants in removal treatments and response to the competition is estimated as the log response ratio of plants in removal and control treatments (see text). Model fitting shown for significant $Cov(E, C)$. Solid line shows least squares fit to raw data. Broken line shows maximum likelihood fit, which corrects for sampling error. Figures next to points in B show numbers of overlapping points.

metacommunity. Here we are not simply interested in the difference of the mean-field model compared to the spatial model but also in the magnitude of the different scale transition components for the two species (equation 13.14), since the relative magnitude of these components determines the importance of the spatial storage effect. For inflorescence number, *Erodium* had a significant positive $Cov(E, C)$, but did not have significant local-scale intraspecific competition (figures 13.4A, 13.5A). Figure 13.4A indicates that the lack of local scale competition was due to a balance between facilitation in poor environments, and competition in favorable environments, which in turn contributed to the positive $Cov(E, C)$. In contrast, *Phacelia* experienced intense interspecific competition from *Erodium*, but did not have a significant $Cov(E, C)$ (figures 13.4B, 13.5B).

Thus, covariance between plant response to the environment and competition, a process occurring only at the scale of the metacommunity, retarded the population growth rate of the dominant competitor, *Erodium*, while local-scale processes did not. While this study did not show that *Erodium* and *Phacelia* will continue to coexist, it demonstrated that, through the spatial storage effect, variation in the environment contributes to the possibility of coexistence, and is therefore important for metacommunity dynamics. Additionally, this study only narrowly considered the spatial storage effect as expressed through one measure of plant yield (inflorescence number). Other forms of environmental response, leading (for example) to variation in survival or germination, will have separate, contributions to the spatial storage effect and potentially provide other avenues for coexistence (Chesson et al., chapter 12).

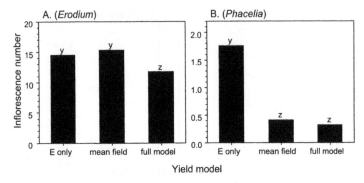

Figure 13.5 Population model fits for per capita inflorescence number production of *Erodium* and *Phacelia*, showing the estimates of different models. The E-only model gives the spatial average yield as the response to the environment in the absence of neighbors. The mean-field model (first term in equation 13.14) gives the spatial average yield in the presence of neighbors, without accounting for variation in the environment or in density dependence. This corresponds to the mean effect of local-scale competition. The full model gives the spatial average yield in the presence of neighbors, accounting for variation in the environment and density dependence (full model in equation 13.13–13.14). Lowercase letters show statistically homogeneous groups.

The metacommunity approach described here contrasts with traditional ANOVA tests of neighborhood competition experiments, which contain the inherent assumption that the environment is uniform because the test compares mean plant responses and does not include the effects of variance. This traditional approach, which corresponds here to the mean-field model, has two problems if we are evaluating competition in a variable environment. First, it does not include the response-enhancing effects of variance (equation 13.14). Second, and most importantly, it does not take into account the effects of $Cov(E, C)$ which may either enhance or retard plant response at the metacommunity scale, depending on the sign of the covariance (equation 13.14). Thus, coexistence predictions from traditional neighborhood competition experiments cannot be scaled up to the metacommunity. However, ANOVA tests can be used to test whether the local-scale effects of competition are significant.

Conclusions

In the three examples above, we have demonstrated the concepts and properties of the scale transition in translating the effects of local scale processes to the scale of the metacommunity. Although concerning different field systems and different processes, the examples share a common three step approach to studying the scale transition: derive a model to scale up, measure nonlinearities, and measure spatial variances and covariances. In each example, the scale transition model identifies the important component mechanisms that change dynamics at the metacommunity scale. The structure of the model for the metacommunity scale is the same in each example, consisting of a mean-field model representing local dynamics modified by new terms to correct for spatial structure. The new terms in each model all represent interactions between nonlinearity and spatial variation. These correction terms, which can be thought of as representing spatial mechanisms, were then quantified by applying standard field experiments and sampling designs. In each example, local nonlinearities were estimated by varying levels of the nonlinear factors in an experiment and fitting models to describe the shape of the nonlinearities. Variances and covariances were measured using standard ANOVA or custom maximum likelihood procedures. In each example components of the scale transition with large magnitude were detected, confirming that metacommunity processes were, or had potential to be, important to the dynamics of the system at the larger scale.

Literature Cited

Addicott, J. F., J. M. Aho, M. F. Antolin, D. K. Padilla, J. S. Richardson, and D. A. Soluk. 1987. Ecological neighborhoods: Scaling environmental patterns. Oikos 49:340–346.

Armsworth, P. R. 2002. Recruitment limitation, population regulation, and larval connectivity in reef fish metapopulations. Ecology 83:1092–1104.

Bertness, M. D., and R. Callaway. 1994. Positive interactions in communities. Trends in Ecology and Evolution 9:191–193.

Bertness, M. D., and E. Grosholz. 1985. Population dynamics of the ribbed mussel, *Geukensia demissa:* The costs and benefits of an aggregated distribution. Oecologia 67:192–204.

Bertness, M. D., and G. H. Leonard. 1997. The role of positive interactions in communities: Lessons from intertidal habitats. Ecology 78:1976–1989.

Bolker, B., and S. W. Pacala. 1997. Using moment equations to understand stochastically driven spatial pattern formation in ecological systems. Theoretical Population Biology 52:179–197.

Burke, R. D. 1986. Pheromones and the gregarious settlement of marine invertebrate larvae. Bulletin of Marine Science 39:323–331.

Chesson, P. 1996. Matters of scale in the dynamics of populations and communities. Pages 353–368 *in* R. B. Floyd, A. W. Sheppard, and P. J. De Barro, eds. *Frontiers of population ecology.* CSIRO Publishing, Melbourne, Australia.

———. 1998a. Making sense of spatial models in ecology. Pages 151–166 *in* J. Bascompte, and R. V. Solé, eds. *Modeling spatiotemporal dynamics in ecology.* Landes Bioscience, Austin, TX.

———. 1998b. Recruitment limitation: A theoretical perspective. Australian Journal of Ecology 23: 234–240.

———. 2000. General theory of competitive coexistence in spatially-varying environments. Theoretical Population Biology 58:211–237.

Cooper, S. D., L. Barmuta, O. Sarnelle, K. Kratz, and S. Diehl. 1997. Quantifying spatial heterogeneity in streams. Journal of the North American Benthological Society 16:174–188.

Cooper, S. D., S. Diehl, K. Kratz, and O. Sarnelle. 1998. Implications of scale for patterns and processes in stream ecology. Australian Journal of Ecology 23:27–40.

De Jong, G. 1979. The influence of the distribution of juveniles over patches of food on the dynamics of a population. Netherlands Journal of Zoology 29:33–51.

Donahue, M. J. 2003. The interaction of dispersal and density dependence: resource variation, competition, and gregarious settlement. Ph.D. dissertation. University of California, Davis.

————. 2004. Size-dependent competition in a gregarious porcelain crab, *Petrolisthes cinctipes* Randall (Anomura: Porcellanidae). Marine Ecology Progress Series 267:195–207.

Downes, B. J., P. S. Lake, and E. S. G. Schreiber. 1993. Spatial variation in the distribution of stream invertebrates: Implications of patchiness for models of community organization. Freshwater Biology 30:119–132.

Gaines, S. D., and J. Roughgarden. 1985. Larval settlement rate: A leading determinant of structure in an ecological community of the marine intertidal zone. Proceedings of the National Academy of Sciences, USA 82:3707–3711.

Goldberg, D. E., T. Rajaniemi, J. Gurevitch, and A. Stewart-Oaten. 1999. Empirical approaches to quantifying interaction intensity: Competition and facilitation along productivity gradients. Ecology 80:1118–1131.

Grosberg, R. K., and D. R. Levitan. 1992. For adults only? Supply-side ecology and the history of larval biology. Trends in Ecology and Evolution 7:130–133.

Hassell, M. P., R. M. May, S. W. Pacala, and P. L. Chesson. 1991. The persistence of host-parasitoid associations in patchy environments. 1. A general criterion. American Naturalist 138:568–583.

Ives, A. R. 1988. Covariance, coexistence and the population dynamics of two competitors using a patchy resource. Journal of Theoretical Biology 133:345–361.

Jensen, G. C. 1989. Gregarious settlement by megalopae of the porcelain crabs *Petrolisthes cinctipes* (Randall) and *Petrolisthes eriomerus* (Stimpson). Journal of Experimental Marine Biology and Ecology 131:223–232.

————. 1991. Competency, settling behavior, and postsettlement aggregation by porcelain crab megalopae (Anomura: Porcellanidae). Journal of Experimental Marine Biology and Ecology 153: 49–62.

Jones, G. P., M. J. Milicich, M. J. Emslie, and C. Lunow. 1999. Self-recruitment in a coral reef fish population. Nature 402:802–804.

Melbourne, B. A., and P. Chesson. Submitted. The scale transition: Scaling up population dynamics with field data.

Nisbet, R. M., S. Diehl, W. G. Wilson, S. D. Cooper, D. D. Donalson, and K. Kratz. 1997. Primary-productivity gradients and short-term population dynamics in open systems. Ecological Monographs 67:535–553.

Pacala, S. W., and S. A. Levin. 1997. Biologically generated spatial pattern and the coexistence of competing species. Pages 204–232 *in* D. Tilman, and P. Kareiva, eds. Spatial ecology: The role of space in population dynamics and interspecific interactions. Princeton University Press, Princeton, NJ.

Pawlik, J. R. 1992. Chemical ecology of the settlement of benthic marine invertebrates. Oceanography and Marine Biology Annual Review 30:273–335.

Searle, S. R., G. Casella, and C. E. McCulloch. 1992. *Variance components*. John Wiley and Sons, New York.

Sears, A. L. W. 2004. Quantifying the effects of spatial environmental variation on dynamics of natural plant populations: Field tests for covariance between the environment and competition. Ph.D. dissertation. University of California, Davis.

Swearer, S. E., J. E. Caselle, D. W. Lea, and R. R. Warner. 1999. Larval retention and recruitment in an island population of a coral-reef fish. Nature 402:799–802.

Swearer, S. E., J. S. Shima, M. E. Hellberg, S. R. Thorrold, G. P. Jones, D. R. Robertson, S. G. Morgan et al. 2002. Evidence of self-recruitment in demersal marine populations. Bulletin of Marine Science 70:251–271.

Thorrold, S. R., G. P. Jones, M. E. Hellberg, R. S. Burton, S. E. Swearer, J. E. Neigel, S. G. Morgan et al. 2002. Quantifying larval retention and connectivity in marine populations with artificial and natural markers. Bulletin of Marine Science 70:291–308.

Underwood, A. J., and M. G. Chapman. 1996. Scales of spatial patterns of distribution of intertidal invertebrates. Oecologia 107:212–224.

PART 4

EMERGING AREAS AND PERSPECTIVES

Marcel Holyoak and Mathew A. Leibold

This section highlights some developing areas of study and synthetic perspectives in ecology and evolution that we believe are likely to contribute fruitfully to future exploration of metacommunity ideas.

One of the largest challenges in studying metacommunities is to link empirically observed patterns with causal mechanisms. The difficulty of this task is made apparent by the recent correspondence about how to test neutral community models (NCMs; e.g., Bell 2001; Hubbell 2001; Chave 2004). Bell stated "the strong version is that the NCM is so successful precisely because it has correctly identified the principal mechanism underlying patterns of abundance and diversity." Furthermore, "the contemplation of pattern is unlikely to succeed in distinguishing between neutral and adaptationist theories of diversity" (Bell 2001, 2418). Hence, Bell advocated testing mechanisms, while rejecting evidence from patterns. Enquist et al. (2001) responded that we can test the NCM by testing for the occurrence of local adaptation and differences in species' demography (i.e., mechanisms). Bell (2002) argued that these would not be sufficient tests of the NCM, but rather we need to test whether adaptation and demography influence local species composition and diversity. Bell (2001) also states that "the neutral model does not deal with the details, only with their consequence," and therefore Enquist et al. (2001) were discussing mechanisms that were not present in Bell's NCM. This debate illustrates the ambivalence many ecologists feel over the value of testing patterns versus mechanisms; it also illustrates how difficult it is to test NCMs.

How to test neutral theory is at this point an open question and one that is poorly understood by many ecologists. Chase et al. (chapter 14) illuminate this issue by placing NCMs into a broader metacommunity perspective. Usefully, they clarify and elaborate on the differences between the four metacommunity paradigms presented by Holyoak et al., chapter 1. A broad range of emergent patterns and differences in assumptions are discussed. Chase et al. discuss empirical evidence for four patterns that have been central to tests of NCMs: (1) the distribution of the relative abundance of species; (2) species composition along environmental and spatial gradients; (3) species composition and diversity in relation to dispersal rates; and (4) patterns of community composition through time both prior to and following disturbances. The evidence serves as a springboard for suggesting future directions in developing more general metacommunity models that could apply to any kind of community.

Building on a theme of testing the neutral perspective, McPeek and Gomulkiewicz describe in chapter 15 methods that are available from population and metapopulation genetics to test neutral model predictions about patterns of species diversity and coexistence. The authors modify genetic diversity and drift metrics to predict temporal drift in species diversity, which parallels Hubbell's (2001) concept of ecological drift. The methods provide quantitative tools that could be used to study and predict temporal change in species composition in the absence of species' fitness differences (an assumption of the neutral community model). The authors distinguish mechanisms capable of causing long term persistence from those that contribute to stability without creating long term persistence. This distinction is useful because it allows us to consider transient dynamics that are an important feature of the NCM (Hubbell 2001) as well as other kinds of spatial models (e.g., Hoopes et al., chapter 2). McPeek and Gomulkiewicz end by discussing evidence for neutral dynamics from their own work on damselfly species in the northeastern United States. Altogether the chapter provides a tantalizing example of neutral dynamics, as well as providing us with methods to study an important aspect of such dynamics.

An area of metacommunity ecology that was touched on in several chapters (e.g., chapters 3 and 5) is the contribution of behavior to metacommunity dynamics. In chapter 16, Resetarits and colleagues use a system consisting of both temporary and permanent ponds to explore the potential importance of habitat selection behavior for local and regional community structure. They present a general theoretical framework for considering habitat selection behavior in metacommunities, and illustrate it using anurans and beetles. Habitat selection behavior is highly relevant to the species sorting and mass effect paradigms, and this chapter does a thorough job of describing the kinds of links that may be found and presents some evidence that suggests species are using both temporary and permanent ponds.

Chapters 1, 6, 7, 9, and 14 of this book involve studying patterns of species diversity. Another species diversity pattern that has intrigued ecologists since Elton published his volume on Animal Ecology in 1927 is the relationship between local and regional species diversity. This pattern has generated a lot of discussion about whether local versus regional species diversity plots can be used to test for the saturation of local communities. Saturation could occur if niches become saturated in local communities and no further species could invade (Terborgh and Faaborg 1980). In chapter 17, Shurin and Srivastva discuss a broad range of theories about mechanisms that can influence these patterns. The review raises points about the importance of spatial scale for both local and regional studies of species diversity, and leads us to articulate which questions can be answered by studying local-regional diversity patterns.

In chapter 18, the metacommunity perspective is expanded in ways that draw parallels with systems concepts developed by Holling and colleagues for entire ecosystems (Holling 1986, 1992, 1994; Allen and Holling 2003). It is not just organisms that move across community boundaries: energy, water, and important biologically-active chemicals do so as well. It is natural therefore to extend the framework of metacommunities to deal with ecosystem processes over space, termed "metaecosystems" (Loreau et al., chapter 18). Spatial coupling of ecosystems leads to critical emergent properties, which the authors illustrate via such topical issues as species diversity and ecosystem productivity, and material fluxes and ecosystem organization at large spatial scales.

This book has primarily illuminated ecological questions about metacommunities. Not surprisingly, metacommunity dynamics have the potential to greatly alter patterns of evolution. In chapter 19 Leibold et al. develop ideas from systems ecology to consider metacommunities as complex adaptive systems. In particular chapter 19 explores the potential for adaptation (and coadaptation) in the four metacommunity perspectives from chapter 1. This is an exciting topic that challenges us to revisit ideas about levels of selection, and we believe that metacommunity evolution has the potential to become a broad area of future study, building on the initial thoughts that are presented here. Developing ideas about metacommunity evolution could be especially valuable for providing a framework for thinking about species' traits and life-history patterns in ways that go beyond things like the competition-colonization trade-offs considered in this volume.

The final chapter by Holt et al. has two main purposes: to discuss a broad range of other topics that could logically contribute to studying metacommunities, and to review some of the emerging themes from this book. Rather than getting into a "metasynthesis" of this book, the review of themes is kept brief and the editors also provide a concluding coda section that gives numbered points summarizing important findings and which could serve as useful material for group discussion.

Literature Cited

Allen, C. R., and C. S. Holling. 2002. Cross-scale structure and scale breaks in ecosystems and other complex systems. Ecosystems 5:315–318.

Bell, G. 2001. Neutral macroecology. Science 293:2413–2418.

———. 2002. Response. Science 295:1836–1837.

Chave, J. 2004. Neutral theory and community ecology. Ecology Letters 7:241–253.

Elton, C. 1927. *Animal ecology.* Macmillan, New York, N.Y.

Enquist, B. J., J. Sanderson, and M. D. Weiser. 2002. Modeling macroscopic patterns in ecology. Science 295:1835–1836.

Holling, C. S. 1986. The resilience of ecosystems: Local surprise and global change. Pages 292-317 *in*

W. C. Clark, and M. R.E., eds. *Sustainable development of the biosphere.* Cambridge, U.K., Cambridge University Press.

———. 1992. Cross-Scale Morphology Geometry and Dynamics of Ecosystems. Ecological Monographs 62:447–502.

———. 1994. Simplifying the complex—the paradigms of ecological function and structure. Futures 26:598–609.

Hubbell, S. P. 2001. *The unified neutral theory of biodiversity and biogeography.* Princeton University Press, Princeton, N.J.

Terborgh, J. W., and J. Faaborg. 1980. Saturation in bird communities in the West Indies. American Naturalist 116:178–195.

CHAPTER 14

Competing Theories for Competitive Metacommunities

Jonathan M. Chase, Priyanga Amarasekare, Karl Cottenie,
Andrew Gonzalez, Robert D. Holt, Marcel Holyoak,
Martha F. Hoopes, Mathew A. Leibold, Michel Loreau,
Nicolas Mouquet, Jonathan B. Shurin, and David Tilman*

Introduction

Four metacommunity frameworks are introduced by Holyoak et al. in chapter 1 and Leibold et al. (2004): the patch dynamic, species sorting, mass effects, and neutral models. These have been utilized to address some of the most fundamental questions that ecologists ask: what factors influence the maintenance of diversity, variation in species composition, and the relative abundance of species at local and regional spatial scales? Hubbell's (2001) treatise on his neutral model has been the focus of much recent interest and debate, most likely because of its success in predicting multiple patterns in natural communities despite its omission of key ecological principles such as differences among species' traits (e.g., Bell 2001, 2003; Whitfield 2002; Norris 2003; Chave 2004). However, each of the other model frameworks, with fundamentally different assumptions (namely differences among species in key ecological traits), can also predict multiple patterns in natural communities (e.g., Hanski and Gyllenberg 1997; Chave et al. 2002; Mouquet and Loreau 2002, 2003; Chase and Leibold 2003; Wilson et al. 2003). In this chapter, we give an overview of the four metacommunity frameworks and the specific assumptions and predictions they make. We then review the empirical evidence available and discuss what is needed to differentiate between the predictions of the various frameworks. Note that throughout, we only consider competitive metacommunities; that is, species interactions only occur through competition. We thus ignore important advances incorporating into metacommunities food web interactions (e.g., Holt 1993, 1996, 1997, 2002; reviewed in Holt and Hoopes, chapter 3) and mutualisms (Amarasekare 2004). Future syntheses of metacommunity ecology will be greatly enhanced by recognizing the importance of species interactions other than just competition.

*Order of authorship after the first is alphabetical

There is a great deal of interest in devising empirical tests that can compare the assumptions and predictions of the various metacommunity model frameworks. However, in many cases several competitive metacommunity models can predict identical patterns even though they make fundamentally different assumptions. To date, most empirical tests have either validated a particular assumption (e.g., the form of a trade-off among species), or used a single pattern (e.g., relative abundance, composition) to support, or more rarely falsify, the various hypotheses for how metacommunities are structured. We believe that a more informative way to evaluate the various frameworks is to recognize two points: (1) Aspects of all of the frameworks are likely to be acting simultaneously, and a reasonable approach is to evaluate the relative importance of the various processes, not whether they are right or wrong as an absolute. (2) A pluralistic perspective should utilize all available evidence concerning the validity of both assumptions (e.g., habitat heterogeneity, dispersal rates) and predictions (e.g., diversity, relative abundance, species composition, invasibility, stability). Although such a pluralistic perspective does not allow the strict falsifiability criterion often desired by ecologists when testing the various metacommunity frameworks (e.g., Bell 2003; McGill 2003), it will be very useful when comparing processes and patterns across a variety of ecosystem types and at large spatial scales. Quantitative tests of model predictions are much stronger than qualitative tests, but they are much more difficult to conduct for metacommunities because the theories often involve many parameters and species. Therefore, we concentrate our discussion on qualitative tests.

In the remainder of this chapter, we will first review the key assumptions and predictions of each of the four model frameworks introduced in chapter 1 by Holyoak et al. Next, we will compare a smattering of available empirical evidence to test the various hypotheses. We will conclude that aspects of the assumptions and predictions of each model framework will be observed in different natural systems and at different spatiotemporal scales.

Key Assumptions and Theoretical Predictions of the Models

We begin with by discussing the key assumptions and predictions of the four model frameworks, which are summarized in table 14.1. We specifically focus on predicted patterns of local and regional diversity, the effects of migration rates on local and regional diversity, the effects of migration rates and distance among localities on β-diversity (a critical scalar between local and regional diversity, indicating the degree of species compositional dissimilarity among local communities [Shurin and Srivastava, chapter 17]), the effects of local- and regional-scale disturbances (density-independent events that cause significant mortality) on the transient and final community structure, as well as how local and regional community structure fluctuates through time in the absence of any corresponding

Table 14.1 Summary of predictions from the four frameworks

Effect	Model Prediction			
	Neutral	Patch dynamics	Species sorting	Mass effects
Overall local diversity	Extinction and colonization balance[a]	Extinction and colonization balance	Depends on species interactions	Depends on species interactions and balance between extinction and colonization
Overall regional diversity	Extinction and speciation balance[a]	Depends on competition-colonization trade-off	Same as above and degree of habitat heterogeneity	Same as above and degree of habitat heterogeneity
Relative species abundance	Zero-sum multinomial (skewed toward rare species)[a]	Variable depending on level of migration and degree of interaction[c]	Variable depending on environmental conditions	Variable depending on level of migration[d]
Dispersal effects: local diversity	Increase	Hump-shaped[d]	No effect	Hump-shaped[e]
Dispersal effects: regional diversity	Decrease	Decrease	No effect	Decrease[5]
Dispersal effects: β-diversity	Decrease[a,b]	Global: no effect Local: decrease	No effect	Global: decrease[e] Local: decrease
Local disturbance	Return immediately	Unpredictable	Return immediately	Return following succession
Regional disturbance	Random walk	Return following succession	Return immediately	Return following succession
Temporal variation: local	Variable	Variable	Static unless environment changes	Static unless environment changes
Temporal variation: regional	Variable	Static unless environment changes	Same as above	Same as above

Note: Predictions without superscripts are speculation not yet backed up by specific theory.

[a]Hubbell 2001, Bell 2001; [b]Chave and Leigh 2002; [c]Chave et al. 2002; [d]Mouquet et al. 2002; [e]Mouquet and Loreau 2003

environmental change. We note that there are a variety of other phenomena predicted by the metacommunity models, such as patterns of range size and community invasibility, but we have left these out for brevity.

At the outset, one of the most striking observations from table 14.1 is that every pattern can be predicted by more than one model framework. Thus, using data from only one pattern will not provide a rigorous test in support or refutation of any of the model frameworks. While in most cases, theoretical models have generated the predictions discussed below, there are not explicit theoretical predictions for some of the responses of each model framework. For example, while Mouquet and Loreau (2003) have theoretically examined many of the patterns expected when metacommunity dynamics are dominated by mass effects, similar patterns are not as well explored for patch dynamic processes. Likewise, although responses to disturbance have been used to differentiate neutral models from niche models (e.g., Hubbell et al. 1999), the specific responses to disturbance for some model frameworks (e.g., the mass effects framework) have not been explored. In these few cases, to be complete, but short of developing novel theoretical predictions, we have speculated as to what the predicted responses are most likely to be for the model frameworks. Future theoretical work should verify these and other related model predictions. Finally, although our discussion will primarily focus on the most basic versions of each framework, a variety of complexities have been added to each metacommunity framework, such as density-dependence to the neutral model (Chave et al. 2002), and heterogeneity to the patch dynamics model (Shurin et al. 2004). However, to compare among the predictions for some circumstances, we include in our discussions some of these alterations that allow them to fit more complex situations.

Neutral Framework

Neutral models assume that individuals of all species have equal net fitness (rates of birth, death, and competitive exclusion; this assumption is often thought to be synonymous with all individuals of each species being identical, but technically is not). The first application of the neutral model to ecological processes was by Caswell (1976), but the majority of the predictions and assumptions we discuss below are derived from Hubbell (2001). Note, however, that all neutral models do not make the same specific assumptions or predictions (see review in Chave 2004). In Hubbell's neutral model, species play a zero sum game, where there are a fixed number of individuals (of all species) that can exist in a metacommunity. Because individuals are neutral with respect to their fitness, there is no stable equilibrium that allows individual species to coexist indefinitely. Instead, each species is on a random walk to extinction. However, at the metacommunity level, so long as immigration and speciation occur on a fast enough time scale, they can balance extinction rates and maintain high levels of species diversity in any given locality. This model also assumes that there is no variation among localities in any sort of environmental conditions that would influence a species' birth or death rates.

Because neutral models predict that species abundances vary through time, many of their predicted community patterns also vary through time. Thus, their predicted patterns are usually considered as a long-term average. Some specific predictions of neutral models (Bell 2001, 2003; Hubbell 2001; Chave and Leigh 2002; Chave et al. 2002; Volkov et al. 2003; Chave 2004) include the following: (1) Local diversity should increase with increasing rates of migration (connectance) among localities. This is because, as with MacArthur and Wilson's (1967) equilibrium theory of island biogeography, increasing immigration rates increases local diversity so long as death (extinction) rates are constant. This implicitly assumes that dispersal is limited. If dispersal were unlimited (that is, if dispersers have an equal probability of reaching any locality within a metacommunity) local diversity would be low because no differences among local communities could develop. (2) Regional diversity should increase with increasing speciation rates in the metacommunity, but at a constant rate of speciation, regional diversity should decrease with increasing rates of dispersal. This is because increased dispersal hastens the time to local extinction, which will decrease regional diversity. This also depends on localized disturbance, whereas if disturbance were global, diversity would again be low. (3) The neutral model predicts that β-diversity will increase at greater distances among localities and with lower dispersal rates. Again, this assumes that dispersal is limited and / or localized (localized dispersal implies "dispersal limitation"). If dispersal were unlimited in a neutral model, there would be no specific effect of dispersal rates or distance on β-diversity. (4) Within a metacommunity, disturbances that are more widespread will tend to lead to more different postdisturbance communities. Composition is expected to change less with disturbance than relative abundance. (5) The relative abundance of any particular species will change through time due to stochastic processes, and be uncorrelated with variation in environmental conditions. Neither diversity nor relative species abundance will vary with variation in environmental conditions, because this model framework assumes that species' dynamics are not responsive to heterogeneous environmental conditions.

Patch Dynamic Framework

Like the neutral model, the patch dynamic framework implicitly assumes that there is no spatially fixed variation in the environmental conditions among patches, at least in the context of what is relevant to the interacting organisms. In addition the patch dynamic framework assumes that each species has a finite rate of extinction in a patch. When there are no differences among species in traits, the patch dynamic framework converges with the neutral model and predicts that species cannot coexist indefinitely (Yu and Wilson 2001; Chave et al. 2002). Several modifications from this limiting case allow coexistence (Amarasekare 2003). For example, species can coexist under many (but not all) parameter values so long as there is a trade-off among species' relative abilities at colonizing patches and competing in patches (Levins and Culver 1971; Hastings 1980; reviewed by Mouquet et al., chapter 10).

The simplest patch dynamic model predicts that any local patch (a microsite;

table 1.1) will be unoccupied (diversity = 0), or occupied by one individual or population of one species (diversity = 1) along the competition-colonization trade-off (see also Mouquet et al., chapter 10); this does not allow us to explore the effects of migration rates, disturbance, et cetera on patterns of local and regional diversity like in the other model frameworks. Thus, we consider a locality to consist of several microsites within a restricted area, and a region to consist of several localities. Specific predictions of the patch dynamic framework, derived from several published sources (Hastings 1980; Tilman 1994; Yu and Wilson 2001; Mouquet et al. 2002; Shurin et al. 2004) as well as some speculation, include the following: (1) Local diversity is a hump-shaped function of dispersal rates within the region; local diversity should increase with rates of migration among localities (connectance) until the point when levels of migration are so high some species are driven extinct from the region. These species can be the poorer dispersing species if the more rapidly dispersing species' advantage is enhanced with higher overall rates of dispersal, or the better dispersing species can be eliminated with increasing dispersal rates if a limit on dispersal speed is reached (Hastings 1980). (2) Regional diversity will decrease with increasing rates of migration because local displacement occurs more rapidly and ultimately fewer species can persist in the metacommunity. (3) β-diversity among localities will not vary with distance (or dispersal rate) among localities. This is because patch dynamic models (to date) explicitly assume that all dispersal is global because they lack a spatially explicit structure (Holyoak et al., chapter 1). If instead, dispersal were more localized, the model's predictions would be more in line with those of neutral models where β-diversity increases with increasing distance among localities. (4) If a locality within a metacommunity is disturbed, it is unpredictable as to which species will recolonize any given microsite, although it is likely that colonization specialists (pioneer species) will exist in more microsites immediately following the disturbance. Alternatively, if most of the region (metacommunity) is disturbed, the system will show transient succession from dominance by colonizing (pioneer) species to eventual coexistence of colonizing and competitive species in the same proportions as their predisturbance configuration (assuming that disturbance has a similar effect on all species). (5) Species' relative abundance will change through time locally due to the stochasticity of colonization-competition processes, but will remain more static through time at the regional scale. Species relative abundances will not, however, change with variation in environmental conditions, because this model framework assumes that species' dynamics are not responsive to heterogeneous environmental conditions.

Species Sorting Framework

In contrast with the previous frameworks, the species sorting framework explicitly assumes that there is heterogeneity in the environment. Furthermore, species persist in the habitats in which their traits and interactions with other species allow them to maintain their populations. That is, species sort themselves so that each persists in its favored environment (e.g., Tilman 1982; Chase and Leibold 2003). In species sorting models, local and regional coexistence depends on types of limiting factors, variation in those limiting factors, and the nature of species trade-offs in competitive abilities. This approach generally ignores the role of dispersal as an explicit process, because dispersal per se does not alter the predictions of the model. It does, however, implicitly assume that dispersal is frequent enough so that all species are able to rapidly reach every locality where they are capable of persistence, even when those localities are rather far apart. An upper limit on dispersal is also assumed, such that dispersal does not perturb abundance or composition away from their within-patch equilibria.

Specific predictions of a species sorting metacommunity (Tilman 1982; Chase and Leibold 2003) are that (1) local and (2) regional diversity will be fairly independent of rates of migration among localities. (3) β-diversity among localities will be less dependent on rates of migration than on the variation in environmental conditions among localities. If environmental conditions are spatially autocorrelated, then β-diversity will be correlated with distance, but if environmental conditions are not spatially autocorrelated, then β-diversity will be independent of distance; this also assumes that dispersal is not localized, which would cause β-diversity to be distance dependent. (4) If a locality or an entire metacommunity is disturbed it will return to its previous state relatively quickly following transient dominance by species that are better colonizers (pioneers). (5) Species' relative abundances will be relatively constant through time, so long as environmental conditions remain constant; they will vary predictably if the environment varies through time.

The Mass Effects Framework

The mass effects framework (Amarasekare 2000; Amarasekare and Nisbet 2001; Mouquet and Loreau 2002, 2003) assumes that there is environmental heterogeneity, and that species trade-off such that they are favored in some habitats but not others. Furthermore, a species can persist as sink populations in patches where they are not favored (if they are maintained by immigration), and that species vary in their relative ability to compete in and colonize habitat patches.

In the mass effects framework, dispersal can influence local and regional diversity, as well as the composition of species. Specific predictions based on previously published sources (e.g., Amarasekare 2000; Amarasekare and Nisbet 2001; Mouquet and Loreau 2002, 2003) and some speculation include the following: (1) Local diversity is a hump-shaped function of dispersal rates within the region; local diversity increases with increasing dispersal because species can persist in habitats where they are not favored due to mass effects until species that are better regional competitors (dispersers) eliminate other species. (2) Regional diversity should remain unchanged from low to intermediate dispersal rates, but decrease with high dispersal rates because locally competitively inferior species become displaced. (3) β-diversity should decrease as localities become closer together or as rates of dispersal increase. This is because with increased rates of dispersal among localities, species that are better colonizers but poorer competitors are favored, regardless of variation in local environmental conditions so long as each species retains a source habitat. (4) If a locality or the entire metacommunity is disturbed it will return to its previous state providing that species do not go extinct from the entire metacommunity. However, localities will go through a successional trajectory such that better colonizing species will dominate immediately after a disturbance but the metacommunity will achieve a configuration identical to predisturbance levels through time. (5) Species' relative abundance will be relatively stable through time at both the local and regional scales so long as environmental conditions remain constant and no other species invade; they will vary predictably if the environment varies through time.

Empirical Support

Real communities clearly do not conform to only one of the above perspectives. One approach to distinguishing the relative roles of different processes is to identify areas where the models make qualitatively distinct predictions that can be subjected to empirical tests. Some recent analyses have explicitly tested the assumptions or predictions of these models, often with particular reference to verifying or refuting the predictions of the neutral model (Condit et al. 2002; Tuomisto et al. 2003; McGill 2003; Volkov et al. 2003; Clark and McClachan 2003). In the next section, we discuss several empirical patterns that can inform the underlying processes that structure metacommunities.

Patterns of Relative Species Abundance

Early thinking on abundance distributions was based on statistical logic rather than mechanistic models. For instance, Fisher et al. (1943) derived a log-series distribution to fit species abundances where the majority of species are in the rarest categories. Alternatively, Preston (1963) supposed that data on species rank-abundances fit a lognormal distribution where the majority of species are rare, but not the rarest in a given community. The first mechanistic model of abundance distribution was MacArthur's "broken stick model" based on niche differentiation, which predicted a pattern of relative species abundance akin to the lognormal pattern (1957, 1960; see also Sugihara 1980). However, none of these ideas combined the patterns of relative species abundance with patterns of species diversity and composition, even though they are obviously intimately linked. Thus, one of the great appeals of neutral models is that they are able to predict patterns of relative species abundance as well as species diversity and composition (Bell 2001, 2003; Hubbell 2001; Chave 2004). Specifically, at local spatial scales, the neutral model predicts a zero-sum multinomial (ZSM) pattern of species rank abundance relationships (the relationship between species abundance and the rank in species abundance). The ZSM is a lognormal-like distribution but predicts fewer common, and rarer species. At larger regional scales, the neutral model can predict a log-series pattern, or a ZSM; the specific shape depends on the nature in which speciation takes place (e.g., point versus allopatric speciation) (Hubbell 2001).

Hubbell (2001) derived the ZSM, and then described several empirical cases that appeared to fit this relationship of species rank abundance better than a log-normal relationship. However, Hubbell's analyses did not rigorously test which hypothesized distribution provided a better statistical fit to the data. In an attempt to remedy this, McGill (2003) calculated a numerically iterative solution to the neutral model in order to derive an expected distribution of the ZSM, which he could statistically compare to a log-normal distribution. Using data from breeding birds and Hubbell's own data from tropical trees on Barro Colorado Island (BCI), Panama, McGill (2003) concluded that in the majority of cases, the data were better predicted by a log-normal relationship, and thus were not consistent with Hubbell's (2001) ZSM predictions based on neutrality. In response, Volkov et al. (2003) derived an analytical solution to the ZSM and rebutted McGill's conclusions by showing that a more rigorous solution to the neutral model's ZSM fit the data on trees from BCI better than the log-normal distribution. In response to this, Etienne and Olff (2004) presented a statistical approach based on individual genealogy and Bayesian statistics to show that the log-normal distribution showed a statistically better fit to the BCI data than the ZSM.

The BCI data are some of the best data available on species rank abundances in such a diverse ecosystem, and yet the debate as to whether they best fit a lognormal or ZSM, and whether those data can provide a definitive test of the different theories remains unclear. The differences between predicted ZSM and lognormal distributions, when compared with the BCI data (figure 14.1) are very subtle. Indeed, Harte (2003) noted that the distributions are nearly indistinguishable, particularly at their tails, and that such minute variation among the model predictions, when compared with the potential measurement error inherent in the dataset, may not provide the sort of definitive test necessary to refute or accept one model over the other.

Even more problematic for testing relative abundance patterns is that nonneutral models can predict species rank abundance distributions that are indistinguishable from the neutral model's predicted ZSM. Chave et al. (2002), in an individual-based model of patch dynamics and trade-offs among species' competition and colonization abilities, derived rank abundance relationships virtually identical to those predicted by the neutral model (figure 14.2; see also Chave

2004). Likewise, Wilson et al. (2003) showed that niche-based Lotka-Volterra models produce a wide range of abundance patterns, including those indistinguishable from the ZSM, depending on the distributions of the underlying parameters. Although the log-normal distribution is most often used as the alternative to the ZSM for purposes of comparison, it is not at all clear that nonneutral theories predict such distributions (Wilson et al. 2003). Finally, the critical feature of the ZSM distribution is the excess of rare species, which likely represent transients or sink populations. The mass effects framework allows for the possibility of many species persisting in population sinks, and thus a few common and many rare species in a pattern similar to the ZSM (Mouquet and Loreau 2003). Indeed, in a long-term survey of marine fish communities, Magurran and Henderson (2003) found that rare or transient species that were present only in some surveys and in some years showed different dynamics than those that were commonly found in all years. Thus, the rank-abundance pattern, in and of itself, is unlikely to be useful for differentiating among the processes operating in a given metacommunity.

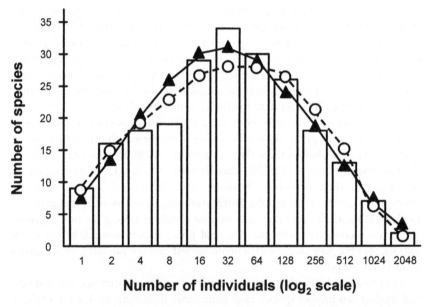

Figure 14.1 Data on tree species abundances from the 50-ha plot at Barro Colorado Island, Panama (21,457 individuals in 225 species). The data are grouped into 12 logarithmic (Log_2) intervals based on Preston's 1948 method. The dotted line with open circles represents the best fit to a lognormal distribution, whereas the solid line with closed triangles represents the best fit to an analytical solution of the neutral model's zero sum multinomial (ZSM). The authors conclude that the solid line ZSM fits the data better than the lognormal. Redrawn from Volkov et al. (2003).

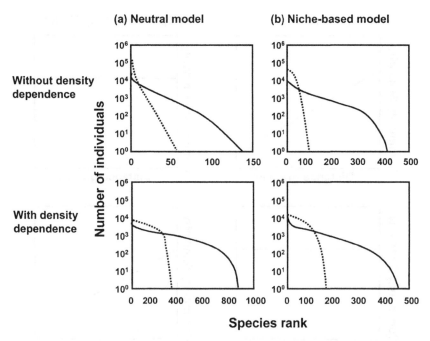

Figure 14.2 Rank abundance curves for a neutral model (a) and a model based on niche differences (b) with (top) and without (bottom) density dependence. Bold lines are when dispersal was global, while dotted lines are when dispersal was local. The figure shows the similarity in predicted patterns in the two very different model structures. Redrawn from Chave et al. (2002).

Patterns of Species Composition along Environmental and Spatial Gradients

Neutral models predict that species composition will vary predictably with space (β-diversity increases with increasing distance among localities), but not the environment (figure 14.3a). Species sorting models predict that species composition varies with environment, but not spatial gradients, so long as environmental variation is not spatially autocorrelated (figure 14.3b). Alternatively, when environments are spatially autocorrelated, species composition should vary with both space and environment. Patch dynamic, species sorting and mass effects models predict a mixture of the above two patterns depending on the assumptions made. When dispersal is global, neither patch dynamic nor mass effects and species sorting models predict specific patterns across spatial gradients, while both mass effects and species sorting models predict variation in species composition along environmental gradients. When dispersal is more localized, both patch dynamic and mass effects models predict variation in species composition along spatial gradients, whereas only mass effects and species sorting models predict variation in species composition along environmental gradients.

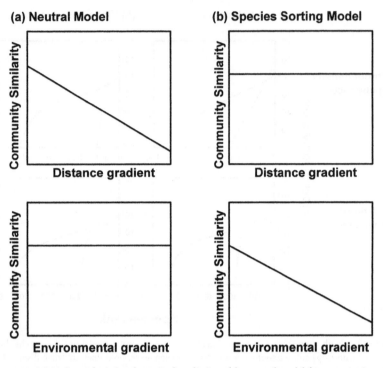

Figure 14.3 (a) A scheme depicting the general prediction of the neutral model that community composition should vary with spatial components (top), but not environmental variation (bottom). (b) A scheme depicting the general prediction of the species sorting model that species composition should not vary with space (top), but will vary with environmental conditions (bottom).

Although comparisons of metacommunity models have reinvigorated studies on species compositional shifts along spatial and environmental gradients (e.g., Condit et al. 2002; Tuomisto et al. 2003), the conceptual foundation for such studies is quite old. Clements (1938) and Gleason (1927) disagreed on whether communities were consistent consortia of species or individuals whose responses to environmental variation randomly overlapped, but they both focused on how species composition would vary as environment varied with little consideration for dispersal limitation. A review of studies by Beard (1955), Whittaker (1975), Chabot and Mooney (1985), and others showed that terrestrial plant species are sorted along major environmental gradients in both tropical and temperate habitats in ways that show a strong correspondence between the pattern of the sorting and plant traits involved in resource competition (Tilman 1988). In a recent review, Leibold and Mikkelson (2002) found that along environmental gradients, a majority of studies from a variety of plant and animal communities showed consistent patterns of turnover in species composition. This is consistent with either species sorting or mass effects models. However, the majority of studies in this review examined patterns of species composition along spatial transects so that the role of space versus environment, and the underlying processes that determine species compositional turnover, could not be disentangled.

Analyses attempting to disentangle the role of space versus environment in determining patterns of species composition have shown mixed support. Condit et al. (2002) and Tuomisto et al. (2003) set out to explicitly examine variation in species composition from site to site as distances among the sites increased in tropical plants (see also Terborgh et al. 1996). Both studies were primarily focused on comparing Hubbell's neutral model, with "niche" models (akin to what we call species sorting). Both found that although spatial distance played a role in determining variation among community composition, there was a considerable amount of variation that could not be explained by distance alone. Tuomisto et al. went a step further, measuring differences in environmental conditions along with spatial distances, to explicitly examine whether the spatial drift processes expected in a neutral model could override variation in environmental conditions, or whether different species' preference for different environmental conditions was the overriding process influencing variation in species composition from site to site. They found that while there appeared to be some spatial correlation in species composition there was a strong overriding effect of site-to-site environmental variation. These results suggest that species sorting played a stronger role than dispersal limitation in shaping communities, and perhaps are most consistent with models that incorporate both environmental variation in species tolerances for different habitats and spatial effects; that is, mass effects models.

Similar analyses have compared the relative roles of environmental conditions versus spatial configuration in other types of systems. Borcard et al. (1992) developed a methodology for detecting the role of environmental versus spatial components in patterns of community structure. To illustrate their procedure, they used patterns from plants, moss invertebrates, and bacteria, and found that both space and environment explained a significant proportion of the variance in species composition. Pinel-Alloul et al. (1995) surveyed zooplankton community structure across a large region of lakes, and also found that both space and environmental conditions influenced patterns of community structure. However, spatial and environmental components were confounded in this study, because communities that were more distant from each other were also more dissimilar in environmental variables. Cottenie and De Meester (chapter 8; Cottenie et al. 2003) compared the role of environmental versus spatial determinants of zooplankton community structure in a series of interconnected ponds and found that although there was a significant effect of space, local environmental conditions played a large role in determining patterns of species composition and diversity. This was despite the fact that rates of dispersal were quite high among ponds (increasing the possibility of mass effects). Similarly, Kunin (1998) examined pat-

terns of plant species composition among habitat patches that were part of a very
long-term experiment of nutrient manipulations in grasslands. He found that although there was evidence for spatial effects, particularly among directly adjacent
patches, environmental variation played the strongest role in determining patterns of species composition. Overall, these examples lend support to the notion
that both space and environment influence species composition. This is inconsistent with the neutral and the species sorting models, but more consistent with the
predictions of the mass effects model.

Patterns of Species Composition and Diversity with Varying Dispersal Rates

Some evidence exists that can be used to evaluate the predictions of different dispersal rates on patterns of local, regional and β-diversity (table 14.1). Harrison
(1997, 1999) compared plant diversity between continuous patches of serpentine
grasslands and isolated fragments that were subject to less dispersal among sites.
She found that both β- and regional diversity were lower in more interconnected
communities. In surveys of wetland amphibians and macroinvertebrates, Chase
(2003) found similar decreases of β- and regional diversity and increases in local
diversity in more closely aligned communities. In addition, experimental studies
on protists inhabiting pitcher-plant communities (Kneitel and Miller 2003; Miller
and Kneitel, chapter 5) and zooplankton in mesocosms (Forbes and Chase 2002)
have experimentally manipulated connectance among local communities and
found increases in community similarity (decreases in β-diversity) and decreases
in regional diversity with increased rates of dispersal. Alternatively, Gonzalez
(chapter 6) found that increased levels of habitat connectivity increased local and
regional microarthropod diversity (Gilbert et al. 1998; Gonzalez et al. 1998). Warren (1996) found similar increases in local diversity with dispersal in experimental microcosms. Again, all of these results, in and of themselves, cannot specifically support one model over others that make similar predictions, but can be
used to eliminate some possible mechanisms. For example, while some of these
results could be used to support neutral, patch dynamic, or mass effects models,
species sorting models make no predictions about the role of varying rates of dispersal on patterns of composition and diversity, and thus are likely to be less
appropriate for these particular systems.

Patterns of Community Composition Through Time and Following Disturbance

Several studies have used data on species diversity and composition from both
small- and large-scale disturbances to discern which of the metacommunity
frameworks is most appropriate (e.g., Hubbell et al. 1999; Vandermeer et al. 2000;
Molino and Sabatier 2001; Schnitzer and Carson 2001; Pitman et al. 2002, reviewed in Brokaw and Busing 2000; Sheil and Burslem 2002). Here, by disturbance, we mean a discrete density independent mortality event that is imposed on
a community, but then relaxed. Hubbell et al. (1999) sampled gaps within the
larger forest matrix at BCI to examine the prediction that species diversity should

vary in disturbed (gap) and undisturbed (nongap) areas. They found that species diversity did not vary between gap and nongap areas when the data were corrected for the higher number of stems found in gap areas, and supported the view that dispersal limitation inherent to neutral models plays a stronger role than niche differentiation. Alternatively, in other tropical forests, Vandermeer et al. (2000) and Molino and Sabatier (2001) showed that the response of tree species composition and diversity to small and large-scale disturbances was more akin to the predictions of niche-based models where species differ in their relative ability to colonize open habitats. Finally, using data from the same site (BCI) as Hubbell et al. (1999) and including data from woody lianas and pioneer trees in addition to shade tolerant species, Schnitzer and Carson (2001) found that gaps appeared to play a strong role in the maintenance of species diversity.

More generally, Mackey and Currie (2001) reviewed the literature on the effects of disturbance on patterns of diversity from a variety of communities. Despite the common notion that disturbance should alter patterns of species diversity, notably by reducing the abundance of competitively superior species and allowing pioneer (colonizing) species to persist, they found that such responses were relatively rare in the literature (less than 20% of studies). Furthermore, they found that approximately 35% of studies showed no effect of disturbance on species richness; a result that Hubbell et al. (1999) attributed to dispersal limitation and neutral processes. Thus, there appears to be little consensus on the responses of communities to disturbance.

Few data exist that were collected over timescales that are appropriate for evaluating the predictions of the metacommunity frameworks on community change through time. However, these sorts of long-term data may be particularly useful in helping to differentiate the neutral model from the other model frameworks. Clark and McLachlan (2003) recently tested these ideas using relative abundance data of temperate North American trees from historic pollen records. They found that following glaciation, populations rapidly stabilized and were unchanged over long temporal scales. This result suggests that these communities are more persistent through time than would be predicted by neutral models (but see Volkov et al. [2004] for criticisms of this analysis).

Conclusions

These model frameworks overlap in their predictions for empirical patterns, and therefore few patterns can definitively differentiate among the models. Furthermore, different systems have shown differential support for each of the frameworks, and thus no single model seems most informative. The challenge is to build a more general, synthetic, and unified model that incorporates appropriate aspects of each of the model frameworks and recognizes that there will be variation among systems.

As a start to such a synthesis, we suggest studies of species diversity in metacommunities may parallel those on genetic diversity in populations (Hastings and Gavrilets 1999; Amarasekare 2000; McPeek and Golmulkiewicz, chapter 15). Several classes of population genetic models deal with the interplay between neutral and nonneutral processes that decrease diversity, such as drift and selection, and processes that increase diversity, such as mutation and migration (Hartl and Clark 1997). These parallel the metacommunity modeling frameworks. For instance, mutation in population genetic models is similar to speciation in Hubbell's (2001) neutral framework, but they operate at different levels of organization (alleles and species). Population genetic models that look at the balance between spatially invariant selection and drift are similar to patch dynamic models where the environment is spatially homogeneous and the species exhibit the same kinds of trade-offs in all locations of the landscape (Amarasekare 2000). Population genetic models that look at the balance between drift and spatially variable selection (Hartl and Clark 1997) are equivalent to species sorting models. Lastly, population genetic models that look at the balance between migration, selection, and drift (Hartl and Clark 1997) are equivalent to mass effects models.

In population genetics there is agreement that genetic drift operates in practically all populations, but its effects are stronger in some (such as when effective population sizes are smaller; Hartl and Clark 1997). By analogy, rare species may be more affected by ecological drift whereas common species may be more affected by competition. This sort of argument provides a way to reconcile the neutral model with niche-based approaches, without having to negate any particular model. In fact, the study of marine fish abundances by Magurran and Henderson (2003) discussed above found a different pattern for core (common) species compared to occasional (rare) species.

Future theoretical work is needed to investigate whether rarer species are more likely to conform to predictions of neutral theory than common species, and whether the patterns of common species follow the predictions of patch dynamics, species sorting, or mass effects models. Future theoretical and empirical work should also continue to explicitly incorporate species interactions other than competition into the metacommunity framework, such as food web interactions (Hoopes et al., chapter 2; Holt and Hoopes, chapter 3) and mutualistic interactions (Amarasekare 2004). These interactions could drastically alter many of the predictions reviewed and evaluated above.

Acknowledgments

We thank R. Law, J. Kneitel, and T. Knight for discussions and comments. This chapter was developed as part of the metacommunities working group at the National Center for Ecological Analysis and Synthesis, a center funded by the NSF, the University of California, and the state of California.

Literature Cited

Abrams, P. A. 2001. A world without competition. Nature 412:858–859.

Amarasekare, P. 2000. The geometry of coexistence. Biological Journal of the Linnean Society. 71: 1–31.

———. 2003. Competitive coexistence in spatially structured environments: A synthesis. Ecology Letters 6:1109–1122.

———. 2004. Spatial dynamics of mutualistic interactions. Journal of Animal Ecology 73:128–142.

Amarasekare, P. and R. Nisbet. 2001. Spatial heterogeneity, source-sink dynamics, and the local coexistence of competing species. American Naturalist 158:572–584.

Beard, J. S. 1955. The classification of tropical American vegetation types. Ecology 36:89–100.

Bell, G. 2001. Neutral macroecology. Science 293:2413–2418.

———. 2003. The interpretation of biological surveys. Proceedings of the Royal Society of London, series B, 270:2531–2542.

Borcard, D., P. Legendre, and P. Drapeau. 1992. Partaking out the spatial component of ecological variation. Ecology 73:1045–1055.

Brokaw, N. and R. T. Busing. 2000. Niche versus chance and tree diversity in forest gaps. Trends in Ecology and Evolution 15:183–188.

Brown, J. H. 2002. Towards a general theory of biodiversity. Evolution 55:2137–2138.

Caswell, H. 1976. Community structure: A neutral model analysis. Ecological Monographs 46:327–354.

Chabot, B. F., and H. A. Mooney. 1985. *Physiological ecology of North American vegetation*. Chapman and Hall, New York, N.Y.

Chase, J. M. 2003. Community assembly: When should history matter? Oecologia 136:489–498.

Chase, J. M. and M. A. Leibold. 2003. Ecological niches: Linking classical and contemporary approaches. University of Chicago Press, Chicago, IL.

Chave, J. 2004. Neutral theory and community ecology. Ecology Letters 7:241–253.

Chave, J. and E. G. Leigh. 2002. A spatially explicit neutral model of beta-diversity in tropical forests. Theoretical Population Biology 62:153–168.

Chave, J., H. Muller-Landau, and S. A. Levin. 2002. Comparing classical community modules: theoretical consequences for patterns of diversity. American Naturalist 159:1–23.

Clark, J. S. and J. S. McLachlan. 2003. Stability of forest biodiversity. Nature 423:635–638.

Clements, F. E. 1938. Nature and structure of the climax. Journal of Ecology 24:252–282.

Condit, R., N. Pitman, E. G. Leigh Jr., J. Chave, J. Terborgh, R. B. Foster, P. Núñez, S. Aguilar, R. Valencia, G. Villa, H. C. Muller-Landau, E. Losos, and S. P. Hubbell. 2002. Beta-diversity in tropical forest trees. Science 295:666–669.

Cottenie, K, E. Michels, N. Nuytten, and L. De Meester. 2003. Zooplankton metacommunity structure: Regional vs. local processes in highly interconnected ponds. Ecology 84:991–1000.

De Mazancourt, C. 2001. The unified neutral theory of biodiversity and biogeography. Science 293: 1772–1772.

Enquist, B. J., J. Sanderson, and M. D. Weiser. 2002. Modeling macroscopic patterns in ecology. Science 295:1835–1836.

Etienne, R. S. and H. Olff. 2004. A novel genealogical approach to neutral biodiversity theory. Ecology Letters 7:170–175.

Fisher, R. A., A. S. Corbet, and C. B. Williams 1943. The relation between the number of species and the number of individuals in a random sample of an animal population. Journal of Animal Ecology 12:42–58.

Forbes, A. E. and J. M. Chase 2002. The role of habitat connectivity and landscape geometry on experimental zooplankton metacommunities. Oikos 96:433–440.

Gilbert, F. S., A. Gonzalez, and I. Evans-Freke. 1998. Corridors maintain species richness in fragmented landscapes of natural microecosystems. Proceedings of the Royal Society London, series B, Biological Sciences. 265:577–582.

Gleason, H. A. 1927. Further views on the succession concept. Ecology 8:299–326.

Gonzalez, A., J. H. Lawton, F. S. Gilbert, T. M. Blackburn, and I. Evans-Freke. 1998. Metapopulation dynamics, abundance, and distribution in a microecosystem. Science 281:2045–2047.

Hanski, and M. Gyllenberg. 1997. Uniting two general patterns in the distribution of species. Science 275:397–400.

Harrison, S. 1997. How natural habitat patchiness affects the distribution of diversity in Californian serpentine chaparral. Ecology 78:1898–1906.

———. 1999. Native and alien species diversity at the local and regional scales in a grazed California grassland. Oecologia 121:99–106.

Harte, J. 2003. Tail of death and resurrection. Nature 424:1006–1007.

Hartl, D. L. and A. G. Clark. 1997. *Principles of population genetics*. 3rd edition. Sinauer Associates, Sunderland, MA.

Hastings, A. 1980. Disturbance, coexistence, history, and competition for space. Theoretical Population Biology 18:363–373.

Hastings, A., and S. Gavrilets 1999. Global dispersal reduces local diversity. Proceedings of the Royal Society of London, series B, Biological Sciences. 266:2067–2070.

Holt, R.D. 1993. Ecology at the mesoscale: The influence of regional processes on local communities. Pages 77–88 *in:* R. Ricklefs and D. Schluter, eds. *Species Diversity in Ecological Communities*. University of Chicago Press, Chicago.

———. 1996. Food webs in space: An island biogeographic perspective. Pages 313–323 *in:* G. Polis and K. Winemiller, eds. *Food Webs: Contemporary Perspectives*. Chapman and Hall.

———. 1997. From metapopulation dynamics to community structure: Some consequences of spatial heterogeneity. Pages 149–164 *in* I. Hanski and M. Gilpin, eds. *Metapopulation biology*. Academic Press, New York, N.Y.

———. 2002. Food webs in space: On the interplay of dynamic instability and spatial processes. Ecological Research 17:261–273.

Hubbell, S. P. 2001. *The unified neutral theory of species abundance and diversity*. Princeton University Press, Princeton, N.J.

Hubbell, S. P, R. B. Foster, S. T. O'Brien, K. E. Harms, R. Condit, B. Wechsler, S. J. Wright, and S. Loo de Lao. 1999. Light-gap disturbances, recruitment limitation, and tree diversity in a neotropical forest. Science 283:554–557.

Kneitel, J. M. and T. E. Miller. 2003. Dispersal rates affect species composition in metacommunities of *Sarracenia purpurea* inquilines. American Naturalist 162:165–171.

Kunin, W. E. 1998. Biodiversity at the edge: A test, of the importance of spatial "mass effects" in the Rothamsted Park Grass experiments. Proceedings of the National Academy of Sciences, USA 95: 207–212.

Legendre, P., R. Galzin, and M. L. Harmelin-Vivien. 1997. Relating behavior to habitat: Solutions to the fourth-corner problem. Ecology 78:547–562.

Leibold, M. A., M. Holyoak, N. Mouquet, P. Amarasekare, J. M. Chase, M. F. Hoopes, R. D. Holt, J. B. Shurin, R. Law, D. Tilman, M. Loreau, and A. Gonzalez, A. 2004. The metacommunity concept: A framework for multi-scale community ecology. Ecology Letters 7:601–613.

Leibold, M. A. and G. M. Mikkelson 2002. Coherence, species turnover, and boundary clumping: Elements of meta-community structure. Oikos 97:237–250.

Levins, R. and D. Culver. 1971. Regional coexistence of species and competition between rare species. Proceedings of the National Academy of Sciences, USA 68:1246–1248.

MacArthur, R. H. 1957. On the relative abundance of species. Proceedings of the National Academy of Sciences, USA 43:293

———. 1960. On the relative abundance of species. American Naturalist 94:25–36.

MacArthur, R. H. and E. O. Wilson, 1967. The theory of island biogeography. Princeton University Press, Princeton, NJ.

Mackey, R. L. and D. J. Currie. 2001. The diversity-disturbance relationship: Is it generally strong and peaked? Ecology 82:3479–3492.

Magurran, A. E. and P. A. Henderson. 2003. Explaining the excess of rare species in natural species abundance distributions. Nature 422:714–718.

McGill, B. J. 2003. A test of the unified neutral theory of biodiversity. Nature 422:881–885.

Molino, J. and D. Sabatier 2001. Tree diversity in tropical rain forests: A validation of the intermediate disturbance hypothesis. Science 294:1702–1704.

Mouquet, N. and M. Loreau. 2002. Coexistence in metacommunities: The regional similarity hypothesis. American Naturalist 149:420–426.

———. 2003. Community patterns in source-sink metacommunities. American Naturalist 162:544–557.

Mouquet, N., J. L. Moore, and M. Loreau. 2002. Plant species richness and community productivity: Why the mechanism that promotes coexistence matters. Ecology Letters 5:56–66.

Norris, S. 2003. Neutral theory: A new, unified model for ecology. Bioscience 2003 53:124–129.

Pinel-Alloul. B., T. Niyonsenga, and P. Legendre 1995. Spatial and environmental components of freshwater zooplankton structure. Ecoscience 2:1–19.

Pitman, N. C. A., J. Terborgh, M. R. Silman, V. Nunez, D. Neill, C. E. Ceron, W. E. Palacious, and M. Aulestia. 2002. Comparison of tree species diversity in two upper Amazonian forests. Ecology 83:3210–3224.

Preston, F. W. 1948. The commonness and rarity of species. Ecology 29:254–283.

———. 1962. The canonical distribution of commonness and rarity. Ecology 43:185–215.

Schnitzer, S. A. and W. P. Carson 2001. Treefall gaps and the maintenance of species diversity in a tropical forest. Ecology 82:913–919.

Sheil, D. F. and R. P. Burslem 2003. Disturbing hypotheses in tropical forests. Trends in Ecology and Evolution 18:18–26.

Shurin, J. B., P. Amarasekare, J. M. Chase, R. D. Holt, M. F, Hoopes, and M. A. Leibold. 2004. Alternative stable states and regional community structure. Journal of Theoretical Biology 227:359–368.

Sugihara, G. 1980. Minimal community structure: An explanation of species abundance patterns. American Naturalist 116:770–787.

Terborgh, J., R. B. Foster, and P. Nunez. 1996. Tropical tree communities: A test of the nonequilibrium hypothesis. Ecology 77:561–567.

Tilman, D. 1982. *Resource competition.* Princeton University Press, Princeton, NJ.

———. 1988. *Dynamics and Structure of Plant Communities.* Princeton University Press, Princeton, NJ.

Tilman, D. 1994. Competition and biodiversity in spatially structured habitats. Ecology 75:2–16.

Tuomisto, H., K. Ruokolainen, and M. Yli-Halla. 2003. Dispersal, environment, and floristic variation of western Amazonian forests. Science 299:241–244.

Vandermeer, J., I. G. de la Cerda, D. Boucher, I. Perfecto, and J. Ruiz. 2000. Hurricane disturbance and tropical tree species diversity. Science 290:788–791.

Volkov, I., J. R. Banavar, S. P. Hubbell, and A. Maritan. 2003. Neutral theory and relative species abundance in ecology. Nature 424:1035–1037.

Volkov, I., J. R. Banavar, A. Maritan, and S. P. Hubbell. 2004. The stability of forest biodiversity. Nature 427:696–696.

Warren, P. H. 1996. The effects of between-habitat dispersal rate on protist communities and metacommunities in microcosms at two spatial scales. Oecologia 105:132–140.

Whitfield, J. 2002. Ecology: Neutrality versus the niche. Nature 417:480–481.

Whittaker, R. H. 1962. Classification of natural communities. Botanical Review 28:1–239.

―――. 1975. *Communities and ecosystems.* Macmillan Press, New York.

Wilson, W. G., P. Lundberg, D. P. Vázquez, J. B. Shurin, M. D. Smith, W. Langford, K. L. Gross, and G. G. Mittelbach. 2003. Biodiversity and species interactions: Extending Lotka-Volterra community theory. Ecology Letters 6:944–952.

Yu, D. W. and H. B. Wilson. 2001. The competition-colonization trade-off is dead: Long live the competition-colonization trade-off. American Naturalist 158:49–63.

CHAPTER 15

Assembling and Depleting Species Richness in Metacommunities

Insights from Ecology, Population Genetics, and Macroevolution

Mark A. McPeek and Richard Gomulkiewicz

The ecological mechanisms that influence levels of species richness remain elusive, particularly in high diversity systems. Ecosystems often harbor many congeneric and ecologically similar species within local areas, and local richness can be spectacular. Hundreds of beetle species (e.g., Didham et al. 1998; Harris and Burns 2000) and hundreds of butterfly species (e.g., DeVries and Walla 2001) can be found in small areas of tropical forests. The African rift lakes each contain hundreds of endemic cichlid species (reviewed in Kornfield and Smith 2000; Turner et al. 2001). Such impressive levels of biodiversity are not restricted to the tropics or to unique "hotspots." At least seventy-eight species in forty-nine genera of chironomid midges inhabit the gravel bottom of one 100 m stretch of stream in southern England, including eight *Orthocladius* and eight *Cricotopus* species (Ruse 1995). Five to twelve *Enallagma* damselfly species co-occur in lakes with fish across eastern North America (Johnson and Crowley 1980; McPeek 1989, 1990, 1998; Shiffer and White 1995; McPeek and Brown 2000). Such examples are easily gleaned from the literature for almost any ecosystem where careful sampling and taxonomic identifications have been done.

In addition to this apparent species richness, molecular phylogenetic studies have identified substantial levels of cryptic local species richness. Several cryptic species of *Chrysoperla* lacewings are often found on the same branch of a bush (Henry et al. 1999). Until very recently, the amphipod *Hyalella azteca* was considered a single species. Recent molecular studies have, however, shown that *H. azteca* is in fact a collection of at least seven species, most of which are morphologically and ecologically indistinguishable and which co-occur in lakes across North America (Witt and Hebert 2000; Witt et al. 2003; G. A. Wellborn, unpublished data). Likewise, nine morphologically indistinguishable rotifer species that were formerly considered *Brachionus plicatilis* co-occur in the lakes of the Iberian Peninsula (Gomez et al. 2002). In both *Hyalella* and *Brachionus*, many of the cryptic species are millions of years old. The literature suggests that such cryptic species richness can be found across the eukaryotes and in almost all ecosystems.

Understanding the mechanisms that promote such high local levels of biodiversity is at the heart of community ecology, because through sheer numbers these are the mechanisms that maintain the bulk of biodiversity on the planet. However, much of the theoretical edifice of community ecology (e.g., Gause's competitive exclusion principle, niche theory) reinforces the belief that such high local diversity of such closely related and ecologically similar species should not be possible. And yet there it is!

How can so many closely related species all co-occur in local communities? Recent reviews have organized the myriad mechanisms that influence species diversity across the landscape into those that promote the indefinite coexistence of species and those that only reduce the rate of species loss (Tilman and Pacala 1993; Chesson and Huntly 1997; Chesson 2000a). In this paper, we restrict our use of the word *coexistence* to situations where persistence is indefinite for multiple species, and we use *co-occurrence* to imply a less restrictive definition in which species are found together regardless of whether persistence is permanent. Chesson (2000a) outlines an extremely useful framework for considering these mechanisms. In this framework, invasibility—a species' population growth rate when at low density and other species are at their normal abundances—is the criterion for assessing coexistence (see also MacArthur 1972; Holt 1977). Mechanisms of species interactions influence the invasibility of species via their *stabilizing* and *equalizing* effects. Stabilizing effects, often resulting from trade-offs, cause species to be limited by different density-dependent factors (e.g., resources, predators, diseases) such that they depress their own population growth rates more than they depress other species' growth rates; these are the fundamental features of mechanisms that promote the coexistence of species. In contrast, equalizing effects are the consequences of species having similar fitnesses to one another. Equalizing effects are capable of prolonging persistence, but not fostering permanent coexistence.

Mechanisms with stabilizing effects can operate on various spatial and temporal scales (Tilman and Pacala 1993; Chesson 2000a). Typically, we first think of coexistence mechanisms as those that regulate a single, local community to a point equilibrium (e.g., Lotka-Volterra models; see Case 2000). However, temporal heterogeneity can also promote coexistence in a single local community through some forms of nonlinear density dependence (e.g., Armstrong and McGehee 1980, Chesson 1994), through endogenously generated cycles (Abrams et al. 2003), or through species differences that lead to the storage effect (Chesson and Warner 1981; Chesson 1994). Within a metacommunity context, spatial mechanisms operating among a number of local communities can also promote coexistence, with or without temporal heterogeneity, to generate regional species coexistence. In the absence of temporal fluctuations, spatial heterogeneity among patches in a metacommunity can often generate nonlinearities or spatial storage effects giving coexistence (Bolker and Pacala 1999; Chesson 2000b; Muko and Iwasa 2000, Amarasekare 2003). Temporal variability (e.g., disturbance or localized individual mortality) and recolonization of sites can promote coexistence among competing species if a trade-off between local competitive ability and dispersal exists (e.g., Hastings 1980; Caswell and Cohen 1991; Tilman 1994; Pacala and Tilman 1994; Hurtt and Pacala 1995; Kinzig et al. 1999; reviewed in Mouquet et al., chapter 10). Predation can have similar effects on prey diversity (Hastings 1977; Caswell 1978).

Multiple species may also inhabit the same metacommunity simply if different species are better adapted to the ecological conditions found in different patches (Shmida and Ellner 1984). Local conditions may allow one species to dominate all others at any particular site, but different species may be favored at different sites (e.g., Pacala and Tilman 1994; see also the species sorting perspective discussed in Holyoak et al., chapter 1). Dispersal among sites could maintain species in patches where they do not dominate, creating a multispecies source-sink system in which each species has a unique but limited number of source patches and each maintains sink populations in the other patches via dispersal. This is termed the "mass effect" perspective and is discussed in Holyoak et al., chapter 1.

Equalizing effects alone can only retard the rate of species loss and not give long-term coexistence, but they do influence the magnitude of stabilizing effects needed for coexistence (Chesson 2000a). Hubbell (2001) has elaborated a neutral model including only equalizing effects (Hubbell and Foster 1986) at a metacommunity scale. If the fitnesses of all species are nearly identical, ecological factors will regulate the summed total abundance of all species and not the abundance of each species individually (i.e., a zero-sum interaction). As a result, species' relative abundances on local and regional scales are not affected by local environmental conditions and will undergo a random walk (Hubbell 2001). The ultimate outcome of this ecological drift is the extinction of all species save one without the continual input of new species via either speciation in member taxa or immigration from outside the system (Hubbell 2001).

Considerations of stabilizing versus equalizing effects of species interactions focus attention on different aspects of the dynamics of species interactions. When we focus on coexistence via the operation of stabilizing effects, we necessarily focus on the ultimate outcome of species interactions—which species can persist indefinitely with one another—and largely ignore how the system approaches that outcome. Also, because of the mathematical tractability of invasibility criteria for predicting which species will be present at the long-term equilibrium, coexistence considerations focus on the dynamics of species when they are rare and increasing in abundance. In contrast, the consequences on equalizing effects of species interactions will only be important to the transient dynamics of an assemblage on the way to its ultimate configuration of coexisting species; thus, this perspective will focus on species as they become rare. If the equalizing effects of species interactions are important to the dynamics of an assemblage, some of the species (i.e., the transient, co-occurring species) are present because the rate at which they are being driven extinct is slow enough to permit them to persist for substantial periods of time in the system before they realize their ultimate fate, and not because local or regional processes foster their indefinite persistence.

Stabilizing effects clearly promote the coexistence of many species in any given local or regional assemblage, but probably not all species in any given assemblage. The stabilizing effects of these mechanisms involving spatial and temporal heterogeneity can allow many more species to coexist in metacommunities than if only local niche partitioning mechanisms dominate (Tilman and Pacala 1993; Chesson 2000a; Abrams et al. 2003; see also reviews in Holt and Hoopes, chapter 3; and Mouquet et al., chapter 10). Moreover, the operation of these mechanisms may involve subtle differences among species that will be hard to document empirically. However, given the numbers of closely related and phenotypically similar species that co-occur, it is untenable to dismiss the possibility that the equalizing effects of their phenotypic similarities play a substantial role in their persistence in ecosystems. Simply assuming that stabilizing effects must be operating to maintain all species in a system is as scientifically dangerous as assuming that any feature of an organism is the product of adaptation (Gould and Lewontin 1979). In addition, given that the ultimate fate of all species is extinction (Raup 1991), the indefinite coexistence of species would seem to be a peculiar theoretical yardstick by which various mechanisms are deemed worthy of consideration. Because of these two empirical facts, we have begun to explore the theoretical interplay between stabilizing mechanisms and ecological drift in a metacommunity context. In this paper, we want to highlight some of the basic theoretical and empirical properties of metacommunities that make ecological drift a potentially forceful mechanism influencing patterns of species co-occurrence in ecosystems.

Ecological Drift in Metacommunities

The processes that shape species richness patterns are highly analogous to the forces of microevolution that shape allelic diversity (see also Amarasekare 2000). The input of new species to a system via speciation or system-level immigration (e.g., species invasions) is analogous to mutation (e.g., Williams 1992). Dispersal of species among patches is analogous to gene flow. Equalizing and stabilizing effects influencing species' coexistence correspond to fitness differences among alleles that generate natural selection. Finally, the dynamics of genetic drift are analogous to those of ecological drift (Hubbell 2001). Clearly, these are merely analogies because evolutionary mechanisms can differ greatly from the ecological analogs. For example, the particulars of natural selection among diploids and of species interactions are very different, although they may be quite similar with asexual genetics (e.g., Gerrish and Lenski 1998). In any case, the analogy establishes what we think is a useful mindset for understanding species coexistence in a dynamically rich setting. Evolutionary theory is typically constructed as a balance of evolutionary forces: allelic diversity within and among populations depends on a balance between mutation and drift (Kimura 1983), or between mutation, selection, and drift (e.g., Bürger 2000), or among selection, drift and gene flow (e.g., Wright 1932). In like manner, we imagine species diversity across the landscape to be governed by a balance of demographic and macroevolutionary forces that shape the distributions and abundances of species on local and regional scales; these forces consist of speciation and system immigration, dispersal, and the stabilizing (coexistence) and equalizing (ecological drift) effects of demography and ecological interactions.

The dynamics of ecological drift are particularly amenable to this analogy, because both the stochastic nature of sampling alleles to pass on to the next generation and the demographic stochasticity that underlies ecological drift are exactly analogous. Hubbell (2001; Hubbell and Foster 1986) uses a birth and death model that closely resembles the Moran (1958) model of genetic drift to model the dynamics of forest trees, illustrating the intimate relationships between this framework and various empirical metrics of species diversity patterns. Here we explore how features of a metacommunity can greatly slow the rate of ecological drift and thus the loss of species from an ecosystem.

Local Community Drift

Rather than use Hubbell's birth and death model (Hubbell 2001), we will introduce an alternative formulation of ecological drift analogous to the haploid version of the workhorse model for genetic drift: the Wright-Fisher model (e.g., Ewens 1979). We do so because the Wright-Fisher model is the more conventional formulation used to explore genetic drift and thus a greater range of previous analyses are available to draw from and motivate our work. Moreover, the Wright-Fisher and Moran models have largely the same stochastic properties in the limit of large population size (Kingman 1982). To begin with, imagine a closed community that contains J individuals drawn from S different species. Every season these individuals die off after producing large numbers of offspring, of which J survive to adulthood to form the new community. Let $N_i(t)$ be the number of individuals of species i present and let $F_i(t)$ be the per capita fecundity for that species in year t. To impose Hubbell's zero-sum assumption (Hubbell 2001), we assume $\sum_{i=1}^{S} N_i(t) = J$ in all years, however ecological drift will occur even if the total community size fluctuates but stays bounded.

If we assume that all offspring compete equally for the J available spaces, then the probability that k members of species i present in year $t + 1$, given the sizes and fecundities of all species in year t, is the binomial probability

$$\Pr[N_i(t + 1) = k \,|\, N_1(t) \ldots N_s(t), F_1(t) \ldots F_s(t)]$$
$$= \frac{J!}{k!(J - k)!} [p_i(t)]^k [1 - p_i(t)]^{J-k}, \tag{15.1}$$

where $p_i(t) = N_i(t)F_i(t)/\sum_{j=1}^{S} N_j(t)F_j(t)$ is the fraction of newborns produced by species i. Ecologists will recognize this as a version of Chesson and Warner's (1981) lottery model with nonoverlapping generations. In the absence of speciation or immigration to the system, species i will ultimately either become extinct or monodominant. The community as a whole will lose all of its original diversity, becoming dominated by one of the original S species.

A fully neutral model akin to Hubbell's assumes that all species are equally fecund, i.e., $F_1(t) = F_2(t) = \cdots = F_S(t)$. Then $p_i(t) = N_i(t)/J$, the local relative abundance of species i. Given the zero-sum assumption, the stochastic dynamics of this community are exactly the same as a neutral Wright-Fisher model with haploid genetics and constant population size J. (Geneticists tend to focus on the dynamics of relative abundance, but this is not required; it is easy enough to convert between absolute and relative sizes.) The stochastic properties of the neutral model are well characterized (e.g., Ewens 1979, Kimura 1983). We highlight just two results here: First, the probability of monodominance for species i is equal to its relative abundance. Second, the mean number of seasons to extinction of a species with relative abundance p is $-2Jp \ln p/(1 - p)$, conditional on its extinction (Kimura and Ohta 1969). This exit time is on the order of J seasons for any moderately abundant species, that is, any species with relative abundance above about 1%. The mean conditional extinction time for rare species is much faster, on the order of log J seasons.

Metacommunity Drift

The Wright-Fisher model of local dynamics extends easily to geographically structured communities. Consider a metacommunity containing C local communities such that the ith local community contains $J_i(t)$ individuals belonging to a subset of our S species. The density of species j in community i is $N_{ij}(t)$ at time t, with $\sum_{j=1}^{S} N_{ij}(t) = J_i(t)$. Assume the zero sum condition applies to the metacommunity as a whole. That is, the total metacommunity size, J_M, is constant: $J_M = \sum_{i=1}^{C} J_i(t) = \sum_{i=1}^{C} \sum_{j=1}^{S} N_{ij}(t)$.

How does geographic structuring affect properties of ecological drift? Some properties are not affected at all, provided no local community is completely isolated from the rest. For example, in the absence of speciation and immigration of new species from outside the system, the metacommunity will eventually lose all diversity and become globally dominated by a single species. Moreover, under the fully neutral model the probability that species i becomes monodominant is equal to its relative abundance across the entire metacommunity. These conclusions follow even when local communities change in size and when dispersal rearranges local species diversities (Maruyama 1970a).

Other aspects of ecological drift can, however, be strongly affected by geographic structuring. Consider the expected time until a species with relative abundance p_M is lost from the entire metacommunity under the neutral model: $-2J_{ME}p_M \ln p_M/(1 - p_M)$ (Kimura and Ohta 1969), where J_{ME} is the effective size of the metacommunity (see below). This implies that the expected time for a moderately abundant species to exit the metacommunity is on the order of J_{ME} seasons. Like the effective population size, which is central to describing random genetic drift in population genetics, we define J_{ME} to be the size of an isolated, unstructured community with the same stochastic features as the metapopulation—the effective metacommunity size.

If dispersal among local communities does not change local community sizes, then the effective size of a subdivided metacommunity can be related to its census size by

$$J_{ME} = J_M/(1 - F_{ST}) \qquad (15.2)$$

(Whitlock and Barton 1997). The parameter F_{ST} accounts for structural aspects of the metacommunity that are critical to ecological drift. It is defined as $F_{ST} = \text{var}(p)/[\bar{p}(1 - \bar{p})]$, where $\bar{p}$ is the mean relative abundance of a particular species, and $\text{var}(p)$ is the variance in relative abundances of the same species across local communities in the metacommunity. F_{ST} lies between zero and one. If the metacommunity is unstructured (i.e., the relative abundances of all species are the same in every local community), $F_{ST} = 0$ and $J_{ME} = J_M$, the metacommunity census size. F_{ST} increases as the relative abundances of species increasingly differ among local communities. The effective size of a structured metacommunity is thus expected never to be less than the total number of individuals that comprise it. Therefore, in the neutral model, the exit of a moderately abundant species is expected to take longer in a structured metacommunity than an undivided one of the same total census size. Moreover, this result emphasizes that the rate of ecological drift is determined by the size of the entire metacommunity: that is, the combined effects of both the number of local communities and the number of individuals in each rather than by just the sizes of the local communities that comprise it.

How does the degree of subdivision affect the rate of extinction? This question can be addressed analytically for certain types of metacommunities, such as island models (Wright 1943; Slatkin 1993), in which all local communities exchange migrants, and stepping-stone models (Maruyama 1970b), in which dispersal is restricted to adjacent communities. For example, the approximation for the island model is $J_{ME} = J_M (1 + [C - 1]^2/[2JmC^2])$ where m is the migration rate, C is the number of local communities, and J is the local community size (Wright 1943; Slatkin 1993). A shared feature is that F_{ST} increases with decreasing migration. Thus the rate at which a species goes extinct in a metacommunity is a declining function of the extent to which the parts of a metacommunity are isolated from one another.

Stabilizing Mechanisms

As explained above, coexistence mechanisms are defined to operate when a new species can invade when rare. With drift, however, the invasion of even highly superior rare species cannot be guaranteed since the probability of immediate extinction may be substantial. One must tease out the coexistence mechanisms by minimizing the effects of drift. This is done by assuming that the local community size is very large so that even a relatively rare species is numerically abundant.

In later publications, we will consider the joint effects of ecological drift and sta-bilizing mechanisms. Here we highlight a few of the applicable insights that have been developed in population genetics and molecular evolution.

The fully neutral model has all equalizing and no stabilizing effects. What spa-tiotemporal patterns of environmental variation and migration will lead to equal-ization, stabilization, or destabilization? An analogous framework for analyzing these types of questions has been developed for molecular evolution by Gillespie (1991). We will use this framework to understand how environmental variability might maintain community diversity.

Consider the Wright-Fisher model of a closed local community containing S species with different fertilities $F_i(t)$. (This is expressly *not* a neutral model.) To eliminate drift, imagine that the community size J is extremely large. It is useful to describe the community in terms of relative abundances, which effectively can take on any value between zero and unity. If $p_i(t)$ is the relative abundance of species i then the change in its relative abundance is equal to its theoretical expectation:

$$\Delta p_i(t) = p_i(t + 1) - p_i(t) = p_i(t)\frac{[F_i(t) - \bar{F}(t)]}{\bar{F}(t)}, \qquad (15.3)$$

where $\bar{F}(t) = \sum_{i=1}^{S} p_i(t)F_i(t)$ is the mean fertility across the metacommunity in sea-son t.

If the fertilities fluctuate from one season to the next, the species with highest geometric mean fertility will become monodominant (Gillespie 1973). If all S species have the same geometric mean fertility, will diversity be maintained in an infinitely-sized community? The answer, it turns out, is not really (Gillespie 1991; Chesson and Huntly 1997). In the long run, the community will always be nearly monodominant, with the remaining species hanging on at extremely low relative abundances, except for occasional rapid switches in the identity of the mono-dominant species. The period between these switches lengthens without bound over time. This suggests that, in a finite community, the rare species will become increasingly vulnerable to extinction by demographic stochasticity. Apparently, temporal variability alone cannot maintain community diversity (Chesson and Huntly 1997).

Can geographical subdivision and spatiotemporal variability maintain com-munity diversity? One way to approach this question is to generalize an ecologi-cally simple model that was originally introduced by Levene (1953) to study anal-ogous questions in population genetics. Imagine a metacommunity with C large local communities. The relative size of the ith community is s_i such that $\sum_{i=1}^{C} s_i = 1$. At the end of the season, the local communities produce a large number of propagules in accordance with local fertilities. These propagules are then freely distributed across the metacommunity landscape.

The dynamics of metacommunity relative abundance for the ith species are described by (suppressing the time variable t)

$$\Delta p_i = \sum_{j=1}^{C} \frac{s_j(\overline{F_j})^\alpha}{\sum_{k=1}^{n} s_k(\overline{F_k})^\alpha} \cdot \frac{p_i[F_{ij} - \overline{F_j}]}{\overline{F_j}}, \tag{15.4}$$

where F_{ij} is the fertility of species i in community j, $\overline{F_j} = \sum_{i=1}^{C} p_i F_{ij}$ is the mean fertility in community j, and α is a parameter that can range from zero to one (Gillespie 1991). The second factor in the summand is the expected change in relative abundance of species i in community j (cf. equation 15.3). The first factor describes the fractional contribution of propagules provided by community j to the metacommunity. At one extreme, $\alpha = 0$, in which case the contribution of community j is proportional to its relative size, s_j. At the other extreme, $\alpha = 1$; the relative contribution by a community is proportional to both its relative size and average productivity, as measured by $\overline{F_j}$. In the population genetics literature, these two extremes are known as soft and hard selection, respectively (Christiansen 1975; Wallace 1970).

It turns out that the extent to which propagule contributions reflect local average productivities is decisive as to whether diversity can be maintained in the metacommunity. Adapting the diffusion methods of Gillespie (1991), it can be shown that when $\alpha < 1/2$, all S species can be maintained under an increasing set of conditions as the degrees of geographic subdivision and heterogeneity increase. This suggests that spatiotemporal variation can serve as a stabilizing mechanism that maintains community diversity. However, when $\alpha > 1/2$, spatial subdivision and variability have just the opposite effect: as they increase, the conditions under which all S species can be maintained become increasingly narrow. We are thus left with the equivocal conclusion that a specific pattern of variability in fitness can be stabilizing or destabilizing, depending on how the metacommunity is structured.

What do these results mean for ecological drift in the face of environmental heterogeneity? Applying Gillespie's results (1991), conditions that support community diversity in the absence of ecological drift (i.e., only stabilizing mechanisms) also slow the rate at which diversity is lost in finite communities compared with a fully neutral model. This means that of the scenarios described above, only spatiotemporal variability with $\alpha < 1/2$ (including the ecological equivalent of soft selection) is consistent with a slower loss of diversity than in a neutral community. By comparison, both temporal and certain kinds of spatiotemporal variability ($\alpha > 1/2$) lead to a more rapid loss of diversity than one would expect under a neutral model.

How Have Real Metacommunities Been Assembled?

The importance of ecological drift and other consequences of demographic stochasticity discussed above in a metacommunity context will depend in large measure on how close real ecological systems lie to their asymptotic deterministic outcomes (i.e., the long term outcomes of species interactions) and the effective metacommunity sizes (J_{ME}) of assemblages. If the metacommunities we study in nature are already at or near their deterministic equilibria, the dynamics of stochastic species losses will be irrelevant and coexistence mechanisms should dominate the dynamics and community structures we see. If, however, real systems are not near their long-term, deterministic outcomes, the dynamics of approach to those outcomes will be the critical features we see in any particular system and not where it will ultimately arrive (see also Hoopes et al., chapter 2 for alternative reasons for the importance of transient dynamics). There is also a much larger philosophical point underlying this choice of how we view systems. The theoretical vantage from which we choose to look will define (a) what features of the system we choose to consider, (b) what dynamics we will a priori define as relevant to addressing our larger goals, and (c) how we will interpret features that may have multiple possible explanations (some explanations we will consider and some we will ignore as irrelevant).

The theoretical utility of invasibility criteria to predict which species will coexist in the long term (Holt 1977; Chesson 2000a) naturally draws our attention to what will happen when a new and rare species invades a system that is currently at its long term deterministic equilibrium. The natural question to ask about the accumulation of species in a system in this theoretical framework is, can a new species increase when it is rare and the others are at their equilibrial abundances? Because the likelihood of long-term persistence of a new, rare species that is nearly ecologically identical to an existing species in the system is exceedingly small, only those species that will deterministically increase when rare (i.e., ones that can coexist with the species already present in the system) would seem to be able to enter the system. Thus from this viewpoint, if only species that can ultimately coexist deterministically have any chance of entering the system, the presence of species that will drift to extinction or those that will ultimately be driven extinct by coexistence mechanisms should be irrelevant to explanations of species diversity and other major features of a multispecies system (e.g., patterns of relative abundance, dominance-diversity relationships, degree of community saturation).

However, this theoretical scenario may not be relevant to many features of real metacommunities if they are not near their long-term deterministic equilibria. Growing evidence from many lines of inquiry suggest that today's ecosystems over large areas of the globe were assembled only recently (on a geological time scale), and that local assemblage membership is quite fluid. This assembly process was shaped by the movements of animals and plants in response to Pleistocene climatic cycles (e.g., see reviews in Davis 1986; Delcourt and Delcourt 1991; Hewitt 1999). Paleobotanical studies have documented rapid changes in local assemblages of trees and range movements of tree species over large areas of most continents. Unpredictable range movements and local assemblage changes have been shown in insects (Coope 1979, 1995; Jost-Stauffer et al. 2001) and rodents (Graham 1986; Graham et al. 1996). In fact, modern North American rodent assemblages "have emerged only in the last few thousand years, and many late Pleistocene communities do not have modern analogs" (Graham et al. 1996). Phylogeographic studies of taxa moving out of refugia to colonize deglaciated continental areas provide a similar picture of faunal upheaval and only recent local and regional assembly (reviewed in Bernatchez and Wilson 1998, Hewitt 1999, and Avise 2000).

In addition to these changing assemblage compositions across the landscape, many new species came into being during this same period because these same biogeographic processes also promote speciation (e.g., Vrba 1985, 1995). In fact, phylogenetic studies show that many different taxonomic groups radiated during the Pleistocene, for example, insects (Henry et al. 1999; McPeek and Brown 2000; Knowles 2000, 2001; Barraclough and Volger 2002), and birds (Arnaiz-Villena et al. 1999; Lovette and Bermingham 1999; Price et al. 2000; see review by Hewitt 1996). Not only were local assemblages being destroyed and reassembled into different configurations, but also many new species were being introduced at the same time. Thus, the metacommunities we study today may be only a few thousand to tens of thousand of years old, with many new species having been introduced to the mix over that same timescale.

On top of all this, we must also consider the degree of ecological differentiation of these new species from their progenitors. Speciation can be accomplished by myriad mechanisms (Dobzhansky 1937; Mayr 1942; Otte and Endler 1989; Howard and Berlocher 1998). Many speciation events are accomplished by processes in which lineage diversification accompanies ecological diversification (e.g., Rosenzweig 1978, 1995; Pimm 1979; Schluter 1993, 1996; McPeek 1996). In these ecological speciation mechanisms (Schluter 1996), speciation is simply a by-product of adaptive ecological differentiation. However, many other mechanisms can generate reproductive isolation with no necessary ecological differentiation. Hybridization clearly plays an important role in plant speciation (Stebbins 1950; Barrett 1989; Rieseberg 1997). Chromosomal rearrangements are a primary reproductive isolating mechanism in many animal taxa (reviewed by King 1993). Also, changes in mate recognition systems and sexual selection are also powerful modes of speciation in many animal taxa (Paterson 1978, 1993; Kaneshiro 1983, 1988, 1989; Kaneshiro and Boake 1987; McKaye 1991; Seehausen et al. 1997; Henry et al. 1999; Boake 2002). From these mechanisms species with varying degrees of ecological differentiation may be produced (Carson 1985): some may produce species that are ecologically nearly identical to one another (e.g., insects changing mating songs [Henry et al. 1999] or genital morphology [Eberhard 1988]), whereas others may produce new species that are ecologically quite different from their progenitors (e.g., hybridization [Rieseberg 1997], ecological speciation [Schluter 1993, 1996; Losos et al. 1998; McPeek 1995, 1999, 2000]), and all variations in between.

Given these recent biogeographic and macroevolutionary dynamics, the commonness of closely related and ecologically similar species that co-occur is not at all surprising within a theoretical framework that acknowledges that systems may not be near their long term, deterministic outcomes. For example, consider the recent history of the *Enallagma* damselflies in North America (McPeek and Brown 2000). Permanent ponds and lakes across North America are today inhabited by *Enallagma* species that are derived from the independent radiation of two progenitors that produced seventeen species within the last 250,000 years (Brown et al. 2000; McPeek and Brown 2000; Turgeon and McPeek 2002). These radiations appear to have occurred both while various lineages were isolated in glacial refuge areas and as lineages recolonized the deglaciated regions of North America (Turgeon and McPeek 2002). Four of the speciation events were driven by ecological differentiation of lineages associated with habitat shifts back and forth between ponds and lakes that have fish as top predators versus large dragonflies as top predators (McPeek 1995, 1999, 2000; McPeek and Brown 2000; Turgeon and McPeek, in preparation). However, most of the speciation events appear to have occurred by differentiation of the mate recognition system to generate new species primarily in the fish-lake habitat with little or no apparent ecological differentiation (McPeek and Brown 2000). As a result, five to twelve ecologically very similar *Enallagma* species can today be found co-occurring at most ponds and lakes with fish across much of North America; the members of the local assemblages differ across the continent, but the general pattern of high species richness does not (Johnson and Crowley 1980; McPeek 1989, 1990, 1998; Shiffer and White 1995; McPeek and Brown 2000). In addition, the current ranges for many of these species would have been completely under ice 15,000 years ago, and so much of the metacommunity these species occupy was only recently assembled.

The metacommunities in which *Enallagma* species are embedded are also huge, and differences in relative abundances among lakes should inflate the effective metacommunity sizes even further. Given that population densities of larvae in lakes range between 200–700 larvae/m^2 in littoral areas (McPeek 1990, 1998), the total abundance of all *Enallagma* species at any given pond or lake containing fish will typically range from 10^4 to 10^6 individuals, depending on the extent of the littoral zone and the overall size of the lake. In addition, many of these species have ranges that encompass literally tens to hundreds of thousands of lakes from the Atlantic Coast to the Rocky Mountains. Dispersal among ponds and lakes within a region (e.g., on the scale of 10^3 km^2; McPeek 1989) or across the entire ranges of species (e.g., on the scale of 10^6 km^2) will link these local systems into a metacommunity with an enormous census metacommunity size (J_M). Also, we can derive an admittedly crude estimate of F_{ST} from the relative abundance data presented in McPeek (1990) among three fish lakes in southwestern Michigan using methods for haploid genetic data (Weir 1996). This estimate of F_{ST} is 0.26 (estimated for $\hat{\theta}$ in Weir [1996, 174]), which suggests that the effective metacommunity size (J_{ME}) will be about $1/(1 - F_{ST}) = 1.35$ times larger than its census value. The enormity of J_M and J_{ME} for these species will make the loss of these ecologically similar species by ecological drift across the landscape exceedingly slow—on the order of perhaps $J_{ME} \approx$ tens to hundreds of millions of years for most species to be ultimately lost from the system and thus leaving one species monodominant, given that *Enallagma* species have one generation per year. Moreover, as illustrated by Gillespie's (1991) results, stabilizing mechanisms could speed or slow these expected exit times, depending on the exact form of fitness variation among ponds and lakes. But in any case, the fact that so many ecologically similar and recently derived species (such as the *Enallagma* in fish lakes) co-occur is not surprising when viewed from this theoretical perspective. Similarly, the persistence of cryptic species diversity in such groups as the *Hyallela* amphipods or *Brachionus* rotifers for ten million years or more may result from similar metacommunity features: local *Hyallela* abundances are typically an order of magnitude higher than those of *Enallagma* in North American ponds and lakes (McPeek, pers. observ.).

Conclusions

Rather than considering the structure of natural metacommunities from the perspective of a rare invader into an established system, these data and what we know about the recent upheavals due to climate that occurred across much of the globe strongly argue that we need to alter our theoretical focus for many systems and taxa away from invasibility to consider how natural metacommunities, which have only recently been assembled, are sorting themselves out on regional and maybe even continental scales. In other words, we need a focus that does not abandon the concepts of coexistence, but that does recognize the potential importance of stochasticity and particularly ecological drift at various scales. When considered from this perspective, the ubiquity of ecologically similar and co-occurring species is not surprising. In many taxa, new species have been introduced recently by mechanisms that do not generate much ecological differentiation among species, and these species are now embedded in huge metacommunities, which will make the loss of the ecologically poorer species from the landscape exceedingly slow.

Thus, the correct theoretical perspective for understanding many natural metacommunities may be one that explores how the interaction of various stabilizing and equalizing effects of species interactions influence the transient dynamics of an assemblage on the way to its ultimate configuration of coexisting species, and one that focuses on species as they become rare. Some, possibly many, species may be present because the rate at which they are being driven extinct is slow enough to permit them to persist for substantial periods of time in the system before they realize their ultimate fates, and not because local or regional processes foster their indefinite persistence. Moreover, it is the very fact that local communities are linked by dispersal into metacommunities that makes ecological drift a potentially important influence on the dynamics of real systems (Hubbell 2001, and results above). Mechanisms with both stabilizing and equalizing effects (Chesson 2000a) influence the dynamics of real metacommunities, and we cannot simply assume that mechanisms promoting coexistence dominate to maintain high biodiversity; many macroevolutionary, historical, and ecological features of high diversity systems argue that it is simply not so. Moreover, the interplay of stabilizing and equalizing effects of species interactions may not be straightforward, as Gillespie's (1991) results illustrate. As population genetics and molecular evolution have done (e.g., Ohta 2002), we need to begin to construct a substantial theoretical edifice that can be applied to real metacommunities to explore the dynamical consequences of the stabilizing and equalizing effects of species interactions.

Acknowledgments

The authors would like to thank Ryan Thum, Alicia Ellis, and three anonymous reviewers for comments on the manuscript. This work was supported by grants from the National Science Foundation to the authors.

Literature Cited

Abrams, P. A., C. E. Brassil, and R. D. Holt. 2003. Dynamics and responses to mortality rates of competing predators undergoing predator-prey cycles. Theoretical Population Biology 64:163–176.

Amarasekare, P. 2000. The geometry of coexistence. Biological Journal of the Linnean Society 71:1–31.

———. 2003. Competitive coexistence in spatially structured environments: A synthesis. Ecology Letters 6:1109–1122.

Armstrong, R. A., and R. McGehee. 1980. Competitive exclusion. American Naturalist 115:151–170.

Arnaiz-Villena, A., M. Álvarez-Tejado, V. Ruíz-del-Valle, C. Garcia-de-la-Torre, P. Varela, M. J. Recio, S. Ferre, and J. Martinez-Laso. 1999. Rapid radiation of canaries (genus *Serinus*). Molecular Biology and Evolution 16:2–11.

Avise, J. C. 2000. *Phylogeography: The history and formation of species.* Harvard University Press, Cambridge, MA.

Barraclough, T. G., and A. P. Vogler. 2002. Recent diversification rates in North American tiger beetles estimated from a dated mtDNA phylogenetic tree. Molecular Biology and Evolution 19:1706–1716.

Barrett, S. C. H. 1989. Mating system evolution and speciation in heterostylous plants. Pages 257–283 *in* D. Otte and J. A. Endler, eds. *Speciation and its consequences.* Sinauer Associates, Sunderland, MA.

Bernatchez, L., and C. C. Wilson. 1998. Comparative phylogeography of Nearctic and Palearctic fishes. Molecular Ecology 7:431–452.

Boake, C. R. B. 2002. Sexual signaling and speciation, a microevolutionary perspective. Genetica 116:205–214.

Bolker, B., and S. W. Pacala. 1999. Spatial moment equations for plant competition: Understanding spatial strategies and the advantages of short dispersal. American Naturalist 153:7575–602.

Brown, J. M., M. A. McPeek and M. L. May. 2000. A phylogenetic perspective on habitat shifts and diversity in the North American *Enallagma* damselflies. Systematic Biology 49:697–712.

Bürger, R. 2000. *The mathematical theory of selection, recombination, and mutation.* John Wiley and Sons, Chichester, UK.

Carson, H. L. 1985. Unification of speciation theory in plants and animals. Systematic Biology 10:380–390.

Case, T. J. 2000. *An illustrated guide to theoretical ecology.* Oxford University Press, New York.

Caswell, H. 1978. Predator-mediated coexistence: A nonequilibrium model. American Naturalist 112: 127–154.

Caswell, H., and J. E. Cohen. 1991. Communities in patchy environments: A model of disturbance, competition, and heterogeneity. Pages 193–218 *in* M. Gilpin and I. Hanski, eds. *Ecological heterogeneity.* Academic Press, London, UK.

Chesson, P. 1994. Multispecies competition in variable environments. Theoretical Population Biology 45:227–276.

———. 2000a. Mechanisms of maintenance of species diversity. Annual Review of Ecology and Systematics 31:343–366.

———. 2000b. General theory of competitive coexistence in spatially-varying environments. Theoretical Population Biology 58:211–237.

Chesson, P., and N. Huntly. 1997. The roles of harsh and fluctuating conditions in the dynamics of ecological communities. American Naturalist 150:519–553.

Chesson, P. L., and R. R. Warner. 1981. Environmental variability promotes coexistence in lottery competitive systems. American Naturalist 117:923–943.

Christiansen, F. B. 1975. Hard and soft selection in a subdivided population. American Naturalist 109: 11–16.

Coope, G. R. 1979. Late cenozoic fossil coleoptera—evolution, biogeography, and ecology. Annual Review of Ecology and Systematics 10:247–267.

———. 1995. Insect faunas in ice age environments: Why so little extinction? Pages 55–74 *in* J. H. Lawton and R. M. May, eds. *Extinction rates.* Oxford University Press, New York.

Davis, M. B. 1986. Climatic instability, time lags, and community disequilibrium. Pages 269–284 *in* J. Diamond and T. J. Case, eds. *Community ecology.* Harper and Row, New York.

Delcourt, H. R., and P. A. Delcourt. 1991. *Quaternary ecology: A paleoecological perspective.* Chapman and Hall, London, UK.

DeVries, P. J., and T. R. Walla. 2001. Species diversity and community structure in neotropical fruit-feeding butterflies. Biological Journal of the Linnean Society 74:1–15.

Didham, R. K., P. M. Hammond, J. H. Lawton, P. Eggleton P, and N. E. Stork. 1998. Beetle species responses to tropical forest fragmentation. Ecological Monographs 68:295–323.

Dobzhansky, T. 1937. *Genetics and the origin of species.* Columbia University Press, New York.

Eberhard, W. G. 1988. *Sexual selection and animal genitalia.* Harvard University Press, Cambridge, MA.

Ewens, W. J. 1979. *Mathematical population genetics.* Vol. 9. Springer-Verlag, New York.

Gerrish, P. J., and R. E. Lenski. 1998. The fate of competing beneficial mutations in an asexual population. Genetica 102/103:127–144.

Gillespie, J. H. 1973. Natural selection with varying selection coefficients—a haploid model. Genetical Research of Cambridge 21:115–120.

———. 1991. *The causes of molecular evolution.* Oxford University Press, Oxford, UK.

Gomez, A., M. Serra, G. R. Carvalho, and D. H. Lunt. 2002. Speciation in ancient cryptic species complexes: Evidence from the molecular phylogeny of *Brachionus plicatilis* (Rotifera). Evolution 56: 1431–1444.

Gould, S. J., and R. C. Lewontin. 1979. The spandrels of San Marcos and the panglossian paradigm: A critique of the adaptationist program. Proceedings of the Royal Society of London, series B, Biological Sciences 205:581–598.

Graham, R. W. 1986. Response of mammalian communities to environmental changes during the late

Quaternary. Pages 300–313 in J. Diamond and T. J. Case, eds. Community ecology. Harper and Row, New York.

Graham, R. W., E. L. Lundelius, Jr., M. A. Graham, E. K. Schroeder, T. S. Toomey III, E. Anderson, A. D. Barnosky, J. A. Burns, C. S. Churcher, D. K. Grayson, R. D. Guthrie, C. R. Harington, G. T. Jefferson, L. D. Martin, H. G. McDonald, R. E. Morlan, H. A. Semken, Jr., S. D. Webb, L. Werdelin, and M. C. Wilson. 1996. Spatial response of mammals to late Quaternary environmental fluctuations. Science 272:1601–1606.

Harris, R. J., and B. R. Burns. 2000. Beetle assemblages of kahikatea forest fragments in a pasture-dominated landscape. New Zealand Journal of Ecology 24:57–67.

Harrison, R. G. 1991. Molecular changes at speciation. Annual Review of Ecology and Systematics 22:281–308.

Hastings, A. 1977. Spatial heterogeneity and the stability of predator-prey systems. Theoretical Population Biology 12:37–48.

———. 1980. Disturbance, coexistence, history, and competition for space. Theoretical Population Biology 18:363–373.

Henry, C. S., M. L. M. Wells, and C. M. Simon. 1999. Convergent evolution of courtship songs among cryptic species of the Carnea group of green lacewings (Neuroptera : Chrysopidae : Chrysoperla). Evolution 53:1165–1179.

Hewitt, G. M. 1996. Some genetic consequences of ice ages, and their role in divergence and speciation. Biological Journal of the Linnean Society 58:247–276.

———. 1999. Post-glacial re-colonization of European biota. Biological Journal of the Linnean Society 68:87–112.

Holt, R. D. 1977. Predation, apparent competition and the structure of prey communities. Theoretical Population Biology 12:197–229.

Howard, D. J., and S. H. Berlocher. 1998. Endless forms: Species and speciation. Oxford University Press, New York.

Hubbell, S. P. 2001. The unified neutral theory of biodiversity and biogeography. Princeton University Press, Princeton, NJ.

Hubbell, S. P., and R. B. Foster. 1986. Biology, chance, and history and the structure of tropical rain forest tree communities. Pages 314–429 in J. Diamond and T. J. Case, eds. Community ecology. Harper and Row, New York.

Hurtt, G. C., and S. W. Pacala. 1995. The consequences of recruitment limitation: Reconciling chance, history and competitive differences between plants. Journal of Theoretical Biology 176:1–12.

Johnson, D. M., and P. H. Crowley. 1980. Habitat and seasonal segregation among coexisting odonate larvae. Odonatologica 9:297–308.

Jost-Stauffer, M., G. R. Coope, and C. Schluchter. 2001. A coleopteran fauna from the middle Wurm (Weichselian) of Switzerland and its bearing on palaeoobiography, palaeoeclimate and palaeoecology. Journal of Quaternary Science 16:257–268.

Kaneshiro, K. Y. 1983. Sexual selection and direction of evolution in the biosystematics of Hawaiian Drosophilidae. Annual Review of Entomology 28:161–178.

———. 1988. Speciation in the Hawaiian Drosophila: Sexual selection appears to play an important role. Bioscience 38:258–263.

———. 1989. The dynamics of sexual selection and founder effects in species formation. Pages 279–296 in L. V. Giddings, K. Y. Kaneshiro, and W. W. Anderson, eds. Genetics, speciation, and the founder principle. Oxford University Press, Oxford, UK.

Kaneshiro, K. Y., and C. R. B. Boake. 1987. Sexual selection and speciation: Issues raised by Hawaiian Drosophila. Trends in Ecology and Evolution 2:207–212.

Kimura, M. 1983. The neutral theory of molecular evolution. Cambridge University Press, Cambridge, UK.

Kimura, M., and T. Ohta. 1969. The average number of generations until fixation of a mutant gene in a finite population. Genetics 61:763–771.

King, M. 1993. *Species evolution: The role of chromosome change.* Cambridge University Press, Cambridge, UK.

Kingman, J. F. C. 1982. Exchangeability and the evolution of large populations. Pages 97–112 *in* G. Koch and F. Spizzichina, eds. *Exchangeability in probability and statistics.* North Holland, Amsterdam, The Netherlands.

Kinsig, A. P., S. A. Levin, J. Dushoff, and S. W. Pacala. 1999. Limiting similarity, species packing, and system stability for hierarchical competition-colonization models. American Naturalist 153:371–383.

Knowles, L. L. 2000. Tests of Pleistocene speciation in montane grasshoppers from the sky islands of western North America (genus *Melanoplus*). Evolution 54:1337–1348.

———. 2001. Did the Pleistocene glaciations promote divergence? Tests of explicit refugial models in montane grasshoppers. Molecular Ecology 10:691–701.

Kornfield, I., and P. F. Smith. 2000. African Cichlid fishes: Model systems for evolutionary biology. Annual Review of Ecology and Systematics 31:163–196.

Levene, H. 1953. Genetic equilibrium when more than one ecological niche is available. American Naturalist 87:331–333.

Losos, J. B., T. R. Jackman, A. Larson, K. de Queiroz, and L. Rodriguez-Schettino. 1998. Contingency and determinism in replicated adaptive radiations of island lizards. Science 279:2115–2118.

Lovette, I. J., E. Bermingham. 1999. Explosive speciation in the New World Dendroica warblers. Proceedings of the Royal Society of London, series B, Biological Sciences 266:1629–1636.

MacArthur, R. H. 1982. *Geographical ecology.* Princeton University Press, Princeton, NJ.

Maruyama, T. 1970a. On the fixation probability of mutant genes in a subdivided population. Genetical Research of Cambridge 15:221–225.

———. 1970b. Stepping stone models of finite length. Advances in applied probability 2:229–258.

Mayr, E. 1942. *Systematics and the origin of species.* Columbia University Press, New York.

McKaye, M. 1991. Sexual selection and the evolution of cichlid fishes of Lake Malawi, Africa. Pages 241–257 *in* M. H. A. Keenleyside, ed. *Cichlid fishes: Behavior, ecology and evolution.* Chapman and Hall, New York.

McPeek, M. A. 1989. Differential dispersal tendencies among *Enallagma* damselflies (Odonata: Coenagrionidae) inhabiting different habitats. Oikos 56:187–195.

———. 1990. Determination of species composition in the *Enallagma* damselfly assemblages of permanent lakes. Ecology 71:83–98.

———. 1996. Linking local species interactions to rates of speciation in communities. Ecology 77:1355–1366.

———. 1998. The consequences of changing the top predator in a food web: A comparative experimental approach. Ecological Monographs 68:1–23.

McPeek, M. A., and J. M. Brown. 2000. Building a regional species pool: Diversification of the *Enallagma* damselflies in eastern North American waters. Ecology 81:904–920.

Moran, P. A. P. 1958. Random processes in genetics. Proceedings of the Cambridge Philosophical Society 54:60–71.

Muko, S., and Y. Iwasa. 2000. Species coexistence by permanent spatial heterogeneity in a lottery model. Theoretical Population Biology 57:273–284.

Ohta, T. 2002. Near-neutrality in evolution of genes and gene regulation. Proceedings of the National Academy of Sciences, USA 99:16134–16137.

Otte, D., and J. A. Endler. 1989. *Speciation and its consequences.* Sinauer and Associates, Sunderland, MA.

Pacala, S. W., and D. Tilman. 1994. Limiting similarity in mechanistic and spatial models of plant competition in heterogeneous environments. American Naturalist 143:222–257.

Paterson, H. E. H. 1978. More evidence against speciation by reinforcement. South African Journal of Science 74:369–371.

———. 1993. Evolution and the recognition concept of species. Johns Hopkins University Press, Baltimore, MD.

Pimm, S. L. 1979. Sympatric speciation: A simulation model. Biological Journal of the Linnean Society 11:131–139.

Price, T., I. J. Lovette, E. Bermingham, H. L. Gibbs, and A. D. Richman. 2000. The imprint of history on communities of North American and Asian warblers. American Naturalist 156:354–367.

Raup, D. M. 1991. *Extinction: Bad genes or bad luck?* W. W. Norton and Co., Inc. New York.

Rieseberg, L. R. 1997. Hybrid origins in plant species. Annual Review of Ecology and Systematics 28:359–389.

Rosenzweig, M. L. 1978. Competitive speciation. Biol. J. Linn. Soc. 10:275–289.

———. 1995. *Species diversity in space and time.* Cambridge University Press, Cambridge, UK.

Ruse, L. P. 1995. Chironomid community structure deduced from larvae and pupal exuviae of a chalk stream. Hydrobiologia 315:135–142.

Schluter, D. 1993. Adaptive radiation in sticklebacks: size, shape and habitat use efficiency. Ecology 74:699–709.

———. 1996. Ecological causes of adaptive radiation. The American Naturalist 148:S40–S64.

Seehausen, O., J. J. M. van Alphen, and F. Witte. 1997. Cichlid fish diversity threatened by eutrophication that curbs sexual selection. Science 277:1808–1811.

Shiffer, C. N., and H. B. White. 1995. Four decades of stability and change in the Odonata populations at Ten Acre Pond in central Pennsylvania. Bulletin of American Odonatology 3:31–41.

Shmida, A. and S. Ellner. 1984. Coexistence of plant species with similar niches. Vegetatio 58:29–55.

Slatkin, M. 1993. Isolation by distance in equilibrium and nonequilibrium populations. Evolution 47:264–279.

Stebbins, G. L. 1950. Variation and evolution in plants. Columbia University Press, New York.

Tilman, D. 1994. Competition and biodiversity in spatially structured habitats. Ecology 75:2–16.

Tilman, D., and S. W. Pacala. 1993. The maintenance of species richness in plant communities. Pages 13–25 *in* R. E. Ricklefs and D. Schluter, eds. *Species diversity in ecological communities: Historical and geographical perspectives.* University of Chicago Press, Chicago, IL.

Turgeon, J., and M. A. McPeek. 2002. Phylogeographic analysis of a recent radiation of *Enallagma* damselflies (Odonata: Coenagrionidae). Molecular Ecology 11:1989–2002.

Turner, G. F., O. Seehausen, M. E. Knight, C. J. Allender, and R. L. Robinson. 2001. How many species of cichlid fishes are there in African lakes? Molecular Ecology 10:793–806.

Vrba, E. S. 1985. Environment and evolution: Alternative causes of the temporal distribution of evolutionary events. South African Journal of Science 81:229–236.

———. 1995. On the connections between paleoclimate and evolution. Pages 24–45 *in* E. S. Vrba, G. H. Denton, T. C. Partridge, and L. H. Burckle, eds. *Paleoclimate and evolution with emphasis on human origins.* Yale University Press, New Haven, CT.

Wallace, B. 1970. Genetic load: Its biological and conceptual aspects. Prentice-Hall, Englewood Cliffs, NJ.

Weir, B. S. 1996. *Genetic Data Analysis II.* Sinauer and Associates, Sunderland, MA.

Whitlock, M., and N. H. Barton. 1997. The effective size of a subdivided population. Genetics 146:427–441.

Williams, G. C. 1992. Natural selection: Domains, levels, and challenges. Oxford University Press, Oxford, U.K.

Witt, J. D. S., and P. D. N. Hebert. 2000. Cryptic species diversity and evolution in the amphipod genus *Hyalella* within central glaciated North America: A molecular phylogenetic approach. Canadian Journal of Fisheries and Aquatic Sciences 57:687–698.

Witt, J. D. S., D. W. Blinn, and P. D. N. Hebert. 2000. The recent evolutionary origin of the phenotypically novel amphipod *Hyalella montezuma* offers an ecological explanation for morphological stasis in a closely allied species complex. Molecular Ecology 12:405–413.

Wright, S. 1932. The roles of mutation, inbreeding, cross-breeding, and selection in evolution. Proceedings of the VI International Congress of Genetics 1:356–366.

Wright, S. 1943. Isolation by distance. Genetics 28:114–138.

CHAPTER 16

Habitat Selection, Species Interactions, and Processes of Community Assembly in Complex Landscapes

A Metacommunity Perspective

William J. Resetarits, Jr., Christopher A. Binckley, and David R. Chalcraft

How are communities assembled? This simple question drives a great deal of theoretical and empirical research in community ecology (reviews by Cody and Diamond 1975; Diamond 1975; Strong et al. 1984; Belyea and Lancaster 1999; Weiher and Keddy 1999), but elucidation of mechanisms or "rules" of community assembly remain a challenge (e.g., Brown et al. 2000; Stone et al. 2000). The majority of community assembly studies focus on the role of internal dynamics (e.g., species interactions, abiotic tolerances) in determining composition of local communities. However, there is increasing awareness that processes operating at larger scales can have important consequences for local and regional dynamics (Danielson 1991; Wilson 1992; Holt 1993; Ricklefs and Schluter 1993; Leibold 1998; Belyea and Lancaster 1999; Mouquet and Loreau 2002; Amarasekare and Nisbet 2001). One such process is the movement of individuals among communities (Tilman et al. 1994; Holt 1997; Holyoak 2000). If local communities are linked by dispersal, mechanisms structuring them must be examined in a larger spatial framework. When communities are linked across space and time new processes and emergent properties may arise from the resulting complex dynamics, hence the metacommunity perspective (Wilson 1992).

The metacommunity perspective represents a logical extension of the metapopulation concept: discrete local populations or communities linked by periodic dispersal. For clarity we identify a local community as the collection of organisms occurring within a discrete habitat patch. Although rates of individual movement among local populations are central to metapopulation and metacommunity models, studies have only recently considered the manner of dispersal among different populations (e.g., Hanski and Singer 2001; Holt and Barfield 2001).

Our work on habitat selection has focused on colonization, the necessary consequence of dispersal. Here we examine ways individuals can select among local communities and how processes of colonization can affect local and regional dynamics. Specifically, we focus on what we call "interactive habitat selection." We have two purposes: to review concepts of habitat selection relevant to metacommunities and to use our ongoing studies of pond communities to illustrate these concepts.

Recent work suggests that habitat selection can dramatically affect both population and community dynamics (Rosenzweig 1991; Brown 1998; Bernstein et al. 1999; Abrams 2000; Remes 2000; Schmidt et al. 2000; Delibes, P. Ferreras, et al. 2001; Delibes, P. Gaona, et al 2001; Hanski and Singer 2001; Heithaus 2001; Holt and Barfield 2001; Grand 2002; Krivan and Sirot 2002; Spencer et al. 2002; Morris 2003). However, few empirical or theoretical studies have examined the effects of habitat selection on local communities composed of more than two species.

Aquatic Mosaic Landscapes as a Model System for Understanding Linkages among Communities

Linkages among communities may be especially critical for persistence of species in complex landscapes where habitats vary both spatially and temporally. Because freshwater habitats typically have more discrete boundaries than their terrestrial or marine counterparts, they are ideal for illustrating many of the processes relating to habitat selection, dispersal, and metacommunities. While we focus here on aquatic habitats as individual communities and linked metacommunities, the concepts apply to a variety of systems.

Freshwater aquatic systems are often composed of isolated patches linked to each other and the surrounding terrestrial matrix by species with complex life cycles. The fauna contains the larval stages of many organisms that are primarily (semi)terrestrial adults, or that must leave the water to complete their life cycle (Merritt and Cummins 1984; Duellman and Trueb 1986; Hutchinson 1993; Schneider and Frost 1996). Persistence of many such species is dependent on seasonal invasion and / or oviposition by dispersing adults. Colonization / oviposition behavior can play a major role in the assembly of individual aquatic communities and link communities across landscapes. The importance of such linkages varies with stability (persistence) of local communities and degree of dispersal. The extent and pattern of spatial variation among communities and temporal variation within and among communities has implications for habitat quality and the dynamics of colonization. If dispersal is minimal, landscape level dynamics are simply the sum of within-community processes across all local community types. If dispersal is substantial, both local and metacommunity dynamics depend on the interaction of dispersal with spatial and temporal variation in habitat quality.

Types of Organisms and Types of Colonization

The causes of dispersal have generated much interest (e.g., Skellam 1952; Gadgil 1971; Maynard-Smith 1972; Hamilton and May 1977; Harper 1977; Parker 1984; Cohen and Levin 1987; Cohen and Mitro 1989; Clobert et al. 2001). Although the impetus for dispersal (e.g., drift, aggression, population density) has implications in a number of domains (Holt 1997), we focus here on the consequences of dispersal for (re)distribution of individuals among habitat patches. It is useful, however, to distinguish between *obligate* and *facultative dispersers*. Obligate dispersers have life cycles that force dispersal at some stage, while facultative dispersers depend on current conditions for cues to initiate dispersal (or not). From a probabilistic perspective obligate dispersers form more consistent links among communities, whereas facultative dispersers may generate more dynamic linkages. Impact of both depends on colonization strategies used by dispersing individuals. Current metacommunity models (e.g., Holyoak et al., chapter 1; Mouquet et al., chapter 10) often follow the lead of metapopulation models in assuming random colonization for simplicity (but see Danielson 1991) and as a starting point (Hanski and Gilpin 1997). Even simple colonization mechanisms can generate complex dynamics (Skellam 1951; Levin 1992); however, the type of colonization may play a central role in determining identity, extent, and strength of linkages among communities. We will use three general types—random colonization, philopatry, and interactive habitat selection—to illustrate how colonization strategies can affect the connectedness of communities.

Random Colonization

Random colonizers have no control over their passive dispersal and settlement patterns, or they actively disperse with random settlement patterns. Random colonization is the mode primarily used in current metacommunity models and gives rise to the notion of the "propagule rain" where all patches have an equal likelihood of receiving propagules from any other patch (figure 16.1a). Random colonization can also be modeled in other ways. For example, proximity and prevailing physical conditions may cause dispersers to have a greater probability of colonizing nearby patches by chance alone (figure 16.1b).

Philopatry

Philopatry is translated as "father loving" and taken to mean "to breed in one's birthplace." However, current usage includes scenarios with very different consequences. Both species breeding at their natal locality because of simple encounter probabilities and those actually exhibiting natal homing are included. In the former case, philopatry represents an extreme form of random colonization where proximity effects or spatial structure dictate that colonization of other patches is unlikely. While this distinction has consequences, it is beyond the scope of our

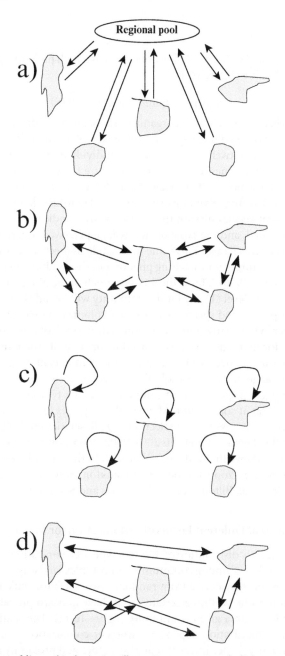

Figure 16.1 Types of dispersal / colonization, illustrating consequences for linkages among local communities: (a) Random dispersal and recolonization in a large, panmictic population. Emigrants enter regional pool and are randomly resorted among original habitats; (b) Random dispersal in a spatially structured (meta)population. Dispersal is random from each node but colonization probabilities are determined by spatial proximity (and patch size) as in internal colonization metapopulation models. (c) Philopatry, in which individuals breed in their natal ponds with probability = 1; all other ponds with probability = 0; (d) IHS where linkages derive from shared habitat traits independent of spatial locations (e.g., closed versus open canopy, fish versus fishless, temporary versus permanent).

current discussion. We simply define philopatry as a high probability of breeding in one's natal locality (figure 16.1c).

Interactive Habitat Selection (IHS)

This is classic habitat selection. An organism assesses habitats during active search and chooses that deemed most appropriate (Baker 1978) (most likely to generate the highest fitness; Fretwell and Lucas 1970). Individuals may either avoid or be attracted to certain patches relative to others (figure 16.1d; Rausher 1993). While avoidance and attraction could be viewed as different descriptions of the same response (e.g., are treefrogs avoiding fish or attracted to fishless habitats?), each can produce different species distributions. "Interactive" indicates that patches are continually assessed, and the same habitat will be perceived differently should it undergo substantive change. This is the most labile of mechanisms and provides the greatest opportunity for matching phenotypes (both adult and propagule) to habitats (Resetarits 1996). This is the focus of much of the discussion below.

At a community level (1) any local community will include both obligate and facultative dispersers and species with various colonization strategies; (2) communities differ in the relative representation of dispersal / colonization types; and (3) different forms of dispersal / colonization operate at different life history stages within a single species. Our aquatic communities contain species exhibiting the entire range of dispersal / colonization types. However, our focal species are mostly obligate dispersers since complex life cycles typically require leaving the aquatic habitat at some point. Many of these species (e.g., certain frogs and beetles) exhibit interactive habitat selection (e.g., Resetarits and Wilbur 1989; Resetarits 2001; Binckley and Resetarits 2002, 2003, 2005), while others appear to be philopatric (e.g., damselflies; McPeek 1990). Still others may be essentially random colonizers (e.g., zooplankton, phytoplankton, aquatic plants, and others with passive dispersal; Bilton et al. 2001; Caceres and Soluk 2002).

Consequences of Different Forms of Colonization for Metacommunities

Clearly the frequency distribution of different types of colonization (as well as dispersal) affects linkages among communities in complex landscapes (figure 16.1). The greatest extent may derive from random dispersal (propagule rain model), where all communities within a certain radius have nonzero probabilities of receiving colonists from a given community (figure 16.1a). The spatial dynamics under random colonization depend on whether colonization is modeled as a propagule rain or whether internal spatial structure and proximity are included but linkages are specifically manifested through changes in population size (mass effects; Shmida and Ellner 1984).

Philopatry reduces connections among communities and fosters isolation and local adaptation. Philopats might be expected to contribute relatively little to dynamic linkages among communities in complex landscapes (figure 16.1c). Only with major disturbance would philopatric populations "reconnect" to others within the landscape, unless philopatry is under dynamic control (which is not indicated by any existing literature). Thus, the extent of philopatry within a community limits the potential for linkages.

The extent of linkages among communities under IHS is intermediate between complete linkage among all patches (propagule rain, figure 16.1a) and no or few linkages (philopatry, figure 16.1c). However, a number of emergent patterns deriving from the process of IHS have implications for metacommunity dynamics. Some of these include the following.

Compression

Species can be compressed into a smaller number of patches via IHS. For example, all patches in a given landscape are ideally suitable for species X before invasion of a subset of patches by a predator, species Y. In the absence of IHS all patches are still colonized resulting in loss of individuals to sinks (figure 16.2a). With IHS the number of patches available to X is compressed because it avoids patches with Y (figure 16.2b). This decreases one set of interspecific interactions (e.g., predation) but can increase the intensity of other intra- and/or inter-specific interactions and affect (meta)population dynamics (Rosenzweig 1991; Blaustein 1999). This is simply habitat loss driven by colonization behavior. All patterns described below potentially involve habitat compression.

Contagion

When contagion occurs, the perceptions of a given habitat "bleed" over into adjacent habitats, affecting their perceived (but not actual) suitability; suitable patches take on the characteristics of the nearby unsuitable habitats in the "eye" of the colonists. This can accelerate habitat loss and generate compression or compromise (the next emergent pattern). In the case of avoidance, contagion causes loss of otherwise favorable habitats perceived as unsuitable due to proximity to truly unsuitable habitats (figure 16.2c). In a changing landscape where processes render habitats unsuitable, contagion accelerates habitat loss by reducing available patches and increasing strain (because of higher densities) on remaining habitats perceived as suitable. In contrast, contagion based on attraction results in colonization of unsuitable habitats perceived as suitable. Regional population size is reduced because these habitats function as persistent sinks. These attractive sinks can have a disproportionate effect on population persistence (Delibes, Ferreras, et al. 2001; Delibes, Gaona, et al 2001; see Hoopes et al. chapter 2).

Compromise

When preferred habitats do not exist, or are co-opted by incompatible species, habitat selection forces species either to local extinction or habitat compromise

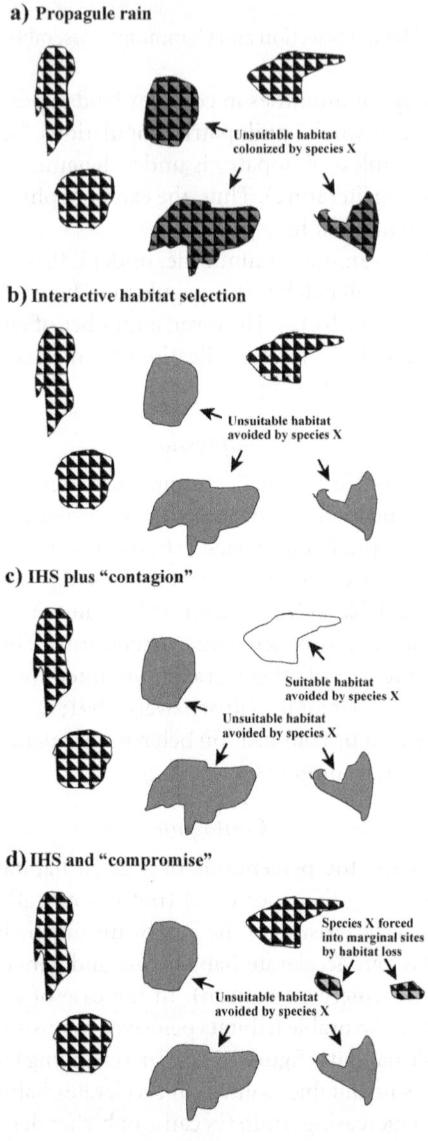

Figure 16.2 Differing modes of dispersal/habitat selection and distribution of hypothetical species X (obligate disperser) in a changing landscape. Initial set of habitats all suitable to X except one. (a) Distribution of colonists of X under passive dispersal/propagule rain model as habitat suitability changes through invasion of incompatible species Y (shaded patches); no change in colonization and propagules placed in unsuitable habitats are lost. (b) Distribution of colonists of X with IHS. X detects and avoids Y as an indicator of unsuitable habitat, propagules redistributed among remaining suitable habitats, increasing densities of X. Increased densities may have further population consequences (see text). (c) Habitat contagion in the context of IHS; suitable habitats in close proximity those invaded by Y are viewed as unsuitable by X, resulting in loss of otherwise suitable habitats and further constriction of the population. Different consequences obtain depending on whether attraction or avoidance is involved (see text). (d) Habitat compromise resulting from habitat loss. Smallest patch initially considered unsuitable by X; after invasion by Y and subsequent loss of habitat, X forced into previously unused habitats as a result of increased densities in remaining high quality habitats (sensu Fretwell and Lucas 1970).

(figure 16.2d); in habitat compromise species select the best remaining habitat, though average fitness may be considerably below that seen in the preferred habitats (sensu the Ideal Despotic Distribution, Fretwell and Lucas 1970) and may even be below replacement rates, resulting in a sink population. This process may also place species into new ecological contexts and into new sets of species interactions.

Covariance

While attraction and avoidance can directly generate patterns of covariance among species (figure 16.3a), they can also generate patterns of secondary covariance via independent species responses to the same or correlated factors. For example, gray (*Hyla chrysoscelis*) and squirrel treefrogs (*H. squirella*) avoid fish, and thus if fishless habitats are limited (and treefrogs do not avoid one another) they will positively covary to a greater extent than if all habitats were fishless and provided other axes across which they could assort (figure 16.3a,b). Alternatively, treefrog A avoids fish, treefrog B avoids shade; if open canopy ponds are less likely to harbor fish, then the two have greater covariance as a result of sharing open canopy, fishless ponds. This could result in competition for enemy-free space (Jeffries and Lawton 1984; Holt and Lawton 1994).

Avoidance (Habitat Loss) and Attraction (Habitat Gain) Cascades

These are similar to cascading effects in other contexts and arise from some combination of processes above. Imagine Z as a predator that excludes X and Y (competitors that have partitioned Z-free habitats along some other axis) (figure 16.3a). Predator Z invades a proportion of habitats suitable to X, forcing X into marginal habitats (for X) normally occupied by Y. If X is either avoided by Y or is a superior competitor, Y's options become very limited (figure 16.3c). Increase in frequency of habitats containing Z results in cascading habitat loss. Cascades can occur as a result of attraction as well. If suitable habitat is defined (in part) by presence of Z, increase in frequency of Z results in cascading habitat gain. We can envision interactions between avoidance and attraction that generate more complex behavioral dynamics (via trait-mediated direct and indirect effects) and also complex dynamics arising from the operation of both density and trait mediated effects over time.

Maladaptive Habitat Selection

Thus far we have assumed that species select habitats in which their performance is enhanced, or in which fitness of individuals selecting among habitats remains constant (Fretwell and Lucas 1970). However, organisms may also make maladaptive choices (Remes 2000; Delibes, Ferreras, et al. 2001; Delibes, Gaona, et al 2001). Random colonization commonly leads to inappropriate habitat matching, and philopatry can fail if habitats change. If cues used to assess habitats under IHS

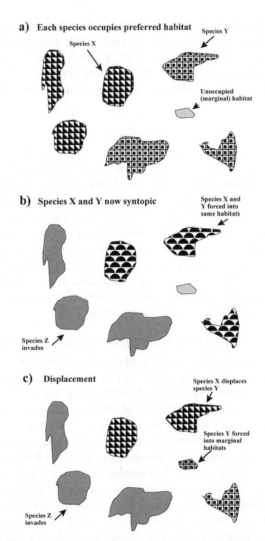

Figure 16.3 Interactive habitat selection and species covariances in dynamic landscapes: (a) two species with allotopic distributions, X (triangles) and Y (squares); (b) Z invades (dark shading). Neither X nor Y can persist with Z, generating negative covariances between Z and X, and Z and Y; X and Y are now forced into syntopy in remaining Z-free habitats (half-circles); Z generates positive covariance between X and Y; (c) Same as in (b) except X is competitively superior and co-opts suitable patches previously occupied by Y, forcing it into the smallest (marginal) patch; Y may even move into even smaller, more marginal habitats (not illustrated). Habitat selection can generate negative covariances among species by avoidance, and positive covariances by attraction (not illustrated) or by shared avoidance and habitat compression.

fail, the consequences can be dramatic. One of the truly puzzling patterns we observed in frogs is oviposition in extremely ephemeral habitats, which often occurs in active agricultural fields adjacent to forested areas. Both forested and open areas contain potential breeding sites; forested sites are cooler, deeper, and less ephemeral; open sites are warm, very shallow, and very ephemeral, often not lasting long enough for hatching much less metamorphosis. What precipitates these seemingly disastrous choices by large numbers of ovipositing females? The strong predilection of females for open canopy sites has been established (Binckley and Resetarits, unpublished data), suggesting that, historically, selection of the most open (warmest, most productive) sites was favored because such sites generated the fastest growth and development. However, rapid anthropogenic change has shifted the frequency of open canopy sites from rare to common, and the upper extreme of the distribution from ponds in forest light gaps to puddles in barren fields. Simple, historically "adaptive" decision algorithms such as, "choose the highest available temperature," now lead to maladaptive preferences for population sinks (Remes 2000; Delibes, Ferreras, et al. 2001; Delibes, Gaona, et al 2001). This consequence is possible whenever the relationship between preference and performance breaks down (Rieger et al. 2004).

Populations, Communities, and Metacommunities

Incorporating habitat selection into population models affects population size (Holt 1985; Smith et al. 2000; Spencer et al. 2002), growth rate, and persistence (Pulliam and Danielson 1991; Sutherland 1996) because individuals actively avoid sinks while selecting source habitats. In dynamic landscapes, where distribution of sinks changes, IHS provides a mechanism for coping with the changing fitness landscape (Pulliam and Danielson 1991; Resetarits 2001). This is because individuals are not immediately lost from the population by mortality or reproductive failure but are initially redistributed spatially and concentrated into remaining source habitat, which can then undergo secondary changes due to increased density, and so on (Resetarits and Wilbur 1989; Abrams 1993; Watkinson and Sutherland 1995; Blaustein 1999; Resetarits 2001, 2005).

Similarly, the presence and specific form of habitat selection can also have important consequences at the community and metacommunity levels. We typically think of linkages among communities being generated by shared species that connect habitats through dispersal. IHS can contribute to this type of linkage in a variety of ways and can even increase the number of species shared as a result of common avoidance or attraction behaviors or as a result of shared responses to habitat loss. IHS also generates another form of linkage. We typically assume species that do not co-occur do not interact, communities lacking shared species are not connected, and species with complimentary distributions do not contribute to metacommunity dynamics. However, our work demonstrates that communities can be linked both by shared species and by species segregation driven by habitat selection. Numerous amphibians and insects avoid fish when colonizing/ovipositing (see "Axes of Habitat Selection: Empirical Perspectives"), and the mutually exclusive distributions seen for squirrel treefrogs and fish, for example (Binckley and Resetarits 2002), derive from behavioral avoidance not from typical face-to-face interactions (e.g., competition and predation). Thus, determination of the distribution of squirrel treefrogs occurs at the colonization stage, and changes in fish distribution affects both treefrog distribution at the landscape scale and their local abundance in available fish-free habitats, linking communities at the landscape level. These phantom interactions (direct but cryptic effects driven by past interactions [sensu Connell 1980]) are a missing element in community ecology and are particularly germane to metacommunity dynamics.

The idea that behavior can generate patterns of negative and positive covariance among species on the landscape scale generates potentially strong linkages among communities that derive from such phantom interactions. Present distributions reflect the past history of species interactions. This idea was once dominant in ecology, spawning Connell's (1980) classic denunciation of the "Ghost of Competition Past"; however the experimental evidence called for by Connell now exists (see "Axes of Habitat Selection: Empirical Perspectives").

Species can have different metapopulation structures and even those structures may be determined by phantom interactions. For example, fish in landscapes such as pine flatwoods in the southeastern United States or other extensive wetlands (e.g., Great Dismal Swamp) typically exist as island-mainland metapopulations with some internal connections (rescue effect). In contrast, fish intolerant species are characterized by internal colonization metapopulations, because larger habitats likely contain fish, precluding an island-mainland structure and restricting spatial dynamics. Landscape level processes, including those driven by IHS, can determine both the structure of individual populations and communities as well as interactions among communities.

Behavioral production of diversity patterns suggests local community structure results from species interactions both within and among communities (at the metacommunity scale). While community ecology has historically emphasized local interactions, recent work (see "Axes of Habitat Selection: Empirical Perspectives" below) suggests that distributions can be determined by species interactions even though species do not co-occur at the local scale. For these species, habitat selection defines the exact nature and extent of interactions occurring within local communities by reducing the frequency of specific deleterious interactions while simultaneously elevating the probability of an entirely different set of local interactions.

Both specific colonization strategies and exchange of individuals among communities can impact metacommunity dynamics (Rosenzweig 1985; McPeek 1989; Bilton et al. 2001; Binckley and Resetarits 2002). When species actively select sites for feeding and reproduction, regional processes (e.g., dispersal and colonization) interact with local processes (e.g., predation, competition) to determine species distributions and community structure. IHS species subdivide regional landscapes into different habitat types, each occurring in a specific spatial configuration, based on shared characteristics that can be quite specific (e.g., species composition). Under IHS priority effects and phenology (e.g., Morin 1984; Alford and Wilbur 1985; Wilbur and Alford 1985; Alford 1989) assume greater significance in the assembly of local communities and regional metacommunities because current occupants can affect habitat choices of later colonists.

Theory emphasizes that habitats differ in their fitness consequences, which are detectable to individuals who match their choice of habitats to these consequences (e.g., Fretwell and Lucas 1970; Fretwell 1972; Holt and Barfield 2001). Habitat types are defined by specific factors that consistently and predictably affect fitness, and may be biotic, abiotic, or a combination of multiple variables (Fretwell and Lucas 1970; Werner and Gilliam 1984; Moody et al. 1997). Habitat quality declines with increased densities of competitors leading to colonization of initially inferior habitats (Fretwell and Lucas 1970). Thus, habitat selection is a context dependent process, where suitability of any patch is relative to all other potential patches. Diversity patterns, patterns of community linkages, and metacommunity dynamics for species exhibiting IHS depend not simply on the relative quality of patches, but on their frequency distribution and spatial configuration in complex landscapes.

Axes of Habitat Selection: Empirical Perspectives

Thus far we have focused on how IHS affects communities and their linkages, but what evidence exists that habitat selection can determine species distributions? Many aquatic insects and amphibians can discriminate and selectively colonize different patches based on a variety of factors (Resetarits and Wilbur 1989; Walton et al. 1990; Crump 1991; Petranka and Fakhoury 1991; Kats and Sih 1992; Blaustein and Kotler 1993; Hopey and Petranka 1994; Sherratt and Church 1994; Holomuzki 1995; Laurila and Aho 1997; Spieler and Linsenmair 1997; Blaustein 1999; Stav et al. 1999; Summers 1999; Marsh and Borrell 2001; Resetarits 2001, 2005; Binckley and Resetarits 2002, 2003, 2005). Selective colonization / oviposition indicates that individuals possess sensory mechanisms and behaviors that facilitate selection of favorable habitats (Rausher 1983, 1993; Singer 1984, 1986; Thompson and Pellmyr 1991; Renwick and Chew 1994; Resetarits 1996; Blaustein 1999). Variation in environmental conditions and resulting success in different local communities (habitats or patches) provides the backdrop for the evolution of interactive habitat selection (Resetarits 1996; Remes 2000). Below we describe studies on habitat selection in treefrogs and aquatic beetles to illustrate the consequences of habitat selection and the range of factors that may affect habitat suitability.

Biotic Factors

Our study of habitat selection traces back to an experiment (Resetarits and Wilbur 1989, 1991) examining the response of ovipositing gray treefrogs to variation in the faunal composition of experimental ponds (figure 16.4a). In this and all subsequent experiments we tested the responses of naturally colonizing populations to experimental conditions presented in as realistic a field setting as possible. We typically set up pools containing base communities that then vary in biotic and / or abiotic conditions (e.g., Morin 1983, Fauth and Resetarits 1991) and allow our focal species to colonize (Resetarits and Wilbur 1989). We assay the responses by removing eggs daily (frogs) or adults and egg cases periodically (beetles).

Our initial goal was to determine whether *H. chrysoscelis* could detect and respond to (avoid) species important in its larval ecology. That experiment established habitat selection as a significant factor in the distribution of larval *H. chrysoscelis* (figure 16.4a). Ovipositing *H. chrysoscelis* detected and avoided fish (*Enneacanthus chaetodon*), salamanders (*Ambystoma maculatum*), and high densities of conspecifics. Not all predators (nor competitors) were equal; other predators (*Notophthalmus viridescens* and *Tramea carolina*) and a potential competitor (*Rana catesbeiana*) were not avoided, either because they could not reliably be detected or did not have significant effects on potential fitness (Resetarits and Wilbur 1989; Resetarits 1996). The critical implication for communities was that behavior in response to the expected levels of predation and competition could determine the distribution of species, even in the absence of what we commonly think of as species interactions. These phantom interactions were the first experimental evidence that past species interactions did in fact determine the present distribution of species (sensu Connell 1980).

Since then we have examined the responses of numerous species to variation in both the density and identity of predatory fish. We have recently returned to the role of competitors in colonization, but predator experiments support the role of conspecific densities, because habitats typically avoided by frogs receive eggs only on nights with considerable oviposition activity (Binckley and Resetarits 2003; Rieger et al. 2004). Since we removed eggs after each oviposition event, this within-night variation in density is the only density variation accessible to frogs. As activity (equals density) increases, preference for preferred habitats weakens, as predicted by the ideal-free distribution (Fretwell and Lucas 1970). This suggests that responses to habitat characteristics may be hierarchical.

We have found that a variety of frogs and aquatic beetles show strong responses to predatory fish, even stronger than that observed in the initial experiment. Given the choice between fish and fishless habitats, *H. squirella* deposits 95% of eggs in fishless pools (Binckley and Resetarits 2002). Even higher avoidance rates (a staggering 99.3%) occurred for *H. femoralis* in an experiment examining

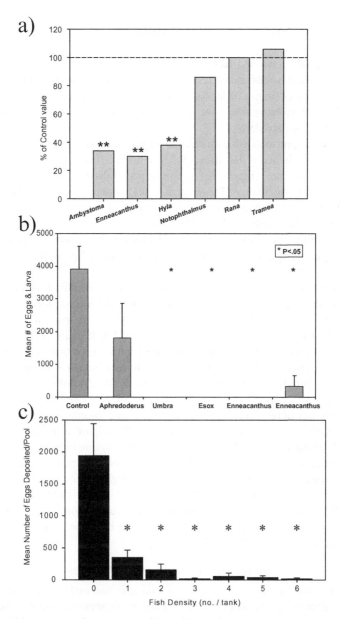

Figure 16.4 (a) Responses of ovipositing *H. chrysoscelis* to predators and competitors. ** indicates significant difference from controls (dashed line) at P < 0.01. Data represents distribution of more than 144,000 eggs into two categories, avoided species (*Enneacanthus chaetodon, Ambystoma maculatum*, and conspecifics) and control equivalents (*Rana catesbeiana, Notophthalmus viridescens* and *Tramea carolina*) (modified from Resetarits and Wilbur 1989). (b) Oviposition responses of a natural population of *H. chrysoscelis* to (nonlethal) presence of five species of predatory fish; only *Aphredoderus sayanus* was not significantly different from controls (modified from Binckley and Resetarits 2003), a result obtained with other taxa as well (unpublished data); monotypic family is only species / family of seven tested not eliciting strong avoidance. (c) Oviposition responses of a natural population of *H. femoralis* to a gradient of (nonlethal) fish density (2 g *Umbra pygmaea*). All treatments significantly differ from controls but not one another. The threshold response density lies below the 1 fish level (<0.53 g/100 L) (modified from Rieger et al. 2004)

habitat selection on two spatial scales (see figure 16.8) (Resetarits 2005). This appears to represent a generalized response to fish, being elicited by species representing six families (in five orders) of freshwater fish (with one interesting exception still under investigation—see figure 16.4b) (Binckley and Resetarits 2003; Binckley, unpublished data).

While the response to fish continues to impress us, more impressive is that the response can be elicited at densities of 0.53g of fish/100 L (a single, 2 g *Umbra pygmaea* in 375 L of water) (figure 16.4c). This response also precisely mirrors the actual effects on performance (Rieger et al. 2004). Interestingly, the four species (figure 16.4b) that elicit avoidance have very different effects on larval anurans (Chalcraft and Resetarits 2003a, 2003b); the nonlethal, phantom effects on species distribution are equivalent among predators, while their lethal effects are very different, providing a complex twist on the concept of functional equivalence (Binckley and Resetarits 2003).

Our work on beetles has revealed similar responses for individual species, plus we have both adult colonization responses and oviposition responses for at least one species. Figure 16.5a shows both responses for *Tropisternus lateralis;* adults colonize fishless habitats with greater frequency and their oviposition directly reflects this (Resetarits 2001). Adults may choose habitats for themselves (egg distribution is a byproduct), their offspring (adult distribution is a byproduct), or they may attempt to optimize both simultaneously. We cannot yet separate these alternatives.

The diversity of beetles has allowed us to examine the effects of fish on multi-species assemblages (Binckley and Resetarits 2005). Figure 16.6 illustrates these effects of habitat selection alone (there is no mortality due to fish), which are almost identical to the results obtained in field surveys: reduction of common species, elimination of rare species, and significantly higher species richness and biomass in the absence of fish (Kenk 1949; Weir 1972; Healey 1984; Fairchild et al. 2000). Thus, the primary filter leading to characteristic communities in fish versus fishless habitats (Wellborn, et al. 1996) may be behavior, rather than predation (Binckley and Resetarits 2003, 2005).

Abiotic Factors and Biotic-Abiotic Interactions

We have recently begun to study effects of abiotic factors on habitat selection and examine interactions between biotic and abiotic factors. Attempts to examine interactions between other factors and predation have been limited by the strong fish response. A similar problem occurs with canopy types (open versus closed); the majority of both treefrogs and beetles prefer open canopy ponds (figure 16.5b). Both preferences are so strong we are unable to test for (meaningful) biological interactions with other factors. For *T. lateralis* we see a significant effect of nutrients in open canopy ponds but no colonization of closed canopy ponds regardless of nutrient level, giving a statistically significant, but not biologically

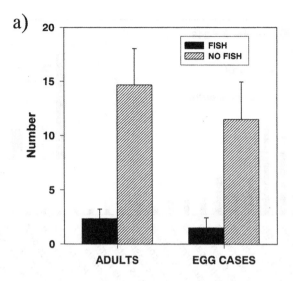

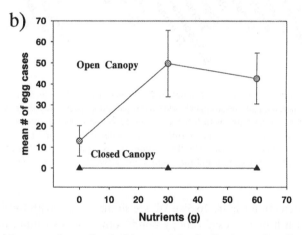

Figure 16.5 (a) Responses of naturally colonizing aquatic beetles, *Tropisternus lateralis,* to (nonlethal) presence of pumpkinseed sunfish (*Lepomis gibbosus*). Number of adults explained 96% of variation in number of egg cases (modified from Resetarits 2001). (b) Responses of *T. lateralis* to open versus closed canopy crossed with nutrient levels showing dramatic response to canopy cover and effects of nutrients only in open canopy ponds (unpublished data).

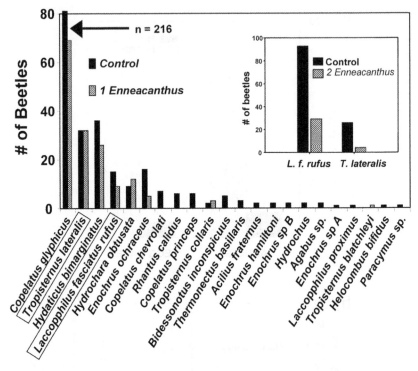

Figure 16.6 Effects of low density fish (nonlethal) presence (1 fish/375 L) on beetles colonizing experimental aquatic habitats. Results match field surveys; common species are reduced and rare species are absent. Eight species found with fish and 21 in fishless pools. Inset: response of two common beetles, *Laccopphilus fasciatus rufus* and *Tropisternus lateralis* at density of 2 fish/375 L, suggesting main graph is conservative estimate of beetle response. This is the strongest evidence to date that IHS can determine species distribution and affect community assembly (Binckley and Resetarits 2005).

relevant, interaction (figure 16.5b)! We see similar results with treefrogs. In the presence of fish or in closed canopy ponds, other factors cannot override the strong primary effects of fish or canopy (figure 16.7).

Spatial Dimensions of Habitat Selection

How habitat selection functions in larger landscapes and at multiple spatial scales impacts the extent of linkages among communities. This is an empirical question, and we have gained some insights into this issue from several relevant experiments. Clearly different organisms will have different capacities to choose among habitats at larger spatial scales. Our treefrogs appear to make decisions at the scale of hundreds of meters, but beyond that we can say little (e.g., Resetarits and Wilbur 1989; Resetarits 1996, 2005). Aquatic beetles vary in flying abilities but certainly outstrip treefrogs and can be expected to cover kilometers or more, depending on species. For other taxa, such as dragonflies, individuals of certain species can cover hundreds of kilometers (or even thousands for *Pantala*), whereas others move very little. This leads to the potential for linkages ranging from local to global (McPeek and Gomulkiewicz, chapter 15). Ovipositioning pinewoods treefrogs, *H. femoralis*, prefer localities consisting of only fishless patches over localities containing one or more patches with fish (regional—illustrating contagion), but also distinguish within localities between fish and fishless patches (local) (Resetarits 2005). The results are quite striking (figure 16.8) and indicate that our experimental arrays capture variation in the landscape on at least two meaningful levels with respect to oviposition site choice and (meta)community assembly.

While the study of habitat selection itself is interesting and productive, it is its role in the distribution and abundance of species that brings it fully into the realm of ecology. The types of behavioral decisions we have observed and their apparent prevalence indicate that choice of habitats by organisms is a dynamic and

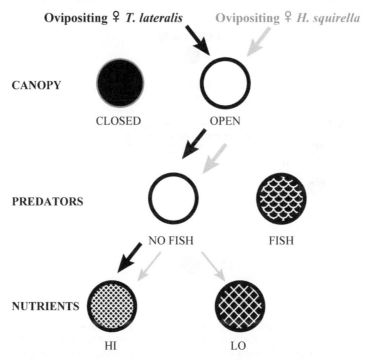

Figure 16.7 Potential effect of shared habitat preferences for open canopy, fishless ponds on covariance in distribution and abundance of *T. lateralis* (black arrows) and *H. squirella* (gray arrows), and effects of variation in preferences (nutrients) on potential covariance. Avoidance generates negative covariance between both species and fish.

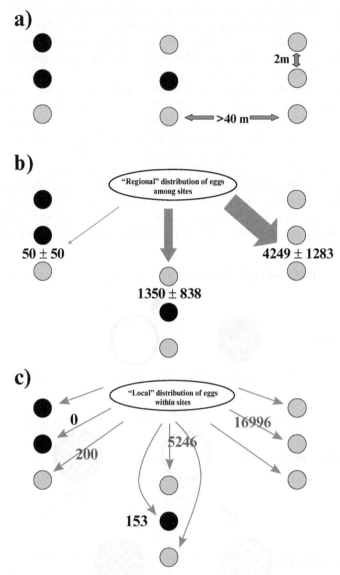

Figure 16.8 Responses of a natural population of ovipositing *H. femoralis* to (nonlethal) presence of predatory fish (*Enneacanthus obesus*) at regional (among localities) and local (among patches within localities) scales. Top panel illustrates a single replicate. At regional scale, females strongly preferred localities of fishless patches, demonstrating a nonlinear response to increasing frequency of fish patches (indicating contagion—see figure 16.2c). At local scale, avoidance of specific patches with fish was nearly complete (only 153 eggs out of 22605 [0.7%]) (modified from Resetarits 2005).

fundamental process in ecology that can serve as a primary filter in determining species distributions and abundances (Binckley and Resetarits 2005). These decisions provide a framework for linkages between communities that has fundamental implications for the emerging field of metacommunity ecology (Resetarits 2005).

Conclusions—Dynamics of Habitat Selection in Complex Landscapes

Metapopulation and metacommunity theory have primarily used random dispersal in modeling landscape level processes. The form of dispersal / colonization can have important consequences for the existence and extent of linkages among local communities (patches, habitats). Whether a landscape is comprised of numerous individual communities or linked into larger metacommunities depends partly on the nature of dispersal and colonization. Interactive habitat selection has the greatest potential to form complex links among populations of multiple species. As we expand the study of community assembly beyond the local scale (Danielson 1991; Wilson 1992; Holt 1993; Resetarits, in review), and from stable to temporally dynamic communities, we must expand our vision beyond individual populations and species pairs, or risk generating a science of metacommunities as simply assembled metapopulations (Holyoak et al., chapter 1). At present most models and empirical studies of habitat selection focus on one or two species. Competition in nature is not constrained to the pairwise construct of the Lotka-Volterra equations, and processes such as diffuse competition and community-wide character displacement have much broader implications for the assembly of natural communities than can be derived from two species models (Leibold 1998; Stevens and Willig 2000; Mouquet and Loreau 2002; Stanton 2003). Similarly, habitat selection has much broader implications at the community and landscape level that affect how we view sets of communities and processes of community assembly in complex landscapes.

We have shown various ways in which habitat selection can influence metacommunities and illustrated the types of decisions species are capable of making. We have also touched on how behavioral decisions can translate into species distributions and compositional patterns in natural communities. Much more work is required before we can begin to understand the role of habitat selection at the population / species and landscape / metacommunity scales. We are continuing our work on habitat selection in aquatic systems along several lines: focusing on issues including the spatial dynamics of habitat selection and trade-offs between habitat quantity and quality; relationships between habitat preference and offspring performance, multiple risks and decision rules (e.g., minimize mortality rate divided by growth rate, minimize adult risk); and effects of habitat alteration and introduced species. It is our hope that this work will lead to a better understanding of habitat selection itself, but more importantly, the consequences of habitat selection for the assembly of communities across complex landscapes.

Acknowledgments

Thanks to M. Holyoak, R. Holt, and M. Leibold for invitations to the ESA Symposium and this book, and for thoughtful editorial comments. J. Shurin, J. Rieger, C. Resetarits, and M. Rosenshield provided helpful comments. We have benefited from discussions of habitat selection with numerous colleagues over the years, including J. Bernardo, L. Blaustein, M. Camara, J. Fauth, M. McPeek, P. Morin, J. Rieger, and E. Werner, and we appreciate their ideas and interest. We appreciate logistical support from Naval Security Group Activity Northwest, College of Charleston, Illinois Natural History Survey, and the Francis Marion National Forest. Funding was provided by INHS-Center for Aquatic Ecology, University of Illinois, Old Dominion University, National Fish and Wildlife Foundation, and NSF (DEB-9727543, DEB-9902188, DEB-0096051) and EPA-STAR (R825795-01-0) grants. CAB was supported by a Dominion Scholarship and DRC as a Postdoctoral Associate at NCEAS.

Literature Cited

Abrams, P. A. 1993. Why predation rates should not be proportional to predator density. Ecology 74:726–733.

———. 2000. The impact of habitat selection on the spatial heterogeneity of resources in varying environments. Ecology 81:2902–2913.

Alford, R. A. 1989. Variation in predator phenology affects predator performance and prey community composition. Ecology 70:206–219.

Alford, R. A. and H. M. Wilbur. 1985. Priority effects in experimental pond communities: Competition between *Bufo* and *Rana*. Ecology 66:1097-1105.

Amarasekare, P., and R. M. Nisbet. 2001. Spatial heterogeneity, source-sink dynamics and the local coexistence of competing species. American Naturalist 158:572–584.

Baker, R. R. 1978. *The evolutionary ecology of animal migration.* Hodder and Stoughton, London, U.K.

Belyea, L. R., and J. Lancaster. 1999. Assembly rules within a contingent ecology. Oikos 86:402–416.

Bernstein, C., P. Auger and J. C. Poggiale. 1999. Predator migration decisions, the ideal free distribution, and predator-prey dynamics. American Naturalist 153:267–281.

Bilton, D. T., J. R. Freeland and B. Okamura. 2001. Dispersal in freshwater invertebrates. Annual Review of Ecology and Systematics 32:159–181.

Binckley, C. A. and W. J. Resetarits, Jr. 2002. Reproductive decisions under threat of predation: Squirrel treefrog *Hyla squirella* responses to banded sunfish *Enneacanthus obesus.* Oecologia 130:157–161.

———. 2003. Functional equivalence of non-lethal effects: Generalized fish avoidance determines distribution of gray treefrog, *Hyla chrysoscelis,* larvae. Oikos 102:623–629.

———. 2005. Habitat selection determines abundance, richness and species composition of beetles in aquatic communities. Biology Letters, in press.

Blaustein, L. 1999. Oviposition site selection in response to risk of predation: Evidence from aquatic habitats and consequences for population dynamics and community structure. Pages 441–456 *in* S. P. Wasser, ed. Evolutionary theory and processes: Modern perspectives. Kluwer, Dordrecht, The Netherlands.

Blaustein, L. and B. P. Kotler. 1993. Oviposition habitat selection by the mosquito *Culiseta longiareolata:* Effects of conspecifics, food and green toad tadpoles. Ecological Entomology 18:104–108.

Brown, J. S. 1998. Game theory and habitat selection. Pages 188–220 *in* L. A. Dugatkin and H. K. Reeve, eds. *Game theory and animal behavior.* Oxford University Press, Oxford, UK.

Brown, J. H., B. J. Fox and D. A. Kelt. 2000. Assembly rules: Desert rodent communities are structured at scales from local to continental. American Naturalist 156:314–321.

Cáceres, C. E. and D. A. Soluk. 2002. Blowing in the wind: A field test of overland dispersal and colonization by aquatic invertebrates. Oecologia 131:402–408.

Chalcraft, D. R. and W. J. Resetarits, Jr. 2003a. Mapping functional similarity on the basis of trait similarities. American Naturalist 162:390–402.

——— 2003b. Predator identity and ecological impacts: Functional redundancy or functional diversity? Ecology 84:2407–2418.

Clobert, J., E. Danchin, A. Dhondt and J. Nichols. 2001. Dispersal. Oxford University Press, Oxford, UK.

Cody, M. L. and J. M. Diamond. 1975. Ecology and evolution of communities. Harvard University Press, Cambridge, MA.

Cohen, D. and S. A. Levin. 1987. The interaction between dispersal and dormancy strategies in varying and heterogeneous environments. Pages 110–122 *in* E. Teramoto and M. Yamaguti, eds. Mathematical topics in population biology, morphogenesis, and neurosciences. Proceedings, Kyoto 1985. Princeton University Press, Princeton, NJ.

Cohen, D. and U. Mitro. 1989. More on optimal rates of dispersal: Taking into account the cost of the dispersal mechanism. American Naturalist 134:659–663.

Connell, J. 1980. Diversity and the coevolution of competitors, or the ghost of competition past. Oikos 35:131–138.

Crump, M. L. 1991. Choice of oviposition site and egg load assessment by a treefrog. Herpetologica 47:308–315.

Danielson, B. J. 1991. Communities in a landscape: The influence of habitat heterogeneity on the interactions between species. American Naturalist 138:1105–1120.

Delibes, M., P. Ferreras and P. Gaona. 2001. Attractive sinks, or how individual behavioural decisions determine source-sink dynamics. Ecology Letters 4:401–403.

Delibes, M., P. Gaona, and P. Ferrera. 2001. Effects of an attractive sink leading into maladaptive habitat selection. American Naturalist 158:277–285.

Diamond, J. M. 1975. Assembly of species communities. Pages 342–444 in M. L. Cody and J. M. Diamond eds. *Ecology and evolution of communities.* Harvard University Press, Cambridge, MA.

Duellman, W. E. and L. Trueb. 1986. *Biology of amphibians.* McGraw-Hill, New York.

Fairchild, G. W., A. M. Faulds, and J. F. Matta. 2000. Beetle assemblages in ponds: Effects of habitat and site age. Freshwater Biology 44:523–534.

Fauth, J. E. and W. J. Resetarits, Jr. 1991. Interactions between the salamander *Siren intermedia* and the keystone predator *Notophthalmus viridescens.* Ecology 72:827–838.

Fretwell, S. D. 1972. *Populations in a seasonal environment.* Princeton University Press, Princeton, NJ.

Fretwell, S. D. and Lucas, H. L. Jr. 1970. On territorial behavior and other factors influencing habitat distribution in birds. I. Theoretical development. Acta Biotheoretica 19:16–36.

Gadgil, M. 1971. Dispersal: Population consequences and evolution. Ecology 52:253–261.

Grand, T. C. 2002. Foraging-predation risk trade-offs, habitat selection, and the coexistence of competitors. American Naturalist 159:106–112.

Hamilton, W. D. and R. M. May. 1977. Dispersal in stable habitats. Nature 269:578–581.

Hanski, I. A. and M. E. Gilpin. 1997. *Metapopulation biology: Ecology, genetics and evolution.* Academic Press, San Diego, CA.

Hanski, I. A. and M. C. Singer. 2001. Extinction-colonization dynamics and host-plant choice in butterfly metapopulations. American Naturalist 158:341–353.

Harper, J. L. 1977. The population biology of plants. Academic Press, London.

Healey, M. 1984. Fish predation on aquatic insects. Pages 255–288 in V. H. Resh and D. M. Rosenberg, eds. *The ecology of aquatic insects.* Praeger, Westport, CT.

Heithaus, M. R. 2001. Habitat selection by predators and prey in communities with asymmetrical intraguild predation. Oikos 92:542–554.

Holomuzki, J. R. 1995. Oviposition sites and fish-deterrent mechanisms of two stream anurans. Copeia 1995:607–613.

Holt, R. D. 1985. Population dynamics in two-patch environments: Some anomalous consequences of optimal habitat distribution. Theoretical Population Biology 28:181–208.

———— 1993. Ecology at the mesoscale: The influence of regional processes on local communities. Pages 77–88 in R. Ricklefs and D. Schluter, eds. Species diversity in ecological communities. University of Chicago Press, Chicago, IL.

———— 1997. From metapopulation dynamics to community structure: Some consequences of spatial heterogeneity. Pages 149–165 in I. A. Hanski and M. E. Gilpin eds. *Metapopulation biology: Ecology, genetics and evolution.* Academic Press, San Diego.

Holt, R. D. and M. Barfield. 2001. On the relationship between the ideal-free distribution and the evolution of dispersal. Pages 83–95 in J. Clobert, E. Danchin, A. Dhondt, and J. Nichols, eds. *Dispersal.* Oxford University Press, Oxford.

Holt, R. D. and J. H. Lawton. 1994. The ecological consequences of shared natural enemies. Annual Review of Ecology and Systematics 25:495-520.

Holyoak, M. 2000. Habitat subdivision causes changes in food web structure. Ecology Letters 3:509–515.

Hopey, M. E. and J. W. Petranka. 1994. Restriction of wood frogs to fish-free habitats: How important is adult choice? Copeia 1994:1023–1025.

Hutchinson, G. E. 1993. *A treatise on limnology.* Volume 4: Zoobenthos. Y. H. Edmondson, ed. Wiley, New York.

Jeffries, M. J. and J. H. Lawton. 1984. Enemy-free space and the structure of ecological communities. Biological Journal of the Linnean Society. 23:269–286.

Kats, L. B. and A. Sih. 1992. Oviposition site selection and avoidance of fish by streamside salamanders *Ambystoma barbouri.* Copeia 1992:468–473.

Kenk, R. 1949. The animal life of temporary and permanent ponds in southern Michigan. Miscellaneous Publications Zoology Museum University of Michigan 71:1–66.

Krivan, V. and E. Sirot. 2002. Habitat selection by two competing species in a two-habitat environment. American Naturalist 160:214–234.

Laurila, A. and T. Aho. 1997. Do female frogs choose their breeding habitat to avoid predation on tadpoles? Oikos 78:585–591.

Leibold, M. A. 1998. Similarity and local co-existence of species in regional biotas. Evolutionary Ecology 12:95–110.

Levin, S. A. 1992. The problem of pattern and scale in ecology. Ecology 73:1943–1967.

Marsh, D. M. and B. J. Borrell. 2001. Flexible oviposition strategies in tungara frogs and their implications for tadpole spatial distributions. Oikos 93:101–109

Maynard-Smith, J. 1972. *On evolution.* Edinburgh University Press, Edinburgh.

McPeek, M. A. 1989. Differential dispersal tendencies among *Enallagma* damselflies (Odonata) inhabiting different habitats. Oikos 56:187–195.

————. 1990. Determination of species composition in the *Enallagma* damselfly assemblages of permanent lakes. Ecology 71:83–89.

Merritt, R. W. and K. W. Cummins. 1984. An Introduction to the Aquatic Insects of North America. Kendall / Hunt, Dubuque, Iowa.

Moody, A. L., W. A. Thompson, B. De Bruijn, A. I. Houston, and J. D. Goss-Custard. 1997. The analysis of the spacing of animals, with an example based on oystercatchers curing the tidal cycle. Journal of Animal Ecology 66:615–628.

Morin, P. J. 1983. Predation, competition, and the structure of larval anuran guilds. Ecological Monographs 53:119–138.

————. 1984. Odonate guild composition: Experiments with colonization history and fish predation. Ecology 65:1866–1873.

Morris, D. W. 2003. Toward an ecological synthesis: A case for habitat selection. Oecologia 136:1–13.

Mouquet, N., and M. Loreau. 2002. Coexistence in metacommunities: The regional similarity hypothesis. American Naturalist 159:420–426.

Parker, G. A. 1984. Evolutionary stable strategies. Pages 30–61 in J. R. Krebs and N. B. Davies eds. *Behavioral ecology: An evolutionary approach.* 2nd edition. Blackwell, Oxford.

Petranka, J. W. and K. Fakhoury. 1991. Evidence for a chemically-mediated avoidance response of ovipositing insects to bluegills and green frog tadpoles. Copeia 1991:234–239

Pulliam, R. H. and Danielson, B. J. 1991. Sources, sinks, and habitat selection: a landscape perspective on population dynamics. American Naturalist 137:S51–S66.

Rausher, M. D. 1983. Ecology of host-selection behavior in phytophagous insects. Pages 223–257 in R. F. Denno and M. S. McClure eds. *Variable plants and herbivores in natural and managed systems.* Academic Press, New York.

————. 1993. The evolution of habitat preference: Avoidance and adaptation. Pages 259–283 in K. C. Kim and B. A. McPheron eds. *Evolution of insect pests.* Wiley, New York.

Remes, V. 2000. How can maladaptive habitat choice generate source-sink population dynamics? Oikos 91:579–582.

Renwick, J. A. A. and F. S. Chew. 1994. Oviposition behavior in lepidoptera. Annual Review of Entomology 39:377–400.

Resetarits, W. J., Jr. 1996. Oviposition site choice and life history evolution. American Zoologist 36:205–215.

————. 2001. Colonization under threat of predation: Non-lethal effects of fish on aquatic beetles, *Tropisternus lateralis* (Coleoptera: Hydrophilidae). Oecologia 129:155–160.

————. 2005. Habitat selection links local and regional scales in aquatic ecosystems. Ecology Letters 8, in press.

————. In review. Ecological character displacement and the nature of species interactions: Intraguild predators in multi-species ensembles.

Resetarits, W. J., Jr. and H. M. Wilbur. 1989. Choice of oviposition site in *Hyla chrysoscelis:* role of predators and competitors. Ecology 70:220–228.

————. 1991. Choice of calling site by *Hyla chrysoscelis:* Effect of predators, competitors, and oviposition site. Ecology 72:778–786.

Ricklefs, R. E. and D. Schluter. 1993. *Species diversity in ecological communities: Historical and geographical perspectives.* University of Chicago Press, Chicago.

Rieger, J. F., C. A. Binckley, and W. J. Resetarits, Jr. 2004. Larval performance and oviposition site preference along a predation gradient. Ecology 85:2094–2099.

Rosenzweig, M. L. 1985. Some theoretical aspects of habitat selection. Pages 517–540 in M. L. Cody, ed. *Habitat selection in birds.* Academic Press, New York.

————. 1991. Habitat selection and population interactions: The search for mechanism. American Naturalist 137:S5–S28.

Schmidt, K. A., J. M. Earnhardt, J. S. Brown and R. D. Holt. 2000. Habitat selection under temporal heterogeneity: Exorcising the ghost of competition past. Ecology 81:2622–2630.

Schneider, D. W. and T. M. Frost 1996. Habitat duration and community structure in temporary ponds. Journal of the North American Benthological Society 15:64–86.

Sherratt, T. N. and S. C. Church. 1994 Ovipositional preferences and larval cannibalism in the Neotropical mosquito *Trichoprosopon digitatum* (Diptera: Culicidae). Animal Behavior 48:645–652.

Shmida, A. and S. Ellner. 1984. Coexistence of plant species with similar niches. Vegetatio 58:29–55.

Singer, M. C. 1984. Butterfly-hostplant relationships: Host quality, adult choice, and larval success. Symposia of the Royal Entomological Society of London 11:81–88.

————. 1986. The definition and measurement of oviposition preference in plant-feeding insects. Pages 65–94 in J. R. Miller and T. A. Miller eds. *Insect-plant relationships*. Springer-Verlag, New York.

Skellam, J. G. 1951. Random dispersal in theoretical populations. Biometrika 38: 196–218.

Smith, C. J., Reynolds, J. D. and Sutherland, W. J. 2000. Population consequences of reproductive decisions. Proceedings of the Royal Society London, series B, 2000:1327–1335.

Spencer, M. L., Blaustein, L. and Cohen J. E. 2002. Oviposition habitat selection by mosquitoes *Culiseta longiareolata* and consequences for population size. Ecology 83: 669–679.

Spieler, M. and K. E. Linsenmair. 1997. Choice of optimal oviposition sites by *Hoplobatrachus occipitalis* (Anura: Ranidae) in an unpredictable and patchy environment. Oecologia 109:184–199.

Stanton, M. L. 2003. Interacting guilds: Moving beyond the pairwise perspective on mutualisms. American Naturalist 162:S10–S23.

Stav, G., L. Blaustein and J. Margalith. 1999. Experimental evidence for predation risk sensitive oviposition by a mosquito, *Culiseta longiareolata*. Ecological Entomology 24:202–207.

Stevens, R. D. and M. R. Willig 2000. Community structure, abundance and morphology. Oikos 88:48–56.

Stone, L., T. Dayan, and D. Simberloff. 2000. On desert rodents, favored states, and unresolved issues: Scaling up and down regional assemblages and local communities. American Naturalist 156:322–328.

Strong, Jr., D. R., D. Simberloff, L. G. Abele, and A. B. Thistle. 1984. *Ecological communities: Conceptual issues and the evidence*. Princeton University Press, Princeton, NJ.

Summers, K. 1999. The effects of cannibalism on Amazonian poison frog egg and tadpole deposition and survivorship in *Heliconia* axil pools. Oecologia 119:557–564.

Sutherland, W. J. 1996. *From individual behaviour to population ecology*. Oxford University Press, Oxford, UK.

Thompson, J. N. and O. Pellmyr. 1991. Evolution of oviposition behavior and host preference in lepidoptera. Annual Review of Entomology 36:65–89.

Tilman, D., R. M. May, C. L. Lehman, and M. A. Nowak. 1994. Habitat destruction and the extinction debt. Nature 371:65–66.

Walton, W. E., N. S. Tietze, and M. S. Mulla. 1990. Ecology of *Culex tarsalis* (Diptera: Culicidae): Factors influencing larval abundance in mesocosms in Southern California. Journal of Medical Entomology 27:57–67.

Watkinson, A. R. and W. J. Sutherland. 1995. Sources, sinks and pseudo-sinks. Journal of Animal Ecology 64:126–130.

Weiher, E. and P. Keddy. 1999. *Ecological assembly rules: Perspectives, advances, retreats*. Cambridge University Press, Cambridge, UK.

Weir, J. S. 1972. Diversity and abundance of aquatic insects reduced by introduction of the fish *Clarias gariepinus* to pools in central Africa. Biological Conservation 4:169–175.

Wellborn, G. A., E. E. Werner, and D. K. Skelly. 1996. Mechanisms creating community structure across a freshwater habitat gradient. Annual Review of Ecology and Systematics 27:337–363.

Werner, E. E. and J. F. Gilliam. 1984. The ontogenetic niche and species interactions in size-structured populations. Annual Review of Ecology and Systematics 15:393–425.

Wilbur, H. M. and Alford, R. A. 1985. Priority effects in experimental pond communities: Responses of *Hyla* to *Bufo* and *Rana*. Ecology:1106–1114.

Wilson, D. S. 1992. Complex interactions in metacommunities, with implications for biodiversity and higher levels of selection. Ecology 73:1984–2000.

CHAPTER 17

New Perspectives on Local and Regional Diversity
Beyond Saturation

Jonathan B. Shurin and Diane S. Srivastava

Introduction

A persistent question in ecology concerns the relative roles of macroscopic processes such as immigration or speciation versus small-scale biotic and abiotic factors as agents of community structure (Ricklefs and Schluter 1993). This issue lies at the heart of several major conceptual debates and many important applied problems. Small-scale and whole-system experimental studies have elucidated the importance of biotic interactions and abiotic constraints for governing the structure and function of ecological communities (Carpenter et al. 1995; Brown et al. 2001). However, broad-scale geographic comparisons have traditionally emphasized the importance of the geologic, migratory, and evolutionary history of biotas for patterns in diversity (Brown 1995; Maurer 1999). This contrast has led to the suggestion that while species interactions may be important within local communities, they offer little mechanistic insight into biogeographic patterns. Instead, larger scale regional processes are often considered dominant in generating differences among communities (Ricklefs 1987; Maurer 1999). The local and regional schools of thought diverge both in scale and the processes they consider most important for influencing community structure. Conceptual and empirical synthesis of the two approaches remains one of the most important challenges in modern ecology.

The study of patterns such as local and regional species richness has been central to testing metacommunity ideas, particularly those associated with neutral theory (e.g., see Chase et al., chapter 14). Here, we provide a detailed exploration of one of these patterns (the correlation between local and regional species richness) as it relates to two issues concerning metacommunities. First we illustrate the need to study both mechanisms and patterns in metacommunities. Second we show the critical importance of the choice of spatial scale for our perception of patterns and processes in metacommunities, and provide guidance on alternatives to choosing specific scales.

One popular approach to estimating the strengths of local and regional processes is to examine the shape of the relationship between local and regional species richness. The technique is based on the hypothesis that strong local interactions place limits on local diversity. In this case, richness within local communities approaches an upper asymptote with increasing regional richness. Saturated patterns are consistent with models where local interactions predominate over dispersal in limiting local diversity. Alternatively, if local richness is unconstrained by local interactions but instead depends on the supply of colonists from the region (the patch dynamics perspective; see Holyoak et al., chapter 1 and Chase et al., chapter 14), then local and regional richness should be positively correlated over their entire range. The approach was first introduced by Terborgh and Faaborg (1980) and has since been applied to data from a broad array of taxa and locations (see reviews in Cornell and Lawton 1992; Srivastava 1999; Hillebrand and Blenckner 2002). This method is attractive in that it offers a simple statistical test of a hypothesis (saturation versus linearity), and that it uses data that are widely available for many types of communities and organisms. However, a number of problems with inferring processes from patterns of local-regional diversity have been pointed out (Srivastava 1999; Fox et al. 2000; Loreau 2000; Shurin and Allen 2001; Hillebrand and Blenckner 2002). These issues generally relate either to the processes underlying patterns of local and regional diversity, or to the appropriate geographic definitions of local and regional scales. Thus, saturated or linear patterns of diversity may not inform us as to the relative importance of local and regional processes.

In this synthesis, we review theoretical foundations for understanding local and regional species diversity and the relationship between the two. We consider two broad categories of issues relating to the control of local and regional diversity: the different types of processes that might reasonably be classified as "local" or "regional," and the appropriate definitions of the two spatial scales. Different types of biotic and abiotic local interactions might have qualitatively different effects on diversity and species coexistence. Regional processes might include dispersal in a metapopulation or biogeographic sense, speciation, source-sink effects or broad-scale climatic or ecosystem processes. Here we review ecological theories that incorporate processes at local and regional scales and assess their predictions regarding the maintenance of local and regional diversity.

The second issue relates to the appropriate definitions of the local and regional spatial scales. A number of studies have pointed out that neither the local or the regional scale can be consistently defined by spatial boundaries that are directly tied to the underlying mechanisms (Westoby 1998; Srivastava 1999; Loreau 2000). We argue that studies should explicitly incorporate spatial scale as a continuous variable rather than drawing arbitrary distinctions (in chapter 9, Kolasa and Romanuk use hierarchy theory to arrive at a similar conclusion). Different definitions of local and regional scales can produce qualitatively different patterns of diversity. Saturation of diversity implies convergence of species-area relations from different regions at small (i.e., local) spatial scales (Westoby 1998). Thus, the slope of the local-regional relationship should increase with the ratio of the local to regional areas if communities are saturated, but not if they are unsaturated. We show using data from a survey of local-regional studies that the slope of the relationship between local and regional richness does indeed increase as the local area becomes larger relative to the regional scale. Thus, although most studies have found linear or unsaturated patterns, the overall pattern of local and regional richness across taxa is consistent with a saturated model of diversity. However, because a diversity of mechanisms can produce either saturated or unsaturated patterns, the distinction between the two is unhelpful for estimating the roles of the two types of processes. Finally, we discuss evidence for the importance of local and regional processes from experimental and observational studies in a number of natural systems. We identify several gaps in our understanding of diversity at multiple spatial scales and propose a number of questions that demand further attention.

Theory: A Continuum of Processes

Here we review several theories of local and regional diversity that have been developed to explicitly incorporate mechanistic local interactions and dispersal among patches within a region (summarized in table 17.1; Levins and Culver 1971; Hubbell 2001; Shurin and Allen 2001). Models based on species-area relations have been applied to questions of local and regional diversity (Westoby 1998; Srivastava 1999; Loreau 2000); however, these are entirely pattern based and do not make predictions regarding the roles of different processes; they will not be considered further here.

The simplest models of the factors that influence local and regional diversity come from island biogeography (MacArthur and Wilson 1967) and metapopulation theory (Levins 1969). If local habitats have no intrinsic limits to their diversity and interspecific interactions do not affect species' distributions, then local diversity should be a constant proportion of the regional species pool (i.e., local richness is a linear function of regional richness). If each species in the region exists in its own metapopulation unaffected by the presence of other organisms, then every species occupies a proportion of the habitat (P_i) equal to $1 - e_i/c_i$, where e_i and c_i are the species-specific extinction and colonization rates, respectively. In this case, local diversity is the sum of the proportional occupancies across all species in the regional pool. That is, mean local diversity for the region (D) is given by $\sum_k P_i$, where k is the number of species in the regional pool. Increasing the size of the regional pool (k) leads to linear increases in the mean number of species found within patches (D) because each species' occupancy is independent of all others. This prediction is the basis for the conclusion that linear, positive relationships between local and regional diversity are indicative of noninteractive local communities (Hugueny and Cornell 2000).

A counterpoint to island biogeography theory is illustrated by models of competition in metapopulations with hard upper limits to local diversity and global

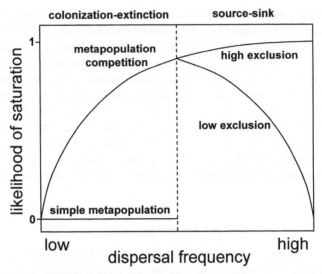

Figure 17.1 A conceptual model of the effects of dispersal rates on patterns of local and regional diversity. The left side of the graph represents patch dynamic models where local interactions occur much faster than dispersal. When local sites have no limits on potential diversity (as in a simple metapopulation), we expect positive relationships between local and regional diversity for all levels of dispersal, although the slope of the function will vary with connectivity. In models with local competitive exclusion, increasing dispersal frequency leads to saturated local-regional functions as local interactions become more intense when more species are supplied from the region. On the right side, dispersal occurs on a similar time scale to demographic rates such that immigration plays an important role in local population dynamics. In this case, low dispersal rates lead to relatively few sink populations and communities are saturated in the patch dynamic sense. Increasing dispersal creates more sink populations and local communities may become regionally enriched, leading to unsaturated patterns of diversity (the low exclusion curve). However, if sink populations exert large competitive influence on nonsink populations, then enhanced dispersal may lead to reduced local diversity due to exclusion. In the case of high exclusion, high dispersal rates may lead to saturation as species in sinks reduce local diversity.

dispersal (Levins and Culver 1971; Hastings 1980; Tilman 1994; Yu and Wilson 2001). Such models are most often motivated by communities of sessile species competing for space where each patch can be occupied by only a single individual. Competition in such a system can be preemptive (i.e., by priority rules) or by dominance (i.e., competitive hierarchies). In this case, species interact by excluding colonists from local sites, not by affecting neighboring individuals, and local diversity within sites has a maximum of one. As outlined in chapter 10 by Mouquet et al., the requirement for regional coexistence of multiple species with dominance competition is a trade-off between dispersal ability and competitive rank (Levins and Culver 1971), while regional coexistence is impossible with preemptive competition (Yu and Wilson 2001). Mean local diversity in the region is

Table 17.1 A summary of theories and their predictions regarding local and regional diversity

Theory	Predictions for local versus regional diversity
Island biogeography—simple metapopulation	Unsaturated at all levels of dispersal (noninteractive).
Metapopulation competition, global dispersal	Unsaturated at low dispersal, saturated at high dispersal if patch is considered local.
Metapopulation competition, local dispersal	Saturated if patches are considered local, possibly unsaturated if the neighborhood is the local scale.
Source-sink (super-saturation)	Unsaturated at low and high dispersal rates, saturated at intermediate levels (a hump-shaped relationship).
Food web, keystone predation	Depends on predator and prey dispersal, predator effects on prey extinction and coexistence.

again the sum of the proportional occupancies among members of the regional pool. However, since we assume absolute local competitive exclusion, the models predict saturated or curvilinear relationships between local and regional diversity at sufficiently high levels of regional richness or dispersal rates (figure 17.1). Linear relationships may occur at low levels of regional richness and slow dispersal, provided there is a very strong trade-off between colonization rate and competitive ability (Shurin and Allen 2001). However, the conditions for proportional sampling of the regional pool become increasingly stringent as more species are added to the region. The contrast between metapopulation models with and without local competition illustrates the two hypothesized relationships between local and regional richness, with strong local exclusion leading to saturating patterns and noninteractive communities giving rise to linear functions. In metapopulation models of competition, saturation occurs at high rates of dispersal, while proportional sampling may be observed when dominant competitors are poor colonizers (figure 17.1, table 17.1).

A third possibility is that species compete locally within patches and dispersal is constrained to occur within a neighborhood of patches as in neutral biogeographic models (Hubbell 2001). Such a system allows for the possibility of nonequilibrial regional coexistence in the case of preemptive competition (Hubbell 2001; Yu and Wilson 2001). Neutral theories also predict saturation of local diversity if a single-occupancy site or patch is considered to be the local scale. However, because of localized dispersal, species are more likely to interact via propagules with nearby species than with those farther away. Thus, the local scale may instead be defined as the neighborhood of patches in close enough proximity to actually exchange propagules due to localized dispersal (Mouquet et al., chapter 10). If we expand our definition of the local scale to the neighborhood of patches, then we may expect linear relationships between local and regional diversity. This occurs because in richer regions, there are more species available to colonize local patches. Thus, whether we expect linear or saturated local-regional patterns in neutral biogeographic models depends on the spatial extent of the local and regional scales (see further discussion of scale issues below).

None of the above models explicitly consider local population dynamics but instead treat patches simply as occupied or unoccupied. This is equivalent to assuming that migrants found in local populations make no significant contribution the size of extant populations, and that local interactions take place on a much faster timescale than dispersal. Another potential role of dispersal in communities is that immigrants may have significant impacts on local population dynamics (Pulliam 1988; Loreau and Mouquet 1999; Amarasekare and Nisbet 2001). If the number of propagules is large relative to local population sizes, then dispersal can sustain populations that would otherwise undergo extinction based on their local birth and death rates (a source-sink system). In such a community, the relationship between local and regional richness again depends on the rates of dispersal of members of the regional pool. If colonization rates are much greater than local extinction, but low enough that sink populations are rarely maintained by immigration, then local exclusion should lead to saturation of local diversity. In this case, dispersal rates are high on the continuum considered by patch dynamic models, but still low relative to local demographic rates. As dispersal rates increase (as in the right half of figure 17.1), local populations become increasingly influenced by immigration and more species are maintained locally. When many species exist as sink populations in some parts of their range, increasing regional diversity may lead to greater numbers of local populations and higher local diversity (figure 17.1, low exclusion scenario). In this case, local diversity is supersaturated, or greater than local conditions would allow in the absence of immigration. Alternatively, if sink populations exert large influence on other species in local communities, then increasing habitat connectivity may lead to exclusion of populations that would otherwise persist (figure 17.1, high exclusion scenario; Amarasekare and Nisbet 2001). Thus, if source-sink dynamics play a significant role in the determining the number of species maintained in a community, then we may expect to see saturation of local diversity at *low* dispersal rates, and either linear or saturated local-regional patterns with *high* dispersal. This prediction contrasts with that of patch-dynamic models where a colonization-extinction balance is considered most important.

All of the above models assume that local interactions serve primarily to place limits on local diversity. However, a wide variety of species interactions can give rise to facilitation and promote colonization by species that would otherwise be excluded. For instance, habitat modification may mitigate physical stress for plants and marine invertebrates (Bertness et al. 1999). Processing chains can also increase resource availability and quality for detritivores (Heard 1994). Indirect interactions mediated via shared competitors or predators can also give rise to facilitation. Shurin and Allen (2001) showed in a metacommunity model that keystone predators that facilitate local coexistence among competitors can promote both local and regional diversity in many situations. The model modifies metapopulation models of competition by incorporating a predator that allows for local coexistence of a dominant and subordinate competitor where the dominant species would otherwise exclude the weaker. If predators facilitate as many or more species than they exclude, then the conditions for linear local-regional richness patterns become much less restrictive than in metapopulation models of competition alone. If keystone predation or other forms of interspecific facilitation play important roles in shaping community structure, then we may expect to find positive relationships between local and regional diversity in the presence of strong local interactions. Thus, mechanisms of local interactions have the potential to affect both local and regional species richness and the shape of the relationship between the two.

Finally, current theoretical foundations for understanding local and regional diversity do not include variation in the local environment among patches. However, empirical evidence and theoretical results indicate that spatial heterogeneity may play an important role in generating both local and regional diversity. For instance, Levine and Rees (2002) found that a colonization-competition trade-off was inadequate to explain relative abundance patterns with respect to seed size and competitive abilities in annual plant communities. Instead, a model that incorporated environmental heterogeneity among patches matched observed patterns where small-seeded species that are good dispersers but poor competitors are most abundant. The model invoked spatial heterogeneity that generated differences in species' competitive abilities among patch types. Second, Shurin et al. (2004) showed that environmental heterogeneity in the supply of two potentially limiting resources can lead to the coexistence of species engaged in alternative stable states. That is, preemptive competition occurs at intermediate supply ratios, while dominance by one species is the rule at high or low ratios. Thus, abiotic variability can lead to the regional persistence of priority effects that would otherwise be lost through extinction. Finally, Chase and Leibold (2002) found that diversity showed hump-shaped relationships with productivity at the local scale and monotonic patterns at the regional scale (see also Loreau et al., chapter 18). This pattern suggests that increasing productivity promotes regional diversity and increases local diversity at low levels while having the opposite effects in very eutrophic systems. Theories to explain the effects of productivity on diversity have focused largely on the local scale (reviewed in Srivastava and Lawton 1998). The reasons for enhanced diversity in highly productive regions are unknown. However, these examples suggest that spatial heterogeneity in local conditions should be considered when attempting to explain local and regional diversity.

Why We Should Consider Continuous Space

The concept of species saturation is typically formulated in terms of local and regional scales. Species interactions are assumed to take place at the local scale and in ecological time. Regional scales contain the species pool from which local sites draw potential colonists. Operationally defining either of these scales involves substantial subjectivity, and neither scale can be easily tied to underlying mechanisms in any particular community. Below we present four arguments for treating space as a continuous variable to avoid the problems of arbitrarily assigning discrete definitions of what is considered local or regional.

First, the shapes of patterns of local and regional richness are very sensitive to the definition of local scale. In simulations of community assembly in a multiple-patch (metacommunity) landscape, the relationship between local and regional richness depends on whether the patch or a neighborhood of patches is considered "local" (Fukami 2004). If the size of the local scale exceeds the scale of local interactions, we expect "local" richness estimates to be increasingly influenced by beta diversity, the turnover in species between local patches, and thus for local-regional richness plots to appear more linear (Huston 1999; Loreau 2000). Empirical results support this supposition. In a meta-analysis of sixty-three local-regional richness plots, Hillebrand and Blenckner (2002) found that studies were more likely to find linear patterns when the local area occupied a greater proportion of the region. Although similar analysis of a single large dataset showed no effect of local scale on the curvilinearity of local-regional richness plots (Caley and Schluter 1997), the smallest local scale (2500 km^2) of this study was probably still too large to reflect the scale of interspecific interactions (Westoby 1998). Because different species likely experience local interactions at distinct spatial scales, defining local scales will always be difficult.

Second, the relationship between local and regional richness is also sensitive to the scale at which regions are defined. This can be easily demonstrated by sampling species-area curves for multiple region sizes. Unsaturated communities will appear to have curvilinear local-regional richness plots when the largest regions have higher regional richness, and when the between-region species-area curve has a higher exponent than the within-region species-area curves (Srivastava 1999). Both of these conditions generally hold in ecological systems (Rosenzweig 1995) so such pseudosaturation is to be expected. For example, in a global comparison of lake zooplankton richness (Shurin et al. 2000), local richness initially appeared to saturate with regional richness. However, after standardizing the size of regions by using the residuals of the species-area relations, local richness was clearly proportional to regional richness. Thus, different definitions of the regional scale can produce qualitatively different patterns of local and regional diversity.

The third reason to treat scale explicitly is to incorporate processes operating at scales other than local and regional. At least two other scales may be important in determining diversity. For many organisms, especially sessile species, individuals only interact with other organisms over very small spatial scales. However, if an individual is considered as the local community (as in many patch dynamic models), then local diversity cannot be greater than one. In this case, localities or clusters of patches that exchange propagules may better represent the local scale, while groups of localities separated over larger scales correspond to metacommunities (Mouquet et al., chapter 10). Groups of metacommunities with distinct geologic or evolutionary histories represent different regions. Thus, there may be two additional scales at which important processes take place, one smaller (the individual or patch) and one larger (the metacommunity) than the local scale. Because patches and metacommunities present all the same problems of definition as local and regional, a continuous treatment of scale may be preferable to a discrete one.

Fourth, processes leading to species saturation are also likely to be scale-dependent. For instance, invasion resistance, which is closely linked to saturation of local communities, can depend on spatial scale. Invasion success depends not only on characteristics of the local community, but also on traits of invading species (Kolar and Lodge 2002). Since these traits are phylogenetically constrained, we expect species from different regions to exhibit different suites of traits. We illustrate this point with two scenarios, one with a potential invader originating within the same biogeographic region as the target community, and one with an invading species originating outside the region. The potential invader originating within a region is more likely to share a common phylogeny with species native to the target community, and thus have traits similar to those already found within the community. If we assume that trait similarity leads to competitive exclusion (the competitive exclusion principle; MacArthur 1972), the potential invader from within the region will have lower invasion success than the species originating outside the region. This leads to the expectation that invasion success will be inversely related to the distance between the target community and the invader's point of origin. There is remarkably little empirical data that can be used to assess this scale-dependence of invasion success. Experiments with pond zooplankton show that local communities resist invasion by most species found within 10 km (Shurin 2000), suggesting saturation at landscape scales. However, zooplankton communities are frequently invaded by exotic species, indicating that they are unsaturated on a global scale (Ricciardi and MacIssac 2000). Many of the most dramatic invasions with the greatest impacts on local communities involve species belonging to novel functional groups from the perspective of the invaded site. Snakes in Guam, zebra mussels in North America, and annual grasses in California may all be examples. These cases illustrate that the apparent saturation of a community depends greatly on the size of the area over which potential invaders are drawn. Many communities are unsaturated or invasible with respect to a global species pool.

Comparative and experimental studies of the relationship between diversity and invasibility also suggest that saturation of diversity is scale dependent. Many small-scale experiments have found a negative relationship between local native diversity and invasion success (Tilman 1997; Stachowicz et al. 1999; Levine 2000; Shurin 2000; Kennedy et al. 2002). By contrast, observational surveys over broad spatial scales often find positive correlations between the numbers of native and introduced species (Stohlgren et al. 1999; Levine 2000; Sax 2002; Shea and Chesson 2002). Sax et al. (2002) found that island chains have experienced many more invasions than extinctions, indicating that diversity at the scale of archipelagos is increasing and apparently not saturated. One explanation for this apparent contradiction may be that experimental studies manipulate local diversity (at scales of less than 1 to a few meters squared) whereas observational studies deal with diversity at broader regional scales (10 m^2 to 100s of km^2). The prediction that diversity confers invasion resistance (Elton 1958) is based on the idea of greater or more complete resource use in more diverse communities, clearly a local-scale phenomenon. However, it is unclear how regional diversity should affect invasibility. If high regional diversity is driven by greater environmental heterogeneity, then more diverse regions may have more available niche space and therefore be more invasible. Alternatively, if more diverse regions have longer evolutionary histories but similar ecological capacities to support species (Stephens and Weins 2003), then more diverse regions may be closer to saturation and less invasible. The relationship between invasibility and regional diversity remains an open question for empirical and theoretical study. However, the contrast between observational and experimental studies of invasions suggests that species saturation via ecological constraints is only expected to occur at relatively small local scales.

How Space Could Be Treated Continuously

Because of the problems of drawing boundaries for discrete scales such as local and regional, introducing space as a continuous variable into models and empirical studies is an important next step. We now describe several ways in which scale can be incorporated into analysis of patterns as a continuous variable.

Earlier, we summarized evidence that the relationship between local and regional diversity changes as local richness is defined for increasingly large scales. One of the simplest ways to predict this change is to resample species-area curves corresponding to saturated and unsaturated communities at increasing "local" scales. The response variable in this analysis is the degree of linearity between local and regional richness, which can be assessed by the exponent of the power function (b) between the two variables ($b = 1$ indicates linearity, $b < 1$ indicates various degrees of downward curvilinearity; see Hillebrand and Blenckner (2002) for justification). The steps in constructing the model are illustrated in figure 17.2,

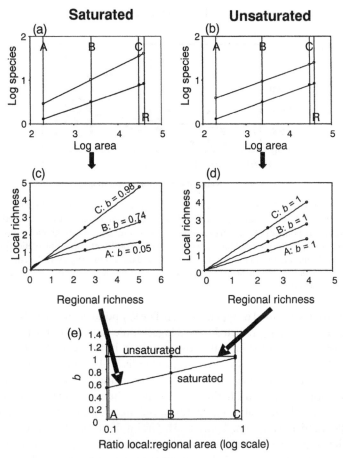

Figure 17.2 Saturated and unsaturated communities can be represented by two different types of log—log species-area plots (a) and (b). In both cases, regions are shown with more (open circles) and less species (closed circles) at every spatial scale. In the saturated community, local richness should be similar in both regions, whereas regional richness should be divergent, yielding nonparallel species-area curves (a). In the unsaturated community, an identical ratio of local to regional richness creates parallel species-area curves in log—log space (b) (Srivastava 1999). Now consider sampling the species-area curves at three different sizes of a local area, denoted A, B and C, but keeping region area (R) fixed. The local—regional plots are predicted to vary between these various local areas, as shown in (c) and (d). In the saturated community (c), local richness will be a power function of regional richness with the exponent b less than 1. However, as the local area approaches the regional area, local and regional richness will become highly correlated, and b will approach one (linearity). In fact, as local area is increased, b will increase as a logarithmic function of the proportion of local area to regional area (e) (appendix 17.1). By contrast, in the unsaturated community, all local-regional plots will remain linear ($b = 1$) even though the slope of this line will depend on local area (d), because the difference between log local and regional richness will be identical for both regions as seen in graph (b). The numerical examples illustrated use the following values: A = 10, B = 30, C = 90, R = 100 area units. The species-area curves for the saturated community are $S = 0.5A^{0.35}$ and $S = 0.5A^{0.5}$. The species-area curves for the unsaturated community are $S = 0.5A^{0.35}$ and $S = 0.8A^{0.35}$. The results can be generalized to multiple regions and any combination of local and regional areas (appendix 17.1).

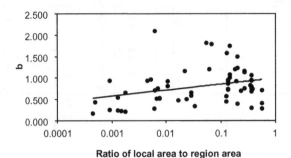

Figure 17.3 Estimates of b, the exponent of a power function between local and regional richness, from sixty-three published studies analyzed by Hillebrand and Blenckner (2002), as a function of the ratio of local to region area (A_L/A_R) in each study. There is a significant logarithmic relationship between b and A_L/A_R (Hillebrand and Blenckner 2002). This is consistent with the theoretical prediction for a saturated community, but not that for an unsaturated community (figure 17.2). The y-intercept of the fitted function is 1, identical to the theoretical prediction (see appendix 17.1). Data were provided by H. Hillebrand.

and the proof is outlined in appendix 17.1. The key point is that, based on species-area curves alone, b is predicted to be a logarithmic function of the ratio of local area to regional area for saturated communities, but unrelated to this ratio in unsaturated communities (figure 17.2e; appendix 17.1). One can now ask which pattern communities follow. Hillebrand and Blenckner's (2002) meta-analysis allows us to address this issue. In this meta-analysis, the exponent (b) is significantly related to the ratio of local to regional area, and follows a logarithmic relationship (H. Hillebrand, pers. comm.; figure 17.3). The implication is that many of the observed linear patterns between local and regional richness may actually represent saturated communities sampled at too large a local area. This conclusion is contrary to previous reviews that, without taking scale into account, showed the predominance of unsaturated patterns (Cornell and Lawton 1992; Srivastava 1999). Thus, although many studies concluded that communities show unsaturated patterns, the literature on local and regional richness collectively shows scale-dependent patterns that are more consistent with saturated, curvilinear patterns at some scales. Although overall the relationship between b and the ratio of local to regional scale is consistent with saturated *patterns,* the relative roles of local and regional *processes* cannot be deduced from this result for the reasons discussed above.

Since the relationship between local and regional diversity is influenced by scale, it is evident that beta diversity (the slope of the local-regional curve) must also be related to scale. There has been some use of hierarchical statistics to quantify how beta diversity changes across scales (e.g., Gering and Crist 2002). Such an approach may be useful in cases where there are strong predictions about the importance of different scales in the hierarchy. However, in the absence of such mechanistic models, results cannot be generalized beyond the particular discrete scales that the researcher chooses to examine. Scale can be incorporated as a continuous variable by considering beta diversity in nested quadrats (Kunin 1998; Harte et al. 1999; Arita and Rodriguez 2002) or between point counts (Condit et al. 2002).

In summary, we argue for a continuous treatment of space in examining patterns of diversity over approaches that depend on untestable definitions of discrete categories such as local and regional. Such patterns can provide critical tests of the importance of different mechanisms provided they are explicitly tied to models of underlying processes (e.g., Condit et al. 2002). However, the distinction between saturated and unsaturated patterns will always provide at best weak tests of the importance of local and regional processes in structuring communities, because many alternate hypotheses can often generate the same patterns (Shurin and Allen 2001). Instead, discontinuities or constancy in patterns of diversity across scales can generate hypotheses that are then subjected to more mechanistic studies (Kunin 1998; Harte et al. 1999; Condit et al. 2002).

Empirical Evidence: Interactions between Local and Regional Processes

Patterns of local and regional diversity have been investigated in a wide range of systems (reviewed in Srivastava 1999; Hillebrand and Blenckner 2002). More mechanistic studies have been conducted in several systems and reveal interesting consequences arising from the joint effects of local and regional processes. Many of these studies indicate that dispersal and local processes are not mutually exclusive, but rather that the two can interact in interesting ways to generate community patterns.

Much of the most compelling evidence for the importance of regional processes comes from studies of sessile species such as terrestrial plants and intertidal invertebrates. A sessile lifestyle poses the problem of dispersal as each individual requires a propagule that colonizes a suitable microsite. However, because of the potential for space and resource competition, biotic interactions can also be of major importance in communities of sessile organisms. Experimental studies indicate that seed limitation in terrestrial plants is a major constraint on species distributions and local diversity (reviewed in Turnbull et al. 2000). Adding seeds of many species, even ones that are present within relatively small regional scales (e.g., 1 km), often results in dramatic increases in diversity and biomass production at small local scales (e.g., within 1 m^2 plots; Tilman 1997; Turnbull et al. 2000). These studies show strong dispersal limitation of local diversity even though more species often fail than succeed in colonizing when introduced as seed (Tilman 1997). Disturbing the local community also results in greatly enhanced invasion (Burke and Grime 1996), indicating that interactions with local species repels many potential invaders. Further evidence for interesting interactions between local competition, dispersal, and environmental heterogeneity comes from metapopulation / metacommunity models discussed in the Theory section above.

The predictions of neutral theories with localized dispersal and preemptive competition have been most extensively tested with tropical forest trees. Although empirical patterns of relative abundance are well described by neutral theories (Hubbell 2001), whether data indicate that the mechanisms envisioned by the model describe the natural processes remains unknown (see Chase et al., chapter 14). Patterns of beta diversity are often inconsistent with neutral models, and instead, similarity among sites appears to correspond to variation in the physical environment more than to the scale of local dispersal. A further complication is that theoretical studies have also shown that neutral and nonneutral models can produce virtually indistinguishable patterns of relative abundance (Chase et al., chapter 14).

Investigations in intertidal invertebrates suggest a shift from strong local interactions to recruitment limitation along an oceanographic gradient of decreasing larval supply along the Pacific coast of North America. A gradient in upwelling and onshore transport results in an increase in recruitment by larval barnacles from California to the Pacific Northwest (Connolly et al. 2001), an increase in cover of adults, and perhaps greater intensity of local competitive and predatory interactions as well (Connolly and Roughgarden 1998). Clear evidence has been shown for both strong local interspecific interactions (e.g., Wootton 1997) and larval supply (Roughgarden et al. 1988) in Pacific intertidal invertebrates. Connolly and Roughgarden (1999) showed in theory that the strength of species interactions increases with greater levels of larval dispersal. They propose that competition and predation are more important at higher latitudes where larval supply is greatest due to onshore advection from oceanic currents. Comparative field experiments to estimate the intensity of species interactions along gradients of larval supply would be extremely valuable in showing such an interaction between processes at local and regional scales.

Fewer examples of mechanistic studies of the roles of local and regional processes are available in communities of motile organisms. Several studies of freshwater microorganisms suggest that the outcome of local interactions depends on the supply of colonists from the surrounding region. Shurin (2001) found that effects of predators on zooplankton diversity and composition varied with the degree of habitat connectivity to the regional pool. In isolated ponds, predator invasion reduced local diversity as a number of large-bodied resident species were unable to coexist with zooplanktivorous fishes and insects. When a diverse group of zooplankton from the region was introduced as rare invaders, a number of small-bodied species invaded in the presence of predators, but not in their absence. This indicates that the occurrence of some species was indirectly facilitated by predators. Fish enhanced local diversity in connected communities, indicating that more species were facilitated than excluded. The results suggest that spatial heterogeneity in predator densities is an important factor promoting regional co-existence in zooplankton, and that effects of predators on local communities are contingent upon habitat connectivity and regional context (see also Cottenie and De Meester, chapter 8). By contrast, in inquiline communities of pitcher plants, invasion by protozoan species is either unaffected or reduced by mosquito predators (Miller et al. 2002). There is evidence that the occurrence of protozoan species in pitcher plants is limited by both dispersal and local interactions (resources and predation; Miller and Kneitel, chapter 5). The persistence times of protozoan and rotifer species in pitcher plants are affected by the interaction of detrital resources (a local process) and mosquito colonization (a regional process; M. K. Trzcinski, S. J. Walde, and P. D. Taylor unpublished results). Results from freshwater communities thus suggest that feedbacks between local and regional processes play important roles in regulating species diversity and composition.

Conclusions

Recent conceptual and empirical advances in understanding local and regional diversity indicate that the distinction between saturated and unsaturated patterns of richness is not a reliable means for estimating the strength of local and regional processes. Current theory suggests that local and regional processes are not mutually exclusive, but may interact in interesting ways to generate patterns in communities (Ricklefs 2004). Empirical examples show that communities that show linear local-regional patterns can be nearly closed to invasion (Shurin 2000; Shurin et al. 2000). A number of reviews have concluded that most communities show linear relationships between local and regional richness; however, these patterns are highly scale dependent. If linear patterns dominate, then the slope of the local-regional curve should be independent of the relative sizes of the local and regional scales. Yet data from the literature survey of Hillebrand and Blenckner (2002) show that the slope increases with the ratio of local to regional area (figure 17.3). Thus, the overall picture from the literature on local and regional diversity is more consistent with saturation, or divergence in species—area relations among areas of different regional richness (figure 17.2). Whether ecological communities are saturated with species or open to invasion from the regional pool remains an important unanswered question. We argue that space should be treated as a continuous variable rather than assigned to discrete categories such as local and regional, as illustrated in figure 17.3. Linking continuous space with models of specific mechanisms of species interactions and dispersal offers much stronger inference than an approach that depends on untestable assumptions about spatial scale.

Acknowledgments

JS was supported by a postdoctoral fellowship from the National Center for Ecological Analysis and Synthesis. DS was supported by a research grant from the Natural Sciences and Engineering Research Council (Canada). We thank Helmut Hillebrand and Thorsten Blenckner for providing raw data from their meta-analysis. Dov Sax, Kurtis Trzcinski, and the reviewers provided valuable comments on the manuscript.

Appendix 17.1

Consider two regions, each with a characteristic species-area relationship. In region 1, arbitrarily the richer of the two regions, the local richness (S_L) and regional richness (S_R) can be predicted from knowledge of the local area (A_L) and region area (A_R) as follows:

$$S_{L_1} = c_1 A_L^{Z_1}$$
$$S_{R_1} = c_1 A_R^{Z_1}$$

where c_1 and Z_1 are constants specific to region 1. In region 2, the appropriate equations are

$$S_{L_2} = c_2 A_L^{Z_2}$$
$$S_{R_2} = c_2 A_R^{Z_2}$$

with c_2 and z_2 being constants specific to region 2.

The relationship between local and regional richness can be expressed as a power function with constants b and k:

$$S_L = k S_R^b$$

Substitution and rearrangement between the above five equations yields the following relationship:

$$b = \frac{(Z_1 - Z_2)\log A_L + \log c_1 - \log c_2}{(Z_1 - Z_2)\log A_R + \log c_1 - \log c_2}.$$

We first examine the form of this relationship for saturated and unsaturated communities. If the community is unsaturated, we expect the species-area curves to be parallel in log–log space, that is, $Z_1 = Z_2$ (figure 2b; Srivastava 1999). In this case $b = 1$, indicating a linear relationship between local and regional richness irrespective of local or region area (figure 17.2d,e). If the community is saturated, then $Z_1 > Z_2$ and $\log c_1 \geq \log c_2$ (figure 17.2a; Srivastava 1999). In this case, $0 \geq b > 1$, indicating a curvilinear relationship between local and regional richness (figure 17.2c).

The relationship can be simplified by considering the two ways in which the ratio of local to region area can be changed. If we change local area but keep region area fixed, and make the further simplifying assumption that a saturated community will have $c_1 \approx c_2$, then the relationship can be written as:

$$b = m \log(A_L/A_R) + 1$$

where m is a constant equivalent to $\log (A_R)^{-1}$. This simple logarithmic relationship is depicted in figure 17.2e. Alternatively, if we change region area but keep local area fixed, then the relationship becomes:

$$b = n \log(A_R)^{-1},$$

where n is a constant equivalent to $\log A_L$. In practice, this inverse logarithmic relationship, when plotted on a log (A_L/A_R) axis, is very similar in shape to the variable local area equation.

Literature Cited

Amarasekare, P., and R. M. Nisbet. 2001. Spatial heterogeneity, source-sink dynamics and the local coexistence of competing species. American Naturalist 158:572–584.

Arita, H. T., and P. Rodriguez. 2002. Geographic range, turnover rate and the scaling of species diversity. Ecography 25:541–550.

Bertness, M. D., G. H. Leonard, J. M. Levine, P. R. Schmidt, and A. O. Ingraham. 1999. Testing the relative contribution of positive and negative interactions in rocky intertidal communities. Ecology 80:2711–2726.

Brown, J. H. 1995. *Macroecology*. University of Chicago Press, Chicago, IL.

Brown, J. H., T. G. Whitham, S. K. M. Ernest, and C. A. Gehring. 2001. Complex species interactions and the dynamics of ecological systems: Long-term experiments. Science 293:643–650.

Burke, M. J. W., and J. P. Grime. 1996. An experimental study of plant community invasibility. Ecology 77:776–790.

Caley, M., and D. Schluter. 1997. The relationship between local and regional diversity. Ecology 78:70–80.

Carpenter, S. R., S. W. Chisholm, C. J. Krebs, D. W. Schindler, and R. F. Wright. 1995. Ecosystem experiments. Science 269:324–327.

Chase, J. M., and M. A. Leibold. 2002. Spatial scale dictates the productivity-diversity relationship. Nature 416:427–430.

Chave, J., H. C. Muller-Landau, and S. A. Levin. 2002. Comparing classical community models: Theoretical consequences for patterns of diversity. American Naturalist 159:1–23.

Condit, R., N. Pitman, E. G. Leigh, J. Chave, J. Terborgh, R. B. Foster, P. Nunez et al. 2002. Beta-diversity in tropical forest trees. Science 295:666–669.

Connolly, S. R., B. A. Menge, and J. Roughgarden. 2001. A latitudinal gradient in recruitment of intertidal invertebrates in the northeast Pacific Ocean. Ecology 82:1799–1813.

Connolly, S. R., and J. Roughgarden. 1998. A latitudinal gradient in northeast Pacific intertidal community structure: Evidence for an oceanographically based synthesis of marine community theory. American Naturalist 151:311–326.

———. 1999. Theory of marine communities: Competition, predation, and recruitment-dependent interaction strength. Ecological Monographs 69:277–296.

Cornell, H. V., and J. H. Lawton. 1992. Species interactions, local and regional processes, and limits to the richness of ecological communities: A theoretical perspective. Journal of Animal Ecology 61: 1–12.

Elton, C. S. 1958. *The ecology of invasions by animals and plants.* Methuen, London.

Fox, J. W., J. McGrady-Steed, and O. L. Petchey. 2000. Testing for local species saturation with non-independent regional species pools. Ecology Letters 3:198–206.

Fukami, T. 2004. Community assembly along a species pool gradient: Implications for multiple-scale patterns of species diversity. Population Ecology 46, in press.

Gering, J. C., and T. O. Crist. 2002. The alpha-beta-regional relationship: Providing new insights into local-regional patterns of species richness and scale dependence of diversity components. Ecology Letters 5:433–444.

Harte, J., A. Kinzig, and J. Green. 1999. Self-similarity in the distribution and abundance of species. Science 284:334–336.

Hastings, A. 1980. Disturbance, coexistence, history, and competition for space. Theoretical Population Biology 18:363–373.

Heard, S. B. 1994. Pitcher-plant midges and mosquitoes—a processing chain commensalism. Ecology 75:1647–1660.

Hillebrand, H., and T. Blenckner. 2002. Regional and local impact on species diversity—from pattern to processes. Oecologia 132:479–491.

Hubbell, S. P. 2001. *The unified neutral theory of biodiversity and biogeography.* Princeton University Press, Princeton, NJ.

Hugueny, B., and H. V. Cornell. 2000. Predicting the relationship between local and regional species richness from a patch occupancy dynamics model. Journal of Animal Ecology 69:194–200.

Huston, M. A. 1999. Local processes and regional patterns: Appropriate scales for understanding variation in the diversity of plants and animals. Oikos 86:393–401.

Kennedy, T. A., S. Naeem, K. M. Howe, J. M. H. Knops, D. Tilman, and P. Reich. 2002. Biodiversity as a barrier to ecological invasion. Nature 417:636–638.

Kolar, C. S., and D. M. Lodge. 2002. Ecological predictions and risk assessment for alien fishes in North America. Science 298:1233–1236.

Kunin, W. E. 1998. Extrapolating species abundance across spatial scales. Science 281:1513–1515.

Levine, J. M. 2000. Species diversity and biological invasions: Relating local process to community pattern. Science 288:852–854.

Levine, J. M., and M. Rees. 2002. Coexistence and relative abundance in annual plant assemblages: The roles of competition and colonization. American Naturalist 160:452–467.

Levins, R. 1969. Some demographic and genetic consequences of environmental heterogeneity for biological control. Bulletin of the Entomological Society of America 15:237–240.

Levins, R., and D. Culver. 1971. Regional coexistence of species and competition between rare species. Proceedings of the National Academy of Sciences 68:1246–1248.

Loreau, M. 2000. Are communities saturated? On the relationship between alpha, beta, and gamma diversity. Ecology Letters 3:73–76.

Loreau, M., and N. Mouquet. 1999. Immigration and the maintenance of local species diversity. American Naturalist 154:427–440.

MacArthur, R. H. 1972. *Geographical ecology: Patterns in the distribution of species.* Princeton University Press, Princeton, NJ.

MacArthur, R. H., and E. O. Wilson. 1967. *The theory of island biogeography.* Princeton University Press, Princeton, NJ.

Maurer, B. A. 1999. *Untangling ecological complexity: The macroscopic perspective.* University of Chicago Press, Chicago, IL.

Miller, T. E., J. M. Kneitel, and J. H. Burns. 2002. Effect of community structure on invasion success and rate. Ecology 83:898–905.

Pulliam, H. R. 1988. Sources, sinks, and population regulation. American Naturalist 132:652–661.

Ricciardi, A., and H. J. MacIssac. 2000. Recent mass invasion of the North American Great Lakes by Ponto-Caspian species. Trends in Ecology and Evolution 15:62–65.

Ricklefs, R. E. 1987. Community diversity: Relative roles of local and regional processes. Science 235: 167–171.

———. 2004. A comprehensive framework for global patterns in biodiversity. *Ecology Letters* 7:1–15.

Ricklefs, R. E., and D. Schluter. 1993. *Species diversity in ecological communities: Historical and geographical perspectives.* University of Chicago Press, Chicago, IL.

Rosenzweig, M. L. 1995. *Species diversity in space and time.* Cambridge University Press, Cambridge, UK.

Roughgarden, J., S. Gaines, and H. Possingham. 1988. Recruitment dynamics in complex life cycles. Science 241:1460–1466.

Sax, D. F. 2002. Native and naturalized plant diversity are positively correlated in scrub communities of California and Chile. Diversity and Distributions 8:193–210.

Sax, D. F., S. D. Gaines, and J. H. Brown. 2002. Species invasions exceed extinctions on islands worldwide: A comparative study of plants and birds. American Naturalist 160:766–783.

Shea, K., and P. Chesson. 2002. Community ecology theory as a framework for biological invasions. Trends in Ecology and Evolution 17:170–176.

Shurin, J. B. 2000. Dispersal limitation, invasion resistance, and the structure of pond zooplankton communities. Ecology 81:3074–3086.

———. 2001. Interactive effects of predation and dispersal on zooplankton communities. Ecology 82: 3404–3416.

Shurin, J. B., and E. G. Allen. 2001. Effects of competition, predation, and dispersal on species richness at local and regional scales. American Naturalist 158:624–637.

Shurin, J. B., P. Amarasekare, J. M. Chase, R. D. Holt, M. F. Hoopes, and M. A. Leibold. 2004. Alternative stable states and regional community structure. Journal of Theoretical Biology 227:359–368.

Shurin, J. B., J. E. Havel, M. A. Leibold, and B. Pinel-Alloul. 2000. Local and regional zooplankton species richness: a scale-independent test for saturation. Ecology 81:3062–3073.

Srivastava, D. S. 1999. Using local-regional richness plots to test for species saturation: pitfalls and potentials. Journal of Animal Ecology 68:1–16.

Srivastava, D. S., and J. H. Lawton. 1998. Why more productive sites have more species: An experimental test of theory using tree-hole communities. American Naturalist 152:510–529.

Stachowicz, J. J., R. B. Whitlatch, and R. W. Osman. 1999. Species diversity and invasion resistance in a marine ecosystem. Science 286:1577–1579.

Stephens, P. R., and J. J. Weins. 2003. Explaining species richness from continents to communities: The time-for-speciation effect in emydid turtles. American Naturalist 161:112–128.

Stohlgren, T. J., D. Binkley, G. W. Chong, M. A. Kalkhan, L. D. Schell, K. A. Bull, Y. Otsuki et al. 1999. Exotic plant species invade hot spots of native plant diversity. Ecological Monographs 69:25–46.

Terborgh, J. W., and J. Faaborg. 1980. Saturation of bird communities in the West Indies. American Naturalist 116:178–195.

Tilman, D. 1994. Competition and biodiversity in spatially structured habitats. Ecology 75:2–16.

———. 1997. Community invasibility, recruitment limitation, and grassland biodiversity. Ecology 78:81–92.

Turnbull, L. A., M. J. Crawley, and M. Rees. 2000. Are plant populations seed-limited? A review of seed sowing experiments. Oikos 88:225–238.

Westoby, M. 1998. The relationship between local and regional diversity: Comment. Ecology 79:1825–1827.

Wootton, J. T. 1997. Estimates and tests of per capita interaction strength: Diet, abundance, and impact of intertidally foraging birds. Ecological Monographs 67:45–64.

Yu, D. W., and H. B. Wilson. 2001. The competition-colonization trade-off is dead; long live the competition-colonization trade-off. American Naturalist 158:49–63.

CHAPTER 18

From Metacommunities to Metaecosystems

Michel Loreau, Nicolas Mouquet, and Robert D. Holt

Introduction

A defining feature of ecology over the last few decades has been a growing appreciation of the importance of considering processes operating at spatial scales larger than that of a single locality, from the scale of the landscape to that of the region (Ricklefs and Schluter 1993; Turner et al. 2001). Spatial ecology, however, has reproduced the traditional divide within ecology between the perspectives of population and community ecology on the one hand and ecosystem ecology on the other hand.

The population and community ecological perspective has focused on population persistence and species coexistence in spatially distributed systems (Hanski and Gilpin 1997; Tilman and Kareiva 1997), and has a strong background in theoretical ecology and simple, generic mathematical models. The metapopulation concept has occupied a prominent role in the development of this perspective (Hanski and Gilpin 1997; Hoopes et al. chapter 2; Mouquet et al. chapter 10). Its strength has been its ability to deliver specific testable hypotheses on the increasingly critical issue of conservation of fragmented populations in human-dominated landscapes. Because local extinction and colonization can be influenced by interspecific interactions such as predation and competition, a natural extension of the metapopulation concept is provided by the metacommunity concept (Holyoak et al., chapter 1). Significant novel insights are being gained from this new approach, as attested by the various contributions in this book.

Another perspective, however, has developed from ecosystem ecology, and is represented by landscape ecology. Landscape ecology is concerned with ecological patterns and processes in explicitly structured mosaics of nearby heterogeneous ecosystems (Turner 1989; Pickett and Cadenasso 1995; Forman 1995; Turner et al. 2001). It has a strong descriptive basis and a focus on whole-system properties, including abiotic processes. Models that address population persistence and conservation from this perspective are usually more detailed; they consider landscape structure and heterogeneity explicitly, and therefore aim to be more realistic and directly applicable to concrete problems than the more general, abstract models of classical metapopulation and community ecology (Gustafson and Gardner 1996; With 1997; With et al. 1997).

The need to integrate the perspectives of community and ecosystem ecology

418

has been increasingly recognized in recent years to understand such fundamental ecological issues as the relationship between biodiversity and ecosystem functioning, the interactions between food web structure and nutrient cycling, and the role of species in ecosystems (DeAngelis 1992; Jones and Lawton 1995; Loreau 2000b; Kinzig et al. 2002; Loreau et al. 2002; Sterner and Elser 2002). Within the field of spatial ecology, there is a similar need to integrate the perspectives of population and community ecology, including the metacommunity approach, on the one hand, and ecosystem and landscape ecology on the other hand. The metacommunity concept has so far had an exclusive focus on the biotic components of ecosystems. Many critical issues at the landscape or larger spatial scales, however, require consideration of abiotic constraints and feedbacks to biotic processes. We have recently proposed the metaecosystem concept as a theoretical framework for achieving this integration of the community and ecosystem perspectives within spatial ecology (Loreau et al. 2003a).

A *metaecosystem* is defined as a set of ecosystems connected by spatial flows of energy, materials, and organisms across ecosystem boundaries. In contrast to the metacommunity concept, which only considers connections among systems via the dispersal of organisms, the metaecosystem more broadly embraces all kinds of spatial flows among systems, including movements of inorganic nutrients, detritus, and living organisms, which are ubiquitous in natural systems. There has been considerable attention to impacts of spatial subsidies on local ecosystems (Polis et al. 1997). Such studies, however, are limited, in that a subsidy entering one local ecosystem must necessarily be drawn from another. Subsidies at one end are losses at another end, and as such should have impacts on both source and target ecosystems. Moreover, flows are rarely completely asymmetrical. Properties of the higher-level system that arise from movements among coupled ecosystems have seldom been considered explicitly.

Expanding the focus from metacommunities to metaecosystems allows one to understand critical functional properties and processes at spatial scales larger than that of the local ecosystems, which have been the object of greatest interest in classical ecology. Metaecosystems can be defined at any scale from that of an ecosystem cluster (Forman 1995), in which the focus is on small-scale spatial processes among contiguous ecosystems, to that of a region or even the entire globe, for some processes such as spatial flows driven by highly mobile organisms or global biogeochemical cycles involving large-scale air or sea currents. The metaecosystem concept provides a theoretical framework for investigating many of the issues that have been addressed form a more empirical perspective in landscape ecology. It focuses on the properties of the higher-level, spatially extended dynamical system that emerges from movement at landscape to global scales. Just as metacommunity theory is giving new concrete insights into the diversity and structure of ecological communities by explicitly considering interactions between local- and regional-scale processes, the metaecosystem concept provides a new tool to understand emergent constraints and properties that arise from spatial coupling of local ecosystems. Emergent properties have been widely discussed within hierarchy theory, which attempts to provide a framework for describing and understanding the spatiotemporal complexity of ecosystems (Allen and Starr 1982; O'Neill et al. 1986). These properties, however, have rarely been studied in a rigorous, quantitative way based on a firm foundation of lower-level interactions.

In this chapter we examine from a theoretical standpoint three examples of emergent properties that arise from spatial coupling of local systems and provide a brief account of the theory that explains the emergence of these properties. Our examples are taken from our recent work at the interface between metacommunities and metaecosystems, and concern three major issues of current interest: (1) the relationship between local and regional species diversity (a community ecology issue); (2) the relationship between species diversity and ecosystem productivity (an issue at the interface between community and ecosystem ecology); and (3) material flows and ecosystem organization at large spatial scales (an ecosystem ecology issue).

Local and Regional Diversity

Species diversity has been studied historically within community ecology from two different, nonoverlapping perspectives: a local perspective, based on niche theory (MacArthur and Levins 1967), and a regional perspective, through island biogeography (MacArthur and Wilson 1967). In the local perspective, interactions between competing species constrain local diversity, and coexistence is viewed either as a function of niche dimensions and resource heterogeneity (MacArthur and Levins 1967) or of differences in species life-history traits (as in the so-called colonization-competition trade-off models; Hastings 1980; Tilman 1994). In the regional perspective, the theory of island biogeography ignores local dynamics and considers local diversity to be the result of regional processes of immigration and extinction. In this theory, there are no limits to diversity except those arising from the size of the regional species pool (continent size) and the constraints on immigration events (continent-island distance). This apparent contradiction was called "MacArthur's paradox" (Schoener 1983; Loreau and Mouquet 1999) because MacArthur's contributions were central in the development of both niche and island biogeography theory.

In reality the dynamics of species diversity at local and regional scales are not independent of one another. Local α diversity and regional γ diversity are mutually dependent through β diversity, the diversity among communities. It is therefore impossible to understand local diversity, regional diversity, and the relationship between them without considering the dynamics that occur across the two scales (Loreau 2000a). Although this mutual dependency of local and regional diversity has been recognized in principle (Cornell and Lawton 1992; Cornell 1993; Rosenzweig 1995), it has generally been ignored in the interpretation of local-regional richness relationships (Cornell and Lawton 1992; Cornell 1993). Even within metacommunity theory (Holyoak et al. chapter 1), the species-sorting perspective (Leibold 1998) and much of the neutral theory (Hubbell 2001) are based on the implicit assumption that local diversity is influenced by regional diversity but there is no feedback of local diversity on regional diversity, just as in the classical theory of island biogeography. In a true metacommunity perspective, local and regional diversity should be emergent properties that arise from the dynamics of species interactions across scales and constrain each other. The source-sink metacommunity perspective that two of us have recently developed (Mouquet and Loreau 2002, 2003) shows precisely this.

Mouquet and Loreau's (2002) metacommunity model concerns sessile organisms with a dispersal stage, such as plants and some marine invertebrates, or territorial animals with natal dispersal. It incorporates spatial structure both within and among communities. At the local scale (within communities) the model considers the environment as a collection of identical discrete sites, each of which can be occupied by a single individual. It uses the classical formalism of metapopulation models (Levins 1969, 1970) applied at the scale of the individual (Hastings 1980, Tilman 1994; see also Mouquet et al. chapter 10). The model assumes exploitation competition for space; once a plant occupies a site, it keeps it until its death (Loreau and Mouquet 1999). There is no direct competitive exclusion because of interference or competition for other resources; a species' competitive ability is determined by its capacity to occupy new sites (reproduction parameter) and keep them (mortality parameter). Thus, the proportion of vacant sites obtained by each species is proportional to the quantity of propagules it produces. This is a simple extension of competitive lottery models as developed by Chesson and Warner (1981). At the regional scale (among communities), dispersal among communities is assumed to occur through a passive immigration-emigration process. Heterogeneity of environmental conditions at the regional scale is obtained by changing species-specific parameters in each community. This assumes that species exhibit different phenotypic responses in different communities as a result of different local environmental factors.

These assumptions are expressed in mathematical terms as follows. Define P_{ik} as the proportion of sites occupied by species i in community k. There are S species that compete for a limited proportion of vacant sites V_k in each community k, and there are N such communities. Each species i is characterized by a set of reproduction-dispersal parameters b_{ilk}, which describe the rate at which new individuals are produced in community l and establish in community k. When $k = l$, b_{ilk} corresponds to local reproduction, and when $k \neq l$, b_{ilk} corresponds to dispersal from community l to community k. Each species i dies in community k at a mortality rate m_{ik}. When a species immigrates into a particular community, it takes the parameters corresponding to that community. This model reads as follows:

$$\frac{dP_{ik}}{dt} = V_k \sum_{l=1}^{N} b_{ilk} P_{il} - m_{ik} P_{ik},$$ (18.1)

where

$$V_k = 1 - \sum_{j=1}^{S} P_{jk}.$$ (18.2)

Mouquet and Loreau (2002) showed that a necessary condition for there to be an equilibrium in this model is $S \le N$. Thus, there cannot be more species than communities in the metacommunity at equilibrium. This rule provides an equivalent to the competitive exclusion principle in a local community (Levin 1970). They further showed that at equilibrium each species satisfies:

$$\overline{R}_i = \frac{\sum_{k=1}^{N} R_{ik} w_{ik}}{\sum_{k=1}^{N} w_{ik}} = 1$$ (18.3)

where

$$R_{ik} = V_k^* r_{ik},$$ (18.4)

$$r_{ik} = \frac{\sum_{l=1}^{N} b_{ikl}}{m_{ik}},$$ (18.5)

$$w_{ik} = \sum_{l=1}^{N} b_{ilk} P_{il}^*.$$ (18.6)

The parameter r_{ik} can be interpreted as the local basic reproductive rate of species i in community k (equation 18.5). Multiplying r_{ik} by the proportion of vacant sites at equilibrium, V_k^*, we obtain the local net reproductive rate of species i in community k at equilibrium, R_{ik} (equation 18.4). Finally, w_{ik} is the total quantity of propagules produced by species i that arrive in community k per unit time at equilibrium (equation 18.6). Consequently, $\overline{R}_i$ is the regional average net reproductive rate of species i, weighted by the total quantity of propagules arriving in each community, at equilibrium (equation 18.3). Clearly, for the metacommunity to reach equilibrium, $\overline{R}_i$ must be equal to one (equation 18.3), that is, each individual of each species must produce one individual on average during its lifetime in the metacommunity as a whole.

Because all the regional average net reproductive rates must be equal at equilibrium, this sets a constraint of regional similarity between coexisting species. Whatever the local net reproductive rates, they have to be equal when averaged at the scale of the region. And since net reproductive rates are simply basic reproductive rates multiplied by the proportion of vacant space in each community, this constrains basic reproductive rates too. The latter must be sufficiently balanced over the region for equation 18.3 to be possible. Local coexistence is then possible in a metacommunity when species are locally different but regionally similar with respect to their reproductive rates. Local coexistence is explained by compensations among species' competitive abilities at the scale of the region. As a corollary, the net reproductive rate, and hence also the basic reproductive rate, of any species cannot be lower than that of any other species in all communities simultaneously. This condition requires habitat differentiation among species, that is, each species should be competitively dominant in at least one community. Thus, in a metacommunity, the number of species that coexist locally and regionally will be highest when species have different niches (habitat differentiation constraint), but similar competitive abilities (regional similarity constraint), at the scale of the region.

These rules place strong constraints on both local and regional species diversity. Within these constraints, however, a wide variation of local and regional diversity is possible, and this variation is driven in particular by changes in dispersal among communities. To demonstrate the effect of dispersal on species diversity, assume for simplicity that a proportion of the total reproductive output remains resident while the rest emigrates through a regional pool of dispersers that are equally redistributed in all other communities, and that the proportions of dispersers (a) and nondispersers ($1 - a$) are equal for all species and all communities. Parameter a may thus also be interpreted as a measure of the relative importance of regional versus local dynamics. With these assumptions,

$$b_{ilk} = (1 - a)c_{il} \text{ for } k = l, \tag{18.7a}$$

$$b_{ilk} = \frac{a}{N-1}c_{il} \text{ for } k \neq l, \tag{18.7b}$$

in equation (18.1). Here c_{il} is the potential reproductive rate of species i in community l, which encapsulates local reproduction, short-distance dispersal and establishment capacities. The model can then be rewritten as

$$\frac{dp_{ik}}{dt} = V_k\left[\frac{a}{N-1}\sum_{l \neq k}^{N} c_{il}P_{il} + (1 - a)c_{ik}P_{ik}\right] - m_{ik}P_{ik}. \tag{18.8}$$

This model was simulated until equilibrium for a metacommunity consisting of twenty species competing in twenty communities. Simulations used an extinction threshold of 0.01, which provides a good approximation of stochastic extinctions at low population size (Loreau and Mouquet 1999), and a matrix of species' local basic reproductive rates corresponding to a deviation of 5% from strict regional similarity. In the case of strict regional similarity the matrix is completely symmetrical with each species being the best competitor in one community

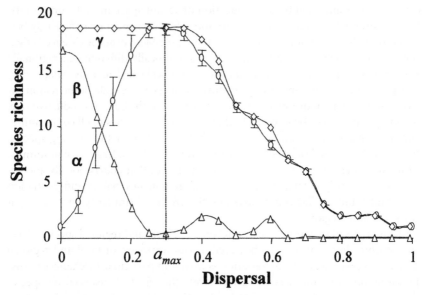

Figure 18.1 Local (α, circles, mean $\pm$ standard deviation across communities), among-community (β, triangles), and regional (γ, diamonds) diversity as functions of dispersal (proportion of dispersers a) in a competitive metacommunity; a_{max} is the dispersal value at which local species diversity is maximal. Modified from Mouquet and Loreau (2003).

(Mouquet and Loreau 2002). The three components of species diversity (α, β and γ) were related through the additive partitioning advocated by Lande (1996) and Loreau (2000a):

$$\gamma = \beta + \overline{\alpha}, \qquad (9)$$

where $\overline{\alpha}$ is the mean alpha diversity of local communities.

Varying dispersal has a dramatic effect on the three components of diversity (figure 18.1; Mouquet and Loreau 2003). When dispersal is zero, local (α) diversity is minimal (1 species) whereas among-community (β) and regional (γ) diversities are maximal; in each community a different species is locally the best competitor. As dispersal increases to an intermediate value a_{max}, an increasing number of species are maintained by immigration above the extinction threshold so that α diversity increases, while at the same time communities become more similar in composition so that β diversity decreases. Regional diversity, however, remains relatively constant. As dispersal increases above a_{max}, both local and regional diversity decrease while β diversity stays close to zero because the best competitor at the scale of the region tends to dominate each community, and other species are progressively excluded. At high dispersal, the metacommunity functions effectively as a single community in which one species outcompetes all others. A hump-shaped relationship between local diversity and dispersal emerges from these constraints.

In the ascending part of the curve, γ diversity is determined by regional environmental heterogeneity, and dispersal acts to transfer its effect from the among-community (β) to the local (α) component of diversity. In the descending part of the curve, dispersal leads to homogenization of the metacommunity, which has a negative effect on regional, and hence also local, diversity.

These results clearly show that dispersal is a major determinant of the relationship between local and regional diversity. To further explore this issue, Mouquet and Loreau (2003) varied maximum regional species richness by varying the degree of regional environmental heterogeneity in the metacommunity for each dispersal value. Variation in environmental heterogeneity was obtained by defining a parameter E_k measuring the environmental condition of community k in a range from 0 to 1, and a parameter H_i measuring the niche preference of species i to environmental conditions, also in a range from 0 to 1. The potential reproductive rate of species i in community k, c_{ik}, was assumed to be greater as its niche optimum was closer to the local environmental condition $\{c_{ik} = (1 - |E_k - H_i|) \times 3\}$. Variation of regional environmental heterogeneity was then generated by varying the distribution of E_k values across communities. Figure 18.2 shows the resulting relationships between local and regional diversity for various levels of

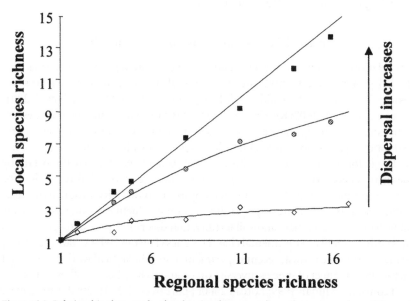

Figure 18.2 Relationships between local and regional species richness for various values of dispersal in a competitive metacommunity ($a = 0.1$, black squares; $a = 0.075$, gray circles; $a = 0.025$, white diamonds). For each dispersal value the gradient of regional species richness was obtained by varying the degree of regional heterogeneity. These results were obtained for low to intermediate dispersal values. At high dispersal, local and regional diversities are equal and the relationship is linear. Modified from Mouquet and Loreau (2003).

dispersal. When dispersal is low, local species richness is limited by the locally dominant species irrespective of regional species richness, and the resulting relationship between local and regional diversity is saturating. When dispersal is higher, local species richness becomes equal to regional species richness and the relationship between local and regional diversity is linear.

Local versus regional diversity plots have often been interpreted as indicative of community saturation; unsaturating linear curves would be typical for unsaturated, noninteractive communities, whereas saturating curves would indicate saturated, interactive communities (Terborgh and Faaborg 1980; Cornell and Lawton 1992; Cornell 1993). The above results show that this interpretation is unwarranted (see also Shurin and Srivastava, chapter 17). Saturation of local-regional richness curves does not tell us anything about community saturation arising from species interactions, but is more fundamentally related to the scale at which a local community is defined and the dispersal properties of the organisms considered (Loreau 2000a). Generally speaking, expanding the scale at which local communities are defined amounts to transferring the environmental heterogeneity that is responsible for the bulk of diversity from the regional to the local scale, hence from β to α diversity. Increasing dispersal across the landscape has a similar effect. The effects of scale and dispersal can be studied quantitatively using modeling frameworks such as that presented here.

Species Diversity and Ecosystem Productivity

The relationship between species richness and ecosystem properties such as productivity has become a central issue in ecological and environmental sciences (see reviews in Tilman 1999; Waide et al. 1999; Loreau 2000b; Mittelbach et al. 2001; Loreau et al. 2001, 2002; Kinzig et al. 2002). It is a unifying fundamental question that requires merging concepts from ecosystem and community ecology. These two subdisciplines have increasingly diverged historically, and merging them is a challenge for modern ecology. This challenge is made particularly important by the current need to understand the potential consequences of biodiversity loss for ecosystem functioning. The traditional approach to diversity-productivity relationships has been to regress species diversity on productivity—or, more exactly, on factors, such as climate and soil fertility, that determine productivity—across sites with different environmental characteristics (Huston 1994; Waide et al. 1999; Grime 2001). In contrast, recent experimental and theoretical work has focused on the specific effect of species diversity on productivity when all other factors are held constant (Tilman 1999; Loreau 2000b; Loreau et al. 2001, 2002; Kinzig et al. 2002). The two approaches have led to different results, which can be reconciled by recognizing that they address different causal relationships at different scales (Loreau 1998, 2000b; Loreau et al. 2001).

Diversity-productivity relationships are also expected to depend strongly on the kind of diversity present in a community, that is, on the coexistence mechanisms that are responsible for the maintenance of diversity within the community (Mouquet et al. 2002). Different coexistence mechanisms involve different environmental and evolutionary constraints on organisms, and these constraints shape both the diversity and productivity of the communities and ecosystems these organisms form. Diversity-productivity relationships then emerge as products of environmental and evolutionary constraints, in which diversity determines productivity as much as productivity determines diversity. What kind of diversity-productivity relationship emerges from source-sink processes in a metacommunity?

Mouquet and Loreau (2003) and Loreau et al. (2003b) explored this issue with two different metacommunity models. The first model is the one presented above, which describes exploitation competition for space. In this model, each species' local productivity is assumed to be determined by its local competitive ability (Tilman et al. 1997; Loreau 1998; Mouquet et al. 2002). Ecosystem productivity in community k, Φ_k, is therefore taken to be the product of the proportion of sites occupied by each species and its local reproductive rate (which is correlated with competitive ability in this model), summed over all species (Loreau and Mouquet 1999):

$$\Phi_k = \sum_{i=1}^{S} c_{ik} P_{ik}. \tag{18.10}$$

Similarly, space occupation by the community as a whole is taken to be simply the summed proportions of sites occupied by all species.

As dispersal increases, average productivity and space occupation across the metacommunity decrease (figure 18.3A). This occurs because the mass effect (Shmida and Ellner 1984), which maintains local species diversity at low to moderate dispersal, also acts to dilute the locally best adapted species—the best competitor—in a mass of locally less adapted species (Loreau and Mouquet 1999). Combining this result with the hump-shaped relationship between local species diversity and dispersal (figure 18.1), a hump-shaped relationship also emerges between average productivity and local species richness (figure 18.3B; note that productivity is on the vertical axis, so that the humped relationship is portrayed in a vertical configuration relative to productivity-diversity plots with productivity on the horizontal axis). At the regional scale, however, the relationship between average productivity and regional species richness is either positive or null (figure 18.3C) because regional species richness is constant or decreases with increasing dispersal (figure 18.1).

These results provide theoretical support for the hypothesis that different diversity-productivity relationships may emerge at different spatial scales, although the mechanisms involved are different from those proposed in other studies (Bond and Chase 2002; Chase and Leibold 2002). Bond and Chase (2002),

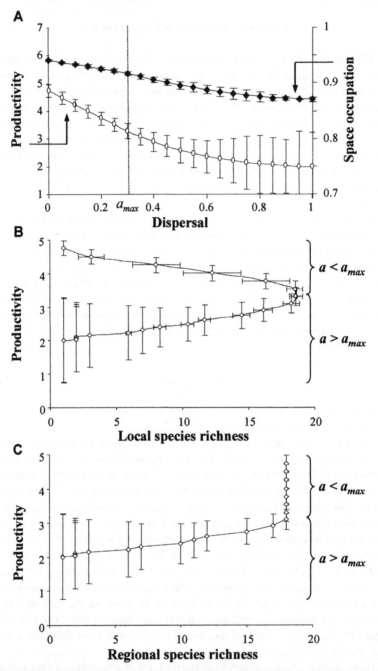

Figure 18.3 (A) Productivity (white circles) and space occupation (solid diamonds) (mean ± standard deviation across communities) as functions of dispersal (proportion of dispersers a) in a competitive metacommunity. (B) The relationship between local species richness and local productivity when dispersal varies. (C) The relationship between regional species richness and local productivity when dispersal varies; a_{max} is the dispersal value at which species richness is maximal. Modified from Mouquet and Loreau (2003).

using a verbal model, suggested that regional complementarity among species could lead to a positive relationship between productivity and regional species richness. In contrast, a hump-shaped relationship would be found at the local scale because local species richness would increase first through local niche complementarity (generating a positive relationship with productivity) and then through a source-sink effect (generating a negative relationship with productivity). Mouquet and Loreau's (2003) results confirm Bond and Chase's intuition, but they involve no local niche complementarity. Both the local hump-shaped and the regional positive diversity-productivity relationships arise from pure source-sink metacommunity processes.

Loreau et al. (2003) developed a more mechanistic consumer-resource model to explore the effects of species diversity on ecosystem productivity and its temporal stability in a metacommunity or metaecosystem under fluctuating environmental conditions. This model makes similar assumptions to the previous one, in particular the fact that dispersal is global and identical for all species, and dispersers are redistributed uniformly across the landscape. The main differences lie in the presence of an explicit consumer-resource local interaction, which allows a more straightforward measurement of productivity, and the presence of environmental fluctuations. The model reads as follows:

$$\frac{dN_{ij}(t)}{dt} = [e_{ij}c_{ij}(t)R_j(t) - m_{ij}]N_{ij}(t) + \frac{a}{M-1}\sum_{k \neq j}^{M}N_{ik}(t) - aN_{ij}(t)$$

$$\frac{dR_j(t)}{dt} = I_j - l_jR_j(t) - R_j(t)\sum_{i=1}^{S}c_{ij}(t)N_{ij}(t) \qquad (18.11)$$

where $N_{ij}(t)$ is the biomass of species i (e.g., a plant) and $R_j(t)$ is the amount of limiting resource (e.g., a nutrient such as nitrogen) in community j at time t. The metacommunity consists of M communities and S species in total. Species i consumes the resource at a rate $c_{ij}(t)$, converts it into new biomass with efficiency e_{ij}, and dies at rate m_{ij} in community j. The resource is renewed locally through a constant input flux I_j, and is lost at a rate l_j. All species disperse at a rate a. Consumption rates $c_{ij}(t)$ vary as local environmental conditions change through time, and are assumed to reflect the matching between species traits and environmental conditions as above. Defining again H_i as the constant trait value of species i and $E_j(t)$ as the fluctuating environmental value of community j, consumption rates are given specifically by:

$$c_{ij}(t) = 1.5 - |H_i - E_j(t)|. \qquad (18.12)$$

Fluctuations of local environmental values are assumed to be sinusoidal with period T:

$$E_j(t) = \frac{1}{2}\left[\sin\left(x_j + \frac{2\pi t}{T}\right) + 1\right], \qquad (18.13)$$

and to be out of phase in the various communities, by choosing x_j such that $E_1(0)$ = 1 and $E_j(0) = E_{j-1}(0) - 1/6$ for j = 2 to 7. With this assumption, there is always a community in which each species is superior, but because of temporal fluctuations, that community shifts in space over time, thus requiring some dispersal for long-term coexistence. The period of fluctuations was chosen to be large enough so that there was competitive exclusion in the absence of dispersal.

Lastly, ecosystem productivity at time t is defined as the production of new biomass per unit time, which, averaged across the metacommunity, is

$$\Phi(t) = \frac{\sum_{i=1}^{S}\sum_{j=1}^{M} e_{ij}c_{ij}(t)R_j(t)N_{ij}(t)}{M}. \tag{18.14}$$

This model leads to the same hump-shaped relationship between local diversity and dispersal as does the previous model (figure 18.4A). In contrast to the previous model, however, average productivity here follows a hump-shaped pattern similar to that of species diversity (figure 18.4B). Similarly, the coefficient of variation of productivity—a common standardized measure of variability (Doak et al. 1998; Ives et al. 1999; Lehman and Tilman 2000; Ives and Hughes 2002)—follows an inverse pattern (figure 18.4C). As a consequence, variations in dispersal rate generate strongly nonlinear, parallel variations in local species diversity, average productivity, and the stability (sensu reduced variability) of productivity.

Differences from the previous model are explained by the specific effects of biodiversity made possible by environmental fluctuations. Biodiversity has been shown to act as biological insurance for local ecosystem functioning by allowing functional compensation among species or phenotypes in time (McNaughton 1977; Yachi and Loreau 1999; Ives et al. 1999; Lehman and Tilman 2000; Norberg et al. 2001). Such insurance effects include an increase in the temporal mean of productivity when there is selection for adaptive responses to environmental fluctuations, and a decrease in productivity's temporal variability because of temporal complementarity among species responses (Yachi and Loreau 1999; Loreau 2000b). Here, however, these effects occur despite the fact that local coexistence is impossible, and thus no temporal insurance effect can occur within a closed system. Therefore, insurance effects shown by this model are entirely generated by the spatial dynamics of the metacommunity. When different systems experience different environmental conditions and fluctuate asynchronously, different species thrive in each system at each point in time, and dispersal ensures that the species adapted to the new environmental conditions locally are available to replace less adapted ones as the environment changes. As a result, biodiversity enhances and buffers ecosystem processes through spatial exchanges among local systems in a heterogeneous landscape, even when such effects do not occur in a closed homogeneous system. This is the spatial insurance hypothesis (Loreau et al. 2003).

As shown by figure 18.4, however, spatial insurance effects are strongly

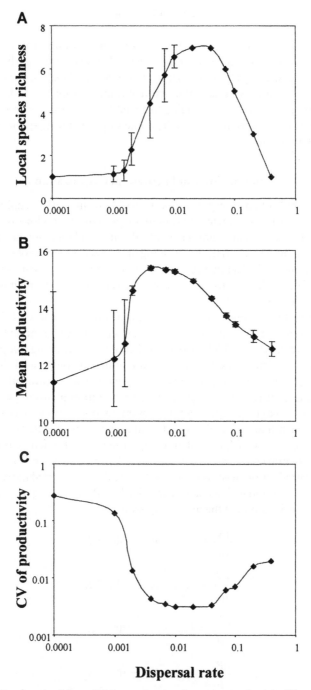

Figure 18.4 Local species richness (A), temporal mean of ecosystem productivity (B), and coefficient of variation of ecosystem productivity through time (C) as functions of dispersal rate (mean ± standard deviation across communities) in a competitive metacommunity under fluctuating environmental conditions. Modified from Loreau et al. (2003b).

dependent on dispersal rate, which determines metacommunity connectivity. Local species diversity and the insurance effects that it generates are highest at an intermediate dispersal rate, and collapse at both low and high dispersal rates. At both ends of the dispersal gradient functional compensations and adaptive shifts between species are prevented, leading to a relatively low average productivity as well as large fluctuations in productivity as the single surviving species tracks environmental fluctuations.

Material Flows and Ecosystem Organization

The last example of an emergent property arising from spatial coupling of local systems is the pattern of material flows in a metaecosystem. Flows of nutrients, whether in the form of inorganic elements, detritus, or living organisms, can exert important influences on the functioning of local ecosystems (Polis et al. 1997). Less appreciated is the fact that these flows may also impose global constraints at the scale of the entire metaecosystem, thereby generating a strong interdependence among local ecosystems.

To highlight these constraints, we concentrate on the simplest possible model of a closed nutrient-limited metaecosystem. Consider two connected local ecosystems, 1 and 2, each of which in turn consists of two interacting compartments, plants (with nutrient stock P) and inorganic nutrients (with stock N). Spatial flows among ecosystems are assumed to occur among similar compartments (i.e., from inorganic nutrient to inorganic nutrient, and from plants to plants). They are also assumed to be independent of local interactions among ecosystem compartments, such that spatial flows and local growth rate are additive in the dynamical equation for each ecosystem compartment. Let F_{Xij} denote the directed spatial flow of nutrient stored in compartment X from ecosystem i to ecosystem j, Φ_i primary production in ecosystem i, and R_i the flow of recycled nutrient within ecosystem i. Local and global mass balance leads to the following set of equations describing the dynamics of the metaecosystem:

$$\frac{dN_1}{dt} = F_{N21} - F_{N12} - \Phi_1 + R_1, \tag{18.15a}$$

$$\frac{dN_2}{dt} = F_{N12} - F_{N21} - \Phi_2 + R_2, \tag{18.15b}$$

$$\frac{dP_1}{dt} = F_{P21} - F_{P12} - \Phi_1 + R_1, \tag{18.15c}$$

$$\frac{dP_2}{dt} = F_{P12} - F_{P21} - \Phi_2 + R_2, \tag{18.15d}$$

This description in terms of directed flows among ecosystems and compartments (figure 18.5A) can be reduced to a simpler description in terms of net flows as follows:

$$\frac{dN_1}{dt} = F_N - G_1, \tag{18.16a}$$

$$\frac{dN_2}{dt} = -F_N + G_2, \tag{18.16b}$$

$$\frac{dP_1}{dt} = F_P - G_1, \tag{18.16c}$$

$$\frac{dP_2}{dt} = -F_P + G_2, \tag{18.16d}$$

where $F_X = F_{X21} - F_{X12}$ is the net spatial flow of nutrient of compartment X from ecosystem 2 to ecosystem 1, and G_i is net local plant growth in ecosystem i.

Note that, as required for closed systems, local mass is conserved in the absence of spatial flows and global mass is conserved with spatial flows. Additional constraints emerge from spatial coupling of local ecosystems as the metaecosystem reaches equilibrium. At equilibrium the left-hand side of equations 18.15 and 18.16 vanishes, which imposes

$$F_N^* = -F_P^* = G_1^* = -G_2^*, \tag{18.17}$$

where asterisks denote functions evaluated at equilibrium. This set of equalities can be interpreted as a double constraint, which can easily be generalized to metaecosystems with an arbitrary number of local ecosystems and an arbitrary number of ecosystem compartments (Loreau et al. 2003a): (1) A *source-sink constraint within* ecosystem compartments. For each compartment, positive growth in some ecosystems must be balanced by negative growth in other ecosystems at equilibrium, which means that some local ecosystems must be sources whereas others must be sinks; and (2) A *source-sink constraint between* ecosystem compartments. The total net spatial flow across the boundaries of each ecosystem must vanish at equilibrium, which means that some compartments must be sources whereas others must be sinks.

In our simple metaecosystem with two ecosystems and two compartments, these constraints result in a global material cycle such that net flows at equilibrium are either in the direction $N_1 \rightarrow P_1 \rightarrow P_2 \rightarrow N_2 \rightarrow N_1$ (figure 18.5B) or in the opposite direction depending on the sign of F_N^* (or any other function) in equation 18.17. In this global cycle, even though production and nutrient recycling occur within each ecosystem (figure 18.5A), one ecosystem acts as a net global pro-

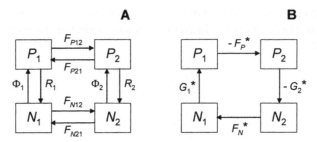

Figure 18.5 Material flows in a closed nutrient-limited metaecosystem consisting of two connected local ecosystems, 1 and 2, each of which in turn consists of two interacting compartments, plants (P) and inorganic nutrient (N): (A) directed nutrient flows, and (B) net nutrient flows at equilibrium.

ducer ($N_1 \rightarrow P_1$) whereas the other acts as a net global recycler ($P_2 \rightarrow N_2$) (figure 18.5B). When there are more than two ecosystem compartments and local ecosystems, the pattern of material circulation in the metaecosystem may be more complex, but all local ecosystems are embedded in a web of material flows constrained by the functioning of the metaecosystem as a whole (Loreau et al. 2003a).

This simple metaecosystem model shows that strong constraints on local ecosystem functioning emerge from spatial coupling of ecosystems. When these constraints can be met, they imply that local ecosystems can no longer be governed by local interactions alone. Instead, by being part of the larger-scale metaecosystem, local ecosystems are constrained to become permanent sources and sinks for different compartments, and thereby to fulfill different functions in the metaecosystem. It is also conceivable, however, that these constraints may be impossible to meet in some cases; during transient dynamics parts of the metaecosystem will then absorb others by progressively depriving them of the limiting nutrient. Specifically, nutrient source-sink dynamics within metaecosystems may drive or accelerate successional changes, until equilibrium is achieved and the final metaecosystem state becomes compatible with global source-sink constraints. Whether energy and material transfers across ecosystem boundaries are strong enough to drive succession, however, depends on their magnitude relative to that of the colonization processes that bring new species into local ecosystems and thereby change their properties. This suggests that combining an explicit accounting of spatial flows of energy and materials with the dynamics of colonization of new patches by organisms in an integrated metaecosystem approach may provide a promising novel perspective on succession theory.

Conclusions

Extending metacommunity theory to a full theory of metaecosystems represents an important and timely development for spatial ecology—a development that has the potential to integrate the perspectives of community, ecosystem, and landscape ecology. At a time when humans are profoundly altering the structure and functioning of natural landscapes, understanding and predicting the consequences of these changes is critical for designing appropriate conservation and management strategies. Metacommunity and metaecosystem perspectives provide powerful tools to meet this goal. By explicitly considering the spatial interconnections among systems, they have the potential to provide novel fundamental insights into the dynamics and functioning of ecosystems from local to regional scales, and to increase our ability to predict the consequences of land-use changes on biodiversity and the provision of ecosystem services to human societies.

In this chapter we have provided some examples of significant emergent properties that arise from spatial coupling of local ecosystems. These range from the coupled dynamics of local and regional diversity, through diversity-productivity relationships at local and regional scales, to patterns of nutrient flows from landscape to global scales. In all these examples, metaecosystem connectivity, as determined by the spatial arrangement of component ecosystems and the movements of organisms, energy, and inorganic substances across these ecosystems, exerts strong constraints on the structure, functioning, and stability of the system at both local and regional scales. It also drives many of the community and ecosystem properties that are traditionally studied at separate scales without consideration of these critical connections among scales. This shows that the metacommunity and metaecosystem perspectives offer a promising theoretical framework to explore hierarchical systems and emergent properties in a spatial context.

We acknowledge that the theoretical models we have reviewed have a number of simplifying assumptions and hence a number of limitations, and that empirical support is still largely missing. The assumptions that are probably most critical in our metacommunity models concern the rules that constrain local interactions (competitive lottery), and the nature of the dispersal process (global, passive dispersal, with no life-history correlates). Modifying these assumptions to make the models better suited for organisms other than plants is likely to change some of the emergent properties we have investigated (such as diversity-productivity relationships), although others (such as local-regional diversity relationships) may be more robust because the mechanisms involved are relatively general. In contrast, our metaecosystem model can be made very general (Loreau et al. 2003a). The challenge here will be to devise more detailed, yet tractable, models capable of providing new insights into more targeted issues for specific systems.

The objective of the models presented in this chapter, however, was not to provide detailed predictions for specific systems, but instead to examine the potential of metacommunity and metaecosystem approaches for exploring new issues and providing new insights into old issues. We feel that, in this respect, these approaches have proven very successful and offer a rich avenue for future theoretical, experimental, and empirical developments. Our hope is also that they will help stimulate the emergence of a mechanistic theoretical landscape ecology.

Literature Cited

Allen, T. F. H., and T. B. Starr. 1982. *Hierarchy: Perspectives for ecological complexity.* University of Chicago Press, Chicago, IL.

Bond, E. M., and J. M. Chase. 2002. Biodiversity and ecosystem functioning at local and regional spatial scales. Ecology Letters 5:467–470.

Chase, J. M., and M. A. Leibold. 2002. Spatial scale dictates the productivity-biodiversity relationship. Nature 416:427–430.

Chesson, P. L., and R. W. Warner. 1981. Environmental variability promotes coexistence in lottery competitive systems. American Naturalist 117:923–943.

Cornell, H. V. 1993. Unsaturated patterns in species assemblages: The role of regional processes in setting local species richness. Pages 243–252 *in* R. E. Ricklefs and D. Schluter, eds. *Species diversity in ecological communities: Historical and geographical perspectives.* University of Chicago Press, Chicago, IL.

Cornell, H. V., and J. H. Lawton. 1992. Species interactions, local and regional processes, and limits to the richness of ecological communities: A theoretical perspective. Journal of Animal Ecology 61: 1–12.

Doak, D. F., D. Bigger, E. K. Harding, M. A. Marvier, R. E. O'Malley, and D. Thomson. 1998. The statistical inevitability of stability-diversity relationships in community ecology. American Naturalist 151:264–276.

DeAngelis, D. L. 1992. Dynamics of nutrient cycling and food webs. Chapman and Hall, London, UK.

Forman, R. T. T. 1995. *Land mosaics: The ecology of landscapes and regions.* Cambridge University Press, Cambridge, UK.

Grime, J. P. 2001. *Plant strategies, vegetation processes and ecosystem properties.* 2nd edition. Wiley, New York.

Gustafson, E. J., and R. H. Gardner. 1996. The effect of landscape heterogeneity on the probability of patch colonization. Ecology 77:94–107.

Hanski, I. A., and M. E. Gilpin. 1997. *Metapopulation biology: Ecology, genetics and evolution.* Academic Press, San Diego.

Hastings, A. 1980. Disturbance, coexistence, history and the competition for space. Theoretical Population Biology 18:363–373.

Hubbell, S. P. 2001. *The unified neutral theory of biodiversity and biogeography.* Princeton University Press, Princeton, NJ.

Huston, M. A. 1994. *Biological diversity.* Cambridge University Press, Cambridge, UK.

Ives, A. R., and J. B. Hughes. 2002. General relationships between species diversity and stability in competitive systems. American Naturalist 159:388–395.

Ives, A. R., K. Gross, and J. L. Klug. 1999. Stability and variability in competitive communities. Science 286:542–544.

Jones, C. G., and J. H. Lawton. 1995. *Linking species and ecosystems.* Chapman and Hall, New York.

Kinzig, A. P., S. W. Pacala, and D. Tilman. 2002. *The functional consequences of biodiversity: Empirical progress and theoretical extensions.* Princeton University Press, Princeton, NJ.

Lande, R. 1996. Statistics and partitioning of species diversity, and similarity among multiple communities. Oikos 76:5–13.

Lehman, C. L., and D. Tilman. 2000. Biodiversity, stability, and productivity in competitive communities. American Naturalist 156:534–552.

Leibold, M. A. 1998. Similarity and local co-existence of species in regional biotas. Evolutionary Ecology 12:95–110.

Levin, S. A. 1970. Community equilibria and stability, and an extension of the competitive exclusion principle. American Naturalist 104:413–423.

Levins, R. 1969. Some demographic and genetic consequences of environmental heterogeneity for biological control. Bulletin of the Entomological Society of America 15:237–240.

———. 1970. Extinction. Pages 77–107 in M. Gerstenhaber, ed. *Some mathematical problems in biology.* American Mathematical Society, Providence, RI.

Loreau, M. 1998. Biodiversity and ecosystem functioning: a mechanistic model. Proceedings of the National Academy of Sciences, USA 95:5632–5636.

———. 2000a. Are communities saturated? On the relationship between α, β and γ diversity. Ecology Letters 3:73–76.

———. 2000b. Biodiversity and ecosystem functioning: Recent theoretical advances. Oikos 91:3–17.

Loreau, M., and N. Mouquet. 1999. Immigration and the maintenance of local species diversity. American Naturalist 154:427–440.

Loreau, M., N. Mouquet, and R. D. Holt. 2003a. Meta-ecosystems: A theoretical framework for a spatial ecosystem ecology. Ecology Letters 6:673–679.

Loreau, M., N. Mouquet, and A. Gonzalez. 2003b. Biodiversity as spatial insurance in heterogeneous landscapes. Proceedings of the National Academy of Sciences, USA 100:12765–12770.

Loreau, M., S. Naeem, and P. Inchausti. 2002. Biodiversity and ecosystem functioning: Synthesis and perspectives. Oxford University Press, Oxford, UK.

Loreau, M., S. Naeem, P. Inchausti, J. Bengtsson, J. P. Grime, A. Hector, D. U. Hooper, M. A. Huston, D. Raffaelli, B. Schmid, D. Tilman, and D. A. Wardle. 2001. Biodiversity and ecosystem functioning: Current knowledge and future challenges. Science 294:804–808.

MacArthur, R. H., and R. Levins. 1967. The limiting similarity, convergence, and divergence of coexisting species. American Naturalist 101:377–387.

MacArthur, R. H., and E. O. Wilson. 1967. *The theory of island biogeography.* Princeton University Press, Princeton, NJ.

McNaughton, S. J. 1977. Diversity and stability of ecological communities: A comment on the role of empiricism in ecology. American Naturalist 111:515–525.

Mittelbach, G. G., C. F. Steiner, S. M. Scheiner, K. L. Gross, H. L. Reynolds, R. B. Waide, M. R. Willig, S. I. Dodson, and L. Gough. 2001. What is the observed relationship between species richness and productivity? Ecology 82:2381–2396.

Mouquet, N., and M. Loreau. 2002. Coexistence in metacommunities: The regional similarity hypothesis. American Naturalist 159:420–426.

———. 2003. Community patterns in source-sink metacommunities. American Naturalist 162:544–557.

Mouquet, N., J. L. Moore, and M. Loreau. 2002. Plant species richness and community productivity: Why the mechanism that promotes coexistence matters. Ecology Letters 5:56–65.

Norberg, J., D. P. Swaney, J. Dushoff, J. Lin, R. Casagrandi, and S. A. Levin. 2001. Phenotypic diversity and ecosystem functioning in changing environments: A theoretical framework. Proceedings of the National Academy of Sciences, USA 98:11376–11381.

O'Neill, R. V., D. L. DeAngelis, J. B. Waide, and T. F. H. Allen. 1986. *A hierarchical concept of ecosystems.* Princeton University Press, Princeton, NJ.

Pickett, S. T. A., and M. L. Cadenasso. 1995. Landscape ecology: Spatial heterogeneity in ecological systems. Science 269:331–334.

Polis, G. A., W. B. Anderson, and R. D. Holt. 1997. Toward an integration of landscape and food web ecology: The dynamics of spatially subsidized food webs. Annual Review of Ecology and Systematics 28:289–316.

Ricklefs, R. E., and D. Schluter. 1993. *Species diversity in ecological communities: Historical and geographical perspectives.* University of Chicago Press, Chicago, IL.

Rosenzweig, M. L. 1995. Species diversity in space and time. Cambridge University Press, Cambridge, UK.

Schoener, T. W. 1983. Rate of species turnover decreases from lower to higher organisms: A review of the data. Oikos 41:372–377.

Shmida, A., and S. Ellner. 1984. Coexistence of plant species with similar niches. Vegetatio 58:29–55.

Sterner, R. W., and J. J. Elser. 2002. *Ecological stoichiometry: The biology of elements from molecules to the biosphere.* Princeton University Press, Princeton, NJ.

Terborgh, J., and J. Faaborg. 1980. Saturation of bird communities in the West Indies. American Naturalist 116:178–195.

Tilman, D. 1994. Competition and biodiversity in spatially structured habitats. Ecology 75:2–16.

———. 1999. The ecological consequences of changes in biodiversity: A search for general principles. Ecology 80:1455–1474.

Tilman, D., and P. Kareiva. 1997. Spatial ecology: The role of space in population dynamics and interspecific interactions. Princeton University Press, Princeton, NJ.

Tilman, D., C. L. Lehman, and K. T. Thomson. 1997. Plant diversity and ecosystem productivity: Theoretical considerations. Proceedings of the National Academy of Sciences, USA 94:1857–1861.

Turner, M. G. 1989. Landscape ecology: The effect of pattern on processes. Annual Review of Ecology and Systematics 20:171–197.

Turner, M. G., R. H. Gardner, and R. V. O'Neill. 2001. *Landscape ecology in theory and practice: patterns and processes.* Springer, New York.

Waide, R. B., M. R. Willig, C. F. Steiner, G. Mittelbach, L. Gough, S. I. Dodson, G. P. Juday, and R. Parmenter. 1999. The relationship between productivity and species richness. Annual Review of Ecology and Systematics 30:257–301.

With, K. A. 1997. The application of neutral landscape models in conservation biology. Conservation Biology 11:1069–1080.

With, K. A., R. H. Gardner, and M. G. Turner. 1997. Landscape connectivity and population distributions in heterogeneous environments. Oikos 78:151–169.

Yachi, S., and M. Loreau. 1999. Biodiversity and ecosystem productivity in a fluctuating environment: The insurance hypothesis. Proceedings of the National Academy of Sciences, USA 96:1463–1468.

CHAPTER 19

Adaptive and Coadaptive Dynamics in Metacommunities

Tracking Environmental Change at Different Spatial Scales

Mathew A. Leibold, Robert D. Holt, and Marcel Holyoak

Introduction

It has long been recognized that community processes operate at numerous scales and that the dynamics of these processes occur differently when they are manifest at these different scales. The metacommunity concept we have championed in this book provides one way of exploring these issues in a general framework that remains conceptually transparent and draws attention to numerous features of multiscale ecological processes.

The work described in the earlier chapters of this book and in previously published works illustrates numerous ways that local and regional dynamics of communities are illuminated by this approach. The effects of metacommunity dynamics can be discerned in many important aspects of community and meta-community structure including patterns of population structure, coexistence, biodiversity, food web architecture, and ecosystem attributes. These studies also show that these various patterns are interrelated; specific patterns predicted by theory and that at times observable in experimental and observational work are contingent on each other. Thus, for example, the observation of patterns that correlate local diversity with regional diversity, despite having strong local saturation of local communities, may be contingent on the dynamics of food web assembly (Shurin and Srivastava, chapter 17). Similarly, observed scaling relations in biomass of organisms at different trophic levels, may also be contingent on how food web assembly processes and community shifts in composition among redundant species occur in metacommunities.

It seems clear that a metacommunity perspective can strongly modify the way we view how communities are regulated and how community structure is related to environmental conditions. In this chapter we evaluate how community structure might be related to environmental conditions by thinking of communities as "complex adaptive systems" (see Levin 1998) and explore how metacommunity dynamics constrain the processes that affect the behavior of such systems. We start with the following propositions that are strongly suggested by work to date on metacommunities:

1) There is feedback between local processes and regional processes. Local community dynamics provide basic rules that interact with dispersal and local extinctions to regulate the dynamics of metacommunities. The metacommunity in turn provides constraints on the attributes and diversity of species that can participate in these local dynamics, and in particular, can coexist. These constraints in the domain of species' traits can modify the dynamics of local communities from those that would be expected with unconstrained species pools. Moreover, dispersal can influence local interactions themselves (e.g., by buffering predator-prey fluctuations and thus reducing extinction risks).

2) Numerous features of the dynamics that result from this interaction of processes at different scales depend on how strongly connected local communities are to each other. Dynamics also depend on how much heterogeneity among local communities there is in local environmental conditions, and the timescale of temporal environmental change (whether gradual or sudden) that might affect local extinctions of species. It is this range of metacommunity attributes that determine the nature of the feedback between local and regional processes described in (1).

3) This interaction of processes results in a continuum of dynamics. They range from situations that can be simplified as a progression through three of the four simple perspectives (patch dynamic, species sorting, and mass effects views) described in the introduction to this book. The fourth perspective— neutral species occupying identical patches—provides a useful null model for evaluating the other three perspectives.

4) The result is that there are scale transition phenomena that can be altered in the dynamics of metacommunities as a function of the factors that drive these dynamics, including the connectivity of the metacommunity, the amount of among patch heterogeneity present, and the frequency and magnitude of environmental change and disturbance in local patches in the metacommunity. Whether such scale transitions are expressed as a continuum in patterns of community organization or instead exhibit discrete breaks and thresholds is an open question.

The relationship between community structure and environmental conditions can be thought of as an adaptive process under constraints that can alter how the ecological attributes of species match up with the environmental conditions that occur where they exist. In the absence of such an adaptive process, there would be no correspondence between environmental conditions and community composition (a null hypothesis for the adaptive process). In contrast, we might expect that the correspondence could be very precise so that community composition is perfectly correlated with environmental conditions. Our point is that many metacommunity processes—identified and discussed in the work done to date on metacommunities—will alter how strong this correlation may be, and may alter the resulting pattern between community composition and environments. In particular, metacommunity theory suggest that the joint effects of dispersal, extinctions, and local environmental change in a metacommunity can strongly determine how closely tied the ecological attributes (or niches) of species in local communities are to local environmental conditions. Here we explore how metacommunity processes alter the behavior of the "complex adaptive systems" that characterize local communities embedded in metacommunities with different levels of connectance to each other.

Abstractly, evolution by natural selection occurs whenever three ingredients are present: there must be variation, the variation must be heritable, and this heritable variation must influence fitness (Lewontin 1970). This can apply at any level of biological organization, including metacommunities (e.g., Wilson 1992; Goodnight 2000; Wade 2004). In our view the genetic composition can include the genetic composition of individual species or it can involve changes in the species composition of the group (Wilson and Swenson 2003). Within a metacommunity, there is variation among species in a number of traits (e.g., niche requirements, dispersal behaviors). Indeed, the magnitude of such variation typically vastly exceeds that available for microevolution within species. Such variation is clearly heritable; species-specific traits often remain reasonably fixed through time. Finally, these differences can affect growth rate, as instantiated for example in Lotka-Volterra community models (and countless descendent theories). Norberg et al. (2001) have recently suggested that evolution of phenotypic traits within a community can be studied using the formalism of quantitative genetics. Expanding this suggestion, one can imagine partitioning shifts in community traits in a metacommunity into within- and between-community processes, analogous to hierarchical levels of selection in a metapopulation (Barton and Whitlock 1997).

Adaptation can be viewed as a process that leads to a match between the genetic and species composition of a group of organisms and the local spatial and temporal conditions present in the environment. For a given set of species and genotypes, adaptation is maximized when every local community is characterized by a composition that is stable and is unaltered by perturbations involving either immigration or emigration (i.e., it is an evolutionarily stable state in relation to all other species and genotypes present). If the environment is fixed, adaptation may take a long time but is eventually expected to stop when there is a balance between selective processes (species sorting and natural selection within species) and other processes that disrupt the process of adaptation (drift, mutation, migration, and sexual selection). In principle, the traits involved in the regulation of communities as complex adaptive systems can involve anything that modifies any of the interactions within or among the component species (see Wade 2004). In much of the following discussion we will focus on consumer resource interactions involving traits such as resource exploitation ability and resistance to consumers. In simple cases the match between species traits and environment can be described by simple and definable criteria. For example, the theory of adaptation in isolated populations indicates that adaptation tends to maximize population growth rate in successional environments, and equilibrial population size when a population tends to a density-dependent equilibrium (Rummel and Roughgarden 1985). As an additional example, theory indicates that adaptation favors minimization of resource requirements in simple systems consisting of consumers on a single resource (the R^* rule of Tilman 1982); at the community level, remaining free resources can be minimized by exploitative competition (MacArthur 1970).

In more complex situations involving multiple limiting factors and enemies, adaptation favors more complicated traits that are context dependent, varying with attributes of the local environment. Furthermore, the final state of the system may not be a point attractor, but could include richer dynamical behaviors (e.g., cycles in character space). The latter situation can arise when there are coevolutionary dynamics, particularly involving predator-prey or host-pathogen interactions (e.g., "red queen" scenarios; Seger 1992; see Steiner and Leibold 2004). Nonetheless, while it may take a long time to equilibrate, the eventual outcome is some sort of broad correspondence between the attributes of the organisms found in a place and the environmental conditions at that place, given that the environment is fixed. This general process, when it involves multiple species interacting among themselves via a variety of direct and indirect interactions has been described as a "complex adaptive system" (Levin 1998).

In environments with fluctuating conditions, however, the degree to which this adaptive process will lead to such a match between the attributes of the organisms present in a local patch and the ambient environmental conditions will depend on how well adaptive dynamics can track environmental change. Below, we evaluate how different factors, including metacommunity dynamics, influence this process of tracking, and refer to this ability as adaptive capacity. If adaptive capacity is high, we predict that there will be a good correspondence between the attributes of the organisms in a community and the environmental conditions there. If adaptive capacity is low, we expect that this will not be as likely and that there will consequently be a less precise fit. Moreover, the observed match will reflect dispersal patterns (as elaborated in more detail below). The term *adaptive dynamics* in some evolutionary literature denotes a specific scenario, where evolution within a community arises via very small changes in an original set of species (Metz et al. 1996). However, in this paper, we use term more broadly to denote the time course of selective processes in general, including at the level of entire communities.

Much of our thinking here parallels corresponding issues in microevolutionary analyses of adaptation within single species as well as coevolution between species, and how such evolutionary processes are influenced by spatial structure and movement (e.g., Barton and Whitlock 1997).

In closed communities (those that consist of isolated local communities and are thus not part of metacommunities), adaptation is limited to changes in species composition that only involve extinctions, or possible cases of sympatric speciation, and to genotypic changes within the remaining species. The range of variation on which selection can act is constrained by the slice of the species pool that was originally captured at the time of isolation. Over sufficiently long timescales, it may be possible for speciation to occur in situ as one form of genotypic change, but this is likely to be a relatively slow process and perhaps one that is unlikely in closed communities (assuming that most speciation occurs in allopatry by vicariance events; the presence of such events within the community would seem to violate the assumption of closed communities). The adaptive capacity of such local communities provides a baseline level of adaptive capacity that can be contrasted with that of metacommunities. Over the long scale, the factors that regulate how fast adaptation can occur in such closed communities in response to changes in the environment include well-studied effects of population size and population structure, mutation rate, genetic constraints, and patterns of selection (e.g., Holt 1997).

A fascinating complication that we will not address is the role of adaptive phenotypic plasticity in the metacommunity context. While recognizing the importance of such plasticity, we suspect that in general differences among species in traits in a community that are in essence "hard-wired" genetically will outweigh the range of variation observed in intraspecific plastic responses to environmental conditions. In any case, to deal adequately with plasticity would mandate our addressing a huge and complex arena of issues, which goes beyond what seems appropriate for this volume.

In metacommunities, in addition to these local processes, there can be influences of dispersal on local adaptation. There is an important and we believe underappreciated parallelism between the impacts of dispersal at the level of individual species and at the level of entire communities. At the level of individual species, dispersal can either facilitate evolution (if gene flow is an important source of local variation, e.g., Gomulkiewicz et al. 1999), or hamper it (as in the classic scenario of gene flow "swamping" local selection). Moreover, and crucially, adaptive capacity at the hierarchical scale of the community is further affected by migration of individuals of different species. Such migration can either promote or hinder the adaptive process matching the community to local conditions. If the local assemblage of organisms (individuals and / or species) has an adaptive capacity that is limited because organisms with attributes that best correspond to the environment are absent, immigration can be a source of such individuals that enhance the adaptive capacity of the system. As noted above, phenotypic variation is usually much greater between species than within a species. Alternatively, if immigration involves large numbers of individuals that do not have such attributes, and if these immigrants reduce the densities of residents that do have the attributes (e.g., because of competition for resources), it can reduce the adaptive capacity of the system. In the context of microevolution, this is gene flow hampering local adaptation. At the community level, this arises, for instance, because of mass effects.

Thus, using closed communities as a baseline, metacommunities may have greater or lesser adaptive capacity to respond to environmental change at the local level. It would be interesting to compare adaptive capacity in local communities differing in their openness, following environmental change. This might perhaps be evaluated by comparisons of rates of recovery of community and ecosystem attributes in closed communities with metacommunities at different levels of connectivity (nonmetacommunity examples of disturbance experiments include Berlow 1997; Shurin 2000; Fukami 2001; Suding and Goldberg 2001; Downing and Leibold 2002). Observed differences would indicate the degree to which metacommunity dynamics might be important in (for instance) determining the relationships between community structure (including biodiversity) and ecosystem attributes and stability.

In short, if we view metacommunities as complex adaptive systems, the precision of this adaptation should be related to the pattern and rate of movement among different local communities. We can now relate this perspective to the simplified views that we have described in this book. We then consider in more detail the effects of heterogeneity of dispersal among species in food webs.

Dispersal and Adaptive Dynamics in Neutral Metacommunities

Metacommunities that conform to the neutral perspective are of course unaffected by the processes described above. Adaptive dynamics in the metacommunity are then synonymous with within-species adaptive evolution in any species (since all are assumed to be equivalent) or uniform plastic responses, with no effect of dispersal rates. A focus on adaptive dynamics highlights the particular nature of the neutrality in such metacommunities by highlighting the simultaneous adaptive dynamics that would have to be present in a neutral metacommunity to avoid disrupting the entire suite of factors that affect the ways species interact with each other, both at the local and regional scales, in ways that involve evolutionary as well as ecological processes.

Adaptive Capacity of Metacommunities: A Species Sorting Perspective

As discussed in chapter 1 and elsewhere in this volume, the species sorting perspective emphasizes that because of resource gradients or heterogeneity in space, different species are superior in different places. Dispersal does not strongly perturb local community structure, yet it is nonetheless crucial for maintaining the adaptive capacity of communities.

High adaptive capacity in metacommunities is enhanced when heterogeneity among patches in local environmental conditions permits patch-type specialization by the component species, and thereby enriches the diversity of the regional biota. In this case, the metacommunity as a whole can serve as a continuous and stable source of species adapted to different local conditions. If there is environmental change in any single patch, coupling between that patch and a heterogeneous metacommunity will ensure that there are appropriately adapted species present somewhere that can colonize the patch, thus enabling the maintenance of a correspondence between species attributes and local environmental conditions.

Imagine a situation where local communities are closed and have had constant environments for long enough that the local assemblage of species shows strong matching between species attributes and environmental conditions. Any subsequent environmental change within a patch can act as a disturbance that will affect the relative fitness of different species in the assemblage. In the absence of immigration, these changes disrupt the outcome of the interactions, potentially favoring some species and eliminating others. When such species elimination occurs, the adaptive capacity of the system has been reduced, because the total variance in ecological traits present in the local community available to respond to future change has been reduced. For instance, if this initial change is reversed, the community may not be able to revert to its prior state because the first change caused extinction of the newly refavored species. Community closure (Lundberg et al. 2000) can thus preclude adaptive response and make the consequences of environmental change irreversible. In the species sorting view of metacommunities, rapid dispersal among sites can greatly enhance the adaptive capacity of the system because it reduces the amount of time needed before these species that have been driven extinct are present again.

Adaptive Capacity in Metacommunities: The Patch Dynamic Perspective

In its most common form, the patch dynamic perspective, assumes that patches are identical. Due to local extinctions, however, some patches are occupied, whereas others are unoccupied; colonization occurs, but with a time lag. The interplay of dispersal with local extinction then becomes important to explaining the persistence of a diverse array of species in the metacommunity. Because dispersal is not rapid, some species may exist as fugitive species; high dispersal rates can allow them to exploit transient opportunities provided by, for instance, temporary absences of locally adapted species. In effect, disturbance creates temporal niches that these species have evolved to exploit. Such fugitive species can enhance the adaptive capacity of the system if they do not inhibit the subsequent colonization by species that are better adapted to local conditions; but if there is niche preemption, they could actually reduce the adaptive capacity if they do inhibit this process. Alternatively, some locally unstable interactions may persist regionally solely because of dispersal (e.g., strong predator-prey or host-parasite inter-actions; see Hoopes et al., chapter 2).

Leaving aside the issue of such fugitive species, as with species sorting, the ability of species to recolonize in patch dynamics should clearly enhance the adaptive capacity of a metacommunity, relative to an array of closed communities experiencing a similar regime of disturbance leading to extinctions even when there is heterogeneity of patches in the metacommunity. This provides an additional way in which the adaptive capacity of the system is reduced because of the lag that results between changes in environmental conditions and the response by changes in community composition.

Adaptive Capacity in Metacommunities: The Mass Effect Perspective

Mass effects can have a variety of impacts on the adaptive capacity of a metacommunity. Mass effects arise when there is heterogeneity among patches, and dispersal is relatively high. Again, as with species sorting, this can permit the rapid reestablishment of species into local patches following appropriate environmental change as described above. However, in heterogeneous systems dispersal can lead to source-sink relations among patches with different environmental conditions. If these mass effects are small, they may enhance the adaptive capacity of the system. This can occur because immigration not only overcomes the limitation to the initial presence of a species, but it may additionally establish a large enough population to facilitate initial growth of this population or to shorten the amount of time spent at very low densities in situations where species show large-amplitude cycles. Coupled with temporal variation, mass effects can greatly increase the average abundance of species (e.g., the "inflationary effect"; Holt 1993; Gonzalez and Holt 2002; Holt et al. 2003). In addition, this facilitation may be important in allowing the initial establishment of species subject to Allee effects or vulnerable to extinction due to demographic stochasticity in small populations.

Mass effects can often have strongly stabilizing effects on locally unstable predator-prey and food web interactions, because an influx of prey that cannot be overexploited (spatial subsidies) is inherently stabilizing, and predator emigration into sinks makes overexploitation in focal productive habitats less likely (e.g., Holt 1985, 1993; Huxel and McCann 1998). Source-sink relations can help stabilize large amplitude oscillations that might also constrain assembly dynamics. For example, Abrams (1998) showed that large-amplitude oscillations can alter numerous aspects of community structure in food webs, including the assembly dynamics of food webs that involve edible and resistant prey when they are involved in oscillations with their consumers in highly productive ecosystems. Source-sink relations among patches with different productivities or among patches that are asynchronously oscillating could greatly reduce the amplitudes of these oscillations and facilitate the development of adaptive assembly dynamics.

Countering these positive effects, however, mass effects and source-sink relations may also reduce the adaptive capacity of the assemblage when dispersal is high. This occurs for two reasons. First, sink populations may depress the abundances of locally adapted species by competition, by preying on them, or by other species interactions (Christiansen and Fenchel 1977; Holt et al. 2003). Another more subtle reason is that emigration from source patches can incur local extinction risks (e.g., the KISS model for minimum plankton patch size; see also Holt 1985). This can endanger the global persistence of a species. Even if this does not occur, emigration reduces the local impact of source species on their environment and makes them more vulnerable to extinction due to disturbance; overall, these effects reduce the adaptiveness of the assemblage in source patches compared to what would obtain if such emigrants were prevented from dispersing and were constrained to remain inside the patch. Of course this last effect will depend on whether the emigrant population is one that would otherwise have lived in the source patch or whether it would otherwise have died (Amarasekare and Nisbet 2001; Mouquet and Loreau 2002, 2003).

If dispersal becomes very high, there results a scale transition in the dynamics such that heterogeneity among patches has greatly reduced effects in allowing for species to be patch type specialists. Under these conditions the mixing effects of dispersal essentially convert the landscape of patches into a single larger community that is essentially homogenous from the perspective of the assemblage. This makes regional coexistence of competitors less likely, essentially because a potential axis for niche differentiation no longer effectively operates (Chase and Leibold 2003), and also makes it more likely for unstable food web interactions to lead to extinctions. High dispersal should lead to biotic impoverishment, and thus a reduced adaptive capacity for coping with environmental change.

These various observations suggest that maximum adaptive capacity occurs at an intermediate level of dispersal—sufficiently high to overcome dispersal limitation in species establishment (permitting a match of traits with local conditions) but not so high that strong source-sink relations compromise the local population dynamics of locally adapted species. This view corresponds well with the species sorting perspective (Leibold 1998; Chase and Leibold 2003) but there may be some possible enhancement if there are weak mass effects that can overcome Allee effects, etc. In this view, metacommunity dynamics are important in maintaining local assembly dynamics, leading to a match between local environments and the attributes of species found there. In effect, weak links between the constituent local communities in a metacommunity facilitate the adaptive capacity of the system, relative to either no linkage at all (no dispersal), or very strong linkages (rapid, homogenizing dispersal).

Leibold and Norberg (2004) use this highly speculative framework to compare and contrast general attributes of zooplankton communities and their roles in limnetic ecosystems. The available (admittedly sparse) evidence indicates that different types of limnetic ecosystems may show a wide range of dynamics that roughly correspond to the various perspectives outlined above. They argue that metacommunities consisting of interconnected ponds in a region may show attributes—low dispersal limitation, resistance to invasion, high compositional turnover among ponds, and strong patterns of trophic structure and scaling of trophic level biomass—that correspond to the species sorting view. In contrast, evidence from deeper lakes (that have a fairly distinct biota from that found in shallower ponds), indicate much more frequent dispersal limitation and sensitivity of local assemblages to trophic structure changes following invasions, corresponding to the patch dynamics view of metacommunities. As a third possibility, Leibold and Norberg argue that strong source-sink relations documented in the population structure of dominant zooplankton species in large water bodies such as very large lakes and in oceans, may explain some elements of the dynamics of these very large systems. (Consistent with this suggestion about oceans, Rex and Etter have recently proposed that large-scale patterns of diversity in oceanic benthic communities as a function of depth may arise from coupling shallow source populations with deep-water sink populations [pers. comm.].) The line of argument we have presented here would suggest that pond metacommunities have greater adaptive capacity than either deeper lakes or ocean systems to respond to changes in local environmental conditions.

Effects of Heterogeneity of Dispersal in Food Webs in Metacommunities

As noted elsewhere in this volume (Holt and Hoopes, chapter 3; Van Nouhuys and Hanski, chapter 4; Miller and Kneitel, chapter 5; Holt et al, chapter 20), food webs often include species that differ greatly in their dispersal biology. This implies that different species in the same community will match different ones of the above idealized scenarios. For example, dispersal in fish-eating birds, fishes, crustaceans, rotifers, algae, bacteria, and insects that interact in lake food webs is very unlikely to be similar to even an order of magnitude. Thus the metacommunity dynamics of these different groups of interacting organisms may be characterized by very different levels of adaptive capacity. How might heterogeneity of dispersal among different sets of species alter the general predictions we made above?

Heterogeneity of dispersal among interacting species (involving competition as well as consumer-resource or other types of relations) is a key attribute of many previously hypothesized components of metacommunity dynamics. Heterogeneity in dispersal plays a key role in the competition-colonization hypothesis of simple patch dynamics models (Levins 1969; Hastings 1980), the coexistence of predators and prey in metacommunities (Huffaker 1958; Caswell 1978; Crowley 1981), the likelihood of coexistence in metacommunities via colonization-competition trade-offs under strong source-sink relations (Amarasekare and Nisbet 2001; Amarasekare et al. 2004), and the possible roles of spatiotemporal dynamics in regulating communities (e.g., Chesson 2000, Chesson et al., chapter 12). If there are recurrent extinctions, it is difficult to maintain long food chains unless top consumers have relatively high colonization rates (e.g., Holt and Hoopes, chapter 3).

Heterogeneity of dispersal among interacting species can also affect the applicability of the metacommunity concept in the first place. The simple definition we have used throughout this book presumes that patches are discrete enough that all interacting species respond somewhat similarly to a given patch structure. This is a useful simplification, but is often a caricature of reality: organisms are so heterogeneous in their dispersal that some perceive a given set of patches as a metapopulation, while other species with which they interact perceive the same set of patches as a single relatively homogenous environment.

How should heterogeneity of dispersal alter the adaptive capacity of metacommunities? Below we use the argument developed above about adaptive capacity to draw attention to the ways that different sets of species might have different levels of co-adaptive capacity. Rather than doing this in full generality we will try to do this by connecting our view of adaptive dynamics in food webs (described above and in Leibold and Norberg 2004) to other work being done on adaptive dynamics of interacting species. We then explore how such differences might affect community and ecosystem features of metacommunities in comparison with closed communities.

Dispersal Heterogeneity and Adaptive Capacity in Metacommunities

It is often useful (and necessary) to aggregate species in communities and food webs into guilds, assemblages, or functional groups (e.g., producers, herbivores, predators). If these different sets of species are heterogeneous in the dispersal rates, but interact both locally and at the metacommunity scales, they may respond to a given landscape of patches in different ways over both ecological and evolutionary timescales. By the arguments sketched above, species assemblages with very low dispersal rates will have their abilities to track local environmental conditions compromised by dispersal limitation, when compared to assemblages with somewhat higher dispersal rates (sufficient to permit species sorting). However, assemblages with very high dispersal rates should also have reduced abilities to track local environmental change, because of mass effects and emigration leading to mismatches between species' traits and local environmental conditions. This means that adaptive capacity may vary substantially among different assemblages or functional groups, with important consequences for both local ecological interactions and evolutionary dynamics in the metacommunity.

As a concrete example illustrating the potential implications of this perspective for population, community, and ecosystem attributes, consider interactions between adjacent trophic levels—that is, interactions between a group of species that serve as common resources for a set of consumer species that therefore potentially compete. Studying the dynamics of consumer-resource interactions has long been at the core of food web ecology. However, most past work has assumed either spatially closed systems (Hairston et al. 1960; Phillips 1974) or examined the local assembly dynamics for food chains and food webs (i.e., not specifically in the context of metacommunities; e.g., Oksanen et al. 1981, Holt 1993). Leibold (1996) began to place this work in the context of metacommunity models based on species sorting. This provides a link between assembly dynamics and ecosystem attributes of local patches, in particular what we will call "trophic structure"—the allocation of biomass to different trophic levels. Importantly, there is evidence that this process occurs in some natural systems and that it results from adaptive dynamics in metacommunities. Leibold (1996) considered metacommunity dynamics involving a single trophic level, which set the stage for examining the consequence of adaptive metacommunity dynamics at multiple trophic levels simultaneously (Steiner and Leibold 2004). This is closely analogous to a process of coevolution (joint sorting of consumers and resources); below we explore this parallel and use it to help generate hypotheses about similar dynamics in metacommunities.

Trophic structure is the allocation of biomass to different trophic levels. It is generally thought that this allocation is reasonably stable and is constrained to a limited domain of ratios (Hairston and Hairston 1993). In many systems, such ratios are constrained despite tremendous variation in productivity and absolute biomass levels. The pattern is best documented in the plant-herbivore link of the food chain; data show that plant biomass and herbivore biomass both increase jointly with ecosystem productivity as driven by environmental conditions such as nutrient levels in lakes and rainfall in arid grasslands.

While this pattern is well documented, it has long been apparent that explaining it with simple food chain and food web models is problematic. For example, the Lotka-Volterra consumer resource equations for plant-herbivore interactions and laissez-faire herbivores predict that steady-state herbivore biomass should increase with ecosystem productivity, but that plants should be unaffected. This implies that there are no constraints on the herbivore to plant biomass ratio. A large literature is devoted to trying to explain this contradiction between observation and simple theory. One set of explanations focuses on how food change length could vary with productivity (Oksanen et al. 1981). Thus when herbivore levels get sufficiently high due to enhanced productivity, this view hypothesizes that carnivores can persist, and thereafter carnivore biomass will increase with increasing productivity, while herbivore biomass will remain fixed, and plant biomass will increase. Over the entire range of productivity, this stair step pattern shows an overall increase in both plant and herbivores. One may interpret this pattern as emerging from a kind of adaptive dynamics, in the sense that it reflects compositional change (i.e., number of trophic levels) arising from changes in productivity. Unfortunately there is scant evidence that the number of trophic levels increases with productivity in the fashion predicted by these simple models (e.g., Holt and Hoopes, chapter 3).

One way to elaborate these models is to permit multiple species at one or more trophic levels. The effect of this on the predictions depends on whether or not systems are closed or open. Phillips (1974) incorporated inedible and resistant plant species, and showed that in simple models of closed systems (i.e., a fixed set of species in the local community) one would not expect both plants and herbivores to respond simultaneously to increased ecosystem productivity. This effect could not alter the prediction that either plants or herbivores should have steady-state biomass that responded to ecosystem productivity but that it was not so for both simultaneously to do so. However, Holt et al. (1994) and Leibold (1996) showed that joint increases could occur if colonization from a species pool was present, permitting a steady progression of species replacements along a productivity gradient toward increasingly resistant plant. As in the situation modeled by Oksanen et al. (1981), the joint responses of plant and herbivore biomass (at steady states) to ecosystem production occurred due to compositional change but here the effects were more subtle because they involved replacements of similar species within a trophic level along the productivity gradient, rather than the addition of trophic levels. There is some evidence that is consistent with this hypothesis (e.g., Leibold et al. 1997; Chase et al. 2000; Steiner 2001). The important point here is that this mechanism involves species sorting, permitting adaptive dynamics of the plant trophic level to grazers (whose abundance in turn is indirectly regulated by productivity).

If an ecosystem undergoes local environmental change resulting in altered productivity, the "adaptive response" (joint increases in both plants and herbivores that prevent either from becoming very large) critically depends in magnitude and rate on which plant species can colonize and modify the local trophic interactions. This of course depends on having a wide range of plants in the species pool that can colonize. Leibold (1996, 1998) considers a metacommunity in which sites vary in productivity, permitting global persistence of a wide range of species. Considering this metacommunity as the source pool for a focal community, he shows that species sorting implies predictable patterns in trophic structure, such as unimodal relations between local species richness and productivity. The metacommunity perspective of species sorting thus leads to predictable patterns that are consistent with much available evidence. It would be instructive to examine how trophic structure is influenced by other metacommunity processes, such as patch dynamics (but see Shurin and Allen 2001) and source-sink relations between habitats (but see Holt 2002).

The trophic structure problem, however, is not completely resolved. Other mechanisms have been proposed for the joint scaling of plant and herbivore biomass with productivity. It is also possible that the effects reflect nonequilibrial behaviors, including oscillations and complex attractors; complex dynamics increase in likelihood with productivity due to the "paradox of enrichment" (Rosenzweig 1971). However, the effects of these various possibilities act locally and do not require colonization from a species pool. This therefore provides a way of evaluating how important metacommunities are in regulating the observed pattern, and by inference in affecting the adaptive dynamics of plant-herbivore communities. Leibold et al. (1997) and Chase et al. (2000) use a meta-analysis of experiments from lakes and in grasslands and show that plant compositional change along productivity gradients may be substantial and important in producing this pattern. This effect is confirmed by a recent experiment (Leibold and Smith, in press) in pond communities, involving both phytoplankton and zooplankton, showing that joint increases in plant and herbivore biomass are much more likely when local communities receive very small inputs of larger biotas present in the region (thus overcoming any dispersal limitation) than when isolated from the regional biota (limiting adaptive responses to those that occur within the original local species set). The results suggest that metacommunity processes can dramatically enhance the adaptive capacity of open systems, compared to closed local communities.

While encouraging, this work raises new issues because compositional change in the community has only been considered to date for plants. What happens if we allow herbivore composition to also change in response to either productivity per se, to the changes in plant composition, to changes in the intensity of predation and the composition of these predators, or to changes in compositions at all levels at once?

Steiner and Leibold (2004, in prep.) studied the assembly of local food webs under different productivity levels, including species sorting from a regional species pool, and contrasted the outcome of simulated assembly histories involving food webs with only one basal trophic level to those with two and three trophic levels. They found that the impact of species sorting was tremendously enhanced by the presence of multiple trophic levels (supporting the implicit conjecture in previous models that food web interactions greatly increase the number of solutions to adaptive dynamics of metacommunities). Moreover, they found that there was much greater variation in herbivore to plant biomass than previously found assuming only plant compositional change. This suggests that adaptive dynamics in such cases are much more complex. (It also makes the previously observed joint scaling of plant and herbivore biomass and the experimental results observed by Leibold and Smith [forthcoming] somewhat puzzling because both plants and herbivores could colonize these experimental systems.)

Comparable issues arise when one compares adaptive evolution within a single species, to adaptive evolution in interacting pairs of species. A critical insight into metacommunities as complex adaptive systems may come from drawing on recent work on adaptive dynamics of plant-herbivore interactions on trophic structure by Loeuille and Loreau (2004, see also Loeuille et al. 2002). In this model, coevolutionary adaptive dynamics of single plant and herbivore species in a closed system were studied. Adaptive evolution involved selection mutation, subject to trade-offs involving plant resistance traits and herbivore's abilities to overcome these resistance traits. They found that the outcome of coevolutionary dynamics for trophic structure were profoundly influenced by the relative adaptive capacity (here determined by the mutation rate) of the plants and herbivores. Two qualitatively different outcomes involving trophic structure were most likely. When plant adaptive capacity was high, relative to that of herbivores, the emergent trophic structure matched models in which plant composition could respond to productivity, whereas herbivore composition was constrained and could not: with higher productivity, both plants and herbivores jointly increased in biomass. Alternatively, when herbivore adaptive capacity was high relative to plants, the pattern more closely corresponded to expectations from the food chain model of Oksanen et al. (1981). With comparable adaptive capacities for both plants and herbivores, the outcome was more variable and hard to predict.

The implications of this study of plant-herbivore coevolution are intriguing. If our suggestion is cogent that adaptive dynamics in metacommunities parallels local coevolutionary dynamics as explored by Loeuille and Loreau (2004)—just with a greater range of variation accessible for selection—then the commonly observed joint scaling of plants and herbivores across productivity gradient could be due to the presence of higher adaptive capacity in plant than in herbivore assemblages. If these adaptive dynamics involve mostly metacommunity processes related to colonization and local extinction (as suggested by Leibold et al. [1997] and Chase et al. [2000]), the differences in adaptive dynamics could be due to differences in how these two trophic levels are regulated by connectivity via dispersal in the metacommunity. Consider pond plankton communities. The zooplankton trophic level may have less adaptive capacity than plants for two distinct reasons. One is that zooplankton may be more dispersal limited than are phytoplankton; thus, plant compositional change can track changes in zooplankton communities more effectively such that zooplankton communities might respond to such changes. Alternatively, pond zooplankton may have stronger source-sink relations than do plants, which hampers their ability to track compositional change in the plant trophic level. Intriguingly, the theoretical study by Loeuille and Loreau (2004) suggests that differences in adaptive capacity between plants and herbivores of either mode can greatly alter the impact of variation in productivity on trophic structure.

Another intriguing pattern that may fit this suggestion comes from a worldwide comparative survey by Milchunas and Lauenroth (1993) on the effect of grazing on aboveground net primary production (ANPP) (Holt 1995). These authors ranked the length of the evolutionary association of large vertebrate grazers and the plants they consumed, and concluded that the difference in

ANPP between grazed and ungrazed sites decreased systematically with an increasing length in shared evolutionary history. Yet, the impact of grazing on local plant species composition increased with the length of shared evolutionary history. One interpretation of this pattern is that with increasing exposure to herbivory over evolutionary timescales, there is an increase in the representation in the species pool of species that are specifically adapted to coping with such herbivory, thus making species sorting among plants more important as a constituent of the overall impact of herbivory at a local scale. Moreover, large herbivores typically have large home ranges and often show migratory behaviors over substantial spatial scales (see figure 20.1b), so it is unlikely that they can develop finely reticulate evolutionary responses to local patterns of variation in plant species composition.

All of these ideas need to be explored more carefully by work specifically focused on metacommunities, but they are highly suggestive of several possibilities:

• Trophic structure (and maybe other similar attributes of communities that result from the interaction of multiple species) can be regulated by the adaptive capacity of consumers and resources involving traits related to food web interactions.
• Variation in trophic structure along productivity gradients may reflect the respective adaptive capacities of different trophic levels.
• Such differences in relative adaptive capacity could be due in part to the local evolutionary potential (not directly related to metacommunity dynamics) of the component species, but it could also be strongly affected by how plants versus herbivores disperse in relation to the patchiness of the system.
• Similar scenarios could describe other types of species interactions. For instance, comparable effects when different groups of competitors for shared resources interact (e.g., the interkingdom competition described by Hochberg and Lawton 1990); many instances of intraguild predation involve taxa with radically different spatial strategies (many examples in Polis et al. 1989). Whichever suite of competitors has an intermediate rate of dispersal in a given landscape will tend to enjoy a species sorting advantage over competitors that either disperse very little, or a great deal.

To summarize the main thrust of our argument to this point:

1) Interacting groups of species in communities and ecosystems have an ability to change in response to environmental change, so they are complex adaptive systems.
2) The ability of complex adaptive systems to change has predictable community and ecosystem level attributes that differ from patterns expected in the absence of compositional and genetic change.
3) Groups of similar species with similar dispersal rates can have variable dynamics matching the four different perspectives we explore in this book.

4) Groups of such species with intermediate dispersal rates (in relation to the connectivity of patches in the metacommunity) will have a greater adaptive capacity than species groups with higher or lower dispersal rates.

5) Groups with intermediate dispersal rates will be able to more effectively track local environmental change (e.g., recover from disturbance).

6) These heterogeneities in adaptive capacity can alter the way the system as a whole operates and may alter numerous aspects of the resulting system ranging from population structure, to community dynamics, to major features of the entire ecosystem.

Connectivity and Scale in Metacommunities and Metaecosystems

The above arguments suggest a rich array of possible behaviors that can emerge when one considers how adaptive compositional change in different groups of organisms can influence large-scale attributes of the ecosystems in which they participate. These consequences are expected to vary as a function of the connectivity of the landscape of patches. Loreau et al. (chapter 18) suggest that this may be captured by the concept of metaecosystems. Here we argue that one crucial element of the dynamics of such metaecosystems may be patch connectivity, because this influences the relative adaptive capacities of different groups of organisms and consequently may govern their participation in ecosystem processes at different spatial scales.

Many authors have observed that environmental heterogeneity scales with distance, so that nearby points are more similar to each other than are more distant points (e.g. Williamson 1988; Bell et al. 1993); sometimes such heterogeneity is fractal in nature and has obvious breakpoints. For purposes of analysis, and to develop a suite of ideas stemming from the metacommunity concept, we have simplified the spatial heterogeneity to two (or three) hierarchical scales. Connectivity among patches can be related to the dissimilarity of local environmental conditions, expressed as a function of distance (figure 19.1). Highly connected metacommunities will have patches that have small distances between them; whereas poorly connected metacommunities will have patches that are more distant from each other. We can compare these distances to the dispersal ability of the species found in the metacommunity and relate these distances to the major metacommunity scenarios we have visited in a number of chapters in this book.

1) Well-mixed: Distances between patches are short enough that individuals in different patches encounter each other with similar likelihood as they do individuals in the same patch over the course of their life cycle.

2) Mass effect: Distances are longer than in a well-mixed scenario but there is sufficient migration that source-sink relations occur between heterogeneous patches.

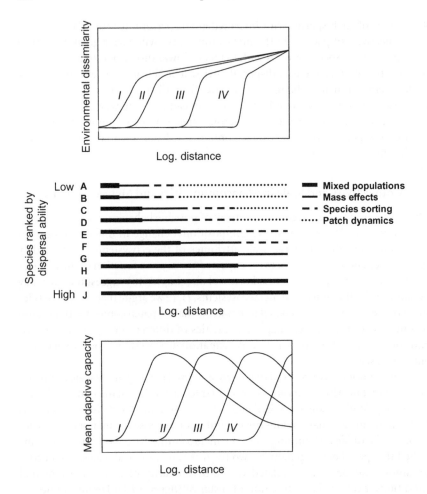

3) Species sorting: Distances are too far to support meaningful source-sink relations, but they are sufficiently small that there is no effective dispersal limitation for compositional change.
4) Patch dynamics: The habitable sites are far enough apart that patches may not be colonized immediately.

In figure 19.1 we portray these effects for four different landscapes (*I–IV*) where patchiness occurs at different levels of connectivity (distances between similar patches). We also compare these distances to the dispersal abilities of a group of ten species (A–J). We have grouped these species into pairs (A and B, C and D, etc.) that have identical dispersal abilities, differing from other such pairs. As depicted, A and B are the poorest dispersers, and I and J are the best (indeed, I

◄   Figure 19.1 The relationship between dispersal ability, environmental variability, and the degree of association between species and their environment (adaptive capacity). The upper panel shows four landscapes (*I–IV*), where patchiness (dissimilar environmental conditions) occurs at different distances. The middle panel shows the kind of dynamics expected for ten species (A–J) with different dispersal abilities, such that A and B are the poorest dispersers and I and J the best. Species are grouped into pairs (A and B, C and D, etc.) that have identical dispersal abilities, differing from other such pairs. Adaptive capacity is the evolutionary association between species and their environment. It is expected to be highest with species sorting, to be reduced with mass effects because of source-sink dynamics, to be further reduced with patch dynamics due to local extinction and inability to disperse to take advantage of favorable conditions, and to be almost zero in well mixed populations. Different groups of species will have the highest adaptive capacity in each of the four landscapes, and consequently mean adaptive capacity across all ten species is shown in the lower panel. In landscape *I*, environmental patchiness (high dissimilarity) occurs at such a small scale that most species (C–J) are evenly mixed at short distances, which reduces their adaptive capacity. In landscape *I*, as spatial scale (distance) increases as more species show mass effects and species sorting dynamics, and net adaptive capacity increases because species sorting is the most frequent type of dynamics. At still larger distances patch dynamics dominate and adaptive capacity is again reduced. By contrast, in landscape *IV*, patchiness occurs only at a much larger spatial scale and there is more heterogeneity in the community in the in adaptive capacity of different species groups: species I and J are in well mixed populations, species G and H show mass effects, species E and F show species sorting, and species A–D show patch dynamics. Landscapes *II* and *III* are bracketed by landscapes *I* and *IV*. The precise shapes of curves are arbitrary and the description is intended to illustrate the patterns reported in the text but not to go beyond the text by making more quantitative predictions.

and J have such high dispersal that they are uniformly distributed over the entire metacommunity).

Different groups of species will have the highest adaptive capacity in each of these four different landscapes. In landscape *I* for example, environmental patchiness occurs at such a small scale that most species (C–J) are evenly mixed. Even species A and B, with the lowest dispersal rates, show mass effects and can exhibit source-sink relations at small spatial scales (distances in figure 19.1). These two species may not show a strong pattern of association with the environment at small scales. In contrast, in landscape *IV*, different groups of species have dispersal rates that result in greater heterogeneity in the adaptive capacity of different species groups at large spatial scales. Species I and J are still evenly distributed over the patches (and thus show no adaptive capacity related to sorting); species G and H have sufficiently high dispersal to show source-sink relations from favorable patches into unfavorable ones; species E and F have the highest adaptive capacity due to moderate dispersal permitting species sorting and so should have the best abilities to track local environmental conditions; and, species A-D are dispersal limited so that they are not able to regularly recolonize favorable patches when they arise. Other landscapes (*II* and *III*) are bracketed by landscapes *I* and *IV*; in each, a different group of species has the greatest adaptive capacity (e.g., greatest for species A and B in landscape *II*, greatest for species C and D in landscape *III*). In the lower panel of figure 19.1 the net adap-

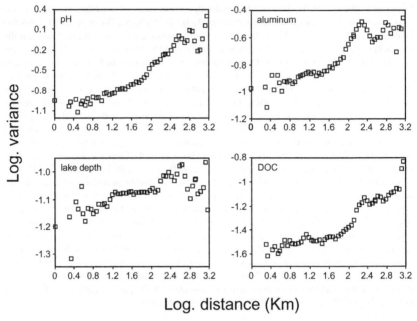

Log. distance (Km)

Figure 19.2 Environmental dissimilarity as a function of distance in Ontario lakes. The graphs show $\log_{10}$ of variance in point measurements against $\log_{10}$ total distance among up to 1236 lakes originally plotted by Bell et al. (1993) using data from the Ontario Ministry of Natural Resources and Ontario Ministry of the Environment (Ontario, Canada). All variables except pH were normalized by log-transformation before analysis. Distance is measured in kilometers, as the maximum distance between lakes at a given spatial scale shown by the horizontal axis. Note that unlike the landscapes in figure 19.1 there is no clear indication that environmental variability reached an upper asymptote. DOC is dissolved organic carbon. Modified from Bell et al. (1993).

tive capacity is shown, which is averaged across all of our ten species in the hypothetical community.

These differences in relative adaptive capacity potentially have important implications in generating patterns in ecological attributes of these systems, as suggested above for trophic structure along productivity gradients. This suggested linking of spatial scale, environmental heterogeneity and dispersal abilities of different organisms provides a potential pathway for linking the metacommunity approach we have championed in this book (which when interpreted literally involves discrete spatial heterogeneity) to analyses of patterns and dynamics in landscapes where heterogeneities are described by continuous variables. Williamson (1988) suggested that spatial variation in environmental conditions often has a fractal structure, and that this scaling could contribute to species-area relationships. Bell et al. (1993) conducted a spatially-distributed array of "common garden" experiments and examined the relationships among plant performance (for a number of species), similarity in environmental conditions, and distance. Some of their results indicate that variation in local environmental quality varies directly as a function of distance between localities (figure 19.2).

We believe that there are manifold potential consequences of extending metacommunity theory to continuous, heterogeneous landscapes along the lines sketched here. In particular, we expect that in different landscapes, different sets of species will have the greatest adaptive capacity for matching patterns of variation, whereas other species will be constrained. The outcome would be a mosaic pattern of responses by species (Thompson 1994) to the spectrum of environmental variability described in the top part of figure 19.1. This could lead to important differences among biomes and systems in responses to environmental perturbations. Holt and Gaines (1992) argued that predicting evolutionary conservatism versus rapid evolution in a given species requires paying attention to the texture in space and time experienced by a species and as modulated by its dispersal behavior. Our suggestion in this chapter is that comparable adaptive phenomena arise when considering interacting species embedded in metacommunities in heterogeneous landscapes.

Conclusions

While highly speculative, we hope that our focus on the adaptive and coadaptive dynamics of metacommunities (including potentially a range of processes, from intraspecific evolution, to coevolution, to species sorting), and the potential consequences of such adaptation for other dimensions of ecology (e.g., control of trophic structure by productivity) will stimulate further work at the interface of community ecology and evolutionary theory. We have outlined how the concept of metacommunities can be linked to the idea of complex adaptive systems, and how metacommunities of different types may have greater or lesser degrees of adaptiveness, depending on the relationship between dispersal and landscape heterogeneity and connectivity in the metacommunity. We suggested an analogy with the concept of adaptive evolution and coevolution in Darwinian models of evolution of single species, and explored the idea of coadaptive dynamics in metacommunity dynamics. This line of thinking suggests that different groups of species may affect each others' distributions in metacommunities in ways that depend on how well they can respond adaptively to variation at different spatial scales. We illustrated this basic idea with the classic issue in community ecology of the relationship between productivity and trophic structure, suggesting the role of adaptive dynamics for aggregate ecosystem properties.

Our approach is different from but complementary to other work done on community genetics (see papers in the 2003 Special Feature in *Ecology* 84:543–601). This work is concerned with the roles of within-species genetic variability on evolutionary dynamics of multispecies communities. The resulting dynamics differ greatly when they occur in subdivided metapopulations and metacommunities (Wilson 1992; Wolf et al. 1998; Wade 2004) because community attributes can then develop strong community-level heritabilities that allow for both altered patterns of individual level selection and for community level selection to act. Our discussion here shows how such heritability may be influenced by the connectivity of the system so that patterns of resulting evolution favor different outcomes. They also suggest that there might be some strong interactions between the evolutionary dynamics that involve within-species evolutionary change with the evolutionary dynamics that involve changes in species composition mediated by assembly dynamics. These interactions will probably be complex in nature. For example De Meester et al. (2002) has proposed a "monopolization hypothesis" in which early colonists to a community can adapt to local conditions rapidly enough to be able to exclude subsequent colonists of other species leading to many alternate stable communities in a landscape and a noisy relationship between local conditions and species composition (even though there would be a good fit between local conditions and the traits distributions of the organisms). However, this seems likely to depend on having slow colonization rates relative to rates of local adaptation and to rates of local environmental change. A broader view of metacommunity dynamics that accounts for different levels of adaptive capacity in different landscapes and of organisms with different levels of dispersal would suggest other possible outcomes as outlined above. Under high dispersal rates for example, local adaptation by individual species may be of relatively little importance because such changes get preempted by the invasion of already better adapted species from the metacommunity (see also Ackerly 2003, 2004).

Loreau et al. (chapter 18; see also Mouquet et al. 2003) have examined how biodiversity might affect ecosystem structure. They argued that the role of biodiversity depends greatly on which type of metacommunity is involved in regulating the biodiversity of the ecosystem. Holt and Loreau (2002) demonstrate that system openness has a dramatic effect on how species sorting maps onto ecosystem functioning. In parallel with our discussion these authors draw attention to the ways that communities affect ecosystem attributes differently when the correspondence between local environment and species composition is maximized rather than disrupted by mass effects or by patch dynamics. Similarly, Moore et al. (2001) examined how invasibility by an outside species is related to biodiversity; they again find that the results depend on how diversity is regulated. These findings indicate some of the ways that thinking of metacommunities and of larger scale relationships between patterns of community structure at different spatial scales can inform our understanding of the interrelations between population dynamics in space and time, community assembly in metacommunities, and the ecosystem dynamics that both emerge and in turn influence these other ecological processes.

The themes that keep rising to the surface in so many of the chapters of this book and in the broader literature on metacommunities often converge on familiar, yet fundamental, themes that permeate much evolutionary and ecological thinking:

1) The centrality of adaptation and coadaptation.
2) Selective forces as a driving element in the dynamics of natural systems.
3) The interplay of disruptive and enhancing processes with these selective forces, including environmental fluctuations, the movement of individual among heterogeneous patches, and the regulation of variability via extinction and colonization episodes.
4) The nonlinear effects of movement in affecting selection by being a force that enhances adaptation at low levels of movement, and disrupting it at high levels.

We suggest that the unique contribution of metacommunity approaches is that they highlight the importance of considering multispecies interactions and spatial dynamics in grappling with the general behavior and features of complex systems. Moreover, because metacommunity dynamics builds on many familiar themes in ecology (metapopulation dynamics, interspecific interactions, etc.) this approach provides a framework for developing hypotheses, experiments, and ultimately a deeper theoretical understanding of natural systems with multiple levels of organization in the biological hierarchy (from genes to ecosystems) and in space (from microsites occupied by single individuals to landscapes, and maybe even further). It is in exploring the action of these features, and in grappling with the particular ways they act at these different scales and in exploring their consequences for novel elements of biological organization, that many of the future challenges in metaecology lie.

Literature Cited

Abrams, P. A. 1998. High competition with low similarity and low competition with high similarity: Exploitative and apparent competition in consumer-resource systems. American Naturalist 152: 114–128.

Ackerly, D. 2003. Community assembly, niche conservatism, and adaptive evolution in changing environments. International Journal of Plant Sciences. 164:S165–S184

———. 2004. Adaptation, niche conservatism and convergence: Comparative studies of leaf evolution in California chaparral. American Naturalist 163:654–671.

Amarasekare, P. 2003. Competitive coexistence in spatially structured environments: A synthesis. Ecology Letters 6:1109–1122.

Amarasekare, P., M. F. Hoopes, N. Mouquet, and M. Holyoak. 2004. Mechanisms of coexistence in competitive metacommunities. American Naturalist164310–326.

Amarasekare, P., and R. M. Nisbet. 2001. Spatial heterogeneity, source-sink dynamics, and the local coexistence of competing species. American Naturalist 158:572–584.

Barton, N. H. and M. C. Whitlock. 1997. The evolution of metapopulations. Pages 183–210 in I. A. Hanski and M. E. Gilpin, eds. Metapopulation biology: Ecology, genetics, and evolution. Academic Press, New York.

Bell, G. , M. J. Lechowicz, A. Appenzeller, M. Chandler, E. DeBlois, L. Jackson, B. Mackenzie, R. Preziosi, M. Schallenberg, and N. Tinker. 1993. The spatial structure of the physical environment. Oecologia 96:114–121.

Berlow, E. L. 1997. From canalization to contingency: Historical effects in a successional rocky intertidal community. Ecological Monographs 67:435–460.

Caswell, H. 1978. Predator mediated co-existence: A non-equilibrium model. American Naturalist 112:127–154.

Chase, J. M., and M. A. Leibold. 2003. *Ecological niches: Linking classical and contemporary approaches.* University of Chicago Press, Chicago, IL.

Chase, J. M., M. A. Leibold, and E. Simms. 2000. Plant tolerance and resistance in food webs: Community-level predictions and evolutionary implications. Evolutionary Ecology 14:289–314.

Chesson, P. 2000. Mechanisms of maintenance of species diversity. Annual Review of Ecology and Systematics 31:343–366.

Christiansen, F. B. and T. M. Fenchel. 1977. *Theories of populations in biological communities.* Springer-Verlag, Berlin, Germany.

Crowley, P. H. 1981. Dispersal and the stability of predator-prey interactions. American Naturalist 118:673–701.

De Meester, L, A. Gomez, B. Okamura, and K. Schwenk. 2002. The monopolization hypothesis and the dispersal-gene flow paradox in aquatic organisms. Acta Oecologica 23:121–235.

Downing, A. L., and M. A. Leibold. 2002. Ecosystem consequences of species richness and composition in pond food webs. Nature 416: 838–841.

Fukami, T. 2001. Sequence effects of disturbance on community structure. Oikos 92: 215–224.

Gomulkiewicz, R., R. D. Holt, and M. Barfield. 1999. The effects of density-dependence and immigration on local adaptation in a black-hole sink environment. Theoretical Population Biology 55: 283–296.

Gonzalez, A., and R. D. Holt. 2002. The inflationary effects of environmental fluctuations in source-sink systems. Proceedings of the National Academy of Sciences, USA 99:14872.

Goodnight, C. J. 2000. Heritability at the ecosystem level. Proceeding of the National Academy of Sciences, USA 97: 9365–9366.

Hairston, N. G., and N. G. Hairston, Sr. 1993. Cause-effect relationships in energy flow, trophic structure, and interspecific interactions. American Naturalist 142:379–411.

Hairston, N. G., F. E. Smith, and L. B. Slobodkin. 1960. Community structure, population control and competition. American Naturalist 94:421–425.

Hastings, A. 1980. Disturbance, coexistence, history, and competition for space. Theoretical Population Biology 18:363–373.

Hochberg, M. E., and J. H. Lawton. 1990. Spatial heterogeneities in parasitism and population dynamics. Oikos 59:9–14.

Holt, R. D. 1985. Population dynamics in two-patch environments: Some anomalous consequences of an optimal habitat distribution. Theoretical Population Biology 28:181–208.

———. 1993. Ecology at the mesoscale: The influence of regional processes on local communities. Pages 77–88 *in* R. E. Ricklefs, and D. Schluter, eds. Species diversity in ecological communities: *Historical and geographical perspectives.* University of Chicago Press, Chicago, IL.

———. 1995. Linking species and ecosystems: Where's Darwin? Pages 273–279 *in* C. G. Jones and J. H. Lawton, eds. *Linking Species and Ecosystems.* Chapman and Hall, New York.

———. 1997. Rarity and evolution: Some theoretical considerations. Pages 210–234 *in* W. E. Kunin and K. J. Gaston, eds. *The biology of rarity: Causes and consequences of rare-common differences.* Chapman and Hall, London, UK.

———. 2002. Food webs in space: On the interplay of dynamic instability and spatial processes. Ecological Research 17:261–273.

Holt, R. D. and M. S. Gaines. 1992. Analysis of adaptation in heterogeneous landscapes: Implications for the evolution of fundamental niches. Evolutionary Ecology 6:433–447.

Holt, R. D., A. Gonzalez, and M. Barfield. 2003. Impacts of environmental variability in open populations and communities: "Inflation" in sink environments. Theoretical Population Biology 64: 315–330.

Holt, R. D., H. Grover, and D. Tilman. 1994. Simple rules for interspecific dominance in systems with exploitative and apparent competition. American Naturalist 144:741–771.

Holt, R. D. and M. Loreau. 2002. Biodiversity and ecosystem functioning: The role of trophic interactions and the importance of system openness. Pages 246–263 in A P. Kinzig, S.W. Pacala, and D. Tilman, eds. *The functional consequences of biodiversity: Empirical progress and theoretical extensions.* Princeton University Press, Princeton, NJ.

Huffaker, C. B. 1958. Experimental studies on predation: Dispersal factors and predator-prey oscillations. Hilgardia 27:343–383.

Huxel, G. R., and K. McCann. 1998. Food web stability: The influence of trophic flows across habitats. American Naturalist 152:460–469.

Leibold, M. A. 1996. A graphical model of keystone predators in food webs: Trophic regulation of abundance, incidence, and diversity patterns in communities. American Naturalist 147:784–812.

———. 1998. Similarity and local co-existence of species in regional biotas. Evolutionary Ecology 12: 95–110.

Leibold, M. A., J. M. Chase, J. B. Shurin, and A. L. Downing. 1997. Species turnover and the regulation of trophic structure. Annual Review of Ecology and Systematics 28:467–494.

Leibold, M. A., and J. Norberg. 2004. Biodiversity in metacommunities: Plankton as complex adaptive systems? Limnology and Oceanography 49:1278–1289.

Leibold, M. A., and V. H. Smith. Forthcoming. Effects of species pool size on the regulation of local trophic structure. Ecology.

Levin, S. A., 1998. Ecosystems and the biosphere as complex adaptive systems. Ecosystems.1:431–436.

Levins, R. 1969. Some demographic and genetic consequences of environmental heterogeneity for biological control. Bulletin of the Entomological Society of America 15:237–240.

Lewontin, R. C. 1970. The units of selection. Annual Review of Ecology and Systematics 1:1–18.

Loeuille, N., M. Loreau, and R. Ferrière. 2002. Consequences of plant herbivore coevolution on the dynamics and functioning of ecosystems. Journal of Theoretical Biology 217:369–381.

Loeuille, N., and M. Loreau. 2004. Nutrient enrichment and food chains: Can evolution buffer top-down control? Theoretical Population Biology 65:285–298.

Lundberg P., Ranta E., Kaitala V. 2000. Species loss leads to community closure. Ecology Letters 3:465–468

MacArthur, R. H. 1970. Species packing and competitive equilibrium for many species. Theoretical Population Biology 1:1–11.

Metz, J. A. J., S. A. H. Geritz, G. Meszena, F. J. A. Jacobs, and J. S. Van Heerwaarden. 1996. Adaptive dynamics, a geometrical study of the consequences of nearly faithful reproduction. Pages 183–231 in S. J. Van Strien and S. M. Verduyn Lunel, eds. *Stochastic and spatial structures of dynamical systems.* North-Holland, Amsterdam, The Netherlands.

Milchunas, D. G. and W. K. Lauenroth. 1993. Quantitative effects of grazing on vegetation and soils over a global range of environments. Ecological Monographs 63:327–366.

Moore, J. L., N. Mouquet, J. H. Lawton, and M. Loreau. 2001. Coexistence, saturation and invasion resistance in simulated plant assemblages. Oikos 94:303–314.

Mouquet, N., and M. Loreau. 2002. Coexistence in metacommunities: The regional similarity hypothesis. American Naturalist 159:420–426.

———. 2003. Community patterns in source-sink metacommunities. American Naturalist 162:544–557.

Mouquet N, P. Munguia, J. M. Kneitel, and T. E. Miller. 2003. Community assembly time and the relationship between local and regional species richness. Oikos 103:618–626

Norberg, J., D. P. Swaney, J. Dushoff, J. Lin, R. Casagrandi, and S. A. Levin. 2001. Phenotypic diversity and ecosystem functioning in changing environments: A theoretical framework. Proceedings of the National Academy of Sciences, USA 98:11376–11381.

Oksanen, L., S. Fretwell, A. Arruda, and P. Niemela. 1981. Exploitation ecosystems in gradients of primary productivity. American Naturalist 118:240–261.

Phillips, O. M. 1974. The equilibrium and stability of simple marine biological systems. II. Herbivores. Archives of Hydrobiology 73:310–333.

Polis, G. A., C. A. Myers, and R. D. Holt. 1989. The ecology and evolution of intraguild predation: Potential competitors that eat each other. Annual Review of Ecology and Systematics 20:297–330.

Rosenzweig, M. L. 1971. The paradox of enrichment: Destabilization of exploitation ecosystems in ecological time. Science 171:385–387.

Rummel, J. D. and J. Roughgarden. 1985. A theory of faunal buildup for competition communities. Evolution 39:1009–1033.

Seger, J. 1992. Evolution of exploiter-victim relationships. Pages 3–25 *in* M. J. Crawley, ed. *Natural enemies: The population biology of predators, parasites, and diseases.* Blackwell Scientific Publications, Oxford, UK.

Shurin, J. B. 2000. Dispersal limitation, invasion resistance, and the structure of pond zooplankton communities. Ecology 81:3074–3086.

Shurin, J. B., and E. G. Allen. 2001. Effects of competition, predation, and dispersal on species richness at local and regional scales. American Naturalist 158:624–637.

Steiner C. F. 2001. The effects of prey heterogeneity and consumer identity on the limitation of trophic-level biomass. Ecology 82:2495–2506.

Steiner, C. F., and M. A. Leibold. 2004. Cyclic assembly trajectories and scale-dependent productivity-diversity relationships. Ecology 85:107–113.

Suding, K. N., and D. Goldberg. 2001. Do disturbances alter competitive hierarchies? Mechanisms of change following gap creation. Ecology 82:2133–2149.

Thompson, J. N. 1994. *The coevolutionary process.* University of Chicago Press, Chicago, IL.

Tilman, D. 1982. *Resource competition and community structure.* Princeton University Press, Princeton, NJ.

Wade, M. J. 2004. Selection in metapopulations: The coevolution of phenotype and context. Pages 259–274 in I. Hanski and O. Gaggiotti, eds. *Ecology, genetics, and evolution of Metapopulations.* Academic Press, New York.

Williamson, M. H. 1988. Relationship of species number to area, distance and other variables Pages 91–115 *in* A. A. Myers, and P. S. Giller, eds. *Analytical biogeography.* Chapman and Hall, London, UK.

Wilson, D. S. 1992. Complex interactions in metacommunities, with implications for biodiversity and higher levels of selection. Ecology 73: 1984–2000.

Wilson, D. S. and W. Swenson. 2003. Community genetics and community selection. Ecology 84:586–588.

Wolf, J. B., E. D. Brodie III, J. M. Cheverud, A. J. Moore and M. J. Wade. 1998. Evolutionary consequences of indirect genetic effects. Trends in Ecology and Evolution 13:64–69.

CHAPTER 20

Future Directions in Metacommunity Ecology

Robert D. Holt, Marcel Holyoak, and Mathew A. Leibold

The rich and diverse contributions gathered in this book champion the utility of the metacommunity as an important concept for understanding the nature of biological diversity, and do so using a variety of theoretical and empirical perspectives. We view the current state of work on metacommunities as being just the beginning of a much broader effort to understand how biological diversity is structured over multiple spatial scales. Numerous issues yet remain to be addressed, and obvious gaps in the perspectives covered in prior chapters are numerous and potentially important. Further, there are many links between metacommunity ecology and other disciplines, such as behavioral ecology, evolutionary ecology, and landscape ecology that need to be explored more fully. Work at these interfaces, we believe, is crucial to a deeper understanding of metacommunity patterns and processes. Here we provide an overview of where we believe future work on metacommunity ecology might go.

Before delving into directions of future growth in metacommunity ecology, it is useful to survey very briefly key insights that have emerged from the theoretical and empirical material presented in the chapters in this book. In chapter 1, we sketched four schematic models for metacommunities, which we labeled as the patch dynamics, species sorting, mass effects, and neutral perspectives (see also Leibold et al. 2004). The empirical contributions in this book reveal that elements of all of these perspectives can be discerned in the workings of natural systems. For instance, Mark McPeek's damselfly system (see chapter 15) may match the assumptions of the neutral theory. Species sorting is evident whenever one examines landscapes with heterogeneous habitats, and often mass effects as well (e.g., the zooplankton system examined by Cottenie and De Meester in chapter 8, and the beetle assemblage in fragmented eucalypt forest studies by Davies et al. in chapter 7). Patch dynamics in its purest sense (colonizations balancing extinctions) is strongly evident in the community modules centered on the Glanville fritillary (see chapter 4 by Van Nouhuys and Hanski). The importance of dispersal in maintaining population abundances and species richness is central to many studies, for instance in the experimental analysis of habitat fragmentation in moss microlandscapes described by Gonzalez (chapter 6). Nearly all these examples also demonstrate the crucial importance of interspecific interactions in determining local community composition, filtering the available pool provided by dispersal processes at the regional scale. Metacommunity ecology thus involves an expansion and enrichment of traditional community ecology, not a replacement for it.

The reviews of theory by Hoopes et al. (chapter 2), Holt and Hoopes (chapter 3) and Mouquet et al. (chapter 10) crystallize basic spatial processes that influence the outcome of traditional community interactions—including the impact of spatial asynchrony in dynamics and heterogeneity in local conditions, patterns of dispersal, and nonlinearities in interactions—in effect permitting an expanded niche theory to be formulated. The relative weighing of species similarity and differences required for robust species coexistence may be played out at large spatial scales, larger than the typical scales of field experiments.

Moreover, the theoretical chapters collectively provide a set of powerful techniques that could be applied in future studies to these and other empirical metacommunity studies; they also provide pointers to critical data that need to be gathered. Holt and Hoopes (chapter 3) sketch some ideas on how extensions of classical island biogeographic theory may be applied to assembly dynamics in metacommunities; this suggests that understanding colonization and extinction rates as a function of measures of local community structure is a key empirical desideratum. Law and Leibold (chapter 11) likewise show that when the time-scales of local dynamics (births and deaths) and regional dynamics (colonizations and extinctions) are reasonably distinct, one can embed assembly dynamics and metapopulation processes in a formal framework. They illustrate their general approach with an example of a nontransitive competitive interaction (which cannot stably persist in any single local community, but robustly persists with recurrent extinctions and colonizations in a metacommunity setting), but the general approach should be much more widely applicable. Chapters 12 and 13 by Chesson et al. and Melbourne et al., respectively, provide an invaluable function by pulling together a set of abstract conceptual tools for analyzing how dynamical rules change as one changes scale in ecological systems, and then applying these tools to several concrete interactions. This approach retains the details of local dynamics, and highlights the role of both local and emergent regional nonlinearities in species' dynamics for understanding their dynamics. The approach also avoids one common limitation of metacommunity theory that it typically abstracts the world into two or three scales (e.g., a patch scale, and a patch ensemble scale). The latter kind of theory will nonetheless continues to play a central role in metacommunity studies because it permits one to isolate in relatively simple fashion some key aspects of species persistence and assemblage dynamics.

Several of the empirical contributions (e.g., the pitcher plant study of Miller and Kneitel in chapter 5, and the rock pool metacommunity studied by Kolasa and Romanuk in chapter 9) in like manner reveal that the spatial dynamics of ecological communities are multiscale in nature. An important direction for future theoretical studies is thus to focus explicitly on multiscale dynamics. To complement the perspective provided by Peter Chesson and his coworkers, in the next two sections we sketch some ideas that can lead to multiscale metacommunity theory. We then discuss issues at the interface of metacommunity ecology and landscape ecology. Both the study of beetle community dynamics described by Davies et al. (chapter 7) and the microlandscapes of Gonzalez (chapter 6) reveal that the detailed spatial structure of landscapes can have crucial effects on metacommunity processes. As noted by Loreau et al. (chapter 18), a consideration of abiotic constraints and feedbacks to population and community processes may be essential in unraveling such landscape effects, because dispersal and other flows lead to energy and material transfers that can drive local processes (see examples in Polis et al. 2004).

There are important temporal dimensions to metacommunities. Metacommunity processes surely will influence the relationship between diversity and stability that has been the focus of much attention (not to mention heat) in the ecological literature. Furthermore, over long timescales, the relationship between regional and local species richness, the strength of dispersal coupling local communities, and the range of interspecific variation present in key traits that determine both interspecific interactions and ecosystem roles all must reflect evolutionary processes. McPeek and Gomulkiewicz (chapter 15) argue that speciation can readily generate species that are near-neutral in their dynamics, so that ecological drift becomes a process of real importance for understanding patterns of species richness at broad biogeographical scales. Grappling with the historical processes that generate the species pool in the first place is of great importance in determining the patterns of the relationship between local and regional abundance (see Shurin and Srivastava, chapter 17). Chapter 19 by Leibold et al. examines the proposition that metacommunities may function as "complex adaptive systems"; the "fit" of organisms to the environment can in principle be examined at any hierarchical scale, and metacommunity processes can help mold the overall distribution of traits in the community in a way that matches species traits to local environments, with or without ongoing microevolution.

Our focus in this volume has been on basic ecological principles and theory. But we fully recognize that in our rapidly changing world, the ideas we put forth could pertain to many significant problems of great human concern. We conclude this chapter with a few comments on potential important applications of these ideas.

The Importance of Spatial Strategies in Metacommunities

Community ecologists often describe broad patterns in communities in terms of frequency distributions of among-species variation—of relative abundances, body size, metabolic rates, and degree of trophic connectance, to name a few. Another important pattern is the distribution among species in the degree to which their local dynamics are spatially open, or closed, (e.g., as measured by the fraction of individuals of each species leaving their natal site, or distances moved). As noted in Holyoak et al. in chapter 1, local communities mix species with a wide variety of spatial strategies, and in effect these species experience the world at different spatial scales (Holt 1993; Ritchie and Olff 1999). Understanding the consequences for community organization of weaving together diverse spatial strategies presents a large challenge to researchers.

There is a growing recognition that even in classic "open" communities such as coral reef fishes, larvae in some species are retained at their natal reef; the community is a blend of species with open patterns of recruitment, and others with relatively little recruitment (Mora and Sale 2002). Conversely, even classic "closed" communities such as oceanic islands are visited by migratory birds. Detailed studies of local communities within a continental setting (e.g., the Eastern Wood bird study by Beven [1976]) typically reveal some species that are persistently present and recruited locally, others that are irregular community members (exhibiting local extinction and recolonization), and some that spill over via mass effects from other distinct habitats. Yet other species occur regularly in the community, but have individual home ranges considerably greater than most other community members. If such a community were suddenly blocked from coupling via dispersal with the larger world, we would expect this heterogeneity in spatial strategies to lead to considerable among-species variation in times to local extinction (Holt 1993).

Differences in dispersal syndromes are often coupled to differences in body size, life history variables, and resource exploitation traits. The interrelationship among key traits is obviously important in mechanisms for coexistence that depend on metacommunity dynamics (e.g., competition-colonization trade-offs; Kneitel and Miller 2004). Relationships among traits may be particularly crucial when considering the food web dimensions of metacommunities, because species engaged in trophic interactions often have radically different spatial strategies and experience the world at different spatial scales. Kolasa and Romanuk (chapter 9) present tantalizing evidence that invertebrate species in a rock pool system have a hierarchy of scales, with entire groups of species having similar abundances and perhaps responding to similar scales of variation in the environment.

Several chapters in this volume begin to relate trophic structure to metacommunity dynamics (Holt and Hoopes, chapter 3; Cottenie and De Meester, chapter 8; Van Nouhuys and Hanski, chapter 4; Miller and Kneitel, chapter 5, Leibold et al. chapter 19). However, much remains to be done. Figure 20.1 depicts a theoretical scheme for how the spatial scaling of population dynamics might depend on trophic level in a community. Figure 20.1A depicts the usual kind of nonspatial food web. In figure 20.1B, species' spatial scale increases with trophic rank. For instance, a vertebrate herbivore with large home ranges may impact several plant species whose dynamics play out on much smaller spatial arenas; in turn, this herbivore might be sustained by a carnivore with a yet larger home range re-

A. Nonspatial food web

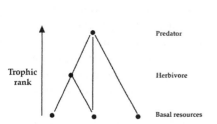

B. Food web in space - mobile top predator

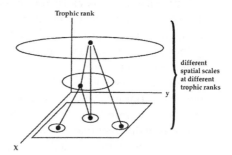

C. Food web in space - mobile herbivore

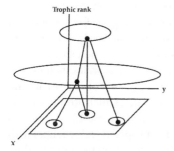

D. Food web in space - mobile basal resources

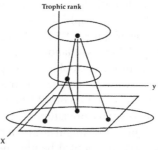

Figure 20.1 A theoretical view of how the spatial scaling of population dynamics might depend on trophic level in a community. Panel A depicts a typical nonspatial food web with species depicted by circles at nodes and lines indicating feeding relationships such that species at higher trophic levels feed on those at lower trophic levels. The same food web is shown in B–D with various spatial scaling relationships for species at particular tropic levels. In B, species' spatial scale increases with trophic rank, such that top predators have the largest area requirements within a generation. In C species at intermediate trophic levels have the largest area requirements. In D basal species are species that are very mobile within generations, such as marine plankton that are carried around by ocean currents. Modified from Holt (1996).

quirement. If trophic rank correlates with body size, and body size defines home range requirements (Schoener 1989), this is the general pattern one expects. Some vertebrate top predators have enormous home ranges (Terborgh et al. 1999). In general, among mammals, for a given body size carnivores have larger home ranges than do omnivores, which in turn have a larger home range than do herbivores (Kelt and Van Vuren 2001). In Arctic tundra ecosystems, mobile avian predators such as jaegers (*Stercorarius* sp.) can aggregate locally due to influx of individuals drawn from large areas; this behavioral response permits them to act as key regulators of vole and lemming population dynamics (Oksanen et al. 1999). The spatial scale of the rodents' dynamics in turn substantially exceeds that of the clonal and edaphically restricted plant species they eat. There is also often heterogeneity in spatial scaling within trophic levels; for instance, larger-bodied species of herbivores may experience spatial variation in food availability in a coarser fashion than do small-bodied species (Ritchie and Olff 1999). Roland and Taylor (1997) describe how within a well-defined functional group of species with similar body size (insect parasitoids), different species respond to changes in forest structure at different spatial scales.

An increase in trophic rank need not always imply an enlarged spatial scale of population dynamics, and figure 20.1B therefore does not portray all communities. For instance, Thies et al. (2003) analyzed a tritrophic interaction among oilseed rape, the oilseed rape beetle, and its parasitoids in an agricultural landscape. The percent noncrop area in surrounding landscapes was correlated (negatively) with both herbivory and parasitism; the correlation coefficient was strongest for both trophic levels for landscapes surrounding the focal points at a spatial scale of about a 1.5 km diameter. Thus, both trophic levels responded to similar spatial scales of landscape structure. In some communities, species at intermediate or lower trophic levels operate over larger spatial scales than do top predators. For instance, the dynamics of the Serengeti plains of East Africa are driven by the migration of wildebeest and other ungulates, which are attacked by relatively sedentary predators such as lions (Sinclair and Arcese 1994). For this system, figure 20.1C might be a better descriptor of how spatial scale changes with trophic rank. In coupled benthic-pelagic aquatic systems, many filter-feeding space occupiers are herbivores or omnivores dependent on a plankton flux of phytoplankton, whose dynamics are determined by water movements at much larger spatial scales. If benthic predators also have in situ recruitment, these communities might be more accurately described by figure 20.1D.

Although figure 20.1 is cast in terms of how trophic rank might lead to variation in the spatial scaling of dispersal and population dynamics within a community, important heterogeneities in dispersal can also arise for other reasons. For instance, many species take a ride on other species to disperse (phoresis; e.g., Pederson and Peterson 2002). These species necessarily have interlocked dispersal rates; a bird-dispersed plant can move no further than the bird it uses, and typically much less. Species experience different mortality risks from disturbances (e.g., due to different microhabitat or resource requirements), which should lead to differences in how dispersal evolves in response to a given disturbance regime.

Metacommunity models to date have only just begun to include such trophic structure heterogeneities in spatial scaling relationships. The challenge is to find general scaling relationships; to recognize the dynamical patterns that they generate (fueling further study); and to identify ways of representing these in quantitative models, so as to gauge the consequences of such heterogeneities in scale for metacommunity dynamics. Leibold et al (chapter 19) provide one view of what some of these implications might be for the evolutionary dimension of meta-communities.

The Interface of Metacommunity and Behavioral Ecology

Species with high mobility can provide key linkages among local communities. Sometimes such mobility is driven by external environmental forces (e.g., winds, ocean currents). However, mobile species can also often adaptively modify their movement among habitats. This observation suggests that quantifying spatial linkages among communities will require a merging of the perspectives and the-ories of community ecology and behavioral ecology. Resetarits et al. (chapter 16) explore some important behavioral dimensions of metacommunity processes (e.g., habitat selection). Here we consider an additional range of issues that war-rant more attention.

There are a wide range of empirical examples of mobile species that are likely to couple local communities and that could select habitats nonrandomly. For ex-ample, lakes appear islandlike, yet at times wide-ranging mergansers (*Mergus* sp.) and other fish-eating birds have substantial effects on fish. In turn, the effects of birds on fish may indirectly influence the remainder of the lake food web. In ma-rine systems frigatebirds (*Fregata* sp.) can range over 200 km in a single day in their search for resource patches (Weimerskirch et al. 2003).

There is likely to be an interaction between the spatial strategy of a species and many aspects of its foraging ecology. A predator that regularly disperses among communities due to localized disturbances might be faced with a shifting array of available prey species, and so could show selection for trophic generalization and labile diet selection. This implies that the magnitude of spatial coupling between different communities can change rapidly, due to behavioral decisions made by such predators. Alternatively, the ability to move may permit a predator to refine specialized tactics required to attack a particular prey species found in a scattered array of patches, so that dispersal could instead foster trophic specialization. An important task for future work is to examine the evolution of trophic specializa-tion, generalization, and diet plasticity in a metacommunity framework.

Incorporating behavioral responses to spatial variability opens up a rich array of potential effects on population and community dynamics, and many of these may lead to more complex effects in metacommunities. For instance, Sih and Wooster (1994) examined a simple model in which prey could emigrate (or not) in response to predators, and showed that in open systems, increasing local pred-ator abundance could at times lead to an *increase* in local prey abundance. This occurred because reduced prey movements more than compensated for increased mortality. Habitat selection behavior can at times weaken the potential impact of metacommunity effects (see Resetarits et al., chapter 16). For instance, Holt (1984; see also Oksanen et al. 1995) considered a simple scenario, in which a predator could limit the abundance of two prey species, each in a distinct habitat. When the predator exhibits ideal-free habitat selection rules (i.e., predators move without cost between habitats and do not interfere with each other), and if the system settles into a demographic equilibrium, the equilibrial abundance of prey in each habitat is expected to be the same as if predators could not move at all. Although species sorting could still occur (e.g., each habitat could contain the prey superior at withstanding predation), the model predicts no mass effect altering local prey species composition or abundance. However, if the predator does not follow ideal-free habitat selection rules, mass effects arise, and prey in lower productivity habitats can be vulnerable to exclusion.

A wide variety of other models have looked at the effects of optimal foraging on metapopulation dynamics (e.g., Sabelis et al. 1991, Krivan 1997, 1998), but there is little work to date on metacommunities. A hint of the rich variety of effects one might expect comes from the work of Schmidt et al. (2000), who examined how coexistence between competitors was influenced by habitat selection when habitats included sinks as well as sources, and dispersal could freely evolve. Temporal variation in local conditions promoted the evolutionary maintenance of dispersal. Schmidt et al. found that if there was a trade-off between competitive ability in the source and mortality in the sink, coexistence could occur that would not be seen in a temporally constant landscape, or if there were fixed movement rules. In this case, because there were reciprocal flows among habitats, a mass effect for both species in the sink indirectly permitted coexistence of both species back in the source, and hence globally. Conversely, if in a stable environment competitive coexistence is permitted by habitat partitioning, temporal variation can select for competitors that behaviorally exploit both habitats, which increases overlap and hence makes coexistence more difficult.

More generally, Bolker et al. (2003) reviewed theoretical studies of community-level effects of behavior and suggested that as a rule of thumb, relative to no behavior, perfect optimization of behavior tended to be more stabilizing for community dynamics, and that suboptimal but adaptive forms of behavior were still stabilizing, but less so. It is not clear whether that generalization would carry across to metacommunities, since there are already both examples that fit and others that do not.

A potential metacommunity effect of habitat selection is seen in work on mobile predators. Mobile predators can selectively move among habitats, permitting the dynamics of different local communities to be indirectly coupled via predator behavior. For instance, outside the breeding season, avian predators can concentrate on areas with high prey density and thereby reduce spatial variability in prey abundance (Norrdahl and Korpimaki 1996, 2000). Kondoh (2003) has argued that adaptive predator foragers (e.g., choosing between prey occurring in different patches) could lead to a positive relationship between food web complexity and community persistence because of environmental fluctuations. This formalizes an idea stemming back to MacArthur (1955), and is related to the notion that predator switching broadly stabilizes predator-prey interactions (Post et al. 2000).

Although appealing, there are also clear counter-examples to the notion that adaptive behaviors are a generic stabilizing influence; in some situations, adaptive behaviors by predators moving among habitats can destabilize population dynamics (Abrams 1992; Schmitz et al. 1997; Abrams and Kawecki 1997). The precise nature of the impact of adaptive behaviors on population dynamics depends closely on behavioral details (e.g., do predators instantaneously achieve adaptive optima, or is there a time lag? Abrams and Matsuda, forthcoming; Holt and Kimbrell, forthcoming). There are some situations where adaptive foraging can be strongly destabilizing. For instance, localized disturbances can generate spatial ripples in plant-herbivore-predator interactions, the magnitude and spatial extent of which depend on the detailed nature of local foraging and patch use behaviors (Holt and Barfield, unpublished results). If a disturbance greatly reduces prey availability in a given patch, sensible mobile consumers will emigrate into other patches, leading to spikes in consumption there. Mobile consumers can thus magnify the spatial extent of otherwise localized disturbances. So, as with dispersal itself, adaptive foraging by mobile consumers is not likely to be a universal stabilizer of metacommunity dynamics.

Landscape Perspectives on Metacommunities

Many issues in landscape ecology are pertinent to metacommunity issues. Metacommunities are hardly ever comprised of cookie-cutter replicate patches, but instead contain considerable variation in local conditions such as productivity and disturbance regimes. Such heterogeneity has profound implications for metacommunity dynamics. One of the simplest but crucial effects is that if different species are differentially superior in different local conditions, the regional species pool can be maintained (and hence local communities enriched) because of the existence of environmental gradients. This is a critical feature of species sorting models (Leibold 1998; introduced in Holyoak et al., chapter 1). In this section we consider two important sets of phenomena that arise from landscape structure: spillover effects and edge phenomena, and spatial patterning and disturbance.

Spillover Effects and Edge Phenomena

An important topic to which we can barely do justice is that metacommunity dynamics may be strongly influenced by the detailed structure of landscapes. In heterogeneous landscapes, this can lead to the juxtaposition of habitats with sharp differences in productivity and species abundances. Spatial fluxes in resources or organisms can then be large and asymmetric, leading to effects such as source-sink dynamics and spatial subsidies (the mass effect perspective). A recent book (Polis et al. 2004) is replete with examples of the importance of such fluxes or subsidies for the dynamics of populations, pairs of interacting populations, and entire community structure. Allochthonous inputs of resources (e.g., prey) can have profound consequences for the stability and persistence of local food webs (e.g., Huxel and McCann 1998, Holt 2002). A trickle of allochthonous prey can stabilize locally unstable interactions; conversely, if mobile predators are sustained by productive prey, those mobile predators can severely overexploit prey elsewhere in the landscape. In a wide-ranging review, Polis et al. (1997) highlighted the profound implications for food webs of movement of resources and organisms across habitat boundaries (for recent examples see Fausch et al. 2002; Henschel et al. 2001; Nakano and Murakami 2001; Power 2001). Such flows typically have strong spatial and landscape signals. For instance, rodent damage in Australian macadamia stands is greater in edges next to forests, than in edges next to grasslands (White et al. 1997). In agricultural landscapes, there is a strong effect of distance from the nest site of avian predators on their impacts on prey assemblages, whereas in forests, which have greater habitat structural complexity, the effects may be weaker and more evanescent (Forsman et al. 2001). Stahl et al. (2002) reported that landscape features (e.g., distance to forest) were excellent predictors of lynx predation on sheep in the French Jura.

In a metacommunity context, the concept of subsidies is a simplification of reality, in that the notion of a subsidy implicitly assumes a donor-controlled supply of allochthonous materials, rather than a true dynamic coupling across habitats (or patches, e.g., Donahue et al. 2003). Loreau et al., chapter 18, introduce the idea of metaecosystems as a fuller view of the spatial dynamics of ecosystem coupling. This topic offers a rich set of possibilities, from theoretical to empirical, for coupling systems more fully and for synthesizing a broad swath of ecology.

Interactions between consumers, their required resources, and their predators play out on complex spatial landscapes. This has a variety of consequences. The individual state or condition of a consumer (influencing both demographic parameters and interaction strengths) integrates landscape features as filtered through an organism's movement rules. For instance, Bommarco (1998) found that the fecundity and internal energy reserves of a generalist carabid beetle in heterogeneous agricultural landscapes were sensitive to the perimeter-to-area ratios of arable fields. Consumer movement among foraging patches can be driven by a number of factors, including localized resource depletion, temporal variation in resource supply rates, shifts in vegetation structure (including those due to impacts of food web interactions), and spatial variation in the risk of predation. Movement permits individuals to exploit spatial variation by moving among habitat patches, thus buffering the effects of temporal variation. But such movements can also lead to enhanced predation risks, and other costs. The magnitude and spatial expression of such risks should depend on the body size of the consumer, and the degree of diet specificity of the predator. Body size influences the size and spatial distribution of suitable foraging patches, and also the nature of predation risk. Generalist predators may exert a relatively fixed background mortality varying across space, whereas specialist predators are likely to define a temporally shifting landscape of mortality risk. Mobile consumers are likely to engage in "foraging games," which at times can exhibit highly complex dynamics.

Local communities coupled by dispersal necessarily have edges. Fagan et al. (1999) classify edge effects on species interactions into four categories: edges as dispersal filters, edge impacts on mortality, edges as arenas for strong spatial subsidy effects (e.g., consumers that nest in one habitat, but forage in adjacent habitats, e.g., Cook et al. 2004), and edges as generators of novel interactions. As an example of the latter, edges of patches of tropical forest are often "hot spots" of plant production, which sustain insect herbivores, which in turn attract insectivorous predators, which in turn forage some distance into the forest patches themselves (Lovejoy et al. 1989, in Fagan et al. 1999). Thus the patches themselves have additional attributes that may have further implications for metacommunity dynamics. Patches that have their own inherent dynamics (e.g., due to recurrent localized disturbances causing extinctions and initiating succession) are likely to exhibit metacommunity dynamics quite different from patches that are fixed in time and space.

In fragmented landscapes, there can be complex responses of trophic interactions to fragmentation (see also Gonzalez, chapter 6). In forest fragments in the Midwestern United States, nest predation increases with degree of fragmentation and proximity to habitat edge (Robinson et al. 1995). By contrast, ground-nesting birds in chaparral fragments in southern California enjoyed lower predation (mainly from snakes) in smaller fragments (Patten and Bolger 2003). Some of the best examples of trophic cascades in terrestrial systems come from fragmented habitats (e.g., Terborgh et al. 2001).

A key distinction at the interface of landscape ecology and food web theory is between species that are habitat specialists, and those that are habitat generalists, which straddle habitats (Waltho and Kolasa 1994; Cook et al. 2002). Spillover effects from a habitat matrix into embedded patches are likely common in anthropogenically-fragmented landscapes (e.g., Davies et al. 2001, Davies et al., chapter 7). Among species that decline in fragments (not able to utilize the matrix), predators at the top of food chains typically decline more than do their prey (e.g., Davies et al. 2000). Such landscape effects have also been suggested to play an important role in driving population cycles in arctic ecosystems. Oksanen et al. (1999) argue that spillover from productive to unproductive habitat patches strongly influences the likelihood of population cycles in lemmings and voles in arctic Fennoscandia. Spillover effects are important in many settings, but are particularly striking when sharply different ecosystems are juxtaposed. For instance, Polis and his colleagues (e.g., Polis and Hurd 1996) found that inputs of marine materials onto the verge of small, unproductive desert islands in the Gulf of California had strong impacts on land communities; such effects were negligible in the interior of large islands. Marine reserves are an applied example of where spillover of fish (and other organisms) from protected areas can potentially enhance the annual catch in adjacent fisheries (Roberts et al. 2001).

Spillover effects challenge us to define the relevant spatial scale for considering meta-ecosystem dynamics. Power and Rainey (2000) coined the term *resource shed* (analogous to *watershed*) to define the spatial scale pertinent to resource availability at given points in space. In like manner, even sedentary consumers can cast spatially delimited resource shadows, altering flows of resources among habitats, and these effects may dissipate with distance. Some of the area effects on food chain length noted in Holt and Hoopes (chapter 3) might actually reflect the differential impact of landscape flows on patches or islands of different sizes (Holt and Post, MS).

Spatial Patterning and Disturbance

Much of the metacommunity theory we have considered has ignored the detailed structure of spatial pattern. The pattern of connectivity in landscapes can have profound consequences for trophic interactions, and thus food web structure (With et al. 2002). This may be particularly important when contrasting systems like streams with prairies and the open ocean. Wilson et al. (1995) carried out studies of an individual-based model for Lotka-Volterra predator-prey interactions, contrasting habitats that differed in dimensionality. For example, one-dimensional habitats might include rivers or coastlines, two-dimensional habitats, an open prairie, and three-dimensional habitat, open bodies of water. Wilson et al. (1995) found strong differences in dynamics as a function of habitat dimensionality. One-dimensional habitats were generally less stable and more prone to extinctions, and phase-locked oscillations in abundance occurred at larger scales than in two- or three-dimensional habitats (Wilson et al. 1995). Although their work dealt with two interacting species, habitat dimensionality is likely to have important implications at the level of entire food webs.

The explicit spatial arrangement of habitats within a landscape can have important consequences for metacommunity dynamics. This volume shows a number of important kinds of dynamics that are possible in spatially explicit models, but not in spatially implicit models (see Hoopes et al., chapter 2, and Holt and Hoopes, chapter 3, for examples). However a great deal remains to be done to explore the consequences of spatially referenced landscape structures. If some species refuse to cross areas with unsuitable habitat, the flow of dispersers across a landscape becomes sensitively dependent on the presence and spatial arrangement of corridors linking patches. Theoretical models of movement suggest the existence of thresholds in degrees of connectivity, below which dispersal is greatly hampered (With and King 1999; With et al. 1997). Empirical studies have shown that corridors can have substantial effects on dispersal rates, with consequences for population persistence and stability (see e.g., Gonzalez, chapter 6; Tewksbury, Levey, et al. 2002). Dispersal rates are often quite sensitive to landscape structure. For instance, Diffendorfer et al. (1999) showed that the rate of distance of movement by rodents in an old field community was sensitive to the degree of habitat fragmentation; in a more highly fragmented landscape, fewer individuals moved, but those that did move went much further in space, than in a comparable but less fragmented landscape. One great challenge in developing spatially explicit, realistic metacommunity models is that estimating dispersal rates is inherently difficult; this difficulty is greatly magnified if dispersal rates are sensitive to the details of landscape composition and structure, and if one must estimate dispersal across an entire community of interacting species.

Heterogeneous landscapes with large-scale disturbances may exhibit interesting patterns reflecting spatial dynamics of food chains and sequential invasions. Many arthropod systems have patterns that are consistent with these processes (e.g., Rey and McCoy 1979). For example, a popular hypothesis to explain why some species become such rampant pests when introduced into novel environments is that they have escaped their natural enemies (Elton 1951; Keane and Crawley 2002). Wolfe (2002) demonstrated that *Silene latifolia* suffered much higher rates of herbivory, particularly by specialists, in Europe (its native range) than in the site of introduction (North America). She suggested that this "escape" hypothesis was more likely to result from leaving behind specialist enemies, than generalists, because the former is unlikely to be encountered in the novel environment. Although the focus of the invasion literature has been on transcontinental scales, comparable phenomena may be observed (if more subtly) at local scales if there are constraints on the rate of colonization of consumers (relative to their prey) following a disturbance (Glasser 1982). These constraints are particularly pronounced when consumers are specialists. In the first place, colonization by such consumers often must await colonization by their required prey. Moreover, even if a required prey species has colonized, its numbers are likely to initially be too low to permit successful establishment of a specialist natural enemy. These two processes together lead to the expectation that during succession, natural enemies (at least those with specialized diets) may be more sluggish in colonization than are their prey. A limitation of this pattern is that succession does not always proceed lockstep up food chains, because of the role of generalist consumers. For example, Hodkinson et al. (2002) observed that often the initial stage of primary succession consists of heterotrophs, whose numbers can be sustained by the allochthonous input of both dead organic matter and living invertebrates, in sufficient numbers to sustain populations of detritivores and predators. Steiner and Leibold (in press) found that assembly models of food web structure involving specialist and generalist consumers could result in strong cyclical assembly trajectories; food chains that build up become vulnerable to invasions by species that compete with low trophic level members of the food chain, that also escape their enemies by immigrating. These successful colonists can then in turn support the assembly of their own food chains, which can be reciprocally invaded by the original low trophic level competitors. Shifts in species composition at one level drive changes in species composition at the adjacent level, which feeds back to alter the species composition of the original level.

Spatial aspects of food chain dynamics can critically influence community responses to disturbance. In 1980, a volcanic eruption at Mount Saint Helens, Washington, extirpated plant and animal communities over a large region, which then underwent succession, a central feature of which was spread of a perennial herb, *Lupinus lapidus,* from remnant patches near the edge. This species plays a key role in ecosystem restoration, because it ameliorates physical conditions and enriches the nitrogen content of the soil. After an initial spurt, reinvasion into available habitat greatly slowed, thereby reducing the rate of succession over a broad landscape. Fagan and Bishop (2000) used experimental manipulations (buttressed by reaction-diffusion models) to show that herbivory by insects was substantially stronger at the invasive edge, than in core regions further from the edge. The magnitude of herbivory was sufficient to greatly hamper the invasive spread of the lupine into empty habitat. Fagan and Bishop argued that the contrast between the core and peripheral zones was likely the result of spatial variation in the impact on insect herbivores of both generalist predators such as spiders (which, though present at the edge, were scarcer there) and specialist parasitoids (which were absent at the edge). In effect, succession occurs up the food chain and because of a lag in response of the top trophic level, herbivory can constrain the spatial spread of crucial producer populations. Indeed, if invasion of the top level is sufficiently slow, there can be a "collapse" of an original range of the basal plant species.

As the chapters in this volume have shown, these various elements—including behavior, heterogeneity of spatial scale among different groups of species, complex attributes of patches and landscapes, and the explicit spatial features of landscapes—all have potentially fascinating implications for refining the views of metacommunity dynamics we have presented in this volume. It is clear that all these factors and the interactions among them have consequences that could potentially modify our views about metacommunity processes. An important challenge for future theoretical and empirical work is to integrate these issues into a deeper theory of metacommunity ecology.

Stability, Complexity, and Metacommunity Dynamics

Metacommunity dynamics have profound implications for our understanding of local interactions. Immigration can lead to the reversal of local competitive dominance and alter both the stability and species richness of predator-prey interactions (Holt 2002; Holt et al. 2003; reviews: Hoopes et al., chapter 2; Mouquet et al., chapter 10).

A central and controversial issue in ecology is the relationship between stability (variously defined) and food web complexity (e.g., May 1973, Polis 1994). Metacommunity dynamics can influence this relationship. First, if there are recurrent weak flows among communities (e.g., adjacent distinct habitats in a landscape). In some circumstances trophic dynamics are directly stabilized (Holt 1984, Closs et al. 1999) so that metacommunity dynamics enrich local communities by reducing local extinction rates (as with allochthonous flows; Polis et al. 1996; Huxel and McCann 1998). Alternatively, if the metacommunity mainly defines a species pool for occasional local colonization episodes, and if local diversity reduces local stability (as in Lotka-Volterra models without strong direct density dependence; May 1973), the realized local stability of communities may be less stable when they occur in metacommunities with richer species pools. Thus the effects of metacommunity species richness on the stability of local communities may depend on the connectivity of the metacommunity patches.

It is unlikely that anything very general can be said about the relationship between rates of external input in complex multispecies webs, and equilibrial population sizes for resident community members. Most ecologists are familiar with the notion of a press perturbation as a kind of experimental manipulation. Less familiar is the recognition that a press perturbation is formally identical to a change in a constant rate of input from external sources. In other words, a press perturbation provides an assessment of the net impact of a small change in external coupling (e.g., the magnitude of a spatial subsidy, such as the immigration rate of a focal species) on all species in the community, including both direct and indirect effects among resident community members. The nature of the problem of predicting equilibrial population sizes can be seen by considering a simple example. Assume that the dynamics of all species in the community are defined by $dN_i/dt = N_i f_i(\{N\}) + I_i$, where N_i is the abundance of species i, $\{N\}$ denotes the vector of abundances, and I_i is the input from external sources of species i. The effect of a small increase in input of species i on the equilibrial abundance of species j can be found by evaluating the inverse of the Jacobian matrix (Yodzis 1988; Nakajima 1992; Higashi and Nakajima 1995), comprised of terms found by evaluating $\partial f_j/\partial N_k$ near equilibrium. In principle, this protocol provides an assessment of how a change in spatial coupling alters local abundances. In practice, Yodzis (1988) showed that a given sign structure of interactions could be compatible with a wide range of quantitatively different impacts, depending on the detailed magnitude of the interaction matrix elements. But in some circumstances (e.g., relatively simple modules) this approach can be used to gauge the relative impact of allochthonous inputs on different community members. Higashi and Nakajima (1995) provide a methodology for partitioning out interaction chains in terms of direct effects along paths and loops, which may be usefully applied to this problem. In metacommunities, however, the situation is much more complex; populations are regulated at different scales so the functions describing growth involve heterogeneous spatial scales. It is unclear what the consequences of this might be for this matrix approach to quantifying spatial effects.

In communities, in principle one could observe a wide variety of sequences of introductions by species to local communities. Both theoretical and empirical studies suggest that the order of colonization events may matter greatly in determining both the composition of the community, and its temporal dynamics in abundance. For instance, Sait et al. (2000) demonstrated that in a host-pathogen-parasitoid system, the dynamical behavior of the system depends on whether or not the pathogen or the parasitoid is first introduced. This is likely to influence both local extinction rates and the average potential output of propagules that could colonize other sites. Sait et al. suggested that one effect could be a long transient, which could be quite relevant to the dynamics of metacommunities with recurrent local extinctions (e.g., Hoopes et al., chapter 2, Law and Leibold, chapter 11).

There have been relatively few attempts to directly examine the implications of spatial dynamics in entire food webs. Keitt (1997) constructed a spatially explicit, individual-based model of species interactions on a lattice, with spatially localized interactions, and compared the persistence of communities with that expected in a comparable Lotka-Volterra mean-field model with global dispersal. He concluded that spatial localization of local interactions typically permitted the persistence of richer and more strongly connected webs. Adding spatial heterogeneity further promoted the persistence of rich, highly connected webs. The internal spatial structure of a metacommunity thus may have profound consequences for its trophic organization. Documented food webs have structures that have complexities that exceed those predicted by stable linear matrix models, similar to those described above. Much work has gone into identifying ecological factors that could explain the stability of these more complex food webs, including behavioral effects such as functional responses, prey switching, and the presence of refuges. However, it may also be that these food webs do not always represent a local community but rather represent a description of species that interact at various spatial scales; or it may be that dispersal from nearby communities stabilizes the dynamics of these complex food webs. In evaluating these ideas, it would be useful to have more experiments that examine food webs, closed off from their prior connections to a broader metacommunity. Useful insights arise from studies of anthropogenically fragmented habitats (e.g., Didham et al. 1998; Crooks et al. 1999), but in such studies there is almost always a blending of disruption of an original pattern of spatial flows within the natural landscape, with an imposition of novel flows (e.g., by invasive species) from matrix habitats (e.g., Cook et al. 2002). Disentangling these effects is a difficult yet essential challenge.

The Temporal Dimension of Metacommunities

The metacommunity perspective inevitably focuses on space, and the consequences of coupling among communities, particularly in heterogeneous landscapes. Yet many important aspects of metacommunity dynamics reflect temporal variation in the environment, both directly and indirectly. This is a theme to which we cannot do justice in the remainder of this chapter, but it is worth noting a few key issues.

First, the patch dynamic perspective (Holyoak et al., chapter 1) includes as a crucial driver rates of local extinctions experienced by different species. Extinctions may arise due to endogenous causes (e.g., due to predator-prey interactions) but also can be caused by disturbances and other causes of temporal spikes in mortality. The relative influence of species sorting versus patch dynamics may reflect the impact of temporal variation in the environment. A crucial desideratum for future work is to integrate metacommunity ecology with disturbance ecology, and, more broadly, nonequilibrial perspectives on community processes (e.g., DeAngelis and Waterhouse 1987).

Second, movement among communities should often vary through time—for many reasons and across many different temporal scales. Many species alternate in their life cycle between relatively sedentary stages and more mobile stages correlated with the annual cycle, which will lead to corresponding variation in spatial coupling among habitats. As a simple example, impacts of migratory birds on small habitat patches or islands located along the migratory route are likely to be seasonally pulsed. If mass effects into a focal habitat arise from emigration from productive habitats, the magnitude of such emigration should fluctuate with temporal variation in abundance of source populations. Mortality during dispersal vary strongly through time (e.g., because of variation in abiotic factors or intensity of predation experienced in transit among patches), so the strength of spatial coupling among patches should also vary. The impact of movement on population size, local coexistence, and stability depends on the strength of local interactions, which themselves can be highly variable (Benedetti-Cecchi 2000); hence, it is likely that the strength of metacommunity processes (e.g., the impact of mass effects, viewed as a press perturbation of a local community) is also quite variable.

Third, unique phenomena may arise when one couples temporal variation with spatial patchiness and heterogeneity. For instance, storage effects promoting coexistence can arise because spatial variation coupled with dispersal permits the retention of local, temporal pulses in production (Chesson et al., chapter 6). Conversely, Holt et al. (2003) have shown that mass effects can be greatly magnified in sink habitats if local growth rates are temporally variable and positively autocorrelated through time. Temporal variation in sinks can greatly increase the average abundance of sink populations, making it difficult for locally superior species to persist in the face of an onslaught of immigrants.

Finally, dispersal itself is an evolved attribute of species. There is an enormous literature on the evolutionary ecology of dispersal (e.g., Clobert et al. 2001), and we will not even attempt to summarize it here. Dispersal should be viewed as part and parcel of an organism's life history, determining where an organism will spend its life. Basically, dispersal is expected to be favored most often when there is temporal variation in the environment, with different patterns in different habitats (e.g., weak spatial autocorrelation). Holt (1997) for instance showed that continued utilization of a sink habitat could be an evolutionarily stable strategy, provided the source habitat experienced temporal variation in fitness. Thus, the magnitude of the mass effect imposed on a focal habitat may indirectly reflect the evolutionary consequences of temporal variation in source habitats. Organisms have many alternative evolutionary responses other than dispersal to temporal variation in the environment, including diapause, the development of perenniality and seed banks, and the maintenance of energy reserves during unfavorable seasons. There will often be trade-offs between dispersal and these alternative mechanisms for coping with temporal variation, so species that utilize these alternatives to dispersal in their life-history mechanisms may experience a reduced magnitude of spatial coupling among habitats. This observation suggests that a valuable direction for future research will be to link metacommunity dynamics to general issues of life-history evolution in temporally and spatially varying environments, as this may define which components of the community are most directly involved in metacommunity processes via dispersal.

Evolution depends on variation. An evolutionary perspective on dispersal in a metacommunity context must consider the implications of intraspecific variation both in dispersal abilities and in niche characteristics. There often will be intraspecific variation in dispersal (e.g., residents versus floaters in bird populations), and dispersers may systematically differ from nondispersers in traits key to local interactions. Immigrants are likely to differ in state from residents (e.g., in body size or age), and the act of dispersal may itself entail energetic costs reflected in body stores, immunological responses, and so on. Such differences may have a wide range of consequences that have been almost entirely ignored in the literature of population and community ecology. For instance, the local population size of a consumer may be a very poor predictor of its attack rates on resources. As an example, Pusenius et al. (2000) showed that immigrant and resident voles inflict very different damage rates on tree seedlings.

If a metacommunity is spatially heterogeneous, a species that can persist in a variety of habitats can also develop local adaptations to those habitats, sometimes very rapidly. This implies that immigrants will often be maladapted relative to residents. In the absence of local adaptation, for single species with direct density-dependence regulating numbers to a stable equilibrium, an increase in immigration will typically increase local population size (the mass effect at the level of single species population dynamics). However, if immigrants are genetically maladapted relative to residents, then an increase in immigration can either have no effect on total population size, or even depress it (Holt 1983). The most severe competitive effect of immigrants may not be exerted on other resident species, but instead on local adapted populations of their own conspecifics. All else being equal, it is likely that this intraspecific effect of dispersal will weaken the importance of mass effects at the community level, with respect to perturbing local competitive dominants. If immigrants are typically weaker and less well-matched to local conditions than are residents, they may be more vulnerable to predation, providing a source of spatial subsidies. Intraspecific variation in traits correlated with dispersal could thus matter greatly in determining the beneficial impact of a supply of allochthonous subsidies of prey for resident predator populations. Mass effects could simultaneously be weak relative to the competitive mechanisms of species sorting in determining local community composition, but key determinants of food web interactions and stability. We feel that a crucial area of future work in metacommunity ecology will be the integration of ecological models of local and regional processes, with microevolutionary theory on the evolution of species' traits such as dispersal and resource, and macroevolutionary theory on diversification (see also Leibold et al., chapter 19).

The Relevance of the Metacommunity Concept to Applied Ecology

This book has focused primarily on fundamental issues in basic ecology, such as understanding the factors maintaining the diversity of ecological communities. However, insights that emerge from the metacommunity perspective clearly have many messages for crucial issues in applied ecology. It would require another large volume to fully address this topic, so here we simply outline some key linkages between metacommunity ecology and applied ecology. The following observations are not meant to be exhaustive, but rather to indicate potential domains of application of metacommunity ideas and theories.

One of the dominant issues in global change is the destruction and fragmentation of natural habitats (Kruess and Tscharntke 1994). These changes can both shrink the regional species pool and reduce connectivity among remnant patches of the original habitat types. Given that in the original, unaltered habitats, metacommunity processes were important in determining the maintenance of local as well as regional species richness, and in determining the development of adaptive matches between organismal traits and the environment, anthropogenic perturbation of these dynamics can obviously entail a corrosion in biodiversity, and the development of species assemblages that are not likely to be resilient to further change.

Another crucial aspect of global change is the homogenization of the world's biota via anthropogenic transport. Introduced nonnative invasive species can potentially impact metacommunity processes in a variety of ways. By reducing population sizes of natives, they may for instance make regional persistence more difficult, even if the invasive species do not directly cause local extinctions. If physical transport or phoretic processes are responsible for connecting habitat patches in a metacommunity, those invasive species that can potentially do the most damage are those that can be carried by those same processes. Coupling between anthropogenic matrix habitats and patches of natural habitats can have serious consequences because of the spillover of invasives into the natural fragments.

Many pest problems and solutions involve metacommunity dynamics. For instance, it is clear that impacts of biological control agents on target species are often quite sensitive to the structure of the landscape in which the control is being attempted (Thies and Tscharntke 1999). For instance, alternative prey present in one habitat may sustain a wide-ranging natural enemy population more effectively than would otherwise be possible, thereby contributing to control against a target pest.

Harvesting of natural populations is also strongly influenced by metacommunity processes. A very active area in fisheries biology, for instance, has been the development of the concept of marine and (more broadly) aquatic reserves (e.g., Roberts et al. 2001). These reserves matter precisely because there is expected to be a spatial coupling between protected reserves and unprotected harvested areas. This spatial coupling is experienced (albeit to different degrees) by all the species in the food web in which harvested species live, so a full understanding of the potential and risks of marine reserves requires a metacommunity perspective.

There is a growing appreciation of the need to apply insights from landscape ecology to the practices of natural resource management (Liu and Taylor 2002). We suggest that a deep understanding of metacommunity dynamics is also an essential ingredient in developing coherent strategies of natural resource management that pay due respect to the complex, multiscale, multispecies dynamics of the natural world.

Conclusions

Understanding how biological processes operate over multiple spatial and temporal scales is a central challenge in contemporary ecology. We suggest that in grappling with this challenge, the ideas we have put forth in this book will play a central role. Understanding the forces that influence the maintenance of biological diversity requires an appreciation of processes operating at multiple scales, from the level of individual interactions as mediated by behavior and life-history traits, through the domain of landscape patterns and processes, to finally the realm of biogeography and marcoevolutionary dynamics. An explicit consideration of metacommunity dynamics provides a bridge between the local processes that have dominated the ecological literature for so long, and the regional to continental scale processes that have long been relegated to biogeography and evolutionary biology. In turn, a mechanistic understanding of metacommunity dynamics will require a fusion of these large-scale pattern and process studies with analyses of individual behavior and life history variables, all in the context of explicit landscape structures. The metacommunity perspective, we suggest, provides a crucial link in the ongoing dialectic among ecologists about how best to relate patterns and processes at different levels in the hierarchy of life.

Acknowledgments

We would like to thank all the participants in the original Ecological Society of America symposium and in the NCEAS working group for providing a stimulating smorgasbord of ideas that made up the raw material from which we could craft this final chapter. RDH would also like to thank NSF, NIH, and the University of Florida Foundation for its support during the final preparation of this chapter, and indeed the entire volume.

Literature Cited

Abrams, P. A. 1992. Adaptive foraging by predators as a cause for predator-prey cycles. Evolutionary Ecology 6:56–72.

Abrams, P. A. and T. J. Kawecki. 1999. Adaptive host preference and the dynamics of host-parasitoid interactions. Theoretical Population Biology 56:307–324.

Abrams, P. A. and H. Matsuda. Forthcoming. Consequences of behavioral dynamics for the population dynamics of predator-prey systems with switching. Population Ecology.

Benedetti-Cecchi, L. 2000. Variance in ecological consumer-resource interactions. Nature 407:370–374.

Beven, G. 1976. Changes in breeding bird populations of an oak-wood on Bookham Common, Surrey over twenty-seven years. Nature 55:23–42.

Bolker, B., M. Holyoak, V. Krivan, L. Rowe, and O. Schmitz. 2003. Connecting theoretical and empirical studies of trait-mediated interactions. Ecology 84:1101–1114.

Bommarco, R. 1998. Reproduction and energy reserves of a predatory carabid beetle relative to agroecosystem complexity. Ecological Applications 8:846–853.

Clobert, J. A., E. Danchin, A. A. Dhondt, and J. D. Nichols. 2001. *Dispersal*. Oxford University Press, Oxford, UK.

Closs, G. P., S. R. Balcombe and M. J. Shirley. 1999. Generalist predators, interaction strength, and food-web stability. Advances in Ecological Research 28:93–126.

Crooks, K. R. and M. E. Soule. 1999. Mesopredator release and avifaunal extinctions in a fragmented system. Nature 400:563–566.

Cook, W. M., R. M. Anderson, and E. W. Schweiger. 2004. Is the matrix really inhospitable? Vole runway distribution in an experimentally fragmented landscape. Oikos 104:5–14.

Cook, W. M., K. T. Lane, B. Foster, and R. D. Holt. 2002. Island theory, matrix effects and species richness patterns in habitat fragments. Ecology Letters 5:619–623.

Davies, K. F., C. R. Margules and J. F. Lawrence. 2000. Which traits of species predict population declines in experimental forest fragments? Ecology 2000:1450–1461.

Davies, K. F., B. A. Melbourne and C. R. Margules. 2001. Effects of within- and between-patch processes on community dynamics in a fragmentation experiment. Ecology 2001:1830–1846.

DeAngelis, D. L. and J. C. Waterhouse. 1987. Equilibrium and nonequilibrium concepts in ecological models. Ecological Monographs 57:1–21.

Didham, R. K., J. H. Lawton, P. M. Hammond, and P. Eggleton. 1998. Trophic structure stability and extinction dynamics of beetles (Coleoptera) in tropical forest fragments. Philosophical Transactions of the Royal Society of London, series B, 353:437–451.

Diffendorfer, J. E., M. S. Gaines, and R. D. Holt. 1999. Patterns and impacts of movement at different scales in small mammals. Pages 63–88 *in* G. W. Barrett and J. D. Peles, eds. *Landscape ecology of small mammals.* Springer-Verlag, Berlin, Germany.

Donahue, M. J., M. Holyoak, and C. Feng. 2003. Patterns of dispersal and dynamics among habitat patches varying in quality. American Naturalist 162:302–317.

Elton, C. S. 1958, *The ecology of invasions by animals and plants.* Methuen and Co. Ltd., London, UK.

Fagan, W. F. and J. G. Bishop. 2000. Trophic interactions during primary succession: Herbivores slow a plant reinvasion at Mount St. Helens. American Naturalist 155:238–251.

Fagan, W. F., R. S. Cantrell and C. Cosner. 1999. How habitat edges change species interactions. American Naturalist 153:165–182.

Fausch, K. D., M. E. Power and M. Murakami. 2002. Linkages between stream and forest food webs: Shigeru Nakano's legacy for ecology in Japan. Trends in Ecology and Evolution 17:425–434.

Forsman, J. T., M. Monkkonen, and M. Hukkanen. 2001. Effects of predation on community assembly and spatial dispersion of breeding forest birds. Ecology 2001:232–244.

Glasser, J. W. 1982. On the causes of temporal change in communities: Modification of the biotic environment. American Naturalist 119:375–390.

Henschel, J. R., D. Mahsberg and H. Stumpf. 2001. Allochthonous aquatic insects increase predation and decrease herbivory in river shore food webs. Oikos 93:429–438.

Higashi, M. and H. Nakajima. 1995. Indirect effects in ecological interaction networks. I. The chain rule approach. Mathematical Biosciences 130:99–128.

Hodkinson, I. D., N. R. Webb, and S. J. Coulson. 2002. Primary community assembly on land: The missing stages: why are the heterotrophic organisms always there first? Journal of Ecology 90:569–577.

Holt, R. D. 1983. Immigration and the dynamics of peripheral populations. Pages 680–694 *in* K. Miyata and A. Rhodin, eds. *Advances in herpetology and evolutionary biology.* Harvard University and the Museum of Comparative Zoology, Cambridge, UK.

———. 1984. Spatial heterogeneity, indirect interactions, and the coexistence of prey species. American Naturalist 124:377–406.

———. 1993. Ecology at the mesoscale: The influence of regional processes on local communities. Pages 77–88 *in* R. E. Ricklefs, and D. Schluter, eds. *Species diversity in ecological communities: Historical and geographical perspectives.* University of Chicago Press, Chicago, IL.

———. 1996. Temporal and spatial aspects of food web structure and dynamics. Pages 255–257 *in* G. A. Polis and K. O. Winemiller, eds. *Food webs: Integration of patterns and dynamics.* Chapman and Hall, New York.

———. 1997. On the evolutionary stability of sink populations. Evolutionary Ecology 11:723–732.

———. 2002. Food webs in space: On the interplay of dynamic instability and spatial processes. Ecological Research 17:261–273.

Holt, R. D., M. Barfield and A. Gonzalez. 2003. Impacts of environmental variability in open populations and communities: "Inflation" in sink environments. Theoretical Population Biology 64:315–330.

Holt, R. D. and T. Kimbrell. Forthcoming. Foraging and population dynamics. *In* D. Stephens, J. Brown, and R. Ydenberg, eds. *Foraging.* University of Chicago Press, Chicago, IL.

Holt, R. D. and D. Post. Manuscript. Spatial controls on food chain length: A review of mechanisms.

Huxel, G. R. and K. McCann. 1998. Food web stability: The influence of trophic flows across habitats. American Naturalist 152:460–469.

Keane, R. M. and M. J. Crawley. 2002. Exotic plant invasions and the enemy release hypothesis. Trends in Ecology and Evolution 17:164–169.

Keitt, T. H. 1997. Stability and complexity on a lattice—coexistence of species in an individual-based food web model. Ecological Modelling 102:243–258.

Kelt, D. and D. Van Vuren. 2001. The ecology and macroecology of mammalian home range area. American Naturalist 157: 637–645.

Kneitel, J. M., and J. M. Chase. 2004. Trade-offs in community ecology: Linking spatial scales and species coexistence. Ecology Letters 7:69–80.

Kondoh, M. 2003. Foraging adaptation and the relationship between food-web complexity and stability. Science 299:1388–1390.

Krivan, V. 1997. Dynamic ideal free distribution: Effects of optimal patch choice on predator-prey dynamics. American Naturalist 149:164–178.

———. 1998. Effects of optimal antipredator behavior of prey on predator-prey dynamics: The role of refuges. Theoretical Population Biology 53:131–142.

Kruess, A. and T. Tscharntke. 1994. Habitat fragmentation, species loss, and biological control. Science 264:1581–1584.

Leibold, M. A. 1998. Similarity and local co-existence of species in regional biotas. Evolutionary Ecology 12:95–110.

Liu, J. and W. W. Taylor. 2002. Integrating landscape ecology into natural resource management. Cambridge University Press, Cambridge, UK.

MacArthur, R. H. 1955. Fluctuations of animal populations, and a measure of community stability. Ecology 36:533–536.

May, R. M. 1973. *Stability and complexity in model ecosystems.* Princeton University Press, Princeton, NJ.

Mora, C. and P. F. Sale. 2002. Are populations of coral reef fishes open or closed? Trends in Ecology and Evolution 17:422–428.

Nakajima, H. 1992. Sensitivity and stability of flow networks. Ecol. Mod. 65:123–133.

Nakano, S. and M. Murakami. 2001. Reciprocal subsidies: Dynamics interdependence between terrestrial and aquatic food webs. Proceedings of the National Academy of Sciences of the USA 98: 166–170.

Norrdahl, K. and E. Korpimaki. 1996. Do nomadic avian predators synchronize population fluctuations of small mammals? A field experiment. Oecologia 107:478–483.

———. 2000. Do predators limit the abundance of alternative prey? Experiments with vole-eating avian and mammalian predators. Oikos 91:528–540.

Oksanen, T., M. E. Power and L. Oksanen. 1995. Ideal free habitat selection and consumer resource dynamics. American Naturalist 146:565–585.

Oksanen, T., M. Schneider, U. Rammul, P. Hamback and M. Aunapuu. 1999. Populations fluctuations of voles in North Fennoscandian tundra: Contrasting dynamics in adjacent areas with different habitat composition. Oikos 86:463–478.

Patten, M. A. and D. T. Bolger. 2003. Variation in top-down control of avian reproductive success across a fragmentation gradient. Oikos 101:479–488.

Pederson, E. J. and M. S. Peterson. 2002. Bryozoans as ephemeral estuarine habitat and a larval transport mechanisms for mobile benthos and young fishes in the north-central Gulf of Mexico. Marine Biology 140:936–947.

Polis, G. A. 1994. Food webs, trophic cascades and community structure. Australian Journal of Ecology 19:121–136.

Polis, G. A., W. B. Anderson and R. D. Holt. 1997. Toward an integration of landscape ecology and food web ecology: The dynamics of spatially subsidized food webs. Annual Review of Ecology and Systematics 28:289–316.

Polis, G. A., R. D. Holt, B. A. Menge, and K. Winemiller. 1996. Time, space and life history: Influence on food webs. Pages 435–460 in Polis, G. A., and K. O. Winemiller, eds. Food webs: Integration of patterns and dynamics. Chapman and Hall, London, UK.

Polis, G. A. and S. D. Hurd. 1996. Linking marine and terrestrial food webs: Allochthonous input from the ocean supports high secondary productivity in small islands and coastal land communities. American Naturalist 147:396–423.

Polis, G. A., G. R. Huxel, and M. Power. 2004. Food webs at the landscape level. University of Chicago Press, Chicago, IL.

Post, D. M., M. E. Conners, and D. S. Goldberg. 2000. Prey preference by a top predator and the stability of linked food chains. Ecology 81:8–14.

Post, D. M., M. L. Pace and N. G. Hairston, Jr. 2000. Ecosystem size determines food-chain length in lakes. Nature 405:1047–1049.

Power, M. E. 2001. Prey exchange between a stream and its forested watershed elevates predator densities in both habitats. Proceedings of the National Academy of Sciences of the USA 98:14–15.

Power, M. E. and W. E. Rainey. 2000. Food webs and resource sheds: Towards spatially delimiting trophic interactions. Pages 291–314 in M. J. Hutchings, E. A. John, and A. J. A. Stewart, eds. The ecological consequences of environmental heterogeneity. Blackwell Science Limited, Oxford, UK.

Pusenius, J., R. S. Ostfeld, and F. Keesing. 2000. Patch selection and tree-seedling predation by resident vs. immigrant meadow voles. Ecology 81:2951–2956.

Rey, J. R. and E. D. McCoy. 1979. Application of island biogeographic theory to pests of cultivated crops. Environmental Entomology 8:577–582.

Ritchie, M. E. and H. Olff. 1999. Spatial scaling laws yield a general theory of biodiversity. Nature 400: 557–560.

Roberts, C. M., J. A. Bohnsack, F. Gell, J. P. Hawkins, and R. Goodridge. 2001. Effects of marine reserves on adjacent fisheries. Science 294:1920–1923.

Robinson, S. K., F. R. Thompson, III, T. M. Donovan, D. R. Whitehead, and J. Faaborg. 1995. Regional forest fragmentation and the nesting success of migratory birds. Science 267:1987–1990.

Roland, J., and P. D. Taylor. 1997. Insect parasitoid species respond to forest structure at different spatial scales. Nature 386:710–713.

Sabelis, M. W., O. Diekmann, and V. A. A. Jansen. 1991. Metapopulation persistence despite local extinction: predator-prey patch models of the Lotka-Volterra type. Biological Journal of the Linnean Society 42:267–283.

Sait, S. M., W. C. Liu, D. J. Thompson, H. C. J. Godfray and M. Begon. 2000. Invasion sequence affects predator-prey dynamics in a multi-species interaction. Nature 405:448–450.

Schmidt, K. A., J. M. Earnhardt, J. S. Brown, and R. D. Holt. 2000. Habitat selection under temporal heterogeneity: Exorcizing the ghost of competition past. Ecology 81:2622–2630.

Schmitz, O. J., A. P. Beckerman, and S. Litman. 1997. Functional responses of adaptive consumers and community stability with emphasis on the dynamics of plant-herbivore systems. Evolutionary Ecology 11:773–784.

Schoener, T. W. 1989. Food webs from the small to the large. Ecology 70:1559–1589.

Sih, A. and D. E. Wooster. 1994. Prey behavior, prey dispersal, and predator impacts on stream prey. Ecology 75:1199–1207.

Sinclair, A. R. E. and P. Arcese. 1995. Serengeti II: Dynamics, management, and conservation of an ecosystem. University of Chicago Press, Chicago, IL.

Stahl, P., J. M. Vandel, S. Ruette, L. Coat, Y. Coat, and L. Balestra. 2002. Factors affecting lynx predation on sheep in the French Jura. Journal of Applied Ecology 39:204–216.

Terborgh, J., J. A. Estes, P. Paquet, K. Ralls, D. Boyd-Heger, B. J. Miller, and R. F. Noss. 1999. The role of top carnivores in regulating terrestrial ecosystems. Pages 39–63 *in* M. E. Soule and J. Terborgh, eds. *Continental conservation: Foundation of regional reserve networks.* Island Press, Washington, D.C.

Terborgh, J., L. Lopez, V. P. Nunez et al. 2001. Ecological meltdown in predator-free forest fragments. Science 294:1923–1926.

Tewksbury, J. J., D. J. Levey, N. M. Haddad, S. Sargent, J. L. Orrock, A. Weldon, B. J Danielson, J. Brinkerhoff, E. I. Damschen, and P. Townsend. 2002. Corridors affect plants, animals, and their interactions in fragmented landscapes. Proceedings of the National Academy of Sciences of the USA 99:12923–12926.

Thies, C. and T. Tscharntke. 1999. Landscape structure and biological control in agroecosystems. Science 285:893–895.

Thies, C., I. Steffan-Dewenter and T. Tscharntke. 2003. Effects of landscape context on herbivory and parasitism at different spatial scales. Oikos 101:18–25.

Waltho, N. and J. Kolasa. 1994. Organization of instabilities in multispecies systems, a test of hierarchy theory. Proceedings of the National Academy of Sciences of the USA. 51:1682–1685.

Weimerskirch, H., O. Chastel, C. Barbraud, and O. Tostain. 2003. Frigatebirds ride high on thermals. Nature 421:333–334.

White, J., J. Wilson, and K. Horskin. 1997. The role of adjacent habitats in rodent damage levels in Australian macadamia orchard systems. Crop Protection 16:727–737.

Wilson, W. G., E. McCauley and A. De Roos. 1995. Effect of dimensionality on Lotka-Volterra predator-prey dynamics: Individual based simulation results. Bulletin of Mathematical Biology 57:507–526.

With, K. A. and A. W. King. Dispersal success on fractal landscapes: A consequence of lacunarity thresholds. Landscape Ecology 14:73–82.

With, K. A., R. H. Gardener, and M. G. Turner. 1997. Landscape connectivity and population distributions in heterogeneous environments. Oikos 78:151–169.

With, K. A., D. M. Pavuk, J. L. Worchuck, R. K. Oaten, and J. L. Fisher. 2002. Threshold effects of landscape structure on biological control in agroecosystems. Ecological Applications 12:52–65.

Wolfe, L. M. 2002. Why alien invaders succeed: Support for the escape-from-enemy hypothesis. American Naturalist 160:705–711.

Yodzis, P. 1988. The indeterminacy of ecological interactions as perceived through perturbation experiments. Ecology 69:508–515.

Coda

Marcel Holyoak, Mathew A. Leibold,
and Robert D. Holt

Below we list the major findings of the book.

1. Four caricatures of the spatial complexity of the world (detailed in chapter 1) seem to capture much of the dynamics of metacommunities (chapters 4–10, 14).
2. There is variation among biomes, landscapes, and taxa in the relative importance of each of these simplified scenarios (chapters 3–9, 14, 20). It remains to be seen if this variation is idiosyncratic or systematic across biomes, landscapes, or taxa.
3. The four scenarios (paradigms in chapter 1) are related to each other by how they deal with various issues: connectivity, spatial habitat heterogeneity, and impacts of temporal variation and disturbance (chapters 1, 10, 12, 13, 20).
4. Metacommunity dynamics are influenced by the ways that individual organisms filter and express temporal and spatial variation via their life histories, niche relationships, and movement patterns.
5. Tentative generalities that emerge from the empirical examples in this book include
   A. Classical patch dynamics are possibly less common than species sorting or mass effects (chapters 4–9, 14);
   B. Some degree of species sorting was found in most systems examined (chapters 4–9, 16);
   C. Mass effects tend to be intermediate in importance and frequency, and in some systems this was the case even in the face of strong dispersal (chapters 5–8, 14);
   D. Some support for neutral community dynamics was provided by two chapters (chapters 9 and 15).
6. Caveats
   A. Because the metacommunity concept is recent, few empirical studies have been designed specifically to test the ideas.
   B. Much of the evidence to date is correlational.
   C. Future empirical studies need to examine processes at multiple spatiotemporal scales and to be explicitly wed to theory development and testing.

7. Metacommunity dynamics affect numerous other aspects of ecology:
   A. The flux of materials among communities and ecosystems, as described by the metaecosystem concept (chapter 18);
   B. Patterns of local and regional species diversity and abundance (for theory see chapters 10, 14, 17, 18; for diversity examples see chapters 6–9; for abundance examples see chapters 5, 6, 8–10, 13, 14).
8. Thinking about metacommunity dynamics elicits new ways of thinking about community ecology. Several techniques stand out as being novel ways of investigating communities and metacommunities:
   A. Scale transition theory provides a rigorous inclusive framework capable of dealing with the full complexities of spatiotemporal structure of the environment and populations (chapters 12 and 13).
   B. Comparative and experimental approaches provide opportunities to assess the relative importance of local habitat factors and spatial dynamics (e.g., chapter 8). These can distinguish between species sorting and the other three metacommunity perspectives.
   C. Mapping community transitions in patch occupancy onto a patch dynamic model framework is an approach that allows analyses of metacommunity dynamics to proceed with relatively small amounts of data (chapter 11). This parallels the use of the incidence function approach in metapopulations.
   D. Considering limiting cases for dispersal in natural systems and creating these by manipulation is key to testing the relevance of the four metacommunity perspectives. Theory is reviewed in chapters 1, 2, 10, 14, and 16. Direct manipulations of dispersal are reported in chapters 5 and 8, and observational or experimental support from variation in connectivity is reported in chapters 4–9.
9. There are many important implications of metacommunity processes for evolutionary biology, including understanding the evolutionary match between the traits of organisms, or assemblages, and their environments (chapters 4, 8, 16, 19, 20).

Contributors

PRIYANGA AMARASEKARE (amarasek@uchicago.edu). Department of Ecology and Evolution, University of Chicago, 1101 East 57th Street, Chicago, Illinois 60637

CHRISTOPHER A. BINCKLEY (cbinc001@odu.edu). Program in Ecology, Evolution, and Integrative Biology, Department of Biological Sciences, Old Dominion University, Norfolk, Virginia 23529

DAVID R. CHALCRAFT (chalcraftd@mail.ecu.edu). Department of Biology, East Carolina University, Greenville, North Carolina 27858

JONATHAN M. CHASE (jchase@biology2.wustl.edu). Department of Biology, Washington University, St. Louis, Missouri 63130

PETER CHESSON (plchesson@ucdavis.edu). Section of Evolution and Ecology, University of California, Davis, One Shields Avenue, Davis, California 95616

KARL COTTENIE (cottenie@nceas.ucsb.edu). National Center for Ecological Analysis and Synthesis, University of California, Santa Barbara, 735 State Street, Suite 300, Santa Barbara, California 93101-3351

KENDI F. DAVIES (kfdavies@ucdavis.edu). Department of Environmental Science and Policy, and Center for Population Biology, University of California, Davis, One Shields Avenue, Davis, California 95616

LUC DE MEESTER (luc.demeester@bio.kuleuven.ac.be). Laboratory of Aquatic Ecology, Katholieke Universiteit Leuven, Ch. De Beriotstraat 32, B-3000 Leuven, Belgium

MEGAN J. DONAHUE (megan.donahue@dartmouth.edu). Department of Biological Sciences, Dartmouth College, Hanover, New Hampshire 03755

RICHARD GOMULKIEWICZ (gomulki@wsu.edu). School of Biological Sciences, P.O. Box 644236, Washington State University, Pullman, Washington 99164

ANDREW GONZALEZ (andrew.gonzalez@mcgill.ca). Department of Biology, McGill University, 1205 Ave. Docteur Penfield, Montreal, Quebec H3A 1B1, Canada

ILKKA HANSKI (ilkka.hanski@helsinki.fi). Metapopulation Research Group, Department of Ecology and Systematics, P.O. Box 65 (Viikinkaari 1), FIN-00014 University of Helsinki, Finland

ROBERT D. HOLT (rdholt@zoo.ufl.edu). Department of Zoology, University of Florida, Gainesville, Florida 32611

MARCEL HOLYOAK (maholyoak@ucdavis.edu). Department of Environmental Science and Policy, University of California, Davis, One Shields Avenue, Davis, California 95616

MARTHA F. HOOPES (mhoopes@mtholyoke.edu). Department of Integrative Biology, University of California, Berkeley, California 94720-3140. Current address: Biological Sciences, 214 Clapp Laboratory, Mount Holyoke College, South Hadley, Massachusetts 01075

JAMIE M. KNEITEL (kneitel@biology2.wustl.edu). Department of Biology, Washington University, Campus Box 1137, St. Louis, Missouri 63130

JUREK KOLASA (kolasa@mcmaster.ca). Department of Biology, McMaster University, 1280 Main Street West, Hamilton, Ontario, Canada L8S 4K1

RICHARD LAW (RL1@york.ac.uk). Department of Biology, University of York, York YO7 1JJ, United Kingdom

JOHN F. LAWRENCE (John.Lawrence@ento.csiro.au). CSIRO Entomology, Box 1700, Canberra ACT 2601, Australia

MATHEW A. LEIBOLD (mleibold@mail.utexus.edu). Section of Integrative Biology, University of Texas, One University Station C0930, Austin, Texas 78712

MICHEL LOREAU (loreau@ens.fr). Laboratoire d'Ecologie, UMR 7625, Ecole Normale Supérieure, 46 rue d'Ulm, 75230 Paris Cedex 05, France

CHRIS R. MARGULES (chris.margules@csiro.au). CSIRO Sustainable Ecosystems and Tropical Forest Research Centre, P.O. Box 780, Atherton, QLD 4883, Australia

MARK A. McPEEK (mark.mcpeek@dartmouth.edu). Department of Biological Sciences, Dartmouth College, Hanover, New Hampshire 03755 USA

BRETT A. MELBOURNE (bamelbourne@ucdavis.edu). Department of Environmental Science and Policy, and Center for Population Biology, University of California, Davis, One Shields Avenue, Davis, California 95616

THOMAS E. MILLER (miller@bio.fsu.edu). Department of Biological Science, Florida State University, Tallahassee, Florida 32306-1100

NICOLAS M. MOUQUET (mouquet@isem.univ-montp2.fr). Equipe Hote-Parasite, ISEM, Université Montpellier II, Place Eugene Bataillon, CC065, 34095, Montpellier Cedex 05, France

WILLIAM J. RESETARITS, JR. (wresetar@odu.edu). Program in Ecology, Evolution, and Integrative Biology, Department of Biological Sciences, Old Dominion University, Norfolk, Virginia 23529

TAMARA N. ROMANUK (romanuke178@rogers.com). Département des sciences biologiques, Université du Québec à Montréal (UQÀM), CP 8888, Succ. Centre Ville, Montréal, QC, Canada H3C 3P8

ANNA L. W. SEARS (alsears@ucdavis.edu). Center for Population Biology, University of California, Davis, One Shields Avenue, Davis, California 95616

JONATHAN B. SHURIN (shurin@zoology.ubc.ca). Department of Zoology, University of British Columbia, 6270 University Boulevard, Vancouver, BC V6T 1ZA Canada

DIANE S. SRIVASTAVA (srivast@zoology.ubc.ca). Department of Zoology, University of British Columbia, 6270 University Blvd., Vancouver, BC, Canada V6T 1Z4

DAVID TILMAN (tilman@umn.edu). Department of Ecology, Evolution and Behavior, University of Minnesota, 1987 Upper Buford Circle, St. Paul, Minnesota 55108

SASKYA VAN NOUHUYS (sdv2@cornell.edu). Department of Ecology and Systematics, University of Helsinki, and Department of Ecology and Evolutionary Biology, Cornell University, Corson Hall, Cornell University, Ithaca, New York 14853

Index

Printed and bound by CPI Group (UK) Ltd, Croydon, CR0 4YY

27/10/2024

14580402-0001